CORE

Introduction to Basic Construction Skills

SIXTH EDITION

Pearson

nccer

NCCER

President and Chief Executive Officer: Boyd Worsham
Vice President of Innovation and Advancement: Jennifer Wilkerson
Core Project Manager: John Esbenshade
Senior Manager of Projects: Chris Wilson
Senior Manager of Production: Erin O'Nora
Testing/Assessment Project Manager: Elizabeth Schlaupitz
Project Assistant: Lauren Corley
Lead Technical Writer: Gary Ferguson
Technical Writers: Troy Staton, Don Congdon, Jeffery Heimgartner, Karen Bouchoux

Pearson

Director of Employability Solutions: Kelly Trakalo
Employability Solutions Coordinator: Monica Perez
Senior Producer: Alexandrina B. Wolf
Content Producers: José Carchi and Alma Dabral
Development Editor: Nancy Lamm
Designer: Mary Siener
Instructor Resources: Emergent Learning
Composition: Integra Software Services
Printer/Binder: LSC Communications
Cover Printer: LSC Communications
Text Fonts: Palatino LT Pro and Helvetica Neue

Cover Image

Cover photo provided by Haskell

16 2024

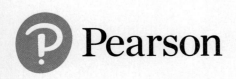

Paperback
ISBN-10: 0-13-748334-1
ISBN-13: 978-0-13-748334-1

Hardcover
ISBN-10: 0-13-748335-X
ISBN-13: 978-0-13-748335-8

PREFACE

To the Trainee

Welcome to the world of construction! You are joining the eight million Americans who have chosen a career in this lucrative field. Construction is one of the nation's largest industries, offering excellent opportunities for high earnings, career advancement, and business ownership.

People with many different talents and educational backgrounds—skilled craftspersons, managers, supervisors, and superintendents—find job opportunities in construction and related fields. As you will learn throughout your training, many other industries depend upon the work you will do in construction. From houses and office buildings to factories, roads, and bridges—everything begins with construction.

New with *Core: Introduction to Basic Construction Skills*

NCCER and Pearson are pleased to present the sixth edition of *Core: Introduction to Basic Construction Skills*. This full-color textbook includes ten modules for building foundational skills in construction. NCCER has added QR codes to videos exploring the different types of jobs available in the industry.

This edition of *Core* features a brand new module called *Build Your Future in Construction*. This new introductory module explores the various careers available in the construction industry and introduces trainees to what it might be like to work on a construction site.

We are also excited to present a significant expansion to the training on construction drawings and construction math. This edition includes some to-scale drawings in the *Introduction to Construction Drawings* module, as well as updated prints throughout to illustrate the parts and pieces of drawings with which trainees must familiarize themselves.

Both the *Introduction to Hand Tools* and *Introduction to Power Tools* modules have been significantly updated to include a variety of new tools, including many power tools that are battery operated.

All 10 modules have updated learning objectives and performance tasks, in-line with the best practices in education.

Core remains aligned with OSHA's 10-hour program. This means that instructors who are OSHA 500 Certified are able to issue 10-hour OSHA cards to their students who successfully complete the program.

Our website, **www.nccer.org**, has information on the latest product releases and training.

Your feedback is welcome. You may email your comments to **curriculum@nccer.org** or send general comments and inquiries to **info@nccer.org**.

NCCER Standardized Curricula

NCCER is a not-for-profit 501(c)(3) education foundation established in 1996 by the world's largest and most progressive construction companies and national construction associations. It was founded to address the severe workforce shortage facing the industry and to develop a standardized training process and curricula. Today, NCCER is supported by hundreds of leading construction and maintenance companies, manufacturers, and national associations. The NCCER Standardized Curricula was developed by NCCER in partnership with Pearson, the world's largest educational publisher.

Some features of the NCCER Standardized Curricula are as follows:

- An industry-proven record of success
- Curricula developed by the industry, for the industry
- National standardization providing portability of learned job skills and educational credits
- Compliance with the Office of Apprenticeship requirements for related classroom training (CFR 29:29)
- Well-illustrated, up-to-date, and practical information

NCCER also maintains the NCCER Registry, which provides transcripts, certificates, and wallet cards to individuals who have successfully completed a level of training within a craft in NCCER's Curricula. *Training programs must be delivered by an NCCER Accredited Training Sponsor in order to receive these credentials.*

Online Badges

Show off your industry-recognized credentials online with NCCER's digital badges!

NCCER is now providing online credentials. Transform your knowledge, skills, and achievements into badges which you can share across social media platforms, send to your network, and add to your resume. For more information, visit **www.nccer.org**.

Cover Image

Norwegian Cruise Line (NCL), one of the largest cruise lines in the world, set a bold vision to provide its guests with the finest global cruise terminal in the industry. Located at Port of Miami, the new terminal would be delivered on an ambitious schedule by a joint venture of two peer companies, NV2A Group and Haskell, in collaboration with renowned architect Bermello Ajamil & Partners (BA). NV2A-Haskell, rather than subcontractors, accepted the unprecedented liability of not requiring field dimensioning. This confidence came from digital modeling and saved over six months on the project schedule. The 3D model gave NV2A-Haskell absolute confidence that the tolerances would be such that all glass would fit, and in fact, it did. The facility was completed in summer 2020. For more information, visit **www.haskell.com**.

CORE FEATURES

Content is organized and presented in a functional structure that allows trainees to access the information where they need it.

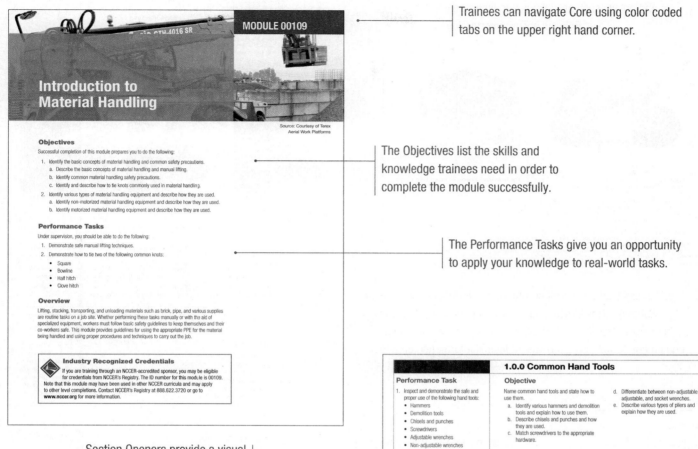

MODULE 00109

Introduction to Material Handling

Source: Courtesy of Terex
Aerial Work Platforms

Objectives
Successful completion of this module prepares you to do the following:

1. Identify the basic concepts of material handling and common safety precautions.
   a. Describe the basic concepts of material handling and manual lifting.
   b. Identify common material handling safety precautions.
   c. Identify and describe how to tie knots commonly used in material handling.
2. Identify various types of material handling equipment and describe how they are used.
   a. Identify non-motorized material handling equipment and describe how they are used.
   b. Identify motorized material handling equipment and describe how they are used.

Performance Tasks
Under supervision, you should be able to do the following:

1. Demonstrate safe manual lifting techniques.
2. Demonstrate how to tie two of the following common knots:
   - Square
   - Bowline
   - Half hitch
   - Clove hitch

Overview
Lifting, stacking, transporting, and unloading materials such as brick, pipe, and various supplies are routine tasks on a job site. Whether performing these tasks manually or with the aid of specialized equipment, workers must follow basic safety guidelines to keep themselves and their co-workers safe. This module provides guidelines for using the appropriate PPE for the material being handled and using proper procedures and techniques to carry out the job.

Industry Recognized Credentials
If you are training through an NCCER-accredited sponsor, you may be eligible for credentials from NCCER's Registry. The ID number for this module is 00109. Note that this module may have been used in other NCCER curricula and may apply to other level completions. Contact NCCER's Registry at 888.622.3720 or go to www.nccer.org for more information.

Trainees can navigate Core using color coded tabs on the upper right hand corner.

The Objectives list the skills and knowledge trainees need in order to complete the module successfully.

The Performance Tasks give you an opportunity to apply your knowledge to real-world tasks.

1.0.0 Common Hand Tools

Performance Task	Objective	
1. Inspect and demonstrate the safe and proper use of the following hand tools: • Hammers • Demolition tools • Chisels and punches • Screwdrivers • Adjustable wrenches • Non-adjustable wrenches • Sockets • Pliers	Name common hand tools and state how to use them. a. Identify various hammers and demolition tools and explain how to use them. b. Describe chisels and punches and how they are used. c. Match screwdrivers to the appropriate hardware.	d. Differentiate between non-adjustable, adjustable, and socket wrenches. e. Describe various types of pliers and explain how they are used.

Construction craftworkers require a combination of knowledge and practical skills to be successful. While there are many things that can be done with hands alone, every craft must use hand tools to get the job done. How craftworkers handle and care for their tools often demonstrates their level of skill and respect for the craft.

Many unique hand tools are used across the crafts—far too many to present in a single module. You will likely be exposed to more hand tools in the early

Section Openers provide a visual organizational structure for the information. Objectives and Performance Tasks are broken out for each section.

1.1.2 Drywall Hammer
A **drywall** hammer (*Figure 5*) is used for driving nails in drywall, also referred to as *sheet rock*. Instead of claws, the opposite end of the hammer is shaped like a small hatchet.

Drywall hammers are lighter than normal hammers, generally weighing 12–14 oz (340–397 g). This light weight helps prevent deep indentations into the drywall. The nailing face is more rounded than other hammers as well, for the same reason. The hatchet-like-tail of the hammer was originally used by plasterers to cut **lath**. However, it is equally valuable for drywall work. It can cut through small woodwork, and the U-shaped area is used to help grab and carry drywall sheets.

Drywall: A large, flat board made of layers of fiberboard and gypsum used primarily for wall construction and finishing. Drywall, also known as *sheet rock* and *wallboard*, is typically manufactured in 4' × 8' panels and readily accepts paint.

Lath: A thin, flat strip of wood used to form latticework or a foundation for wall plaster.

Trade Terms appear on the page adjacent to the text where they are first presented.

1.2.0 Adding and Subtracting Whole Numbers

Sum: The result of an addition problem. For example, in the problem 7 + 8 = 15, 15 is the sum.

To add means to combine the values of two or more numbers together into one sum or total. To add whole numbers, perform the following steps:

723 + 84	**Step 1** Line up the digits in the top and bottom numbers by place value columns.
723 + 84 7	**Step 2** Beginning at the right side, add the numbers in the ones column (3 and 4) together first.
1 723 + 84 07	**Step 3** Continue adding the digits in each column, moving from right to left, one column at a time. In this example, adding the 2 and 8 in the tens column gives 10. This requires carrying the 1 from the tens column over to the next column on the left. To do so, place the 0 in the tens column and carry the 1 over to the top of the hundreds column as shown. Always add the carried-over number to the rest of the digits in that column.
1 723 + 84 807	**Step 4** Add the 7 already in the hundreds column to the 1 carried over. The resulting sum is 807.

Step-by-step math equations help make the concepts clear and easy to grasp.

QR codes link trainees directly to videos that highlight current content.

SUCCESS STORIES: *Credentials Matter: Start Building Your Career Today*

Scan this code using the camera on your phone or mobile device to view this video.

Important information is highlighted, illustrated, and presented to facilitate learning.

Placement of images near the text description and details such as callouts and labels help trainees absorb information.

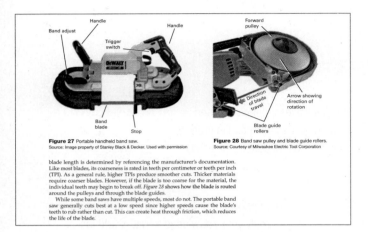

Figure 27 Portable handheld band saw.
Source: Image property of Stanley Black & Decker. Used with permission

Figure 28 Band saw pulley and blade guide rollers.
Source: Courtesy of Milwaukee Electric Tool Corporation

blade length is determined by referencing the manufacturer's documentation. Like most blades, its coarseness is rated in teeth per centimeter or teeth per inch (TPI). As a general rule, higher TPIs produce smoother cuts. Thicker materials require coarser blades. However, if the blade is too coarse for the material, the individual teeth may begin to break off. *Figure 28* shows how the blade is routed around the pulleys and through the blade guides.

While some band saws have multiple speeds, most do not. The portable band saw generally cuts best at a low speed since higher speeds cause the blade's teeth to rub rather than cut. This can create heat through friction, which reduces the life of the blade.

Preparing Drills with Keyless Chucks

Most cordless drills use a keyless chuck. While the steps for preparing a cordless drill are similar, there are some small differences. Follow the steps below when preparing to use drills with keyless chucks:

Step 1 Disconnect the drill from its power source by removing the battery pack before loading a bit.

Step 2 As shown in (*Figure 7A*), open the chuck by turning it counterclockwise until the jaws are wide enough to insert the bit shank.

Step 3 Insert the bit shank into the chuck opening (*Figure 7B*). Keeping the bit centered in the opening, turn the chuck by hand until the jaws grip the bit shank.

Step 4 Tighten the chuck securely with your hand so that the bit does not move (*Figure 7C*). You are now ready to use the cordless drill.

(A) Insert the Bit Shank.   (B) Keep Bit Straight and Partially   (C) Tighten the Chuck Securely.
Tighten the Chuck.

Figure 7 Loading the bit on a keyless chuck.
Source: Cianbro Corporation

New boxes highlight safety and other important information for trainees. Warning boxes stress potentially dangerous situations, while Caution boxes alert trainees to dangers that may cause damage to equipment. Notes boxes provide additional information on a topic.

WARNING!
A portable band saw always cuts in the direction of the user. For that reason, workers must be especially careful to avoid injury when using this type of saw. Always wear appropriate PPE and stay focused on the work.

CAUTION
Never assume anything. It never hurts to ask questions, but disaster can result if you don't ask. For example, do not assume that an electrical power source is turned off. First ask whether the power is turned off, then check it yourself to be completely safe.

NOTE
This training alone does not provide any level of certification in the use of fall arrest or fall restraint equipment. Trainees should not assume that the knowledge gained in this module is sufficient to certify them to use fall arrest equipment in the field.

Did You Know?
Louis Henry Sullivan, an American architect in the late 19th century, created a new style of architecture that resulted in buildings that were tall but still considered beautiful, a unique concept at the time. Called the "Father of Skyscrapers," he is most known for his design of the Wainwright Building in St. Louis.

These boxed features provide additional information that enhances the text.

Around the World
GOST
While OSHA serves to protect workers by setting safety standards in the United States, other systems are used internationally. One such set of technical standards used on a regional basis is known as GOST. GOST standards are more far-reaching than OSHA standards, as they cover a much broader range of topics than worker safety alone. The first set of GOST standards were published in 1968 as state standards for the former Soviet Union. After the Soviet Union was dismantled, GOST became a regional standard used by many previous members of the Soviet Union. Although other countries may also have some standards of their own, countries such as Belarus, Moldova, Armenia, and Ukraine continue to use GOST standards as well. The standards are no longer administered by Russia, however. Today, the standards are administered by the Euro-Asian Council for Standardization, Metrology and Certification (EASC).

Orthographic Drawings
Orthographic drawings are used for elevation drawings. They show straight-on views of the different sides of an object with dimensions that are proportional to the actual physical dimensions. In orthographic drawings, the designer draws lines that are scaled-down representations of real dimensions. Every 12 inches, for example, may be represented by ¼ inch on the drawing. Similarly, in an example using metric measurements with a ratio of 1:2, every 30 millimeters may be represented by 15 millimeters on the drawing.

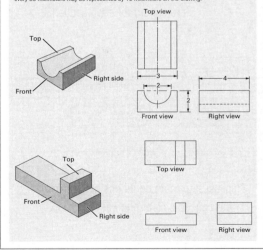

Going Green
Biodiesel

Cranes and other equipment used in rigging operations consume lots of fuel—just like all the other pieces of equipment at a typical job site. Most large trucks and construction equipment run on diesel fuel. These vehicles and machines could go green and use biodiesel instead. Biodiesel is a plant oil based fuel made from soybeans, canola, and other waste vegetable oils. It is even possible to make biodiesel from recycled frying oil from restaurants. Biodiesel is considered a green fuel since it is made using renewable resources and waste products. Biodiesel can be combined with regular diesel at any ratio or be run completely on its own. This means any combination of biodiesel and regular petroleum diesel can be used or switched back and forth as needed.

But what benefits does biodiesel have over traditional fuels?

• It's environmentally friendly. Biodiesel is sustainable and a much more efficient use of our resources than diesel.

• It's non-toxic. Biodiesel reduces health risks such as asthma and water pollution linked with petroleum diesel.

• It produces lower greenhouse gas emissions. Biodiesel is almost carbon-neutral, contributing very little to global warming.

• It can improve engine life. Biodiesel provides excellent lubricity and can significantly reduce wear and tear on your engine.

Think about the environmental impact that would occur if every vehicle and piece of equipment at every job site were converted to biodiesel. The use of biodiesel also continues to increase in Europe, where Germany produces the majority of these fuels. However, even tiny countries such as Malta and Cyprus have some level of production.

Going Green looks at ways to preserve the environment, save energy, and make good choices regarding the health of the planet.

Review questions at the end of each section and module allow trainees to measure their progress.

1.0.0 Section Review

1. The person primarily responsible for your safety is _____.
   a. your foreman
   b. your instructor
   c. yourself
   d. your employer

2. The color commonly used for informational signs is _____.
   a. green
   b. red
   c. yellow
   d. blue

3. The SDS for any chemical used at a job site must be available _____.
   a. at the job site
   b. online
   c. at the contractor's office
   d. at the nearest hospital

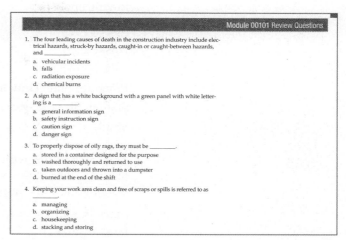

Module 00101 Review Questions

1. The four leading causes of death in the construction industry include electrical hazards, struck-by hazards, caught-in or caught-between hazards, and _____.
   a. vehicular incidents
   b. falls
   c. radiation exposure
   d. chemical burns

2. A sign that has a white background with a green panel with white lettering is a _____.
   a. general information sign
   b. safety instruction sign
   c. caution sign
   d. danger sign

3. To properly dispose of oily rags, they must be _____.
   a. stored in a container designed for the purpose
   b. washed thoroughly and returned to use
   c. taken outdoors and thrown into a dumpster
   d. burned at the end of the shift

4. Keeping your work area clean and free of scraps or spills is referred to as _____.
   a. managing
   b. organizing
   c. housekeeping
   d. stacking and storing

NCCERconnect

This interactive online course is a unique web-based supplement that provides a range of visual, auditory, and interactive elements to enhance training. Also included is a full eText. Visit **www.nccerconnect.com** for more information!

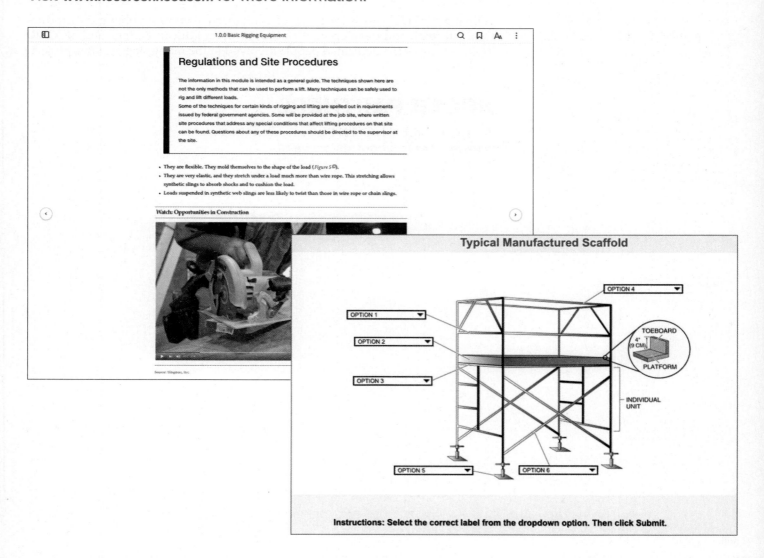

ACKNOWLEDGMENTS

This curriculum was revised as a result of the vision and leadership of the following sponsors:

MCTA

Haskell

ABC San Diego

Pittsburg State University

Stanley Black & Decker/DeWalt

Sundt

Knox School District

River Valley Technical Center

TIC – The Industrial Company

Industrial Management & Training Institute Inc.

This curriculum would not exist without the dedication and unselfish energy of those volunteers who served on the Authoring Team. A sincere thanks is extended to the following:

Mark Bonda

Ryan Camer

Jason Decker

Josh Garland

Harold Heintz

Erin Hunter

Jon Jones

Heidi Pratt

Fernando Sanchez

John Stronkowski

Chad Sutton

Tim Thompson

Darry Welker

A final note: This book is the result of a collabortive effort involving the production, editorial, and development staff at Pearson Education, Inc., and NCCER. Thanks to all of the dedicated people involved in the many stages of this project.

NCCER PARTNERS

To see a full list of NCCER Partners, please visit **www.nccer.org/about-us/partners**.

CONTENTS

1.0.0 Overview of the Construction Industry 2

  1.1.0 **What Is Construction?** 2

    1.1.1 State of the Industry 3

    1.1.2 Growing Demand for Skilled Craft Professionals 3

    1.1.3 Common Construction Industry Myths 4

  1.2.0 **History of Construction** 7

    1.2.1 Famous Construction Examples 8

    1.2.2 Important Construction Innovations and Inventions 8

    1.0.0 Section Review 9

2.0.0 Benefits of a Career in Construction 10

  2.1.0 **Making a Difference** 10

    2.1.1 Providing Essential Services 10

    2.1.2 Working to Help the Environment 10

  2.2.0 **Construction Career Benefits and Opportunities** 11

    2.2.1 Competitive Salaries 11

    2.2.2 Starting Out Debt-Free 11

    2.2.3 Career Advancement and Mobility 12

    2.0.0 Section Review 13

3.0.0 Construction Craft Careers 13

  3.1.0 **Construction Industry Sectors** 14

  3.2.0 **Types of Construction Careers** 16

    3.2.1 Carpenter 16

    3.2.2 Electrician 17

    3.2.3 Welder 18

    3.2.4 Pipefitter 19

    3.2.5 Heavy Equipment Operator 20

    3.2.6 Crane Operator 22

    3.2.7 Ironworker 23

    3.2.8 HVAC Technician 23

    3.2.9 Project Manager 24

    3.2.10 More Construction Careers 25

    3.0.0 Section Review 25

4.0.0 Starting Your Construction Career 26

  4.1.0 **Career and Technical Education** 26

  4.2.0 **Craft Training and Apprenticeships** 28

  4.3.0 **Community Colleges and Universities** 29

    4.3.1 Community College 29

    4.3.2 Four-Year Universities 29

    4.0.0 Section Review 30

MODULE 00100 Review Questions 30

Module 00101 Basic Safety (Construction Site Safety Orientation)

1.0.0 Safety and Hazard Recognition **34**

1.1.0 Incidents and Accidents **36**

1.1.1 Incident and Accident Categories 36

1.1.2 Costs 37

1.2.0 Incident and Accident Causes **38**

1.2.1 Failure to Communicate 38

1.2.2 At-Risk Work Habits 40

1.2.3 Alcohol and Drug Abuse 41

1.2.4 Lack of Skill 41

1.2.5 Intentional Acts 42

1.2.6 Unsafe Acts 42

1.2.7 Rationalizing Risk 42

1.2.8 Unsafe Conditions 43

1.2.9 Poor Housekeeping 44

1.2.10 Management System Failure 44

1.3.0 Hazard Recognition, Evaluation, and Control **44**

1.3.1 Hazard Recognition 45

1.3.2 Job Safety Analysis (JSA) and Task Safety Analysis (TSA) ... 46

1.3.3 Risk Assessment 46

1.3.4 Reporting Injuries, Incidents, and Near-Misses 48

1.3.5 Safety Data Sheets 48

1.0.0 Section Review 51

2.0.0 Elevated Work and Fall Protection **51**

2.1.0 Fall Types and Hazards **52**

2.1.1 Walking and Working Surfaces 52

2.1.2 Unprotected Sides, Wall Openings, and Floor Holes 52

2.2.0 Fall Arrest **54**

2.2.1 Anchor Points 54

2.2.2 Harnesses 55

2.2.3 PFAS Inspection 58

2.2.4 Lanyards 58

2.2.5 Lifelines.. 59

2.2.6 Guardrails 61

2.2.7 Safety Nets 61

2.3.0 Ladders and Stairs **62**

2.3.1 Straight Ladders 63

2.3.2 Extension Ladders 65

2.3.3 Stepladders 67

2.3.4 Inspecting Ladders 67

2.3.5 Stairways 67

2.4.0 Scaffolds **68**

2.4.1 Types of Scaffolds 68

2.4.2 Inspecting Scaffolds 70

2.4.3 Using Scaffolds 71

2.0.0 Section Review 71

3.0.0 Struck-By and Caught-in-Between Hazards 72

3.1.0 Struck-By Hazards 72

3.1.1 Falling Objects 72

3.1.2 Flying Objects 72

3.1.3 Vehicle and Roadway Hazards 73

3.2.0 Caught-In and Caught-Between Hazards 74

3.2.1 Trenches and Excavations 74

3.2.2 Tool, Machine, and Equipment Guards 78

3.2.3 Cranes and Heavy Equipment 79

3.0.0 Section Review 80

4.0.0 Energy Release Hazards 80

4.1.0 Electrical Safety Guidelines 80

4.1.1 Grounding 81

4.1.2 Ground Fault Circuit Interrupters 82

4.1.3 Summary of Electrical Safety Guidelines 83

4.1.4 Working Near Energized Electrical Equipment 83

4.1.5 If Someone Is Shocked 84

4.2.0 Lockout/Tagout Requirements 84

4.2.1 Pressurized or High-Temperature Systems 85

4.0.0 Section Review 86

5.0.0 Personal Protective Equipment 86

5.1.0 PPE Items 87

5.1.1 Hard Hats 88

5.1.2 Eye and Face Protection 88

5.1.3 Hand Protection 89

5.1.4 Foot and Leg Protection 90

5.1.5 Hearing Protection 91

5.2.0 Respiratory Hazards and Protection 92

5.2.1 Silica Standard 92

5.2.2 Other Respiratory Hazards 93

5.2.3 Types of Respirators 95

5.2.4 Wearing a Respirator 96

5.2.5 Selecting a Respirator 96

5.0.0 Section Review 96

6.0.0 Job-Site Hazards 97

6.1.0 Job-Site Exposure Hazards 97

6.1.1 Lead 97

6.1.2 Bloodborne Pathogens 98

6.1.3 Chemical Splashes 98

6.1.4 Container Labeling 99

6.1.5 Radiation Hazards 100

6.1.6 Biological Hazards 101

6.1.7 Evacuation 101

6.2.0 **Environmental Extremes** **101**

    6.2.1 Heat Stress 101

    6.2.2 Heat Cramps 101

    6.2.3 Heat Exhaustion 102

    6.2.4 Heat Stroke 102

    6.2.5 Cold Stress 103

    6.2.6 Frostbite 103

    6.2.7 Hypothermia 103

6.3.0 **Hot Work Hazards** **104**

    6.3.1 Arc Welding Hazards 105

    6.3.2 Oxyfuel Cutting, Welding, and Brazing 105

    6.3.3 Transporting and Securing Cylinders 106

    6.3.4 Hot Work Permits 107

6.4.0 **Fire Hazards and Fire Fighting** **109**

    6.4.1 How Fires Start 109

    6.4.2 Combustibles 109

    6.4.3 Fire Prevention 109

    6.4.4 Basic Fire Fighting 110

    6.4.5 Using a Fire Extinguisher 112

6.5.0 **Confined Spaces** **113**

    6.0.0 Section Review 116

MODULE 00101 Review Questions **117**

Module 00102 Introduction to Construction Math

1.0.0 **Whole Numbers** **122**

  1.1.0 **Place Values of Whole Numbers** **123**

    1.1.1 Study Problems: Place Values of Whole Numbers 123

  1.2.0 **Adding and Subtracting Whole Numbers** **124**

    1.2.1 Study Problems: Adding and Subtracting Whole Numbers 125

  1.3.0 **Multiplying and Dividing Whole Numbers** **125**

    1.3.1 The Order of Operations 127

    1.3.2 Study Problems: Multiplying and Dividing Whole Numbers 128

    1.0.0 Section Review 128

2.0.0 **Fractions** **129**

  2.1.0 **Equivalent Fractions and Lowest Common Denominators** **129**

    2.1.1 Finding Equivalent Fractions 129

    2.1.2 Reducing Fractions to Their Lowest Terms 130

    2.1.3 Comparing Fractions and Finding Lowest Common Denominators 130

    2.1.4 Study Problems: Finding Equivalent Fractions 131

  2.2.0 **Improper Fractions and Mixed Numbers** **132**

    2.2.1 Study Problems: Changing Improper Fractions to Mixed Numbers... 133

  2.3.0 **Adding and Subtracting Fractions** **133**

    2.3.1 Study Problems: Adding and Subtracting Fractions 134

2.4.0 Multiplying and Dividing Fractions 135

2.4.1 Study Problems: Multiplying and Dividing Fractions 136

2.0.0 Section Review 136

3.0.0 The Decimal System 137

3.1.0 Decimals 137

3.1.1 Rounding Decimals 138

3.1.2 Comparing Decimals with Decimals 138

3.1.3 Study Problems: Working with Decimals 139

3.2.0 Adding, Subtracting, Multiplying, and Dividing Decimals 140

3.2.1 Adding and Subtracting Decimals 140

3.2.2 Multiplying Decimals 140

3.2.3 Dividing with Decimals 141

3.2.4 Using the Calculator to Add, Subtract, Multiply,
and Divide Decimals 142

3.2.5 Study Problems: Decimals 143

3.3.0 Converting Decimals, Fractions, and Percentages 145

3.3.1 Converting Decimals to Percentages
and Percentages to Decimals 145

3.3.2 Converting Fractions to Decimals 146

3.3.3 Converting Decimals to Fractions 147

3.3.4 Converting Inches to Decimal Equivalents in Feet 147

3.3.5 Study Problems: Converting Different Values 148

3.3.6 Practical Applications 150

3.0.0 Section Review 151

4.0.0 Measuring Length 152

4.1.0 Reading English and Metric Rulers 152

4.1.1 The English Ruler 152

4.1.2 The Metric Ruler 153

4.1.3 Study Problems: Reading Rulers 153

4.2.0 The Measuring Tape 154

4.2.1 The English Measuring Tape 154

4.2.2 The Metric Measuring Tape 154

4.2.3 Using a Tape Measure 155

4.2.4 Study Problems: Reading Measuring Tapes 155

4.0.0 Section Review 157

5.0.0 Metric and Inch-Pound Measurement Systems 157

5.1.0 Units of Length Measurement 158

5.1.1 Inch-Pound System Units of Length 158

5.1.2 Metric System Units of Length 159

5.1.3 Converting Length Units Between Systems 160

5.1.4 Study Problems: Converting Measurements 160

5.2.0 Units of Weight Measurement 161

5.2.1 Inch-Pound Units of Weight 161

5.2.2 Metric Units of Weight 161

5.2.3 Converting Weight Units Between Systems 162

5.2.4 Study Problems: Converting Weight Units 162

5.3.0 Units of Volume Measurement 162

5.3.1 Inch-Pound Units of Volume 163

5.3.2 Metric Units of Volume 163

5.3.3 Converting Volume Units Between Systems 163

5.3.4 Study Problems: Converting Volume Units 164

5.4.0 Temperature Units 164

5.4.1 Study Problems: Converting Temperatures 167

5.0.0 Section Review 167

6.0.0 Introduction to Geometry 167

6.1.0 Angles 168

6.2.0 Shapes 169

6.2.1 Rectangle 169

6.2.2 Square 169

6.2.3 Triangle 170

6.2.4 Circle 171

6.3.0 Calculating the Area of Shapes 171

6.3.1 Study Problems: Calculating Area 172

6.4.0 Volume of Three-Dimensional Shapes 172

6.4.1 Three-Dimensional Rectangles 173

6.4.2 Cubes 173

6.4.3 Cylinders 174

6.4.4 Triangular Prisms 174

6.4.5 Study Problems: Calculating Volume 174

6.4.6 Practical Applications Using Volume 175

6.0.0 Section Review 176

MODULE 00102 Review Questions 176

Module 00103 Introduction to Hand Tools

1.0.0 Common Hand Tools 182

1.1.0 Hammers and Demolition Tools 183

1.1.1 Claw Hammers 183

1.1.2 Drywall Hammer 185

1.1.3 Ball Peen Hammer 185

1.1.4 Sledgehammers 185

1.1.5 Nail Pullers 186

1.1.6 Safety and Maintenance 188

1.2.0 Chisels and Punches 189

1.2.1 Chisels 189

1.2.2 Punches 190

1.2.3 Safety and Maintenance 190

1.3.0 Screwdrivers 191

1.3.1 Safety and Maintenance 192

1.4.0 **Wrenches** **193**

    1.4.1 Non-Adjustable Wrenches 193

    1.4.2 Adjustable Wrenches 194

    1.4.3 Socket Wrenches 195

    1.4.4 Safety and Maintenance 195

1.5.0 **Pliers and Wire Cutters** **196**

    1.5.1 Slip-Joint Pliers 197

    1.5.2 Long-Nose Pliers 197

    1.5.3 Lineman's Pliers 197

    1.5.4 Tongue-and-Groove Pliers 197

    1.5.5 Locking Pliers 197

    1.5.6 Safety and Maintenance 198

    1.0.0 Section Review 198

2.0.0 Measurement and Layout Tools **199**

2.1.0 **Rules and Other Measuring Tools** **199**

    2.1.1 Steel Rule 199

    2.1.2 Tape Measure 200

    2.1.3 Wooden Folding Rule 200

    2.1.4 Laser Measuring Tools 200

    2.1.5 Safety and Maintenance 201

2.2.0 **Levels and Layout Tools** **201**

    2.2.1 Spirit Levels 201

    2.2.2 Digital Levels 202

    2.2.3 Laser Levels 202

    2.2.4 Level Safety and Maintenance 203

    2.2.5 Squares 204

    2.2.6 Use and Maintenance 205

    2.2.7 Plumb Bob 206

    2.2.8 Chalk Lines 206

    2.0.0 Section Review 207

3.0.0 Other Common Hand Tools **207**

3.1.0 **Saws** **208**

    3.1.1 Handsaws 209

    3.1.2 Safety and Maintenance 210

3.2.0 **Clamps** **210**

    3.2.1 Safety and Maintenance 212

3.3.0 **Files and Utility Knives** **212**

    3.3.1 Files and Rasps 212

    3.3.2 Utility Knives 214

    3.3.3 Safety and Maintenance 215

3.4.0 **Shovels and Picks** **216**

    3.4.1 Safety and Maintenance 216

    3.0.0 Section Review 217

MODULE 00103 Review Questions **217**

Module 00104 Introduction to Power Tools

1.0.0 Power Drills/Drivers **222**

1.1.0 Basic Power Tool Safety Guidelines **223**

1.2.0 Types of Power Drills/Drivers **224**

1.2.1 Power Drills and Bits 224

1.2.2 Cordless Tools and Batteries... 225

1.2.3 Power Drill Safety and Maintenance Guidelines 227

1.2.4 Preparing to Use a Power Drill 228

1.2.5 Operating Power Drills... 230

1.2.6 Electromagnetic Drill Presses 230

1.3.0 Hammer Drills and Impact Drivers **232**

1.3.1 Hammer Drill Safety and Maintenance.................................... 233

1.3.2 Operating Hammer Drills 234

1.3.3 Impact Drivers... 235

1.3.4 Impact Driver Safety and Maintenance 235

1.3.5 Operating Impact Drivers.. 236

1.4.0 Pneumatic Drills and Impact Wrenches **236**

1.4.1 Pneumatic Drills 236

1.4.2 Impact Wrenches... 236

1.4.3 Pneumatic Tool Safety and Maintenance Guidelines 237

1.4.4 Operating Pneumatic Drills and Impact Wrenches................. 237

1.0.0 Section Review 238

2.0.0 Power Saws **239**

2.1.0 Circular Saws.. **239**

2.1.1 Circular Saw Safety Guidelines and Maintenance.................. 240

2.1.2 Operating Circular Saws 241

2.2.0 Jigsaws and Reciprocating Saws **242**

2.2.1 Jigsaws.. 242

2.2.2 Reciprocating Saws... 243

2.2.3 Jigsaw and Reciprocating Saw Safety and Maintenance....... 243

2.2.4 Operating Jigsaws and Reciprocating Saws 244

2.3.0 Portable Band Saws **244**

2.3.1 Band Saw Safety and Maintenance 245

2.3.2 Operating a Portable Band Saw 246

2.4.0 Miter and Cutoff Saws **246**

2.4.1 Power Miter Saws 246

2.4.2 Abrasive Cutoff Saws 247

2.4.3 Miter and Cutoff Saw Safety and Maintenance 247

2.4.4 Operating Miter and Cutoff Saws.. 247

2.5.0 Table Saws 248

    2.5.1 Table Saw Safety and Maintenance 248

    2.5.2 Ripping Wood with a Table Saw 248

    2.0.0 Section Review 249

3.0.0 Grinders and Oscillating Multi-Tools 250

    3.1.0 Grinders 250

        3.1.1 Grinder Safety and Maintenance 251

        3.1.2 Operating Grinders 253

    3.2.0 Grinder Attachments and Accessories 254

    3.3.0 Oscillating Multi-Tools 255

        3.3.1 Oscillating Multi-Tool Safety and Maintenance 255

        3.0.0 Section Review 256

4.0.0 Miscellaneous Power Tools 256

    4.1.0 Power Nailers 256

    4.1.1 Power Nailer Safety and Maintenance 258

    4.1.2 Operating Power Nailers 259

    4.2.0 Hydraulic Jacks 260

        4.2.1 Hydraulic Jack Safety and Maintenance 260

        4.0.0 Section Review 261

MODULE 00104 Review Questions 261

Module 00105 Introduction to Construction Drawings

1.0.0 Construction Drawings and Their Components 266

    1.1.0 Basic Components of Construction Drawings 267

        1.1.1 Title Block 268

        1.1.2 Border 268

        1.1.3 Drawing Area 268

        1.1.4 Revision Block 268

        1.1.5 Legend 269

        1.1.6 North Arrow 269

    1.2.0 Drawing Elements 269

        1.2.1 Lines of Construction 269

        1.2.2 Abbreviations, Symbols, and Keynotes 270

        1.2.3 Using Gridlines to Identify Plan Locations 274

    1.3.0 Dimensions and Drawing Scale 274

        1.3.1 Dimensions 274

        1.3.2 Drawing Scale 275

1.4.0 Measuring Scales **276**

1.4.1 Architect's Scale 276

1.4.2 Metric Scale (Metric Architect's Scale) 277

1.4.3 Engineer's Scale 278

1.5.0 Six Types of Construction Drawings **278**

1.5.1 Civil Plans 279

1.5.2 Architectural Plans 279

1.5.3 Structural Plans 284

1.5.4 Mechanical Plans 289

1.5.5 Plumbing/Piping Plans 296

1.5.6 Electrical Plans 297

1.5.7 Other Drawings and Documents 304

1.0.0 Section Review 304

MODULE 00105 Review Questions **305**

Module 00106 Introduction to Basic Rigging

1.0.0 Basic Rigging Equipment **308**

1.1.0 Slings **309**

1.1.1 Sling Tagging Requirements 309

1.1.2 Synthetic Slings 310

1.1.3 Alloy Steel Chain Slings 314

1.1.4 Wire Rope Slings 315

1.2.0 Sling Inspection **316**

1.2.1 Synthetic Sling Inspection 316

1.2.2 Alloy Steel Chain Sling Inspection 318

1.2.3 Wire Rope Sling Inspection 319

1.3.0 Rigging Hardware **321**

1.3.1 Shackles 321

1.3.2 Eyebolts 323

1.3.3 Lifting Clamps 324

1.3.4 Rigging Hooks 326

1.4.0 Hoists **327**

1.4.1 Operation of Chain Hoists 328

1.4.2 Hoist Safety and Maintenance 328

1.5.0 Hitches **329**

1.5.1 Vertical Hitch 330

1.5.2 Choker Hitch 330

1.5.3 Basket Hitch 332

1.5.4 The Emergency Stop Signal 333

1.0.0 Section Review 334

MODULE 00106 Review Questions **334**

Module 00107 Basic Communication Skills

1.0.0 Communication **338**

   1.1.0 The Communication Process **339**

      1.1.1 Nonverbal Communication 339

      1.1.2 Listening and Speaking Skills 341

   1.2.0 Active Listening on the Job **343**

      1.2.1 Barriers to Listening 344

   1.3.0 Speaking on the Job **345**

      1.3.1 Placing Telephone Calls 346

      1.3.2 Receiving Telephone Calls 347

      1.0.0 Section Review 348

2.0.0 Reading and Writing **348**

   2.1.0 The Importance of Reading and Writing Skills **349**

   2.2.0 Reading on the Job **349**

   2.3.0 Writing on the Job **353**

      2.3.1 Emails 357

      2.3.2 Texting 359

      2.0.0 Section Review 360

MODULE 00107 Review Questions **361**

Module 00108 Basic Employability Skills

1.0.0 Opportunities in the Construction Industry **364**

   1.1.0 The Construction Business **364**

   1.2.0 Entering the Construction Workforce **366**

      1.0.0 Section Review 369

2.0.0 Critical Thinking and Problem Solving **370**

   2.1.0 Critical Thinking and Barriers **370**

      2.1.1 Barriers to Problem Solving 370

   2.2.0 Solving Problems Using Critical Thinking **371**

      2.2.1 Defining the Problem 372

      2.2.2 Analyzing the Alternatives 372

      2.2.3 Choose a Solution and Plan 372

      2.2.4 Implement and Monitor 373

      2.2.5 Evaluate the Final Result 373

   2.3.0 Planning and Scheduling Problems **373**

      2.3.1 Materials 373

      2.3.2 Equipment 374

      2.3.3 Tools 374

      2.3.4 Labor 374

2.3.5 Handling Delays 375

2.0.0 Section Review 375

3.0.0 Relationship and Social Skills 376

3.1.0 Personal and Social Skills 376

3.1.1 Personal Habits 376

3.1.2 Work Ethic 377

3.1.3 Tardiness and Absenteeism 377

3.2.0 Conflict Resolution 378

3.2.1 Resolving Conflicts with Co-Workers 379

3.2.2 Resolving Conflicts with Supervisors 379

3.3.0 Giving and Receiving Criticism 379

3.3.1 Offering Constructive Criticism 380

3.3.2 Receiving Constructive Criticism 380

3.3.3 Destructive Criticism 381

3.4.0 Social Issues in the Workplace 381

3.4.1 Harassment 382

3.4.2 Drug and Alcohol Abuse 383

3.5.0 Teamwork and Leadership 384

3.5.1 Leadership Skills 385

3.0.0 Section Review 388

MODULE 00108 Review Questions 389

Module 00109 Introduction to Material Handling

1.0.0 Material Handling 392

1.1.0 Material Handling Basics 392

1.1.1 Pre-Task Planning 392

1.1.2 Personal Protective Equipment 393

1.1.3 Proper Lifting and Lowering Procedures 393

1.2.0 Material Handling Safety 395

1.2.1 Stacking and Storing Materials 395

1.2.2 Working from Heights 396

1.3.0 Knots for Material Handling 397

1.3.1 The Square Knot 397

1.3.2 The Bowline 398

1.3.3 The Half Hitch 399

1.3.4 The Clove Hitch 400

1.0.0 Section Review 402

2.0.0 Material Handling Equipment 402

2.1.0 Non-Motorized Material Handling Equipment 402

2.1.1 Material Carts 403

2.1.2 Hand Trucks 403

2.1.3 Cylinder Carts 404

2.1.4 Drum Dollies and Carts 404

2.1.5 Roller Skids 405

2.1.6 Wheelbarrows 405

2.1.7 Pipe Mule 405

2.1.8 Pipe Transport 406

2.1.9 Pallet Jack 406

2.2.0 Motorized Material Handling Equipment **406**

2.2.1 Powered Wheelbarrow 407

2.2.2 Concrete Mule 408

2.2.3 Freight Elevator 408

2.2.4 Industrial Forklift 409

2.2.5 Rough Terrain Forklift 411

2.2.6 Hand Signals 413

2.0.0 Section Review 413

MODULE 00109 Review Questions **413**

Appendix A Answer Keys 417

Appendix B 00101 Basic Safety 429

Appendix C 00102 Construction Math 433

Appendix D 00105 Construction Drawings 439

Glossary 443

References 451

Index 453

CORE

Introduction to Basic Construction Skills

Build Your Future in Construction

Source: Smileus/123RF

Objectives

Successful completion of this module prepares you to do the following:

1. Describe the construction industry.
   a. Define construction and summarize the current and future outlook for jobs.
   b. Identify some of construction's more prominent contributions in history.
2. Explain the benefits of a construction career.
   a. Recognize and describe how construction careers make a difference in the community.
   b. Describe the financial and professional benefits of pursuing a construction career.
3. Describe the typical career path for craft professionals.
   a. Describe industry sectors and the progression path for construction careers.
   b. Identify different construction careers and the types of skills they require.
4. Identify ways to pursue a career in the construction industry.
   a. Explain the benefits of career and technical education programs.
   b. Describe the advantages of craft training programs and their relationship with apprenticeships.
   c. Summarize the path to a construction career through community colleges and universities.

Performance Task

This is a knowledge-based module. There are no performance tasks.

Overview

Construction is an exciting, well-paying industry that offers an abundance of career opportunities. With a growing need for individuals who are ready to learn while getting paid, it provides a great fit for people of all backgrounds, skills, and strengths. Carpenter, pipefitter, welder, electrician, and crane operator are just a few of the construction professions in high demand. This module will help you understand the state of the industry, the job opportunities that currently exist, and the training options that will lead you on a path to your new construction career.

Industry Recognized Credentials

If you are training through an NCCER-accredited sponsor, you may be eligible for credentials from NCCER's Registry. The ID number for this module is 00100. Note that this module may have been used in other NCCER curricula and may apply to other level completions. Contact NCCER's Registry at 1.888.622.3720 or go to **www.nccer.org** for more information.

1.0.0 Overview of the Construction Industry

Performance Tasks	**Objective**

There are no performance tasks in this section.	Describe the construction industry.
	  a. Define construction and summarize the current and future outlook for jobs.
	  b. Identify some of construction's more prominent contributions in history.

When you think of construction, what images come to mind? If you are like many people, you might imagine someone hammering a nail or driving a bull-dozer. While these are important activities within a specific construction trade, the opportunities within the entire industry are much broader than you may imagine. The construction industry employs millions of skilled individuals working in trades that offer rewarding jobs, excellent benefits, and great opportunities to have a positive impact on society and the world. Whatever your skills or interests, construction has a place for everyone.

1.1.0 What Is Construction?

At its most basic level, the term *construction* means to build. As an industry, construction includes the creation of residential and commercial buildings, roads, bridges, dams, industrial facilities, and more. It also includes lesser-known activities related to designing, managing, planning, repairing, maintaining, and even demolishing these structures.

Craft/trade: A general term referring to one of the many construction specialties.

Craft professionals: A trained and skilled individual who designs and builds things. Craft professionals conform to the technical and ethical standards of their trade.

Construction is both an art and a science and involves individuals skilled in a particular **craft/trade**. While **craft professionals** must rely on their technical knowledge, they often combine it with creativity and artistic skills. Creativity can mean many things, from using an artistic eye to looking at problems in an original way. It takes imagination to look at a job site or a piece of wood and know how to transform it into a finished product.

Think about the work performed by carpenters, welders, and masons. In most cases, it is critical that the products they build are structurally sound. In many instances, they are also visually appealing. If you have ever admired a building, a stained-glass window, a bridge (*Figure 1*), or even someone's renovated kitchen, then you understand the value of construction and the different trades. The term *trade* is a general way of referring to one of the many construction specialties or crafts. From skyscrapers to small homes, highways to pipelines, and

Figure 1 The Golden Gate Bridge.
Source: Travel Stock/Shutterstock

Figure 2 Craft professionals impact all parts of society.
Source: Icon Sportswire/Getty Images

power lines to ocean liners, everything in society was built by individuals using a wide variety of materials, tools, and skills.

Finding a way to follow one's passion in construction is easy. Restoring historic structures, like the Statue of Liberty or the United States Capitol, is its own exciting specialization of the construction field. Even building a professional football stadium (*Figure 2*) or basketball arena takes hundreds of trade professionals with a wide variety of expertise. Imagine the pride in creating lasting structures used by thousands of individuals and the opportunity to say you helped build it.

1.1.1   State of the Industry

The construction industry is currently thriving. According to the Associated General Contractors of America (AGC), the industry builds over $1.3 trillion in structures per year and includes over 700,000 construction companies that employ more than 7 million total employees. In fact, construction contributed over $887 billion to the US economy in 2019 alone, which equates to about 4 percent of the total value of all goods and services produced in the US.

Despite recent challenges in the overall US economy, the construction market is steadily growing. In addition to continued private spending, more construction growth is projected because of future plans to invest in the country's aging **infrastructure**. These are the transportation, water, electrical, and telecommunications systems that allow countries to operate.

Infrastructure: The transportation, water, electrical, and telecommunications systems that allow communities and countries to operate.

1.1.2   Growing Demand for Skilled Craft Professionals

Opportunities abound for individuals entering the construction field. In fact, there simply are not enough skilled craft professionals to meet the industry's existing or anticipated demands in the coming years. Studies show that 80 percent of construction companies intend to hire more people over the next few years to meet the growing needs of the market. Many companies have encountered difficulties filling these positions and employee shortages are now a top concern.

Since experienced professionals are leaving the industry and fewer new employees are replacing them, there is a significant workforce shortage and a high demand for talented craft professionals. In fact, construction companies are trying to hire and train individuals immediately for the projected future growth in construction. For those seeking to enter the construction industry, this means there are great jobs available and they can earn a good income while they are being trained (*Figure 3*).

Figure 3 The construction industry offers many opportunities to today's craft professionals.
Source: Mr Twister/Shutterstock

There are several reasons for this growing labor shortage. By 2030, an estimated 40 percent of the current construction workforce is expected to retire. Baby Boomers, the portion of the US population that is nearing or past retirement age, currently make up a large portion of the construction workforce. This group of highly experienced craft professionals have built their careers for decades and now hold leadership positions. They will leave behind a large gap in the number of people needed to perform the work in many of the trades.

At the same time, fewer young people are entering into construction. Instead, many individuals over the past few decades have opted to earn a degree at a four-year university. According to the website *Statista*, in 2018, 34 percent of men and 35 percent of women in the US had received at least four years of college education. In 1968, these numbers were much lower. Only 13 percent of men and 8 percent of women had attended a four-year college. So, in just 50 years, college enrollment has nearly tripled. With a continued national emphasis on attending college, many young people are simply unaware of the abundant construction career opportunities.

1.1.3 Common Construction Industry Myths

Despite its essential role in keeping the world's infrastructure running smoothly, the current construction industry is widely misunderstood. These misconceptions may prevent some people from pursuing careers in construction. The following list of myths about construction explains why they are not true.

Myth: People in Construction Are Unskilled

Construction is often mistakenly thrown into the category of unskilled labor; that is, work that can be performed without any special training or knowledge. This misconception likely stems from the fact that a person does not need a college degree to get a construction job. In reality, craft professionals perform extremely detailed, specialized work that requires a high level of expertise gained through years of training, both in the classroom and on the job. Many professionals must also earn specialized licenses, credentials, and/or certifications that prove their expertise.

Myth: Construction Employees Make Minimum Wage

According to a 2018 survey by *U.S. News & World Report*, the lowest-paid 25 percent of the construction workforce made almost double the minimum wage amount,

Success Story: Holley Thomas

Holley Thomas became interested in welding in 2007 while earning a degree in robotics from Central Alabama Community College. She discovered that not only did she have a great talent for welding, but she also loved the craft. She was recruited by KBR, a global engineering and construction firm, as part of their outreach and engagement with the college's technical programs.

Over the next 11 years, Thomas took advantage of KBR's training offerings and grew as an experienced craftsperson. She started as a welder, worked her way into supervision, and moved into a quality inspector role. She is the first woman welder to have taken top honors (gold) in welding at the Associated Builders and Contractors (ABC) National Craft Championship in 2010. In 2015, Thomas won the ABC's National Craft Professional of the Year award.

Thomas shares, "Being a skilled craft professional, and furthermore a female in the industry, helps me leave a great legacy. I've been able to show other women in the industry that you can start from the bottom and work your way up to where I am today."

over $28,520 per year. The highest 25 percent of employees made almost double that amount. The average annual salary is over $40,000 and rising. Salaries do not include overtime or bonus pay, which means potential incomes are even higher.

Myth: Construction Is Not for Women

For most of history, men dominated the world of construction. Now, greater focus on refined skills and knowledge of specific crafts means that the most successful and best-paid construction professionals are those who can do the best work, regardless of gender.

According to the National Association of Women in Construction (NAWIC), 1.1 million women work in construction industry jobs that include field, office, and management positions. With women making up only 10 percent of the construction workforce (*Figure 4*), many construction companies see this underrepresented demographic as an opportunity to fill important job vacancies.

In addition, women are paid more equally in construction than in many other industries. The US Bureau of Labor Statistics (BLS) reports that women earn more than 99 cents for every dollar a man earns in construction jobs, compared to the US average of 81 cents.

Myth: Construction Does Not Use Technology

Construction has embraced technology and invested in it. According to the website **techcrunch.com**, funding in US-based construction technologies increased

Figure 4 Construction opportunities abound for men and women.
Source: Jhorrocks/E+/Getty Images

Figure 5 Many construction professionals use technology on the job.
Source: AzmanL/E+/Getty Images

by about 320 percent in recent years and is expected to increase even more. Examples of technology used in construction include:

- *Computers, mobile apps, and IoT devices* — Allow workers to share data, collaborate, and communicate more quickly (*Figure 5*). Examples of IoT devices include smartphones, tablets, residential and commercial sensors, and many other items.
- *Drones* — Survey job sites and identify potential hazards in risky or hard-to-see areas.
- *Robots* — Lift and move heavy objects on job sites.
- *Simulators* — Create realistic sensations of heights, stress, and hazards, allowing trainees and craft professionals to experience what it feels like to operate equipment and practice safety procedures before actually doing so on the job.
- *Software* — Create and update construction designs easily and seamlessly.

Myth: Construction Is Dangerous

Jobs involving heavy machinery, power tools, and large components have inherent risks. But the construction world has made great strides to create a safer and healthier work environment for craft professionals. Working in construction is safer than ever before due to many factors:

- *Higher government safety standards* — Since the formation of the **Occupational Safety and Health Administration (OSHA)** in 1971, construction injuries have dropped significantly. The mission of OSHA is to establish and enforce safe working conditions for the US workforce. Protective equipment requirements and dedicated safety managers on job sites are two of the most influential reasons for safer work environments.
- *More emphasis on safety in training* — Many injuries can be avoided if craft professionals are well-trained and understand the proper procedures. Modern construction programs place a significant focus on teaching safe practices in the workplace.
- *Improved technology* — Digital technology has enabled construction companies to work more safely and efficiently. For example, drones allow construction crews to more easily spot potential hazards and to monitor work sites remotely.
- *Safety matters in winning bids* — A construction company's safety records are a major factor in winning bids and being selected for projects.

Occupational Safety and Health Administration (OSHA): An agency of the US Department of Labor with the mission to establish and enforce safe working conditions for the US workforce.

Did You Know?

President Nixon played an important role in the history of construction by signing the Occupational Safety and Health Act in 1970, which established the government agency called OSHA.

Figure 6 Solar panels at Walt Disney World Resort.
Source: Engdao Wichitpunya/123RF

Myth: Construction Is Bad for the Environment

The construction industry is leading the charge on **sustainability**—the practice of minimizing negative impacts to Earth's climate and natural environment. Today, construction companies are attempting to reduce impacts on the environment by recycling materials, installing renewable energy sources, and using **green construction**, which involves creating structures that adhere to sustainability principles. In 1993, the US Green Building Council (USGBC) was created with a mission to promote sustainability-focused practices in the building industry. The council created the green building rating system known as Leadership in Energy and Environmental Design (LEED), which has become an international standard for certifying environmentally sound buildings.

Examples of construction practices that are good for the environment include:

- Preassembly of structures off site, which allows for more material reuse
- Wind power technologies, which use wind to generate energy
- Solar panels, which harness sunlight to provide energy
- Water reclamation

An example of green construction in use is the Walt Disney World Resort in Orlando, Florida, which recently installed a solar panel facility on 270 acres of land that will reduce annual emissions of greenhouse gas by more than 50,000 tons (*Figure 6*).

Sustainability: The practice of minimizing negative impacts to Earth's climate and natural environment.

Green construction: Construction practices that involve the creation of structures that adhere to sustainability principles.

1.2.0   History of Construction

Construction had its start thousands of years ago. History is documented through the built environment—think about the architectural wonders like the Great Pyramid of Giza in Egypt, the Great Wall of China (*Figure 7*), or Machu Picchu in Peru. It's easy to forget that human beings created these amazing structures. Imagine the many aspects of work that were required to bring these structures to completion; the design, planning, transporting of heavy materials, building, and so much more, all without the tools and technology available today.

Construction began by necessity, with humans building personal shelters using their hands or simple hand tools. Over time, people gained expertise in certain tasks to become specialists such as carpenters, bricklayers, and blacksmiths, allowing societies to build more complex structures. As steam-powered machinery came about in the 19th century, more advanced structures could be built, and could be built much faster.

With technology advancements that came with the Information Age, computer systems have revolutionized construction faster than in any other time period.

Did You Know?

American sculptor, inventor, and businessperson Pat Billings invented the product Geobond®, an indestructible plaster product that was patented in 1997. The most significant characteristic of Geobond® products is that they are extremely resistant to heat, able to withstand temperatures reaching over 2,000°F. Not even a rocket engine can burn the product.

Figure 7 Construction on the Great Wall of China began around 700 BC.
Source: Aphotostory/Shutterstock

Improvements in accuracy, consistency, speed, and complexity of construction exist today in ways never even imagined just 50 years ago. Construction has evolved into a high-tech business. Innovative individuals who tried new ideas, invented new products and processes, and manufactured easy-to-use materials have helped transform construction into a much more efficient and technology-driven business.

1.2.1   Famous Construction Examples

All the famous and iconic man-made structures around the world would not be here today without the skilled people who built them. Consider these feats of construction that were groundbreaking at the time they were built:

- *Eiffel Tower, in Paris, France* — Made of wrought iron in the late 1800s, it is still the second tallest building in France.
- *Empire State Building, in New York, NY* — The world's tallest building upon its completion in 1931.
- *Golden Gate Bridge, in San Francisco, CA* — The longest suspension bridge in the world upon its completion in 1937.
- *Hoover Dam, on the Nevada/Arizona border* — A dam built with a workforce that peaked at more than 5,000 people in 1934.
- *Burj Khalifa, in Dubai, United Arab Emirates* — A skyscraper completed in 2009, it is still the tallest artificial structure in the world at 2,722 feet (*Figure 8*).

1.2.2   Important Construction Innovations and Inventions

It is hard to imagine the world without inventions that make construction projects work more efficiently. The following are just a few examples of the construction industry's important innovations and inventions:

- **Personal protective equipment (PPE)** — This equipment or clothing is designed to prevent or reduce injuries on a job site. Examples include hard hats, safety glasses, face shields, safety-toed footwear, gloves, and earplugs (*Figure 9*).
- **Computer-aided design (CAD)** — Before this invention in the 1960s, construction designs had to be drawn on paper by hand, using a tedious and time-consuming process. With CAD, users employ a more automated approach to sketching two-dimensional or three-dimensional designs, allowing other people to more clearly see how aspects of a structure would look and work together when built.
- **Building information modeling (BIM)** — In this process, software reads computer files containing information to get virtual pictures of the construction

Figure 8 The Burj Khalifa in Dubai is the tallest building in the world.
Source: Iain Masterton/Alamy Stock Photo

Personal protective equipment (PPE): Equipment or clothing designed to prevent or reduce injuries.

Computer-aided design (CAD): The use of computers to digitally sketch construction designs, allowing other people to see how aspects of a structure would look and work together when built.

Building information modeling (BIM): A process in which software reads computer files containing building data to create a virtual picture of a facility before it is built. BIM objects represent many dimensions of construction, and allow builders to adjust the plans prior to beginning actual construction.

Figure 9 Craft professionals wear PPE to stay safe on the job.
Source: Xavierarnau/E+/Getty Images

of a facility before it is built. BIM objects represent many dimensions of construction such as time, cost, operation, and safety to simulate the construction process and anticipate any issues that may arise. As a result, schedules, site layouts, and other details can be adjusted for the best outcome, long before the physical building begins.

- **Virtual reality (VR)** — Like CAD and BIM, virtual reality takes what used to be paper drawings and information and allows users to visualize a structure before it is created. VR provides a 3D model of a construction site and allows users to directly immerse themselves into the virtual space. VR enables multiple people to envision the final product, feel what it is like to be in the space, and identify any issues before work begins.

Virtual reality (VR): Provides a 3D model of a construction site and allows users to directly immerse themselves into the virtual space. VR enables multiple people to envision the final product, feel what it is like to be in the space, and identify any issues before work begins.

1.0.0 Section Review

1. A major reason for the projected shortage of trained craft professionals is _____.
   a. construction salaries are too low
   b. technology has eliminated many construction careers
   c. demand for construction projects is declining
   d. the retirement of Baby Boomers

2. For every dollar made by men in construction, women make approximately _____.
   a. 50 cents
   b. 68 cents
   c. 87 cents
   d. 99 cents

3. During the 19th century, many structures could be built faster due to the invention of _____.
   a. hand tools
   b. steam-powered machinery
   c. self-healing concrete
   d. virtual reality

2.0.0 Benefits of a Career in Construction

Performance Tasks

There are no performance tasks in this section.

Objective

Explain the benefits of a construction career.

a. Recognize and describe how construction careers make a difference in the community.

b. Describe the financial and professional benefits of pursuing a construction career.

When you choose a career in construction, you are choosing an industry that has a lasting impact on people and communities. At the same time, it provides an opportunity to combine knowledge, skills, and artistic expression. There are limitless opportunities for career mobility and advancement, plus the financial benefits that come from an industry filled with high-paying occupations. With so many specialties from which to choose, the benefits of a career in construction are many.

2.1.0 Making a Difference

In many ways, the state of a society's infrastructure depends on the skills and expertise of craft professionals. The end result of their work is what allows society to function and thrive. In short, craft professionals make a lasting difference in the world.

Craft professionals who work in construction take pride in being part of something greater than their individual contributions. They understand that it often takes the skills of many individuals working together to achieve great things.

2.1.1 Providing Essential Services

When you wake up and flip the light switch on, that switch is connected to wires that ultimately lead to a power plant that is miles away that craft professionals built and maintain. When you turn on the faucet and clean water flows, that water comes through pipes that connect to a water station miles away that craft professionals built and maintain. The roads you drive on and the buildings in the schools you attend were all built by craft professionals. These services and structures are only possible because of skilled individuals who used their talents to install the components and systems that make activities like these a regular part of our daily lives.

During natural disasters or other catastrophes, craft professionals often act as first responders. Hurricanes, tornados, floods, and earthquakes can create destruction that interrupts many essential services. Restoring essential services and rebuilding structures in a timely manner are critical to minimizing the impact of these emergencies. Examples of essential services include restoring utilities such as water and electricity and rebuilding residential structures that may have been destroyed in a disaster.

2.1.2 Working to Help the Environment

The current construction workforce has become increasingly focused on minimizing environmental impacts of construction. As a result, the construction industry is changing, and the professionals who work in the industry are making positive differences for the environment. In addition to employing different building techniques, they are actually installing devices and systems that help businesses and homeowners conserve energy, reduce environmental impacts, and save money.

2.2.0 Construction Career Benefits and Opportunities

Construction careers can provide competitive salaries, a debt-free start, and financial security. With the sizeable shortage of construction professionals, the increasing demand for craft professionals has resulted in rising salaries and enticing benefits designed to attract talented individuals.

2.2.1 Competitive Salaries

Salaries for craft professionals are extremely competitive. Every other year, NCCER releases construction craft salary survey results based on the national average of base salaries for skilled craft professionals. The survey represents more than 350,000 industrial and commercial construction employees from across the US. Results for some of the construction trades are shown in *Figure 10*. Construction salaries for all of the crafts included in this survey range from $47,000 to $92,000. The salaries shown do not include overtime, bonuses, or other incentives. When these are combined, the overall benefits of construction jobs are even higher.

2.2.2 Starting Out Debt-Free

College tuition costs and student loans are the second-highest source of debt in the United States. In fact, college tuition debt is second only to mortgage debt. *Forbes* reports that the total amount of student debt is over $1.5 trillion, which is equivalent to the value of about 500 National Football League franchises.

College tuition costs continue to rise at a fast rate. While some aspiring craft professionals may choose to obtain a bachelor's degree—particularly if they want to become civil engineers, project managers, or architects—the majority of construction careers do not require a degree. In fact, 7 out of every 10 jobs in the US economy require less than a four-year degree (*Figure 11*).

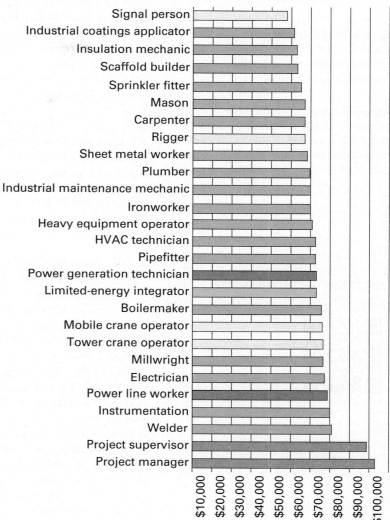

Construction Craft Salary Survey Results

Figure 10 NCCER construction craft salary survey results.

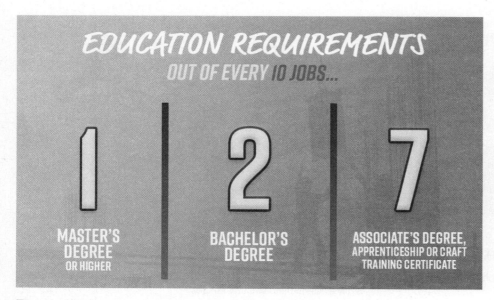

Figure 11 Education and construction career requirements.

Earn as You Learn: An approach to education that pays trainees wages for work completed as part of an on-the-job learning (OJL) program.

Apprenticeship/craft training: A program in which you gain valuable skills on the job under the instruction of more experienced professionals, often making money as you obtain training.

SUCCESS STORIES: *Get Started in Construction: Progress in Your Field*

Scan this code using the camera on your phone or mobile device to view this video.

Skilled craft training is available through technical schools, community colleges, associations, or contractor programs. These programs are shorter and far less expensive per year than the cost of attending a university.

With craft training programs (*Figure 12*), trainees learn both on the job and in the classroom under the instruction of experienced craft professionals. Craft training typically follows an **Earn as You Learn** model, which allows trainees to earn wages for their work in the field with pay increases as their skills advance. In addition, many construction companies will cover any additional education costs, allowing trainees to complete the training program debt-free and with a growing bank account.

In fact, successful completion of an **apprenticeship/craft training** program provides immediate marketability and acquisition of highly valuable skills—all while earning an income as you are being trained. This type of training makes it possible to avoid the high costs of attending a university. On the other hand, despite spending several years and considerable funds to earn a college diploma, university graduates often end up working in jobs unrelated to their degree. A study by the Federal Reserve Bank of New York found that only 27 percent of graduates work in a field related to their major.

2.2.3   Career Advancement and Mobility

Construction careers typically offer clear paths to advancement and craft professionals have excellent opportunities to progress in their careers. Many construction jobs begin with individuals learning how to use their tools. As they become proficient and develop their expertise, they can obtain additional licenses and certifications to increase their chances for promotion.

In addition to core craft skills, working in construction often requires leadership skills such as good communication, teamwork, and decision-making (*Figure 13*). These critical leadership skills allow advancement from trainee to craft professional to higher managerial and executive roles. Some craft professionals ultimately gain the knowledge and skills necessary for starting their own businesses and leading their own teams of employees.

During the development of their skills, these same individuals are often given the opportunity to travel or relocate to different areas because the skills they possess are in such high demand. Excellent construction jobs exist in cities and states all across the US and the world, meaning that skilled craft professionals are free to travel or move to locations as the demand allows. In some cases, companies even send their employees to construction sites in foreign countries.

Figure 12 Craft training provides on-the-job learning (OJL).
Source: Sturti/E+/Getty Images

Figure 13 Leadership skills help you advance in construction.
Source: kzenon/istock/Getty Images

Success Story: Boyd Worsham, from Builder to President/CEO

As a boy growing up, Boyd Worsham always liked to build things. The carpentry program at his high school solidified his desire to enter the construction industry. He was inspired by his carpentry teacher, who taught not only skills in carpentry but also responsibility and how to be employable. His teacher introduced Boyd to a job opening as a carpentry helper, which he started in 1980, right out of high school.

As he developed his skills over the next 38 years, Worsham worked his way up, first as a journeyman carpenter, then carpenter foreman, assistant superintendent, superintendent, and, finally, to vice president of construction support for The Haskell Company. Haskell is a leader in design-build project delivery. At Haskell, he was involved in major projects throughout the US. Haskell even paid for Worsham to obtain his Master of Business Administration from Jacksonville University, located in Jacksonville, Florida. Although he did not have a bachelor's degree, his 25 years of industry experience was counted in place of a degree. Ultimately, in September 2020, Worsham was appointed president and chief executive officer of NCCER, an international education foundation for the construction industry.

Source: Courtesy of Boyd Worsham

2.0.0 Section Review

1. During disasters, craft professionals are considered first responders because they perform work to restore essential services and structures.
   a. True
   b. False

2. A major benefit of craft training is that _____.
   a. you have no financial responsibilities until you finish training
   b. you are able to earn while you learn
   c. you eventually make enough to pay off your training debt
   d. you have your apprenticeship after one year of training

3. One reason construction careers provide craft professionals with mobility is _____.
   a. subcontractors often give company automobiles to employees
   b. most craft professionals must travel due to the lack of work
   c. their skills are often in demand everywhere
   d. most good contractors demand that their employees travel long distances

3.0.0 Construction Craft Careers

Objective

Describe the typical career path for craft professionals.
   a. Describe industry sectors and the progression path for construction careers.

   b. Identify different construction careers and the types of skills they require.

Performance Tasks

There are no performance tasks in this section.

Skilled construction professionals are part of a growing industry that offers a wide array of career opportunities. While some people still hold an outdated view that lumps all construction jobs together, today's reality is much different. The construction industry offers many career paths and specializations. In fact, the National Center for Construction Education and Research (NCCER) offers specialized training programs for more than

45 construction crafts. This means individuals can choose from a large variety of construction jobs, many of which offer tremendous earning potential and great opportunity for advancement.

3.1.0   Construction Industry Sectors

Sector: A category of construction distinguished by specific types of work, materials, equipment, and skills. The four sectors of construction are residential, commercial, industrial, and heavy civil/infrastructure.

The construction industry comprises four primary sectors. Each **sector** includes specific types of construction, materials, equipment, and skills. The four sectors are:

- *Residential* — Includes design, construction, and maintenance of single-family homes, multi-family homes such as apartment buildings, public housing developments, and even separate garages and sheds.
- *Commercial* — Includes design, construction, and maintenance of schools, government buildings, medical facilities, hotels, sports arenas and stadiums, shopping centers, and large office buildings.
- *Industrial* — Includes construction of public works such as manufacturing plants, oil refineries, electrical generating plants, chemical processing plants, and large mills.
- *Heavy civil/infrastructure* — Often referred to as horizontal construction, this sector includes professionals who build bridges, roadways, airports, tunnels, and dams.

Each construction sector has its own approach to initiating and paying for projects, selecting the required equipment, and ensuring its workforce is adequately trained and prepared for the job.

Although each sector and construction craft has specific requirements for skills and training, the progression path for construction careers is usually the same as the one shown in *Figure 14*.

Contractor: Oversees many parts of a construction project, including resource management, budgeting, code adherence, quality assurance, and construction materials.

Project owner: The initiator of the project who usually finances the building endeavor.

Subcontractor/specialty contractor: Typically employs craft professionals who have specialized skills needed to complete a particular part of the construction project (e.g. HVAC, electrical, plumbing).

Crew leader: Also called *foreman*, this supervisory role oversees a crew of craft professionals. It is the crew leader's job to make sure that work is completed correctly and on time. Crew leaders are responsible for the safety and work of those under them.

Site superintendent: Manages day-to-day activities on a construction site and supervises work performed by the subcontractors.

Project manager: Oversees the planning and delivery of construction projects on time and within budget.

A **contractor** oversees many parts of a construction project, including resource management, budgeting, code adherence, quality assurance, and construction materials. A contractor typically works directly with the construction **project owner** or the owner's representative. The project owner is the initiator of the project and usually finances the building endeavor. A contractor hires a **subcontractor/specialty contractor** to complete specific construction tasks. In most cases, subcontractors employ craft professionals with expertise and skills that can be used across multiple construction sectors. For example, while some craft professionals are responsible for tasks related to erecting a structure, others are involved in installing and maintaining electrical, communication, HVAC, and plumbing systems.

Many construction professionals start their careers working for a subcontractor and may continue there or move up to become general contractors. As they refine their skills and develop their leadership capabilities, they may be promoted to **crew leader** or superintendent, or even progress to higher senior management positions. The crew leader, sometimes called the *foreman*, manages a team or crew of craft professionals assigned to work on a construction project (*Figure 15*). A **site superintendent** manages day-to-day activities on a construction site and supervises work performed by the subcontractors. A superintendent can be a craft professional who has been a crew leader and has completed the necessary supervisory training, or he or she can be a person who graduated from a university construction management program. On the other hand, a **project manager** plans, schedules, and executes projects within a set budget, but is not always on the construction site. A project manager typically holds a university degree.

As in many other industries, opportunities for advancement in construction are directly related to workers' skill level, their continued willingness to grow and develop, and the quality of their work.

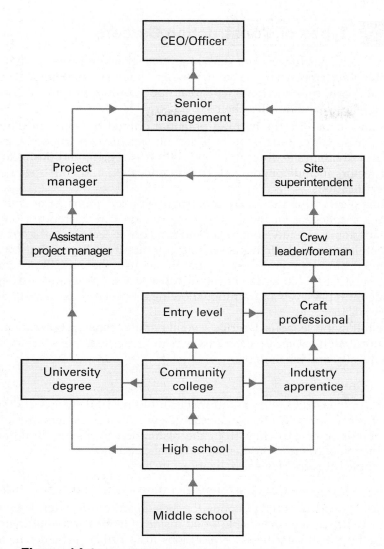

Figure 14 Career path for construction careers.

Figure 15 Construction project team.

3.2.0	**Types of Construction Careers**

One of the biggest benefits of pursuing a construction career is the variety of available jobs. Depending on your personal interests, professional goals, and desire to learn, construction offers great opportunities for practically everyone willing to learn and develop their skills.

Another benefit is the earning potential offered by many of the construction careers. Salary ranges differ based on specific construction career paths and an individual's skill level, quality of work, region, and desire to succeed. Other factors include ongoing classroom education and craft-specific training and certifications.

While performing the role of construction project manager or civil engineer requires a college education, most construction careers require only a high school diploma or equivalent to get started. Formal training programs are generally offered through technical schools, community colleges, and construction-oriented training organizations, and many use the NCCER curricula. An image of NCCER is shown in *Figure 16*. Most NCCER programs are designed to provide multiple levels of training, allowing a trainee to learn and obtain craft credentials in stages.

The following sections describe several popular construction careers, providing information related to each career path's responsibilities, salary potential, needed skills, and job outlook.

3.2.1 Carpenter

Carpenters construct, repair, and install building frameworks and structures made from wood and other materials (*Figure 17*). Carpenters are involved in many kinds of construction from the restoration of historic buildings to the construction of homes, commercial buildings, and more.

Examples of carpentry specializations include:

- *Rough carpenter* — Does the framing, formwork, and other structural work.
- *Finish carpenter* — Puts finishing touches on structures after they are almost fully built. Most of the visible wood inside of homes and buildings is placed there by finish carpenters. Typical jobs for a finish carpenter include window and door trim, crown molding, bookshelves, baseboards, and wooden stairways.

SUCCESS STORIES: *SkillsUSA 2018 Carpentry Competition Highlights*

Scan this code using the camera on your phone or mobile device to view this video.

Figure 16 National Center for Construction Education and Research.

Figure 17 Carpenters build both large and small structures.
Source: Stevecoleimages/E+/Getty Images

Salary

The average salary for a carpenter is approximately $57,000 per year. The top 10 percent of carpenters earn more than $84,690.

Job Growth Estimates

The demand for carpenters in the US is expected to remain the same, with about 90,000 new openings per year expected through the next decade.

Skills

To be successful, a carpenter should possess the following traits or skills:

- Physically fit
- Sense of balance
- Good eye-hand coordination
- Detail oriented
- Ability to read blueprints and follow instructions for installing certain products
- Basic math skills (to calculate sizes and amounts of materials accurately)
- Good problem-solving skills

Specific Qualifications

In some cases, carpenters are expected to earn an apprenticeship, which requires a specific number of hours on the job and in the classroom. NCCER offers a four-year carpentry training course that can be used to fulfill the classroom instruction requirement of an apprenticeship.

3.2.2 Electrician

Electricians install and maintain the electrical and power systems for homes, businesses, and factories (*Figure 18*). In large factories, electricians usually do maintenance work that is more complex. These kinds of electricians may repair motors, transformers, generators, and electronic controllers on machine tools and industrial robots. They also advise management regarding electrical hazards.

Another related electrical career is focused more on installation and maintenance of low-voltage electronic cables and devices. Some of the systems

Figure 18 Electricians install and maintain high-voltage power systems.

installed by these professionals include audio and video, security, fire protection, residential and commercial building networks, CCTV, and access control systems. These electronics professionals are known as limited-energy integrators or electronic systems technicians.

Salary

The average salary for an electrician is approximately $67,000 per year. The top 10 percent of electricians earn more than $96,580.

Job Growth Estimates

The demand for electricians in the US is expected to grow by almost 10 percent during the next decade, which equates to almost 80,000 new openings per year over the same time period.

Skills

To be successful, an electrician should possess the following traits or skills:

- Physically fit
- Good color vision (to recognize electrical wires by color)
- Detail-oriented
- Good math skills
- Good problem-solving skills
- Customer service skills
- Skilled with hands

Specific Qualifications

Credentials needed by electricians vary by state. In most states, they must be licensed. To earn a license, an electrician must pass a test with questions related to national, state, and local codes. These codes are in place to ensure safe installation of electrical wiring and equipment.

Electricians are generally required to take continuing education courses related to safety, code changes, and product training in order to maintain their licenses. To work in a specialty area, electricians are often required to obtain additional training.

3.2.3   Welder

Welders join metals together using a high-intensity electrical arc at temperatures between 6,000°F and 10,000°F. By joining metal together, welders create a variety of structures, including buildings, ships, bridges, automobiles, and other smaller items. They also cut steel using oxyfuel, air carbon-arc, or plasma-arc equipment. A strong focus on precision and safety is key for success in this craft. Examples of welding specializations include:

- *Pipe welder* — As the name implies, pipe welders construct and repair sections of pipe and related components (*Figure 19*).
- *Stainless steel welder* — Uses either TIG welding, spot welding, or MIG welding to join stainless steel metals. The method used typically depends on the qualities of the stainless steel being welded.
- *Weld inspector* — Assures the quality of existing welds by using high-tech testing methods.
- *Underwater welder* — While wearing full underwater diving gear, this professional performs different types of welding using a power source supplied through cables and hoses connected to welding equipment.
- *Robotic welder* — Programs robots and computer-controlled machines to perform welding.

SUCCESS STORIES: *Discover Success: Following in Footsteps*

Scan this code using the camera on your phone or mobile device to view this video.

Did You Know?

Two men of the 19th century are credited with making significant advancements in the welding field. French engineer Auguste de Meritens was the first to use heat from electricity to weld together lead plates. Around the same time, Russian inventor Nikolay Benardos discovered a similar method of welding using carbon rods.

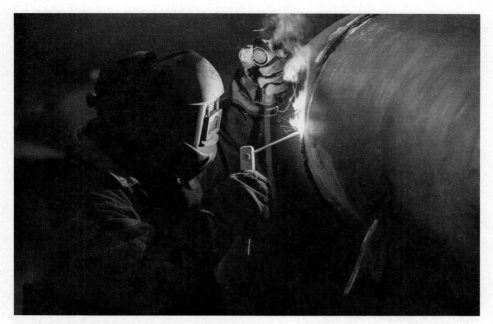

Figure 19 Pipe welders join sections of pipe.
Source: Imantsu/istock/Getty Images

Salary

The average salary of a pipe welder is approximately $69,000. Structural welders make an average salary of approximately $56,160. When other advanced specialty welding salaries are combined, the average salary of an experienced welder rises to $71,067.

Job Growth Estimates

The demand for welders in the US is expected to grow by about 3 percent during the next decade. As the nation's infrastructure ages, the expertise of welders will be needed to rebuild roads, bridges, and buildings. Aspiring welders have more opportunities if they are open to traveling or relocating to work on what are often short-term projects.

Skills

To be successful, a welder should possess the following traits or skills:

- Detail-oriented
- Hand-eye coordination
- Physically fit
- Strength in reading two- and three-dimensional drawings
- Basic understanding of electricity

Specific Qualifications

Some welding positions require a general welding certification. The American Welding Society (AWS) offers the Certified Welder credential. To become a specialist, a welder must obtain a certificate focused on a particular area, such as pipe welding. Underwater welders must also have a diver's certification.

3.2.4 Pipefitter

Pipefitters install, assemble, fabricate, maintain, and repair mechanical piping systems. They work with many kinds of pipe in manufacturing, commercial, and industrial settings such as chemical plants, oil refineries, food processing plants, and paper mills. They work with piping systems that carry water, gases, liquid chemicals, solids, and fuels (*Figure 20*). Pipefitters need to have a good understanding of blueprints, basic math, welding, pipe bending, fittings, and valves.

SUCCESS STORIES: *Discover Success: Earn More Through Welding Jobs*

Scan this code using the camera on your phone or mobile device to view this video.

Figure 20 Pipefitters connect and repair pipe in a variety of industrial settings.

Salary

The average salary for a pipefitter is approximately $63,000 per year. The top 10 percent of pipefitters earn more than $97,170 annually.

Job Growth Estimates

The demand for pipefitters in the US is expected to grow approximately 4 percent in the next decade, which works out to approximately 50,000 new openings per year.

Skills

To be successful, a pipefitter should possess the following traits or skills:

- Mechanical abilities
- Physical fitness
- Problem-solving skills
- Mathematical skills, particularly in geometry
- Ability to measure, cut, thread, and assemble pipe
- Ability to sketch piping systems

Specific Qualifications

Pipefitters typically go through a four-year craft training or apprenticeship program to become a journey-level fitter. Several states require pipefitters to be licensed when working on gas lines.

3.2.5 Heavy Equipment Operator

Heavy equipment operators drive and control heavy construction equipment to move construction materials, dirt, and other heavy objects at construction sites, large mills, mining operations, and distribution centers. On large construction

Figure 21 Heavy equipment operators drive large construction machinery.
Source: Ewg3D/istock/Getty Images

sites, heavy equipment may be used to clear, grade, and lift equipment prior to and during the construction of roads, buildings, bridges, airports, and power generation facilities.

Examples of heavy equipment include backhoes, bulldozers, dump trucks, forklifts, and excavators. Almost all operators choose to specialize in one or more of these types of equipment (*Figure 21*).

Salary

The average salary for a heavy equipment operator is approximately $61,000 per year. In a recent survey, the top 10 percent of people in this career earned more than $84,650. Professionals operating more complex equipment typically get paid higher salaries.

Job Growth Estimates

The demand for heavy equipment operators in the US is expected to grow by about 4 percent in the next decade. Metropolitan areas usually offer the most job opportunities, and operators trained to work on multiple types of equipment typically have the best opportunities.

Skills

To be successful, a heavy equipment operator should possess the following traits or skills:

- Physical strength
- Good sense of balance
- Hand-eye-foot coordination
- Mechanical skills

Specific Qualifications

Several OSHA standards apply to operating different types of heavy equipment. OSHA's *1910.178(l)(3)* standard requires operators of powered industrial trucks (forklifts) to be licensed or certified. When operators are tasked with hauling their own equipment between job sites, they may also be required to obtain a commercial driver's license.

SUCCESS STORIES: *James Snyder: I Built This*

Scan this code using the camera on your phone or mobile device to view this video.

3.2.6 Crane Operator

Crane operators use state-of-the-art machinery to hoist heavy construction materials above and around a job site. Joysticks, levers, and pedals allow operators to control the equipment. Professional crane operators are regularly tasked with using trigonometry and physics equations to calculate maximum load size based on the configuration of the crane and load.

There are many different types of cranes, including tower cranes that are seen high in the sky over tall buildings in cities around the world. Tower crane operators frequently control equipment that remains in place but grows taller as the building progresses upward. In some cases, cranes can rise many stories above their original starting point (*Figure 22*).

Salary

The average salary for a tower or mobile crane operator is approximately $66,000 per year. In a recent survey, the top 10 percent of people in this career earned more than $91,840.

Job Growth Estimates

The demand for crane operators in the US is expected to grow by about 8 percent in the next decade, which is above the average growth for all occupations. Metropolitan areas usually have the largest need for crane operators and pay the highest salaries. Because crane operators have unique skills, job prospects are extremely good for this profession.

Skills

To be successful, a crane operator should possess the following traits or skills:

- Endurance
- Agility
- Physical coordination
- Good sense of balance
- Ability to judge distance
- Hand-eye-foot coordination
- Good math skills
- Comfortable with technology

Figure 22 Crane operators hoist heavy materials off the ground.
Source: Hxdyl/istock/Getty Images

Specific Qualifications

According to OSHA standard *1926.1400(a)*, crane operators must be certified or licensed, with the type of certification based on the type or capacity of the crane. Licensing usually consists of a combination of written exams and physical tests in which individuals must demonstrate safe operating practices.

3.2.7   Ironworker

Ironworkers place and install iron or steel beams, columns, and other construction materials to form and reinforce structures. Contrary to the name, ironworkers primarily work with steel and are often referred to as *structural ironworkers.* These professionals primarily build structural frameworks that support large construction (*Figure 23*).

Salary

The average salary for an ironworker is approximately $60,000 per year. In a recent survey, the top 10 percent of people in this career earned more than $89,790.

Job Growth Estimates

The demand for ironworkers in the US is expected to grow by about 5 percent in the next decade. Because of the ongoing need to repair aging highways and bridges, an ironworker's skills should remain in high demand for years to come.

Figure 23 Ironworkers build steel structural frameworks.

Skills

To be successful, an ironworker should possess the following traits or skills:

- Physical fitness
- Agility
- Good sense of balance
- Hand-eye coordination
- Ability to work at heights
- Mechanical skills
- Math skills
- Problem-solving skills

SUCCESS STORIES: *Discover Success: Looking for Something Different*

Scan this code using the camera on your phone or mobile device to view this video.

Specific Qualifications

Ironworkers with a welding certification often have more job prospects.

3.2.8   HVAC Technician

HVAC technicians install, maintain, and repair heating, ventilation, and air conditioning systems. HVAC work includes both mechanical and electronic systems such as motors, pumps, fans, thermostats, and computerized switches that control systems in residential, commercial, and industrial structures (*Figure 24*). Technicians can also specialize in specific equipment, such as water-based heating systems, or commercial refrigeration.

Salary

The average salary for an HVAC technician is approximately $62,000 per year. In a recent survey, the top 10 percent of people in this career earned more than $77,920.

Job Growth Estimates

Demand for HVAC technicians in the US is expected to grow by about 4 percent over the next decade. Increased opportunities exist for technicians willing to move to higher-growth areas.

Figure 24 HVAC technicians install and repair air conditioning systems.
Source: Sturti/E+/Getty Images

Skills

To be successful, an HVAC technician should possess the following traits or skills:

- Good hand-eye coordination
- Mechanical aptitude
- Basic math skills
- Good problem-solving skills

Specific Qualifications

Some states and localities require technicians to be licensed; others require technicians to be certified when working with different types of chemicals.

3.2.9 Project Manager

Project managers are essential to completing construction projects on time and on budget. Their duties include planning, coordinating, budgeting, and supervising projects from beginning to end. They typically prepare cost estimates, explain project contracts to other professionals, manage personnel, resolve project issues, and ensure that projects adhere to safety codes and regulations. Project managers usually spend most of their time on site, where they monitor projects and ensure construction activities stay on schedule (*Figure 25*).

SUCCESS STORIES: *No One Day Is the Same—Discover Success*

Scan this code using the camera on your phone or mobile device to view this video.

Figure 25 Project managers monitor project schedules and activities.
Source: Kali9/E+/Getty Images

Salary

The average salary for a construction project manager is approximately $92,000 per year. In a recent survey, the top 10 percent of people in this career earned more than $164,790. Project managers may also earn bonuses based on their performance.

Job Growth Estimates

The demand for construction project managers in the US is expected to grow by about 8 percent over the next decade, which is well above the average growth for all occupations. As construction processes and mobile technology become more sophisticated, the need for project managers who stay current with technological changes is expected to increase.

Skills

To be successful, a project manager should possess the following traits or skills:

- Good oral and written communication skills
- Ability to make sound decisions
- Business acumen
- Ability to manage time and oversee multiple ongoing activities
- Strong customer service skills
- Good leadership skills

Specific Qualifications

Unlike most construction careers, project managers typically need a bachelor's degree from an accredited college or university. Large construction companies tend to seek candidates who have a bachelor's degree in a construction-related field and direct construction experience.

Some states require construction project managers to be licensed. Additionally, certifications in construction management and project management can increase the chances of landing a project management job.

3.2.10   More Construction Careers

While some of the more popular construction careers have been highlighted, NCCER offers training on more than 45 crafts, and you can discover more about them on NCCER's Build Your Future website, **www.byf.org**. This site offers information about salaries by craft, career paths, labor demand by state, and much more.

3.0.0 Section Review

1. Which construction sector includes schools, government buildings, hotels, and shopping centers?
   - a. Residential
   - b. Commercial
   - c. Public
   - d. Industrial

2. The construction role that typically reports to the project owner is the
   _____.
   - a. superintendent
   - b. foreman
   - c. subcontractor
   - d. contractor

3. To start your career path in most construction crafts, you generally need a _____.
   - a. high school diploma or equivalent
   - b. bachelor's degree
   - c. master's degree
   - d. state permit

4. The construction craft professional regularly tasked with using trigonometry and physics equations to calculate maximum load sizes is the _____.

   a. heavy equipment operator

   b. ironworker

   c. crane operator

   d. pipefitter

4.0.0 Starting Your Construction Career

Performance Tasks

There are no performance tasks in this section.

Objective

Identify ways to pursue a career in the construction industry.

   a. Explain the benefits of career and technical education programs.

   b. Describe the advantages of craft training programs and their relationship with apprenticeships.

   c. Summarize the path to a construction career through community colleges and universities.

You have several training options to consider if you are interested in pursuing a construction career. If you are still in high school, you should find out if any local schools offer additional construction classes. You can also look for summer jobs with local contractors. The Internet is a great resource for learning about construction crafts while you earn your diploma.

After earning your high school diploma or equivalent, choose which training path works best for you. Available training options include:

- Technical school
- Apprenticeship program
- Community college or university
- Industry training program

Most craft professionals are taught through a combination of technical schools, on-the-job learning (OJL), craft training programs, and/or apprenticeships.

4.1.0 Career and Technical Education

Career and technical education (CTE): Programs offered in high school and technical education centers that allow students to explore career options available in construction.

Career and technical education (CTE) classes are generally offered at high schools and technical education centers. Some schools start earlier by offering middle school classes that teach introductory concepts in electronics, engineering, carpentry, or other construction-related crafts. In either case, students attending these programs have the opportunity to explore career options available for each of the construction crafts. In some cases, students attending high school CTE classes have the opportunity to develop their hands-on skills through construction competition events like SkillsUSA. School guidance counselors can provide information about which CTE courses are offered and how students can enroll.

CTE programs place an emphasis on career readiness and hands-on learning (*Figure 26*). These programs emphasize the following approaches:

- *Real-life application* — Being able to practice skills learned has been shown to increase students' retention and understanding. For many students, being able to link traditional lessons directly to their future careers means they are more committed to succeeding in their education. Being able to do so in a low-pressure environment makes learning easier.

Figure 26 CTE programs focus on hands-on learning.

- *Teaching nontechnical skills* — Having a combination of craft-specific technical skills along with other important but nontechnical skills increases a student's employability. Some of these nontechnical skills include effective communication, time management, attention to detail, critical thinking, and a customer service attitude.
- *Career exploration* — By learning about careers early on, and from teachers with industry experience (*Figure 27*), students can discover their own strengths and interests, find a career that fits their passions, and set a path to achieve their goals.

After completing CTE classes, students generally have the qualifications needed to progress to one of these next steps:

- Industry apprenticeship
- Community college
- University degree

Figure 27 Students learn from industry experts in CTE programs.
Source: Istock/Getty Images

Success Story: Construction in The Villages

A growing retirement community called The Villages is located in north central Florida. In 2017, The Villages Charter School (TVCS) added a construction management academy to introduce students to construction, one of the most prevalent career fields in their community.

Each year, Bruce Haberle, a construction teacher at the school, has his senior class help construct a home for Habitat for Humanity. This project gives students a platform to practice the skills they are learning with real-life applications like reading blueprints, working on a schedule, and keeping an efficient pace. In their senior year, students are able to see what a career in construction is really like. Most recently, students worked to construct a 1,200-square-foot, 3-bedroom, 2-bathroom home.

Haberle is still refining the program for his students but says he has three primary goals for his students: prepare them, get them in the industry, and set them up for success in their careers.

4.2.0   Craft Training and Apprenticeships

The craft training and apprenticeship approach enables students to learn skills from experienced construction professionals. Regardless of students' previous exposure in high school, many construction companies provide training themselves, either through in-house programs or by paying for trainees to attend classes at a local construction association or community college. Ultimately, craft training programs and apprenticeships are excellent opportunities to be paid while working on the job.

A craft training or apprenticeship program features a split between learning in the classroom (typically 20 percent of learning) and on the job (typically 80 percent of learning) (*Figure 28*). It usually takes two to four years and requires at least 2,000 hours of OJL and 144 hours of classroom instruction during each year of the apprenticeship. According to the US Department of Labor, there are five components to an apprenticeship that are also included in craft training:

- Business involvement
- Structured OJL
- Related instruction
- Rewards for skill gains
- Nationally recognized credentials

Figure 28 Apprenticeships and craft training offer paid work while learning.

Craft training programs and apprenticeships offer many benefits, including full-time, paid work while learning skills on the job. This model of education not only gives trainees needed experience, but it keeps them from incurring high education debt. It also allows them to make valuable connections in the industry through relationships built with mentors.

You can find craft training programs and apprenticeships in your area in the following ways:

- Search the national apprenticeship database at **www.apprenticeship.gov**.
- Contact local construction associations like Associated Builders and Contractors (ABC) as well as Associated General Contractors (AGC).
- Search online job boards or other social media job search sites.
- Reach out to local employers.
- Research information on websites for national construction organizations.

4.3.0　Community Colleges and Universities

More and more colleges and universities offer construction-related degree programs. These institutions offer a variety of construction-oriented academic degrees. Community colleges usually provide two-year programs and offer associate degrees, while four-year universities are more often used for bachelor's degrees and beyond.

4.3.1　Community College

Community colleges often collaborate with contractors and construction companies by helping facilitate their training programs. Rather than teach employees in-house, these companies can send trainees to a community college. Doing so allows companies to fulfill the knowledge-based component of the program by taking advantage of the college's dedicated teaching staff and classroom facilities.

Earning an associate degree from a community college allows you to receive college credit as well as industry-recognized certifications and credentials upon graduation. Options for the next steps include:

- Industry apprenticeship
- Entry-level job
- University degree

4.3.2　Four-Year Universities

An increased number of four-year universities offer degree programs in construction-related areas such as design, construction management, estimating, human resources, architecture, engineering, and safety (*Figure 29*).

Degrees signify formal academic achievement and offer in-depth learning. A bachelor's degree focuses on an area of study (called a "major") and typically requires at least 120 hours of credit work. Typical degree programs will consist of general education courses (science, math, history, and English) along with specific construction courses.

A big advantage of a college degree is that it may allow you to move up the construction career ladder more quickly after gaining initial experience in your craft. You may also be more qualified for higher-paying roles that involve decision-making responsibilities.

However, there are financial challenges related to obtaining a four-year degree. The tuition can be very expensive and it can be difficult to work a full-time job while attending classes. In addition, a college degree is not a guarantee that you will be offered a specific construction job, and unfortunately, unless you have on-the-job construction experience, you may still have to start your career in an entry-level position.

SUCCESS STORIES: *How Many Jobs Need a 4-Year Degree*

Scan this code using the camera on your phone or mobile device to view this video.

Did You Know?

Louis Henry Sullivan, an American architect in the late 19th century, created a new style of architecture that resulted in buildings that were tall but still considered beautiful, a unique concept at the time. Called the "Father of Skyscrapers," he is most known for his design of the Wainwright Building in St. Louis.

Figure 29 Many universities offer construction-related degrees.
Source: JanPietruszka/istock/Getty Images

4.0.0 Section Review

1. CTE programs place an emphasis on preparing students for construction careers through _____.
   a. hands-on learning
   b. lectures and quizzes
   c. virtual lessons
   d. group projects

2. Depending on the particular craft, an apprenticeship or craft training program can generally be completed within _____.
   a. six months
   b. two to four years
   c. six to eight years
   d. ten years

3. Earning a construction degree guarantees you will be offered a construction job upon graduation.
   a. True
   b. False

Module 00100 Review Questions

1. What percentage of construction companies anticipate having to hire more people in the coming years to replace retiring craft professionals?
   a. 60 percent
   b. 70 percent
   c. 80 percent
   d. 90 percent

2. What percentage of construction field and office positions are currently filled by women?
   a. 10 percent
   b. 20 percent
   c. 30 percent
   d. 40 percent

3. The international standard used for certifying environmentally sound buildings is known as _____.

   a. LEED
   b. OSHA
   c. BIM
   d. NCCER

4. A developing technology that allows craft professionals to gather and share data and reports is _____.

   a. IoT devices
   b. robots
   c. wireless tools
   d. drones

5. The Empire State Building is important in construction history because, upon completion, it was the tallest building in _____.

   a. New York City
   b. the United States
   c. the Western Hemisphere
   d. the world

6. The fact that craft professionals restore essential services and structures following disasters is one reason they can be classified as _____.

   a. EMTs
   b. first responders
   c. public servants
   d. OSHA workers

7. Recent surveys indicate construction careers for all craft professionals typically pay salaries that fall between _____.

   a. $18,000 and $31,000
   b. $27,000 and $47,000
   c. $37,000 and $72,000
   d. $47,000 and $92,000

8. In addition to possessing specific craft skills, what other traits are essential for construction professionals to advance in their careers?

   a. Being friends with the contractor.
   b. Having good math and drafting skills.
   c. Expertise and decision-making skills.
   d. Willingness to travel and work off the clock.

9. The construction of single-family homes falls under the construction sector called _____.

   a. small structure
   b. infrastructure
   c. industrial
   d. residential

10. The construction role that is sometimes referred to as the *foreman* is also called the _____.

    a. site superintendent
    b. crew leader
    c. project owner
    d. contractor

11. The construction career that typically requires a four-year college education is a(n) _____.
    a. project manager
    b. finish carpenter
    c. underwater welder
    d. limited-energy integrator

12. One of the biggest advantages of pursuing a construction career is the _____.
    a. amount of vacation
    b. ability to make your own schedule
    c. large variety of jobs
    d. lack of needed skills for most crafts

13. A craft professional who creates structures by joining metals using high heat and pressure is called a(n) _____.
    a. heavy equipment operator
    b. ironworker
    c. crane operator
    d. welder

14. A craft profession that requires a good knowledge of blueprint reading and geometry is a(n) _____.
    a. electrician
    b. pipefitter
    c. HVAC technician
    d. welder

15. Apprenticeship programs are typically structured so that learning is _____.
    a. 10 percent in the classroom and 90 percent on the job
    b. 20 percent in the classroom and 80 percent on the job
    c. 30 percent in the classroom and 70 percent on the job
    d. 40 percent in the classroom and 60 percent on the job

Answers to Section Review Questions and Module Review Questions are found in *Appendix A*.

Basic Safety
(Construction Site Safety Orientation)

Source: Cleanfotos/Shutterstock

Objectives

Successful completion of this module prepares you to do the following:

1. Explain the benefits of safety, the cost of workplace incidents, and ways to reduce related hazards.
   a. Describe the types of workplace incidents along with physical and monetary impacts.
   b. Summarize the causes and consequences of common incidents.
   c. Explain how to recognize, evaluate, and control workplace hazards.
2. Describe common fall hazards and methods to prevent them.
   a. Summarize the most common types of construction fall hazards.
   b. Describe components of effective fall arrest systems and how they prevent or halt falls.
   c. Explain how to use ladders and stairs safely.
   d. Identify key steps to ensuring scaffolds are assembled and used safely.
3. Recognize and avoid struck-by and caught-in-between hazards.
   a. Describe struck-by hazards and how to avoid them.
   b. Describe common caught-in/caught-between hazards and steps that can prevent them.
4. Identify common electrical hazards and how to avoid them.
   a. Summarize basic job-site electrical safety guidelines.
   b. Explain the importance of disabling equipment as well as basic lockout/tagout procedures.
5. Associate personal protective equipment (PPE) with the hazards they reduce or eliminate.
   a. Explain how PPE is used to protect craftworkers from different types of injuries.
   b. Explain how respirators protect craftworkers from respiratory dangers.
6. Describe safety practices used with other common job-site hazards.
   a. List other types of hazards craftworkers may encounter.
   b. Describe common environmental hazards and how craftworkers should respond to them.
   c. Summarize hazards associated with hot work.
   d. Identify fire hazards and describe basic fire fighting procedures.
   e. Name different types of confined spaces and how to avoid related hazards.

Performance Tasks

Under supervision, you should be able to do the following:

1. Properly set up and climb/descend an extension ladder, demonstrating proper three-point contact.

2. Inspect the following PPE items and determine if they are safe to use:

- Eye protection
- Hearing protection
- Hard hat
- Gloves
- Fall arrest harnesses
- Lanyards
- Connecting devices
- Approved footwear

3. Properly don, fit, and remove the following PPE items:

- Eye protection
- Hearing protection
- Hard hat
- Gloves
- Fall arrest harness

4. Inspect a typical power cord and GFCI to ensure their serviceability.

Overview

Work at construction and industrial job sites can be hazardous. Most job-site incidents are caused by at-risk behavior, poor planning, lack of training, or failure to recognize the hazards. To help prevent incidents, every company must have a proactive safety program. Safety must be incorporated into all phases of the job and involve employees at every level, including management.

Industry Recognized Credentials

If you are training through an NCCER-accredited sponsor, you may be eligible for credentials from NCCER's Registry. The ID number for this module is 00101. Note that this module may have been used in other NCCER curricula and may apply to other level completions. Contact NCCER's Registry at 888.622.3720 or go to **www.nccer.org** for more information.

1.0.0 Safety and Hazard Recognition

Performance Tasks	Objective
There are no performance tasks in this section.	Explain the benefits of safety, the cost of workplace incidents, and ways to reduce related hazards. a. Describe the types of workplace incidents along with physical and monetary impacts. b. Summarize the causes and consequences of common incidents. c. Explain how to recognize, evaluate, and control workplace hazards.

When you take a job, you have an obligation to your employer, co-workers, family, and yourself to work safely. You also have an obligation to make sure anyone you work with is working safely. Your employer is likewise obliged to maintain a safe workplace for all employees. The ultimate responsibility for on-the-job safety, however, rests with you; safety is part of everyone's job. In this module, you will learn to ensure your safety, and that of the people you work with, by obeying the following rules:

- Follow safe work practices and procedures, both regulatory and corporate.
- Inspect safety equipment before use.
- Use safety equipment properly.

To take full advantage of the wide variety of training, job, and career opportunities the construction industry offers, you must first understand the importance of safety. Successful completion of this module will be your first step toward achieving this goal. Later modules offer more detailed explanations of safety procedures, along with opportunities to practice them.

On a typical job site, there are often many workers from many trades in one place. These workers are all performing different tasks and operations. As a result, the job site is constantly changing and hazards are continually emerging. These hazards can jeopardize your safety. Your employer should make every effort to plan safety into each job and to provide a safe and healthful job site. Ultimately, however, your safety is in your own hands.

Safety training is provided to make you aware that hazards exist all around you every day. The time you spend learning and practicing safety procedures can save your life and the lives of others.

Safety is a learned behavior and attitude. It is a way of working that must be incorporated into the company as a culture. A **safety culture** is created when all the workers at a job site or in an organization see the value of a safe work environment and support it through their actions. Creating and maintaining a safety culture is an ongoing process that includes a sound safety structure and attitude, and relates to organizations as well as individuals. Everyone in the company, from management to laborers, must be responsible for safety every day they come to work.

Safety culture: The culture created when the whole company sees the value of a safe work environment.

There are many benefits to having a safety culture. Companies with strong safety cultures usually have the following characteristics:

- Fewer at-risk behaviors
- Lower **incident** and **accident** rates
- Less turnover
- Lower absenteeism
- Higher productivity

Incident: Per the US Occupational Safety and Health Administration (OSHA), an unplanned event that does not result in personal injury but may result in property damage or is worthy of recording.

Accident: According to the US Occupational Safety and Health Administration (OSHA), an unplanned event that results in personal injury and/or property damage.

A strong safety culture can also improve a company's safety record, which leads to winning more bids and keeping workers employed. Contractors with poor safety records are sometimes excluded from bidding, so good safety performance is essential. Factors that contribute to a strong safety culture include the following:

- Embracing safety as a core value
- Strong leadership
- Establishing and enforcing high standards of performance
- The commitment and involvement of all employees
- Effective communication and commonly understood and agreed-upon goals
- Using the workplace as a learning environment
- Encouraging workers to have a questioning attitude and empowering them to stop work when faced with potential hazards
- Good organizational learning and responsiveness to change
- Providing timely responses to safety issues and concerns
- Continually monitoring performance
- Positive reinforcement when proper safety practices are demonstrated by employees

Did You Know?

Safety First

Safety training is required for all activities. Never operate tools, machinery, or equipment without prior training. Always refer to the manufacturer's instructions.

Around the World
GOST

While OSHA serves to protect workers by setting safety standards in the United States, other systems are used internationally. One such set of technical standards used on a regional basis is known as GOST. GOST standards are more far-reaching than OSHA standards, as they cover a much broader range of topics than worker safety alone. The first set of GOST standards were published in 1968 as state standards for the former Soviet Union. After the Soviet Union was dismantled, GOST became a regional standard used by many previous members of the Soviet Union. Although countries may also have some standards of their own, countries such as Belarus, Moldova, Armenia, and Ukraine continue to use GOST standards as well. The standards are no longer administered by Russia, however. Today, the standards are administered by the Euro-Asian Council for Standardization, Metrology and Certification (EASC).

Occupational Safety and Health Administration (OSHA): An agency of the US Department of Labor. Also refers to the Occupational Safety and Health Act of 1970, a law that applies to more than 130 million workers and 7 million job sites in the United States.

1.1.0   Incidents and Accidents

Incidents and accidents can occur at any job site. Both at-risk behavior and poor working conditions can cause these undesirable events. You can help prevent such events by using safe work habits, understanding what causes them, and learning how to prevent them.

The terms *incident* and *accident* are often used interchangeably. However, according to the US **Occupational Safety and Health Administration (OSHA)**, an incident is an unplanned event that may or may not result in property damage. However, an incident is worthy of being documented so that steps can be taken to prevent it from recurring. When an incident occurs, no personal injury has occurred.

An accident is defined as an unplanned event that results in personal injury and/or property damage. Therefore, an event that results in property damage alone could be considered an incident or an accident. If personal injury or a fatality has occurred, the event is definitely an accident.

There are varying opinions on the use of these two terms, however. The US National Safety Council defines an incident as an unplanned, undesired event that adversely affects the completion of a task. In this definition, there is no mention of injury or property damage. Other safety organizations across the globe are likely to have their own definitions of these terms as well, or may use completely different terms.

The most important thing to understand is that both incidents and accidents are undesirable events that have a negative effect on both projects and workers. Do not be surprised when you hear the terms used interchangeably by other workers. The definitions provided by OSHA are used here to provide context for these terms as they are used throughout this module.

The lessons you will learn in this module will help you work safely. You will be able to spot and avoid hazardous conditions on the job site. By following safety procedures, you will help keep your workplace free from incidents and accidents, and protect yourself and others from injury or even death.

1.1.1   Incident and Accident Categories

Incidents and accidents are often categorized by their severity and impact, as follows:

- *Near-miss* — An unplanned event in which no one was injured and no damage to property occurred, but during which either could have happened. Near-miss incidents are warnings that should always be reported rather than overlooked or taken lightly.

- *Property damage* — An unplanned event that results in damage to tools, materials, or equipment, but no personal injuries.

- *Minor injuries* — Personnel may have received minor cuts, bruises, or strains, but the injured workers returned to full duty on their next regularly scheduled work shift.

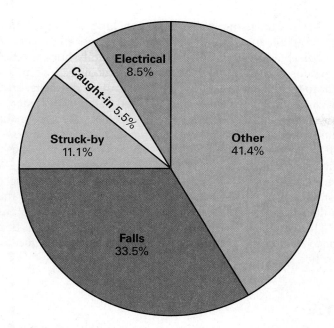

Figure 1 The four high-hazard areas.
Source: US OSHA, 2018

- *Serious or disabling injuries* — Personnel received injuries that resulted in temporary or permanent disability. Included in this category would be lost-time incidents, restricted-duty or restricted-motion cases, and those that resulted in partial or total disability.

- *Fatalities* — Deaths resulting from unplanned events.

Studies have shown that for every serious or disabling injury, there were 10 injuries of a less serious nature and 30 property damage incidents. Further study showed that 600 near-miss incidents occurred for every serious or disabling injury.

There are four leading causes of death in construction work. These are often referred to as the "big four," the "fatal four," or the "focus four." They include falls; struck-by hazards; caught-in or caught-between hazards; and electrical hazards (*Figure 1*). Deaths from falls far exceed all other causes.

Here are explanations of the four leading hazard groups:

- Falls from elevation are incidents involving failure of, failure to provide, or failure to use appropriate fall protection.

- Struck-by accidents involve unsafe operation of equipment, machinery, and vehicles, as well as improper handling of materials, such as through unsafe rigging operations.

- Caught-in or caught-between accidents involve unsafe operation of equipment, machinery, and vehicles, as well as improper safety procedures at **trench** sites and in other **confined spaces**.

- Electrical shock accidents involve contact with overhead wires; use of defective tools; failure to disconnect power source before repairs; or improper **ground fault** protection.

1.1.2 Costs

Incidents and accidents cost billions of dollars each year and cause much needless suffering. The National Safety Council estimates that the organized safety movement has saved more than 4.2 million lives since it began in 1913. This section examines why incidents and accidents happen and how you can help prevent them.

Trench: A narrow excavation made below the surface of the ground that is generally deeper than it is wide, with a maximum width of 15 feet (4.6 m). Also see *excavation*.

Confined spaces: Work areas large enough for a person to work, but arranged in such a way that an employee must physically enter the space to perform work. Confined spaces have a limited or restricted means of entry and exit. They are not designed for continuous work. Tanks, vessels, silos, pits, vaults, and hoppers are examples of confined spaces. Also see *permit-required confined space*.

Ground fault: Incidental grounding of a conducting electrical wire.

Accidents that result in injury or death can have a lasting effect not only on the victims, but on their families, co-workers, and employers. An injured worker who is disabled by an accident faces potentially huge medical bills. On top of those expenses, the worker's family faces loss of the income they rely on. It's convenient to think that insurance will take care of the costs, but it may not. Workers who are injured because they violated established safety rules may have their insurance claims denied and may be dismissed from the company because of the violation. A worker who is injured in a fall because he or she was not using fall protection could be refused compensation. How these issues are handled varies dramatically among companies and countries.

Employers and co-workers can be affected because many contract awards are based, in part, on a company's safety record. Therefore, incidents and accidents can also result in the loss of future jobs, which affects the company's financial position. This can mean layoffs, hiring freezes, or inability to purchase new equipment or tools. In this way, these events affect not only injured employees and their families, but everyone on the job site.

The Fatal Four

Out of 4,779 worker fatalities in US private industry during the 2018 calendar year, 1,008, or 21.1 percent, were in construction. In other words, one out of five worker deaths were in construction. The leading causes of worker deaths on construction sites were falls, followed by struck-by-object, electrocution, and caught-in/between. These fatal four were responsible for nearly 3 out of 5 (58.6 percent) construction worker deaths in 2018 reports, as reported by the US Bureau of Labor Statistics. Overall, a total of 591 workers died in 2018 as a result of the fatal four.

1.2.0   Incident and Accident Causes

When an incident or accident occurs on the job site, it can often be attributed to one of the following causes:

- Failure to communicate
- At-risk work habits
- Alcohol or drug abuse
- Lack of skill
- Intentional acts
- Unsafe acts
- Rationalizing risks
- Unsafe conditions
- Poor housekeeping
- **Management system** failure

Each of these causes is discussed further in the sections that follow.

1.2.1   Failure to Communicate

Many incidents happen because of a lack of communication. For example, you may learn how to do things one way on one job, but what happens when you go to a new job site? You need to communicate with the people at the new job site to find out whether they do things the way you have learned to do them. If you do not communicate clearly, incidents can happen. Remember that different people, companies, and job sites do things in different ways.

Making assumptions about what other workers know and what they will do can cause incidents. Don't assume, for example, that all workers understand what you are saying; some workers have limited language skills, especially

Management system: The organization of a company's management, including reporting procedures, supervisory responsibility, and administration.

Half-Measures

Most workers who die from falls are wearing harnesses but failed to tie off properly. Always follow the manufacturer's instructions when wearing a harness. Know and follow your company's safety procedures when working on roofs, ladders, and other elevated locations, and make sure you have an adequate anchor point at all times.

Information Sign

Caution Sign

Warning Sign

Safety Sign

DANGER

HIGH VOLTAGE

Danger Sign

Figure 2 Communication tags and signs.
Source: Courtesy of Accuform.com LLC

outside of their native language. Also, don't use terms or jargon that other people may not understand.

All work sites have specific markings and signs to identify hazards and provide emergency information (*Figure 2*). Learn to recognize these types of signs:

- Informational
- Safety
- Caution
- Danger
- Temporary warnings

Informational markings or signs provide general information. These signs are blue. The following are considered informational signs:

- No Admittance
- No Trespassing
- For Employees Only

Safety signs give general instructions and suggestions about safety measures. The background on these signs is white; most have a green panel with white letters. These signs tell you where to find such important areas as the following:

- First aid stations
- Emergency eye wash stations
- Evacuation routes
- **Safety Data Sheet (SDS)** stations
- Exits (usually have white letters on a red field)

Caution markings or signs tell you about potential hazards or warn against unsafe acts. When you see a caution sign, protect yourself against a possible

CAUTION

Never assume anything. It never hurts to ask questions, but disaster can result if you don't ask. For example, do not assume that an electrical power source is turned off. First ask whether the power is turned off, then check it yourself to be completely safe.

Toolbox Talks

Toolbox talks are one way to effectively keep all workers aware and informed of safety issues and guidelines. Toolbox talks are 5- to 10-minute meetings that review specific health and safety topics. These are very common at construction sites of all types.

Safety Data Sheet (SDS): A document that must accompany any hazardous substance. The SDS identifies the substance and gives the exposure limits, the physical and chemical characteristics, the kind of hazard it presents, precautions for safe handling and use, and specific control measures.

TABLE 1 Tags and Signs

Basic Stock (background)	Safety Colors (ink)	Message(s)
White	Red panel with white or gray letters	Do Not Operate Do Not Start
White	Black square with a red oval and white letters	Danger Unsafe Do Not Use
Yellow	Black square with yellow letters	Caution
White	Black square with white letters	Out of Order Do Not Use
Yellow	Red/magenta (purple) panel with black letters and a radiation symbol	Radiation Hazard
White	Fluorescent orange square with black letters and a biohazard symbol	Biological Hazard

Respirators: Devices that provide clean, filtered air for breathing, no matter what is in the surrounding air.

Flammable: Capable of easily igniting and rapidly burning; used to describe a fuel with a flash point below 100°F (38°C).

hazard. Caution signs are yellow and have a black panel with yellow letters. They may give you the following information:

- Hearing and eye protection are required
- **Respirators** are required
- Smoking is not allowed

Danger markings or signs tell you that an immediate hazard exists and that you must take certain precautions to avoid an incident. Danger signs are red, black, and white. They may indicate the presence of the following:

- Defective equipment
- **Flammable** liquids and compressed gases
- Safety barriers and barricades
- Emergency stop button
- High voltage

Safety tags are temporary warnings of immediate and potential hazards. They are not designed to replace signs or to serve as permanent means of protection. Learn to recognize the standard incident and accident prevention signs and tags (*Table 1*).

1.2.2 At-Risk Work Habits

Some examples of at-risk work habits are procrastination, carelessness, and horseplay. Procrastination (putting things off) is a common cause of incidents. For example, delaying the repair, inspection, or cleaning of equipment and tools can cause incidents and even accidents. If you try to push machines and equipment beyond their operating capacities, you risk injuring yourself and your co-workers.

Machines, power tools, and even a pair of pliers can hurt you if you don't use them safely. It is your responsibility to be careful; tools and machines don't know the difference between wood or steel, and flesh and bone.

Work habits and work attitudes are closely related. If you resist taking orders, you may also resist listening to warnings. If you let yourself be easily distracted, you won't be able to concentrate. If you aren't concentrating, you could cause a significant problem.

Your safety is affected not only by how you do your work, but also by how you act on the job site. This is why most companies have strict policies for employee behavior. Horseplay and other inappropriate behavior are forbidden. Workers who engage in such behavior on the job site may be fired.

These strict policies are provided for the worker's protection. There are many hazards on construction sites. Each person's behavior—at work, on a break, or at lunch—must follow the principles of safety.

A person pulling a practical joke on a co-worker could consider it just having fun, but in fact, it could cause the co-worker serious, even fatal, injury. If you horse around on the job, play pranks, or don't concentrate on what you are doing, you are showing a poor work attitude that can lead to a serious incident.

1.2.3 Alcohol and Drug Abuse

Alcohol and drug abuse costs the construction industry millions of dollars a year in incidents, accidents, lost time, and lost productivity. The true cost of alcohol and drug abuse is much more than just money, of course. Substance abuse can cost lives. Just as drunk driving kills thousands of people on our highways every year, alcohol and drug abuse kills on the construction site.

Using alcohol or drugs creates a risk of injury for everyone on a job site. Many states have laws that prevent workers from collecting insurance benefits if they are injured while under the influence of alcohol or illegal drugs.

Would you trust your life to a crane operator who was under the influence of drugs? Would you bet your life on the responses of a co-worker under the influence of alcohol or drugs? Alcohol and drug abuse have no place in the workplace. A person on a job site who is under the influence of alcohol or drugs is an incident or accident waiting to happen—possibly a fatal accident.

People who work while using alcohol or drugs are at risk of incident or injury; their co-workers are at risk as well. That's why most employers have a formal substance abuse policy that you should be aware of and follow. Avoid any substances that can affect your job performance; the life you save could be your own.

You do not have to be abusing drugs such as marijuana, cocaine, or heroin to create a job hazard. Many prescription and over-the-counter drugs, taken for legitimate reasons, can affect your ability to work safely. Amphetamines, barbiturates, and antihistamines are only a few of the prescription-controlled legal drugs that can affect the ability to work or operate machinery safely. The main thing is to understand and follow your company's substance abuse policy.

1.2.4 Lack of Skill

Every worker needs to learn and practice new skills under careful supervision. Never perform new tasks alone until you have been checked out by a supervisor. Lack of skill can cause incidents and accidents quickly. For example, suppose you are told to cut some boards with a circular saw, but you aren't skilled with that tool.

CAUTION

If your doctor prescribes any medication that you think might affect your job performance, ask about its effects. Your safety and the safety of your co-workers depend on everyone being alert on the job.

Did You Know?

Stress Effects

Stress creates a chemical change in your body. Although stress may heighten your hearing, vision, energy, and strength, long-term stress can harm your health. Not all stress is job-related; some stress develops from the pressures of dealing with family and friends and daily living. In the end, your ability to handle and manage your stress determines whether stress hurts or helps you. Use common sense when you are dealing with stressful situations. For example, consider the following:

- Keep daily occurrences in perspective. Not everything is worth getting upset, angry, or anxious about.

- When you have a particularly difficult workday scheduled, get plenty of rest the night before.

- Manage your time. The feeling of always being behind creates a lot of stress. Waiting until the last minute to finish an important task adds unnecessary stress.

- Talk to your supervisor. Your supervisor may understand what is causing your stress and may be able to suggest ways to manage it better.

A basic rule of circular saw operation is never to cut without a properly functioning guard. Because you haven't been trained, you don't know this. You find that the guard on the saw is slowing you down, so you jam the guard open. The result could be a serious accident. Proper training can prevent this from happening.

Never operate a power tool or machine until you have been trained to use it. Some power tools require that a worker be trained and certified in their use. You can greatly reduce the chances of incidents and accidents by learning the safety rules for each task you perform.

1.2.5 Intentional Acts

When someone purposely causes property damage or injury, it is called an intentional act. An angry or dissatisfied employee may purposely create a situation that leads to an incident or accident. If someone you are working with threatens to get even or pay back someone, let your supervisor know at once.

Unfortunately, terrorism must also be a consideration on the job site. The level of concern is typically dependent on the site itself and the nature of the structure. Pipelines, for example, may represent a significant target due to their size, importance, location (sections of pipeline are in remote areas), and potential for destruction. Controversial sites also tend to attract attention, perhaps from terrorist groups operating within a country that oppose the construction or what the structure represents.

As an individual worker, the most important thing you can do is to pay attention. Become familiar with people on the job site and look for strangers that do not seem to fit. A terrorist attack can come from any direction and be delivered in many different ways. It can begin with a single individual gaining access to the job site.

1.2.6 Unsafe Acts

An unsafe act is a change from an accepted, normal, or correct procedure that can cause an incident or accident. It can be any conduct that causes unnecessary exposure to a job-site hazard or that makes an activity less safe than usual. Here are examples of unsafe acts:

Personal protective equipment (PPE): Equipment or clothing designed to prevent or reduce injuries.

- Failing to use required **personal protective equipment (PPE)**
- Failing to warn co-workers of hazards
- Lifting improperly
- Loading or placing equipment or supplies improperly
- Making safety devices (such as saw guards) inoperable
- Operating equipment at improper speeds
- Operating equipment without authority
- Servicing equipment in motion
- Taking an improper working position
- Using defective equipment
- Using equipment improperly

1.2.7 Rationalizing Risk

Everybody takes risks every day. When you get in your car to drive to work, you know there is a risk of being involved in an accident. Yet when you drive using all the safety practices you have learned, you know that there is a very good chance you will arrive at your destination safely. Driving is an acceptable risk because you have some control over your own safety and that of others.

Some risks are not acceptable. On the job, you must never take risks that endanger yourself or others just because you think you can justify doing so. This is called rationalizing risk, which means ignoring safety warnings and practices. For example, because you are late for work, you might decide to run a red light. Perhaps you feel the risk is worth the time saved. This is not only a dangerous driving habit, but this kind of thinking is unacceptable on the job site.

Self-Care

Practicing good self-care not only benefits you personally, but also increases your chances of a long and successful construction career. Self-care strategies can be done in simple, cost-effective ways and include the following:

- Quality sleep — Sleep deprivation negatively impacts every system in your body and can result in significant errors in judgment, reduced ability to safely operate machinery, and increased risk of becoming distracted. These consequences are extremely dangerous on a job site. Quality sleep is enhanced by other kinds of self-care and limiting screen time before bed.

- Healthy eating — The high energy levels and physical strength required by construction work are supported by a healthy diet filled with lean proteins, vegetables, fruits, and whole grains. If your grocery store has limited fresh produce, look for plain, frozen fruits and vegetables. Dozens of apps and blogs exist to assist with budgeting and recipes. Hydration, especially in hot weather, is also critical.

- Exercise — In addition to supporting the physical requirements of construction work, regular exercise aids in regulating sleep and reducing stress. Many workouts can be done at home using just your body weight.

- Massage therapy — Regular massage, particularly for the upper back, shoulder, and neck, keeps your muscles and connective tissue flexible, which helps prevent and heal injuries.

- Chiropractic care — Many people who suffer from chronic low back pain benefit from chiropractic care, which involves manipulation of joints with a focus on the spine.

- Acupuncture — This ancient practice has become increasingly popular as a minimally invasive, drug-free way of addressing chronic pain and stress management.

Most schools offering massage therapy, acupuncture, or chiropractic programs have student clinics that provide these services at a heavily discounted rate. Insurance also covers many self-care strategies.

Make self-care a priority as you embark on your career path. Your long-term health and career longevity depend on it.

The following are common examples of rationalized risks on the job:

- Crossing barricades or boundaries because there is no apparent activity
- Not wearing gloves because it's easier to do the job with bare hands
- Removing your hard hat because you are hot and you cannot see anyone working overhead
- Not tying off your fall protection because you only have to lean over by about a foot

Think about the job before you do it. If you think that it is unsafe, then it is unsafe. Stop working until the job can be done safely. Bring your concerns to the attention of your supervisor. Your health and safety, and that of your co-workers, make it worth taking extra care.

1.2.8 Unsafe Conditions

An unsafe condition is a physical state that is different from the acceptable, normal, or correct condition found on the job site. It usually results in an incident or accident. An unsafe condition can be anything that reduces the degree of safety normally present. The following are some examples of unsafe conditions:

- Congested workplace
- Defective tools, equipment, or supplies
- Excessive noise
- Fire and explosive hazards
- Hazardous atmospheric conditions (such as gases, dusts, fumes, and vapors)
- Inadequate supports or guards
- Inadequate warning systems
- Cluttered work area
- Poor lighting
- Poor ventilation
- Radiation exposure
- Unguarded moving parts such as pulleys, drive chains, and belts

All employees should be given the authority to stop work when an unsafe condition is observed.

Figure 3 Container for oily rags and waste.
Source: Courtesy of Justrite Mfg. Co. LLC

Combustible: Capable of easily igniting and rapidly burning; used to describe a fuel with a flash point at, or above, 100°F (38°C).

Tool Blades

Dull blades cause more accidents than sharp ones. If you do not keep your cutting tools sharpened, they won't cut very easily. When you have a hard time cutting, you exert more force on the tool. When that happens, something is bound to slip. And when something slips, you can be injured.

Hazard Communication (HAZCOM) Standard: The Occupational Safety and Health Administration standard that requires contractors to educate employees about hazardous chemicals on the job site and how to work with them safely.

1.2.9   Poor Housekeeping

Housekeeping means keeping your work area clean and free of scraps, clutter, or spills. It also means being orderly and organized. Store your materials and supplies safely and label them properly. Arranging your tools and equipment to permit safe, efficient work practices and easy cleaning is also important.

If the work site is indoors, make sure it is well-lit and ventilated. Don't allow aisles and exits to be blocked by materials and equipment. Make sure that flammable liquids are stored in safety cans. Oily rags must be placed only in approved, self-closing metal containers (*Figure 3*).

Remember that the major goal of housekeeping is to prevent incidents. Good housekeeping reduces the chances for slips, fires, explosions, and falling objects. Here are some good housekeeping rules:

- Remove from work areas all scrap material and lumber with nails protruding.
- Clean up spills to prevent falls.
- Remove all **combustible** scrap materials regularly.
- Make sure you have containers for the collection and separation of refuse. Containers for flammable or harmful refuse must have covers.
- Dispose of wastes often.
- Store all tools and equipment in the proper location and condition when you are finished with them.
- Keep all aisles and walkways clear of materials and tools.

Another term for good housekeeping is pride of workmanship. If you take pride in what you are doing, you won't let trash build up around you. The saying "a place for everything and everything in its place" is the right attitude on the job site.

1.2.10   Management System Failure

Sometimes the cause of an incident or accident is failure of the management system. The management system should be designed to prevent or correct the acts and conditions that can create safety hazards. If the safety management system is not functioning properly, problems are likely to occur.

What traits could mean the difference between a management system that fails and one that succeeds? A company implementing a good management system will do the following:

- Put safety policies and procedures in writing
- Distribute written safety policies and procedures to each employee
- Review safety policies and procedures periodically
- Enforce all safety policies and procedures fairly and consistently
- Evaluate supplies, equipment, and services to see whether they are safe
- Provide regular, periodic safety training for employees

1.3.0   Hazard Recognition, Evaluation, and Control

The process of hazard recognition, evaluation, and control is the foundation of an effective safety program. When hazards are identified and assessed, they can be addressed quickly, reducing the hazard potential. Simply put, the more aware you are of your surroundings and the dangers in them, the less likely you are to be involved in an incident.

There is a standard rule in the United States that affects every worker in most industries. It is often referred to as HAZCOM, which is short for **Hazard Communication Standard**. The US HAZCOM also aligns with the Globally Harmonized System of Classification and Labeling of Chemicals (GHS). HAZCOM may also be called the Right-To-Know requirement. It requires all US contractors to educate their employees about the hazardous chemicals they may be exposed to on the job site. Employees must be taught how to work safely around these materials. Other countries have similar programs.

Many people think that there are very few hazardous chemicals on construction job sites. That is simply not true. In practice, the term *hazardous chemical* applies to solvents, paint, concrete, and even wood dust, along with more obvious substances such as acids.

As an employee, you have the following responsibilities under HAZCOM:

- Know where SDSs are on your job site.
- Report any hazards you spot on the job site to your supervisor.
- Know the physical and health hazards of any hazardous materials on your job site, and know and practice the precautions needed to protect yourself from these hazards.
- Know what to do in an emergency.
- Know the location and content of your employer's written hazard communication program.

The final responsibility for your safety rests with you. Your employer must provide you with information about hazards, but you must know this information and follow safety rules.

1.3.1 Hazard Recognition

There are many potential hazard indicators. The best approach in determining if a situation or equipment is potentially hazardous is to ask these questions:

- How can this situation or equipment cause harm?
- What types of energy sources are present that can cause an incident?
- What is the magnitude of the energy?
- What could go wrong to release the energy?
- How can the energy be eliminated or controlled?
- Will I be exposed to any hazardous materials?

Before you can fully answer these questions, you need to know the different types of incidents and accidents that can happen and the energy sources behind them. Some of the different types of events that can cause injuries include the following:

- Falls on the same elevations or falls from elevations
- Being caught in, on, or between equipment
- Being struck by falling objects
- Contact with acid, electricity, heat, cold, radiation, pressurized liquid, gas, or toxic substances
- Being cut by tools or equipment
- Exposure to high noise levels
- Repetitive motion or excessive vibration

You might recognize the first four high-hazard conditions as the so-called fatal four previously discussed. Remember, these four types of accidents cause more than half of all construction fatalities.

When equipment is the cause of an incident or accident, it is usually because there was an uncontrolled release of energy. The different types of energy sources that can be released include the following:

- Mechanical
- **Pneumatic**
- **Hydraulic**
- Electrical
- Chemical

Pneumatic: Powered by air pressure, such as a pneumatic tool.

Hydraulic: Powered by fluid under pressure.

- Thermal (heat or cold)
- Radioactive
- Gravitational
- Stored energy

There are a number of ways to recognize hazards and potential hazards on a job site. Some techniques are more complicated than others. In order to be effective, they all must answer this question: What could go wrong with this situation or operation? No matter what hazard recognition technique you use, answering that question in advance will save lives and prevent equipment damage.

1.3.2   Job Safety Analysis (JSA) and Task Safety Analysis (TSA)

Performing a job safety analysis (JSA), also known as job hazard analysis (JHA), is one approach to hazard recognition. Another common technique is performing a task safety analysis (TSA), also called a task hazard analysis (THA).

In a JSA, the task at hand is broken down into its individual parts or steps and then each step is analyzed for its potential hazards. Once a hazard is identified, certain actions or procedures are recommended that will correct that hazard. For example, during a JSA, it is determined that using a chain hoist to install a pump motor in a tight space would be safer than having a worker do it manually. By using the chain hoist, the chance that the worker's hand would get crushed during installation is reduced. Using the JSA process saved the worker from injury. *Figure 4* shows an example of a form used to conduct a JSA.

JSAs can also be used as pre-planning tools. This helps to ensure that safety is planned into the job. You may be asked to take part in a JSA during job planning. When JSAs are used as pre-planning tools, they contain the following information:

- Tools, materials, and equipment needs
- Staffing or manpower requirements
- Duration of the job
- Quality concerns

Task safety analysis is similar to job safety analysis in that both require workers to identify potential hazards and needed safeguards associated with a job they are about to do. The difference is the form used to report the hazards. During a TSA, a pre-printed, fill-in-the-blank checklist is often used to document any hazard found during analysis. Before work begins, the first-line supervisor or team leader should discuss the conclusions found during the TSA with the crew. Some companies require workers to sign the completed TSA forms or checklists before they start work. This helps companies document the hazards and ensure that workers have been told of the potential hazards and safety procedures.

1.3.3   Risk Assessment

Whether an action is considered safe is often a matter of evaluating risk. Risk is a measure of the probability, consequences, and exposure related to an event. Probability is the chance that a given event will occur. Consequences are the results of an action, condition, or event. Exposure is the amount of time and/or the degree to which someone or something is exposed to an unsafe condition, material, or environment.

A safe operation is one in which there is an acceptable level of risk. This means there is a low probability of an incident and that the consequences and exposure risk are all acceptable. For example, climbing a ladder has risk that is considered to be acceptable if the proper ladder is being used as intended, if it is set up correctly, and if it is in good condition. The probability

JOB SAFETY ANALYSIS

TITLE OF JOB OR TASK

TASK	START	END	HAZARDS	CONTROLS
1.				
2.				
3.				
4.				
5.				
6.				
7.				
8.				

Required Training:

Required Personal Protective Equipment (PPE):

Weekly Vehicle Check List:

Tire Pressure _____   Transmission Fluid _____
Oil _____   Lights _____
Air Filter _____   Wkly Mileage _____

Job Name:
Job Number:
Supervisor:
Date:

Names of Employees:	PRINT NAME	SIGN NAME	TOTAL HOURS

Figure 4 Job safety analysis form.
Source: L.J. LeBlanc

of exposure to a hazard and its potential consequences are all low. If any one of these conditions were different, climbing the ladder would have an unacceptable level of risk.

1.3.4   Reporting Injuries, Incidents, and Near-Misses

All on-the-job incidents and accidents, no matter how minor, must be reported to your supervisor. Some workers think they will get in trouble if they report minor injuries, but that is not the case. Small injuries, like cuts and scrapes, can later become big problems because of infection and other complications.

US employers with more than 10 employees are required to maintain a log of significant work-related injuries and illnesses using specific OSHA forms and documents. Employee names can be kept confidential in certain circumstances. A summary of these injuries must be posted at certain intervals, although employers do not need to submit it to OSHA unless requested. Employers can calculate the total number of injuries and illnesses and compare the result with the average national rates for similar companies. By analyzing incidents and accidents, companies and OSHA can improve safety policies and procedures. By reporting an incident, you can help keep similar events from happening in the future. For details on the operation of OSHA and its important mission, refer to the *Appendix B.*

1.3.5   Safety Data Sheets

SDSs are fact sheets prepared by the chemical manufacturer or importer. Each product used on a construction site must have an SDS. An SDS describes the substance, along with its hazards, safe handling, first aid, and emergency spill procedures. The HAZCOM standard requires new SDSs (formerly known as material safety data sheets or MSDSs) to be in a standardized format. The sections of the form include the following:

- *Section 1, Product identification, manufacturer contact information, recommended uses and restrictions*
- *Section 2, Hazard identification*
- *Section 3, Composition/information on ingredients*
- *Section 4, First aid measures*
- *Section 5, Fire fighting measures*
- *Section 6, Accidental release measures*
- *Section 7, Handling and storage*
- *Section 8, Exposure controls/personal protection*
- *Section 9, Physical and chemical properties*
- *Section 10, Stability and reactivity*
- *Section 11, Toxicological information*
- *Section 12, Ecological information*
- *Section 13, Waste disposal considerations*
- *Section 14, Transport information*
- *Section 15, Regulatory information*
- *Section 16, Other information*

Figure 5A and *Figure 5B* shows a sample SDS for a PVC solvent cement. The most important things to look for on an SDS are the specific hazards, personal protection, handling procedures, and first aid information. Most SDSs have a 24-hour emergency-response number.

Using *Figure 5A* and *Figure 5B*, try to find the information you would need to use the cement described on the sample SDS. First, locate the hazards; section 2 of the SDS shows that the adhesive is flammable, and also an eye and skin irritant that can cause respiratory irritation, dizziness, and drowsiness.

WELD·ON® GHS SAFETY DATA SHEET

WELD-ON® 705™ Low VOC Cements for PVC Plastic Pipe

Date Revised: JUN 2018
Supersedes: NOV 2017

SECTION 1 - PRODUCT AND COMPANY IDENTIFICATION

PRODUCT NAME: WELD-ON® 705™ Low VOC Cements for PVC Plastic Pipe
PRODUCT USE: Low VOC Solvent Cement for PVC Plastic Pipe
SUPPLIER:

MANUFACTURER: IPS Corporation
17109 South Main Street, Gardena, CA 90248-3127
P.O. Box 379, Gardena, CA 90247-0379
Tel. 1-310-898-3300

EMERGENCY: Transportation: CHEMTEL Tel. 800.255.3924, +1 813-248-0585 (International) **Medical:** CHEMTEL Tel. 800.255.3924, +1 813-248-0585 (International)

SECTION 2 - HAZARDS IDENTIFICATION

GHS CLASSIFICATION:

Health		Environmental		Physical	
Acute Toxicity:	Category 4	Acute Toxicity:	None Known	Flammable Liquid	Category 2
Skin Irritation:	Category 3	Chronic Toxicity:	None Known		
Skin Sensitization:	NO				
Eye:	Category 2				

GHS LABEL:

Signal Word: Danger

WHMIS CLASSIFICATION: CLASS B, DIVISION 2
CLASS D, DIVISION 1B

Hazard Statements	Precautionary Statements
H225: Highly flammable liquid and vapor	P210: Keep away from heat/sparks/open flames/hot surfaces – No smoking
H319: Causes serious eye irritation	P261: Avoid breathing dust/fume/gas/mist/vapors/spray
H332: Harmful if inhaled	P280: Wear protective gloves/protective clothing/eye protection/face protection
H335: May cause respiratory irritation	P304+P340: IF INHALED: Remove victim to fresh air and keep at rest in a position comfortable for breathing
H336: May cause drowsiness or dizziness	P403+P233: Store in a well ventilated place. Keep container tightly closed
H351: Suspected of causing cancer	P501: Dispose of contents/container in accordance with local regulation
EUH019: May form explosive peroxides	

SECTION 3 - COMPOSITION/INFORMATION ON INGREDIENTS

	CAS#	EINECS #	REACH Pre-registration Number	CONCENTRATION % by Weight
Tetrahydrofuran (THF)	109-99-9	203-726-8	05-2116297729-22-0000	25 - 50
Methyl Ethyl Ketone (MEK)	78-93-3	201-159-0	05-2116297728-24-0000	5 - 36
Cyclohexanone	108-94-1	203-631-1	05-2116297718-25-0000	15 - 30

All of the constituents of this adhesive product are listed on the TSCA inventory of chemical substances maintained by the US EPA, or are exempt from that listing.
* Indicates this chemical is subject to the reporting requirements of Section 313 of the Emergency Planning and Community Right-to-Know Act of 1986 (40CFR372).
# indicates that this chemical is found on Proposition 65's List of chemicals known to the State of California to cause cancer or reproductive toxicity.

SECTION 4 - FIRST AID MEASURES

Contact with eyes:	Flush eyes immediately with plenty of water for 15 minutes and seek medical advice immediately.
Skin contact:	Remove contaminated clothing and shoes. Wash skin thoroughly with soap and water. If irritation develops, seek medical advice.
Inhalation:	Remove to fresh air. If breathing is stopped, give artificial respiration. If breathing is difficult, give oxygen. Seek medical advice.
Ingestion:	Rinse mouth with water. Give 1 or 2 glasses of water or milk to dilute. Do not induce vomiting. Seek medical advice immediately.
Likely Routes of Exposure:	Inhalation, Eye and Skin Contact

Acute symptoms and effects:

Inhalation:	Severe overexposure may result in nausea, dizziness, headache. Can cause drowsiness, irritation of eyes and nasal passages.
Eye Contact:	Vapors slightly uncomfortable. Overexposure may result in severe eye injury with corneal or conjunctival inflammation on contact with the liquid.
Skin Contact:	Liquid contact may remove natural skin oils resulting in skin irritation. Dermatitis may occur with prolonged contact.
Ingestion:	May cause nausea, vomiting, diarrhea and mental sluggishness.
Chronic (long-term) effects:	Category 2 Carcinogen

SECTION 5 - FIREFIGHTING MEASURES

			HMIS	NFPA	
Suitable Extinguishing Media:	Dry chemical powder, carbon dioxide gas, foam, Halon, water fog.				0-Minimal
Unsuitable Extinguishing Media:	Water spray or stream.	Health	2	2	1-Slight
Exposure Hazards:	Inhalation and dermal contact	Flammability	3	3	2-Moderate
Combustion Products:	Oxides of carbon, hydrogen chloride and smoke	Reactivity	0	0	3-Serious
		PPE	B		4-Severe
Protection for Firefighters:	Self-contained breathing apparatus or full-face positive pressure airline masks.				

SECTION 6 - ACCIDENTAL RELEASE MEASURES

Personal precautions:	Keep away from heat, sparks and open flame.
	Provide sufficient ventilation, use explosion-proof exhaust ventilation equipment or wear suitable respiratory protective equipment. Prevent contact with skin or eyes (see section 8).
Environmental Precautions:	Prevent product or liquids contaminated with product from entering sewers, drains, soil or open water course.
Methods for Cleaning up:	Clean up with sand or other inert absorbent material. Transfer to a closable steel vessel.
Materials not to be used for clean up:	Aluminum or plastic containers

SECTION 7 - HANDLING AND STORAGE

Handling:	Avoid breathing of vapor, avoid contact with eyes, skin and clothing.
	Keep away from ignition sources, use only electrically grounded handling equipment and ensure adequate ventilation/fume exhaust hoods.
	Do not eat, drink or smoke while handling
Storage:	Store in ventilated room or shade below 44°C (110°F) and away from direct sunlight.
	Keep away from ignition sources and incompatible materials: caustics, ammonia, inorganic acids, chlorinated compounds, strong oxidizers and isocyanates.
	Follow all precautionary information on container label, product bulletins and solvent cementing literature.

SECTION 8 - PRECAUTIONS TO CONTROL EXPOSURE / PERSONAL PROTECTION

EXPOSURE LIMITS:

Component	ACGIH TLV	ACGIH STEL	OSHA PEL	OSHA STEL	OSHA PEL-Ceiling	CAL/OSHA PEL	CAL/OSHA Ceiling	CAL/OSHA STEL
Tetrahydrofuran (THF)	50 ppm	100 ppm	200 ppm	N/E	N/E	200 ppm	N/E	250 ppm
Methyl Ethyl Ketone (MEK)	200 ppm	300 ppm	200 ppm	N/E	N/E	200 ppm	N/E	300 ppm
Cyclohexanone	20 ppm	50 ppm	50 ppm	N/E	N/E	25 ppm	N/E	N/E

Engineering Controls:	Use local exhaust as needed.
Monitoring:	Maintain breathing zone airborne concentrations below exposure limits.

Personal Protective Equipment (PPE):

Eye Protection:	Avoid contact with eyes, wear splash-proof chemical goggles, face shield, safety glasses (spectacles) with brow guards and side shields, etc. as may be appropriate for the exposure.
Skin Protection:	Prevent contact with the skin as much as possible. Butyl rubber gloves should be used for frequent immersion.
	Use of solvent-resistant gloves or solvent-resistant barrier cream should provide adequate protection when normal adhesive application practices and procedures are used for making structural bonds.
Respiratory Protection:	Prevent inhalation of the solvents. Use in a well-ventilated room. Open doors and/or windows to ensure airflow and air changes. Use local exhaust ventilation to remove airborne contaminants from employee breathing zone and to keep contaminants below levels listed above. With normal use, the Exposure Limit Value will not usually be reached. When limits approached, use respiratory protection equipment.

Figure 5A Solvent cement SDS (1 of 2).
Source: Weld-On Adhesives, Inc., a division of IPS Corporation

GHS SAFETY DATA SHEET

WELD-ON® 705™ Low VOC Cements for PVC Plastic Pipe

Date Revised: **JUN 2018**
Supersedes: **NOV 2017**

SECTION 9 - PHYSICAL AND CHEMICAL PROPERTIES

Appearance:	Clear or gray, medium syrupy liquid
Odor:	Ketone
pH:	Not Applicable
Melting/Freezing Point:	-108.5°C (-163.3°F) Based on first melting component: THF
Boiling Point:	66°C (151°F) Based on first boiling component: THF
Flash Point:	-20°C (-4°F) TCC based on THF
Specific Gravity:	0.9611 @23°C (73°F)
Solubility:	Solvent portion soluble in water. Resin portion separates out.
Partition Coefficient n-octanol/water:	Not Available
Auto-ignition Temperature:	321°C (610°F) based on THF
Decomposition Temperature:	Not Applicable
VOC Content:	When applied as directed, per SCAQMD Rule 1168, Test Method 316A,VOC content is: ≤ 500 g/l.

Odor Threshold:	0.88 ppm (Cyclohexanone)
Boiling Range:	66°C (151°F) to 156°C (313°F)
Evaporation Rate:	> 1.0 (BUAC = 1)
Flammability:	Category 2
Flammability Limits:	**LEL:** 1.1% based on Cyclohexanone
	UEL: 11.8% based on THF
Vapor Pressure:	129 mm Hg @ 20°C (68°F)based on THF
Vapor Density:	>2 (Air = 1)
Other Data: Viscosity:	Medium bodied

SECTION 10 - STABILITY AND REACTIVITY

Stability:	Stable
Hazardous decomposition products:	None in normal use. When forced to burn, this product gives off oxides of carbon, hydrogen chloride and smoke.
Conditions to avoid:	Keep away from heat, sparks, open flame and other ignition sources.
Incompatible Materials:	Oxidizers, strong acids and bases, amines, ammonia

SECTION 11 - TOXICOLOGICAL INFORMATION

Toxicity:	LD$_{50}$	LC$_{50}$	Target Organs
Tetrahydrofuran (THF)	Oral: 2842 mg/kg (rat)	Inhalation 3 hrs. 21,000 mg/m^3 (rat)	STOT SE3
Methyl Ethyl Ketone (MEK)	Oral: 2737 mg/kg (rat), Dermal: 6480 mg/kg (rabbit)	Inhalation 8 hrs. 23,500 mg/m^3 (rat)	STOT SE3
Cyclohexanone	Oral: 1535 mg/kg (rat), Dermal: 948 mg/kg (rabbit)	Inhalation 4 hrs. 8,000 PPM (rat)	

Reproductive Effects	Teratogenicity	Mutagenicity	Embryotoxicity	Sensitization to Product	Synergistic Products
Not Established	Not Established	Not Established	Not Established	Not Established	Not Established

SECTION 12 - ECOLOGICAL INFORMATION

Ecotoxicity:	None Known
Mobility:	In normal use, emission of volatile organic compounds (VOC's) to the air takes place, typically at a rate of ≤ 500 g/l.
Degradability:	Not readily biodegradable
Bioaccumulation:	Minimal to none.

SECTION 13 - WASTE DISPOSAL CONSIDERATIONS

Follow local and national regulations. Consult disposal expert.

SECTION 14 - TRANSPORT INFORMATION

Proper Shipping Name:	Adhesives
Hazard Class:	3
Secondary Risk:	None
Identification Number:	UN 1133
Packing Group:	PG II
Label Required:	Class 3 Flammable Liquid
Marine Pollutant:	NO

EXCEPTION for Ground Shipping
DOT Limited Quantity: Up to 5L per inner packaging, 30 kg gross weight per package.
Consumer Commodity: Depending on packaging, these quantities may qualify under DOT as "ORM-D" .

TDG INFORMATION	
TDG CLASS:	FLAMMABLE LIQUID 3
SHIPPING NAME:	ADHESIVES
UN NUMBER/PACKING GROUP:	UN 1133, PG II

SECTION 15 - REGULATORY INFORMATION

Precautionary Label Information:	Highly Flammable, Irritant, Carc. Cat. 2 Ingredient Listings: USA TSCA, Europe EINECS, Canada DSL, Australia
Symbols:	F, Xi AICS, Korea ECL/TCCL, Japan MITI (ENCS)
Risk Phrases:	R11: Highly flammable.
	R20: Harmful by inhalation.
	R36/37: Irritating to eyes and respiratory system.

R66: Repeated exposure may cause skin dryness or cracking
R67: Vapors may cause drowsiness and dizziness

Safety Phrases:	S9: Keep container in a well-ventilated place.
	S16: Keep away from sources of ignition - No smoking.
	S25: Avoid contact with eyes.

S26: In case of contact with eyes, rinse immediately with plenty of water and seek medical advice.
S33: Take precautionary measures against static discharges.
S46: If swallowed, seek medical advise immediately and show this container or label.

SECTION 16 - OTHER INFORMATION

Specification Information:	
Department issuing data sheet:	IPS, Safety Health & Environmental Affairs
E-mail address:	<EHSinfo@ipscorp.com>
Training necessary:	Yes, training in practices and procedures contained in product literature.
Reissue date / reason for reissue:	6/21/2018 / Updated GHS Standard Format
Intended Use of Product:	Solvent Cement for PVC Plastic Pipe

All ingredients are compliant with the requirements of the Europea
Directive on RoHS (Restriction of Hazardous Substances).

This product is intended for use by skilled individuals at their own risk. The information contained herein is based on data considered accurate based on current state of knowledge and experience. However, no warranty is expressed or implied regarding the accuracy of this data or the results to be obtained from the use thereof.

Figure 5B Solvent cement SDS (2 of 2).
Source: Weld-On Adhesives, Inc., a division of IPS Corporation

Next, find out how to minimize these hazards. Section 7 gives general handling and storage information. It indicates that ventilation is needed to reduce hazardous vapors, which can be a fan in an open window. If ventilation is not enough, respiratory protection is needed. Section 8 tells you how to protect your eyes and skin.

Section 4 lists the first aid measures for eye contact, skin contact, or inhalation. Section 5 explains fire hazards and fire fighting measures. Now you have the information you need in case of an emergency.

The SDSs must be kept in the work area and be readily accessible to all workers. The company's safety officer or **competent person** should review the SDS before a hazardous material is used. Ask your supervisor to point out where the SDSs are located if you are not sure. Have him or her point out the sections that relate to your job; the health and safety of you and your co-workers may depend on it.

Competent person: A person who is capable of identifying existing and predictable hazards in the surroundings or working conditions that are unsanitary, hazardous, or dangerous to employees, and who has authorization to take prompt corrective measures to eliminate them.

1.0.0 Section Review

1. The person primarily responsible for your safety is _____.
    a. your foreman
    b. your instructor
    c. yourself
    d. your employer

2. The color commonly used for informational signs is _____.
    a. green
    b. red
    c. yellow
    d. blue

3. The SDS for any chemical used at a job site must be available _____.
    a. at the job site
    b. online
    c. at the contractor's office
    d. at the nearest hospital

2.0.0 Elevated Work and Fall Protection

Objective

Describe common fall hazards and methods to prevent them.
  a. Summarize the most common types of construction fall hazards.
  b. Describe components of effective fall arrest systems and how they prevent or halt falls.
  c. Explain how to use ladders and stairs safely.
  d. Identify key steps to ensuring scaffolds are assembled and used safely.

Performance Tasks

1. Properly set up and climb/descend an extension ladder, demonstrating proper three-point contact.
2. Inspect the following PPE items and determine if they are safe to use:
   • Fall arrest harnesses
   • Lanyards
   • Connecting devices
3. Properly don, fit, and remove the following PPE item:
   • Fall arrest harness

Falls from elevated areas are the leading cause of fatalities in the workplace. Falls from elevated heights account for about one-third of all deaths in the construction trade. Approximately 85 percent of the injured workers lose time from work; approximately 33 percent require hospitalization; some never return to the job.

While the risk of falls is high in construction and some other trades, there are many things that workers can do to safeguard themselves. Using the appropriate PPE; following proper safety procedures; practicing good housekeeping habits; and staying alert at all times will help you stay safe when working at an elevation. Employers are required to provide for both fall prevention and fall arrest, and to make sure workers are trained and certified in the use of fall protection equipment. Fall prevention consists of covered floor openings, climbing aids, barricades, and guardrails that are designed to protect against falls. Fall arrest consists of equipment such as body harnesses, **lanyards**, connection devices, lifelines, and safety nets that are intended to protect a worker in case a fall occurs. All these topics are covered in this section.

Lanyards: Short sections of rope or strap, one end of which is attached to a worker's safety harness and the other to a strong anchor point above the work area.

2.1.0 Fall Types and Hazards

Falls are classified into two groups: falls from an elevation and falls from the same level. Falls from an elevation can happen during work from **scaffolds**, work platforms, decking, concrete forms, ladders, stairs, and work near **excavations**. Falls from an elevation often result in death unless the fall is arrested. Falls on the same level are usually caused by tripping or slipping. Sharp edges and pointed objects, such as exposed concrete reinforcing bars (rebar), could cut and otherwise harm a worker. Other bodily injuries are also common results of tripping or slipping.

Scaffolds: Elevated platforms for workers and materials.

Excavations: Any man-made cuts, cavities, trenches, or depressions in an earth surface, formed by removing earth. They can be made for anything from basements to highways. Also see *trench*.

In the United States, fall protection is required for platforms or work surfaces with unprotected sides or edges that are 6 feet (≈2 m) or above the ground or the level below. This is commonly referred to as the **six-foot rule**. However, some international regulations and company policies may require fall protection for heights less than 6 feet.

Six-foot rule: A rule stating that platforms or work surfaces with unprotected sides or edges that are 6 feet (≈2 m) or higher than the ground or level below it require fall protection.

2.1.1 Walking and Working Surfaces

Slips, trips, and falls on walking and working surfaces cause 15 percent of all incidental deaths in the construction industry. Some incidents occur due to environmental conditions, such as snow, ice, or wet surfaces. Others happen because of poor housekeeping and careless behavior, such as leaving tools, materials, and equipment out and unattended. You can avoid slips, trips, and falls by being aware of your surroundings and following the rules on your site. Remember these general walking and working surface guidelines to avoid incidents:

- Keep all walking and working areas clean and dry. If you see a spill or ice patch, clean it up, or barricade the area until it can be properly attended to.
- Keep all walking and working surfaces clear of clutter and debris.
- Run cables, extension cords, and hoses overhead or through crossover plates so that they will not become tripping hazards.
- Do not run on scaffolds, work platforms, decking, roofs, or other elevated work areas.

2.1.2 Unprotected Sides, Wall Openings, and Floor Holes

Any opening in a wall or floor is a safety hazard. There are two types of protection for these openings: (1) they can be **guarded** or (2) they can be covered. Cover any hole in the floor when possible. Hole covers must be clearly marked. When it is not practical to cover the hole, use barricades. If the bottom edge of a wall opening is less than 39 inches (1 m) above the floor and would allow someone to fall 6 feet (≈2 m) or more, then place guards around the opening.

Guarded: Enclosed, fenced, covered, or otherwise protected by barriers, rails, covers, or platforms to prevent dangerous contact.

The types of barriers and barricades used vary from one job site to another. There may also be different procedures for when and how barricades are put up. Learn and follow the policies at your job site. Several different types of guards are commonly used:

- Railings are used across wall openings or as a barrier around floor openings to prevent falls (*Figure 6A*).

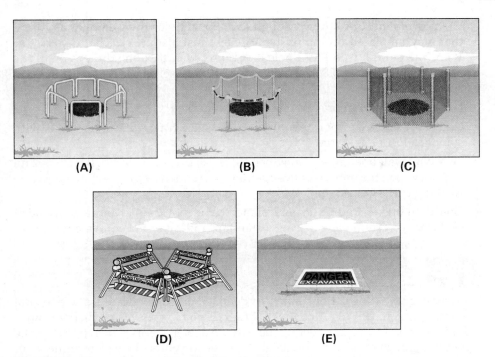

Figure 6 Common types of barriers and barricades.

- Warning barricades alert workers to hazards but provide no real protection (*Figure 6B*). Warning barricades can be made of plastic tape or rope strung from wire or between posts. The tape or rope is color-coded as follows:

  - *Red means danger.* No one may enter an area with a red warning barricade. A red barricade is used when there is danger from falling objects or when a load is suspended over an area.

  - *Yellow means caution.* You may enter an area with a yellow barricade, but be sure you know what the hazard is, and be careful. Yellow barricades are used around wet areas or areas containing loose dust. Yellow with black lettering warns of physical hazards such as bumping into something, stumbling, or falling.

  - *Yellow and purple together mean radiation warning.* No one may pass a yellow and purple barricade without authorization, training, and the appropriate PPE. These barricades are often used where piping welds are being X-rayed.

- Protective barricades give both a visual warning and protection from injury (*Figure 6C*). They can be wooden posts and rails, posts and chain, or steel cable. People should not be able to get past protective barricades.

- Blinking lights are placed on barricades so they can be seen at night (*Figure 6D*).

- Hole covers are used to cover open holes in a floor or in the ground (*Figure 6E*).

WARNING!

Never remove a barricade unless you have been authorized to do so. Follow your employer's procedures for putting up and removing barricades.

Follow these guidelines when working near unprotected sides, floor holes, and wall openings:

- Hole covers must be cleated, wired, or otherwise secured to prevent them from slipping sideways or horizontally beyond the hole.

- Covers must extend adequately beyond the edge of the hole.

- Hole covers must be strong enough to support twice the weight of anything that may be placed on top of them. They must also be clearly marked. Use ¾-inch (2 cm) plywood as a hole cover, provided that one dimension of the opening is less than 18 inches (46 cm); otherwise, 2-inch (5-cm) lumber is required.
- Never store material or equipment on a hole cover.
- Guard all stairway floor openings, with the exception of the entrance, with standard railing and **toeboards**.
- Guard all wall openings from which there is a drop of more than 6 feet (≈2 m) and for which the bottom of the opening is less than 39 inches (1 m) above the working surface.
- Guard all open-sided floors and platforms 6 feet (≈2 m) or more above adjacent floor or ground level, using a standard railing or the equivalent.

Toeboards: Vertical barriers at floor level attached along exposed edges of a platform, runway, or ramp to prevent materials and people from falling.

2.2.0 Fall Arrest

The key to preventing serious injury or death should a fall occur is the personal fall arrest system (PFAS). A complete PFAS (*Figure 7*) consists of anchor points, a body harness, and connecting devices. Anchor points are related to the structure, and the type and availability of anchor points help determine what other equipment should be chosen. The body harness comprises the system of belts, rings, or hooks worn by the worker. Connecting devices and lanyards are used to maintain attachment between anchor points and the PFAS, and include lanyards and various pieces of hardware.

NOTE

This training alone does not provide any level of certification in the use of fall arrest or fall restraint equipment. Trainees should not assume that the knowledge gained in this module is sufficient to certify them to use fall arrest equipment in the field.

2.2.1 Anchor Points

There are both permanent anchor points and temporary, reusable anchor points (*Figure 8*). Anchor points must be rated at or equal to 5,000 pounds (2,267 kg) breaking or tensile strength, or twice the intended load.

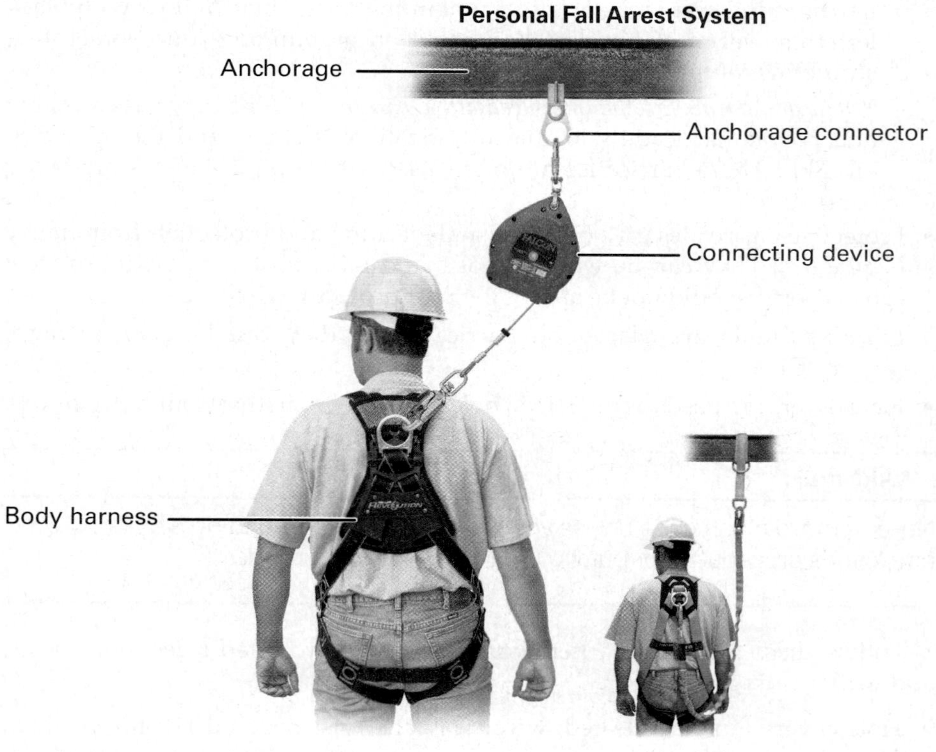

Personal Fall Arrest System

Anchorage ———

——— Anchorage connector

——— Connecting device

Body harness ———

Figure 7 Personal fall arrest system.
Source: Fall protection materials provided courtesy of Miller Fall Protection, Franklin, PA

(A) Permanent Anchor

(B) Concrete Anchor

(C) Beam Anchor

(D) Tie-Back Lanyard

Figure 8 Anchor points.
Sources: Fall protection materials provided courtesy of Miller Fall Protection, Franklin, PA (8A); Honeywell Safety Products (8B–8D)

Figure 9 Back D-ring.
Source: Fall protection materials provided courtesy of Miller Fall Protection, Franklin, PA

Ideally, the fall arrest anchor point will be located directly above the back D-ring (*Figure 9*) in order to minimize any swing-zone hazards, as well as the free-fall distance. The maximum free-fall distance is 6 feet (≈2 m). To understand swing zones, picture a tied-off worker standing 3 feet (1 m) away from a point directly under the anchor point. If the worker falls, gravity will cause his body to swing toward the anchor point axis and momentum will cause the body to swing past that point. If there is a wall or other solid object close by, he may strike it and be injured. Swing zones are minimized when the anchor point is directly above the worker when he falls. Serious injury and damage can occur when a human body strikes an immovable object while swinging as a pendulum. Although the PFAS may do its job by preventing the worker from falling a great distance, serious injury or death can still occur by striking an object in the swing zone.

Anchor points are often needed to secure the position of the worker, leaving the hands free to accomplish a task. Ideally, workers will select anchor points to maintain a potential fall distance of no more than 2 feet (61 cm). For example, a worker placing reinforcing bars in a concrete wall form may use two short positioning lanyards connected to the hip rings on the harness. The fall protection lanyard would be connected to the rear D-ring.

Positioning connections cannot be considered the primary anchor. Positioning anchor points are required to be rated at a 3,000-pound (1,360-kg) strength instead of the 5,000-pound (2,268-kg) rating for primary fall arrest. Remember that positioning lanyards, connected to D-rings on the harness other than the back or front chest D-ring, are fall restraints rather than fall arresting connections. A positioning lanyard does not take the place of a fall arrest lanyard or anchor point.

2.2.2 Harnesses

Full body harnesses, like the one shown in *Figure 10,* are available in sizes that are based on the height and weight of the user. The back D-ring is the only one used to connect the harness to the anchor point for primary fall arrest purposes unless you are climbing a ladder. When climbing a ladder, the front chest D-ring is the likely choice for connecting the fall arrest harness. D-rings located at the hips are used for positioning and fall restraint only. D-rings mounted to shoulders are often used for rescue situations. All of them can be used for fall restraint, but the back D-ring is the primary connection for fall arrest.

A body harness must fit correctly to ensure that it will provide proper protection. Do not place additional holes or openings in harness components under any circumstances. No field modifications to a body harness or lanyard should be attempted. Installation, maintenance, and inspection instructions are provided for every harness, and it is the responsibility of the worker to read and understand the details regarding his or her personal equipment. Workers must inspect their body harness each day that it is in use.

Harness straps are generally designed with some stretch to help absorb some of the potential force of a fall. This means good, taut installation on the body is essential so that the worker cannot fall out of the harness in the event of a fall.

Figure 11 shows the proper procedure for donning a common full body harness. The most important adjustments to be made include the chest straps,

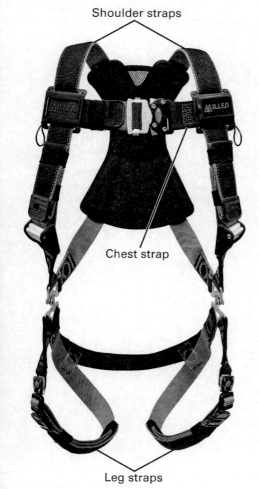

Figure 10 Full body harness.
Source: Fall protection materials provided courtesy of Miller Fall Protection, Franklin, PA

6 Easy Steps That Could Save Your Life
How To Don A Harness

1 Hold harness by back D-ring. Shake harness to allow all straps to fall in place.

2 If chest, leg and/or waist straps are buckled, release straps and unbuckle at this time.

3 Slip straps over shoulders so **D-ring is located in middle of back between shoulder blades.**

4 Pull leg strap between legs and connect to opposite end. Repeat with second leg strap. If belted harness, connect waist strap after leg straps.

5 **Connect chest strap and position in midchest area.** Tighten to keep shoulder straps taut.

6 Snug Fit

After all straps have been buckled, **tighten all buckles so that harness fits snug but allows full range of movement.** Pass excess strap through loop keepers.

Figure 11 Installing the body harness.
Source: Fall protection materials provided courtesy of Miller Fall Protection, Franklin, PA

Figure 12 Suspension trauma strap use.
Source: Photo courtesy of Capital Safety (DBI-SALA and PROTECTA)

the groin straps, and the final position of the back D-ring. The related details that follow must be considered during the fitting and wearing of a full body harness:

- The back D-ring location is vital to proper fall arrest. Position this ring between the shoulder blades. If it is too low, it will tend to cause the body to hang in a more horizontal position during fall arrest, increasing pressure on the diaphragm and affecting breathing. If it is positioned too high or with too much slack, the D-ring may strike the worker's head at the base of the fall, and the shoulder straps may be pulled too tightly into the neck and restrict blood flow. The impact at the base of the fall arrest can be dramatic, so the force must be spread all around the body to prevent injury to any one portion.

- Chest straps generally form either an "H" pattern or an "X" pattern. Adjust "H" pattern straps to land between the bottom of the sternum and the belly button. This helps ensure the horizontal portion of the "H" does not contact the throat during a fall, choking the worker. Some harness designs may not allow for this adjustment, with the final position of the "H" being based solely on a properly sized harness.

- The position of the chest straps is also crucial for "X" pattern harnesses. Position the "X" at or just below the sternum.

- Leg straps are an integral and required part of a PFAS. Adjust the groin straps for a good, snug fit. Too much slack here will cause extreme discomfort in a fall, when the impact snatches them up tight and you are left suspended this way.

- A suspension trauma strap (*Figure 12*) is recommended as part of the PFAS gear. The suspension trauma strap is stored in a pouch connected to the harness within easy reach. This is done by either one end of the strap being permanently sewn to the harness (by the manufacturer); one end of the strap attached to the harness with a carabiner (*Figure 13*); or by choking the pouch around a harness strap or hip D-ring. The strap can then be quickly removed and used without any possibility of the user dropping it. Once connected, the strap allows the worker to stand up in the harness, relieving suspended weight and pressure from the hips and groin. This helps open the path for blood flow from the legs back to the heart, preventing blood from pooling in the lower extremities.

- A separate waist or tool belt, while not considered a necessary component of the fall arrest system, must be fitted properly. It is best used for body positioning with the D-rings precisely located at the hips, rather than in the front

NOTE

These guidelines are general in nature, and the instructions provided for specific equipment by the manufacturer must always take precedence. It is the worker's responsibility to be intimately familiar with the duty of each and every ring and strap on a given harness.

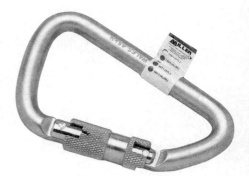

Figure 13 A carabiner.
Source: Fall protection materials provided courtesy of Miller Fall Protection, Franklin, PA

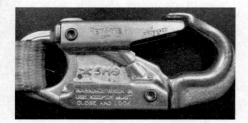

Figure 14 Double-locking snap hook.

(A) Shock-Absorbing

(B) Retractable

Figure 15 Fall arrest lanyards.
Source: Fall protection materials provided courtesy of Miller Fall Protection, Franklin, PA

or rear. Do not adjust it in a way that could apply pressure to the kidneys or lower back. If the waist belt is not an integral part of the harness, it must not be worn on the outside of the harness—put it on first, then add the harness over it. This is also true of any added tool belts.

- Saddles, like waist belts, are also not considered an integral or required part of the PFAS. They are optional and often detachable. They are generally used by workers to allow a seated, suspended position when the task may require long periods in the same location.

It is important to note that not all hardware will qualify as a component of a PFAS. Some hardware is to be used only for attaching tools and equipment to the worker or to structures. Hardware used as connecting devices as part of the PFAS must be drop-forged steel, and have a corrosion-resistant finish resistant to salt spray per ANSI standards. Any type of hook or carabiner must be equipped with safety gates or keepers to prevent the hooked object from being disconnected incidentally. In most cases, these safety gates will be required to be two-step, also called double action. Designs for these features vary, but those designed so that both movements required to open the gate can be done with a single hand are generally better. Using both hands to manipulate a single connector can be a hazard in itself.

Double-locking snap hooks (*Figure 14*) are usually curved and have an opening to allow connection to a line that can then be securely closed. They are usually not as consistent in appearance as carabiners, and come in somewhat different shapes. When used as part of a PFAS, the security closure should be automatic. Snap hooks are designed to connect to D-rings primarily, not to each other. In most cases, snap hooks are already connected to a lanyard or rope to ensure the integrity of the connection, and are not purchased separately.

2.2.3   PFAS Inspection

Always inspect your lanyard, harness, and any other fall protection gear prior to use. Never use damaged equipment. Treat a safety harness as if your life depends on it, because it does! To maintain their service life and high performance, all belts and harnesses should be inspected frequently. Damage to fall arrest systems includes burns, hardening due to chemical contact, and excessive wear. When inspecting a harness, check that the buckles and D-ring are not bent or deeply scratched. Check the harness for any cuts or rough spots.

The PFAS should be inspected monthly by a competent person. This requirement should be established through your company's safety program. The competent person has the authority to impose prompt corrective measures to eliminate any hazards. If there is any question about a defect, no matter how small it may seem to be, err on the side of caution. Take the fall arrest component(s) out of service for testing or replacement.

2.2.4   Lanyards

Lanyards consist of some primary material of construction (rope, webbing, aircraft cable, etc.) with a connecting device attached to the ends. Lanyards used for fall arrest are either shock absorbing or self-retracting (*Figure 15*). Non-shock absorbing lanyards, such as the one shown in *Figure 16*, are used for positioning and fall restraint. Lanyards used for positioning are not considered part of the fall arrest system; they are fall restraints and will be attached to D-rings on the harness other than the back D-ring. Since fall restraint is all about preventing a fall from happening, lanyards used for positioning should not allow a fall or movement greater than 2 feet (0.61 m). Two such lanyards are usually required to permit movement from one place to another.

The retractable fall arrest lanyard shown in *Figure 15B* is rapidly becoming the preferred device. Since it automatically retracts or feeds lanyard as the worker moves about, there is never a great deal of slack that can pose a risk in itself. In addition, the potential fall distance is significantly reduced.

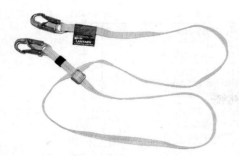

Figure 16 Non-shock absorbing lanyard.
Source: Fall protection materials provided courtesy of Miller Fall Protection, Franklin, PA

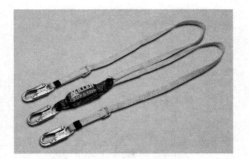

Figure 17 Y-configured shock absorbing lanyard.
Source: Fall protection materials provided courtesy of Miller Fall Protection, Franklin, PA

Lanyards for fall restraint or arrest must never be field-fabricated, and must never be connected together to increase their length. Depending on the use, padding may be added during fabrication or added in the field to protect against sharp edges. Never tie a knot in a lanyard, as knots can severely reduce the load limit. Never wrap a lanyard around a structure and then choke it back into itself unless it is designed for this use. A special large D-ring is usually attached to the lanyard for this purpose. It is important to remember that whenever you are wearing a PFAS, 100 percent tie-off is required. A PFAS is absolutely useless unless you are tied off.

The D-ring used on the PFAS for fall arrest may only be connected to one live connection at a time. This can be challenging when trying to move from one point to another, especially horizontally. The user must be able to reach back and disconnect a lanyard, while maintaining one lanyard connected at all times. A Y-configured lanyard (*Figure 17*) can be used for this purpose. A Y-configured lanyard has a single point of attachment at the D-ring that is used to accommodate two lanyards. They are also referred to as double-leg or tieback lanyards.

Before using a shock absorbing or self-retracting lanyard, the potential fall distance is determined. Then the proper equipment is selected to meet available fall clearance (*Figure 18*). Note in the figure that the fall distance with the self-retracting lanyard is significantly shorter than it is with a shock absorbing lanyard; hence its increasing popularity. Failure to select proper equipment and calculate fall distance may result in serious personal injury or death. These calculations must be done by experienced and qualified personnel on the job site. If a personal fall arrest system is actuated due to a fall, it must be inspected by a competent person before it can be used again. In most cases, replacement will be necessary.

2.2.5   Lifelines

A lifeline is a flexible line such as a cable or rope connected vertically to an anchorage at one end (vertical lifeline), or horizontally to an anchorage at both ends (horizontal lifeline). It serves as a means for connecting other components of a personal fall arrest system to the available anchorage.

Vertical lifelines are suspended from a fixed anchorage. A fall arrest device such as a rope grab (*Figure 19*) or retractable lanyard is attached to the lifeline. A beam grab, or beamer (*Figure 20*), is sometimes used as an anchorage for a lifeline. Vertical lifelines must have a minimum breaking strength of 5,000 pounds (2,267 kg). Workers must use separate vertical lifelines. A vertical lifeline must have a termination on the end unless it extends to the ground or to the next lower level of the structure.

Horizontal lifelines (*Figure 21*) are connected between two fixed anchorages. A lanyard is attached to the lifeline. Horizontal lifelines must be designed, installed, and used under the supervision of a competent person. The required breaking strength of any horizontal lifeline is determined by the number of workers that will be attached to it.

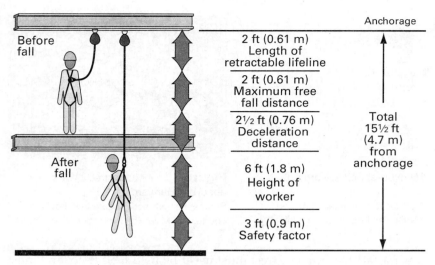

Anchorage

2 ft (0.61 m)
Length of
retractable lifeline

2 ft (0.61 m)
Maximum free
fall distance

2½ ft (0.76 m)
Deceleration
distance

Total
15½ ft
(4.7 m)
from
anchorage

6 ft (1.8 m)
Height of
worker

3 ft (0.9 m)
Safety factor

Using a Self–Retracting Lanyard

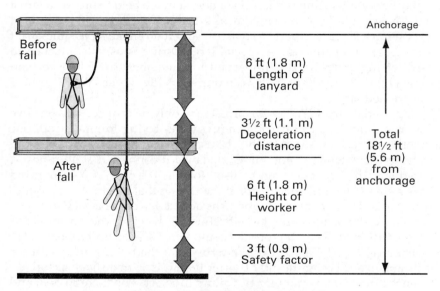

Anchorage

6 ft (1.8 m)
Length of
lanyard

3½ ft (1.1 m)
Deceleration
distance

Total
18½ ft
(5.6 m)
from
anchorage

6 ft (1.8 m)
Height of
worker

3 ft (0.9 m)
Safety factor

Using a Shock Absorbing Lanyard

Figure 18 Potential fall distances for self-retracting and shock absorbing lanyards.

Figure 19 Rope grab.
Source: Fall protection materials provided
courtesy of Miller Fall Protection, Franklin, PA

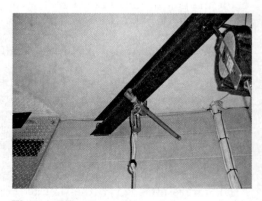

Figure 20 Beam grab.

Figure 21 Horizontal lifeline.
Source: Honeywell Safety Products

WARNING!

Keep in mind that horizontal lifelines are rated for a maximum number of connected workers. Check with your supervisor or the manufacturer before connecting to a lifeline that is being used by other workers.

2.2.6 Guardrails

Guardrails (*Figure 22*) are a common type of fall prevention. They protect workers by providing a barrier between the work area and the ground or lower work areas. They may be made of wood, pipe, steel, or wire rope and must be able to support 200 pounds (90 kg) of force applied in any direction to the top rail and 150 pounds (68 kg) for the **midrail**. A guardrail must be 42 inches ±3 inches (106 ±8 cm) high to the top rail and have a toeboard that is a minimum of 4 inches (10 cm) high. This helps to prevent the inadvertent loss of tools or material through the bottom rail. The toeboard must be securely fastened with not more than ¼-inch (5 mm) clearance above the floor level.

2.2.7 Safety Nets

Safety nets are used for fall protection on bridges and similar projects. They must be installed not more than 30 feet (9 m) beneath the work area. There must be enough clearance under a safety net to prevent a worker who falls into it from hitting the surface below it. There must also be no obstruction between the work area and the net.

Midrail: Mid-level, horizontal board required on all open sides of scaffolds and platforms that are more than 14 inches (35 cm) from the face of the structure and more than 10 feet (3 m) above the ground. It is placed halfway between the toeboard and the top rail.

Figure 22 Guardrails.
Source: Guardian Fall Protection

Depending on the actual vertical distance between the net and the work area, the net must extend 8 to 13 feet (2 to 4 m) beyond the edge of the work area. Mesh openings in the net must be limited to 36 square inches (232 sq cm) and 6 inches (15 cm) from the side. The border rope must have a 5,000-pound (2,267 kg) minimum breaking strength, and connections between net panels must be as strong as the nets themselves. Safety nets must be inspected at least once a week and after any event that might have damaged or weakened them. Worn or damaged nets must be removed from service.

2.3.0   Ladders and Stairs

Ladders are used daily to perform work in elevated locations. Any time work is performed above ground level, there is a risk of incidents. This risk can be reduced by carefully inspecting ladders before you use them and by using them properly.

Overloading means exceeding the **maximum intended load** of a ladder. Overloading can cause ladder failure, which means that the ladder could buckle, break, or topple. The maximum intended load is the total weight of all people, equipment, tools, materials, loads that are being carried, and other loads that the ladder can hold at any one time. Check the manufacturer's specifications to determine the maximum intended load. Ladders are usually given a duty rating that indicates their load capacity, as shown in *Table 2*. Note that ladders designed for the metric market are not usually direct equivalents to ladders built for the American market. Capacity ratings of 130 kilograms (286 lbs) and 150 kilograms (330 lbs) are the most common for the trades. Ladder heights will also be stated in metric units and may not be exactly the same size as those designed for the American market.

There are different types of ladders to use for different jobs (*Figure 23*). Selecting the right ladder for the job at hand is important to complete a job as safely and efficiently as possible. Ladder types include portable straight ladders, extension ladders, and stepladders.

Maximum intended load: The total weight of all people, equipment, tools, materials, and loads that a ladder can hold at one time.

WARNING!

Use ladders only for their intended purposes. Ladders are not interchangeable; incorrect use of a ladder can result in injury or damage.

When using a ladder, be sure to maintain three-point contact with the ladder while ascending or descending. Three-point contact means that either two feet and one hand or one foot and two hands are always touching the ladder.

Drop-Testing Safety Nets

Safety nets should be drop-tested at the job site after the initial installation, whenever relocated, after a repair, and at least every six months if left in one place. The drop test consists of a 400-pound (181 kg) bag of sand of 29 to 31 inches (74 to 79 cm) in diameter that is dropped into the net from at least 42 inches (107 cm) above the highest walking/working surface at which workers are exposed to fall hazards. If the net is still intact after the bag of sand is dropped, it passed the test.

TABLE 2 Ladder Duty Ratings and Load Capacities

Duty Rating	Load Capacities
Type IAA	375 lbs., extra-heavy duty/professional use
Type IA	300 lbs., extra-heavy duty/professional use
Type I	250 lbs., heavy duty/industrial use
Type II	225 lbs., medium duty/commercial use
Type III	200 lbs., light duty/household use

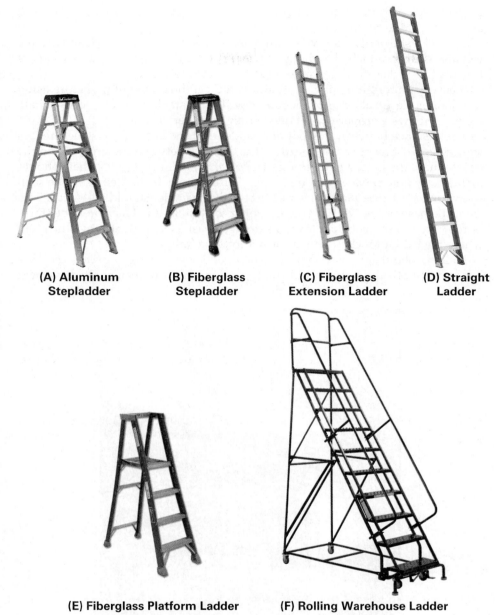

(A) Aluminum Stepladder   **(B) Fiberglass Stepladder**   **(C) Fiberglass Extension Ladder**   **(D) Straight Ladder**

(E) Fiberglass Platform Ladder   **(F) Rolling Warehouse Ladder**

Figure 23 Different types of ladders.
Source: Courtesy of Louisville Ladders

2.3.1 Straight Ladders

Straight ladders consist of two rails, rungs between the rails, and safety feet on the bottom of the rails (*Figure 24*). Straight ladders are generally made of aluminum, wood, or fiberglass.

Metal ladders conduct electricity and should never be used around electrical equipment. Any portable metal ladder must have "Danger! Do Not Use Around Electrical Installations" stenciled on the rails in 2-inch red letters. Ladders made of dry wood or fiberglass, neither of which conducts electricity, should be used around electrical equipment. Check that any ladder, especially a wooden ladder, is completely dry before using it; even a small amount of water will conduct electricity.

Different types of ladders are intended for use in specific situations. Aluminum ladders are corrosion-resistant and can be used where they might be exposed to the elements. They are also lightweight and can be used where they must often be lifted and moved. Fiberglass ladders are very durable, so they are useful where some amount of rough treatment is unavoidable. Wooden ladders, which are heavier and sturdier than fiberglass or aluminum ladders, can be used where heavy loads must be moved up and down. However, wooden ladders are subject

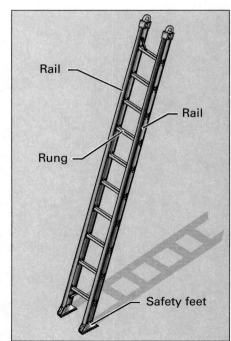

Rail

Rail

Rung

Safety feet

Figure 24 Portable straight ladder.

Figure 25 Ladder safety foot.

to more rapid deterioration as the wood swells and shrinks. Both fiberglass and aluminum are easier to clean than wood.

Figure 25 shows the safety feet attached to a straight ladder. Make sure the feet are securely attached and that they are not damaged or worn down. Do not use a ladder if its safety feet are not in good working order.

It is very important to place a straight ladder at the proper angle before using it. A ladder placed at an improper angle will be unstable and could cause a fall. *Figure 26* shows a properly positioned straight ladder.

The distance between the foot of a ladder and the base of the structure it is leaning against must be one-fourth of the distance between the ground and the point where the ladder touches the structure. Stated another way, there should be a 4-to-1 ratio between the distances. For example, if the height of the wall shown in *Figure 26* is 16 feet (4.9 m), the base of the ladder should be 4 feet (1.2 m) from the base of the wall. If you are going to step off a ladder onto a platform or roof, the top of the ladder should extend at least 3 feet (0.9 m) above the point where the ladder touches the platform, roof, side rails, etc.

Ladders should be used only on stable and level surfaces unless they are secured at both the bottom and the top to prevent any incidental movement

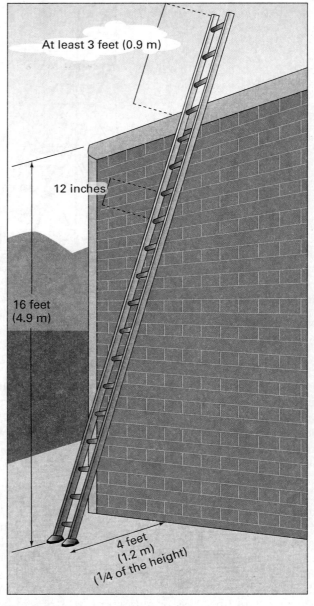

At least 3 feet (0.9 m)

12 inches

16 feet (4.9 m)

4 feet (1.2 m) (1/4 of the height)

Figure 26 Proper positioning of a straight ladder.

Bottom Secured   **Top Secured**

Figure 27 Securing a ladder.

(*Figure 27*). Never try to move a ladder while you are on it. If a ladder must be placed in front of a door that opens toward the ladder, the door should be locked or blocked open. Otherwise, the door could be opened into the ladder.

Ladders are made for vertical use only. Never use a ladder as a work platform by placing it horizontally. Make sure the ladder you are about to climb or descend is properly secure before you do so. Check to make sure the ladder's feet are solidly positioned on firm, level ground. Also check to make sure the top of the ladder is firmly positioned and in no danger of shifting once you begin your climb. Remember that your own weight will affect the ladder's steadiness once you mount it. It is important to test the ladder first by putting some of your weight on it without actually beginning to climb. This way, you can be sure that the ladder will remain steady as you climb.

When climbing a straight ladder, keep both hands on the rails or rungs (*Figure 28*). Maintain three points of contact at all times, as shown in the figure. This can be two feet and one hand, or one foot and two hands. Always keep your body's weight in the center of the ladder between the rails. Face the ladder at all times. Never go up or down a ladder while facing away from it.

To carry a tool while you are on the ladder, use a **hand line** or tagline attached to the tool. Climb the ladder and then pull up the tool. Don't carry tools in your hands while you are climbing a ladder.

2.3.2 Extension Ladders

An extension ladder is actually two straight ladders connected so that the overlap between them can be altered to increase or decrease the length of the ladder (*Figure 29*).

Extension ladders are positioned and secured following the same rules as straight ladders. When adjusting the length of an extension ladder, always reposition the movable section from the bottom, not the top, to ensure that the rung locks (*Figure 30*) are properly engaged after you make the adjustment. Check to make sure the section locking mechanisms are fully hooked over the desired rung. Also check to make sure that all ropes used for raising and lowering the extension are clear and untangled.

Extension ladders are positioned and secured following the same rules as straight ladders. There are, however, some safety rules that are unique to extension ladders:

Figure 28 Moving up or down a ladder.

Hand line: A line attached to a tool or object so a worker can pull it up after climbing a ladder or scaffold.

Aluminum   Fiberglass

Figure 29 Examples of extension ladders.
Source: Werner Co.

Figure 30 Rung locks.
Source: Werner Co.

> **WARNING!**
>
> Extension ladders have a built-in extension stop mechanism. Do not remove this mechanism. If the mechanism is removed, it could cause the ladder to collapse under a load.
>
> Haul materials up on a line rather than hand carrying them up an extension ladder.

- Make sure the extension ladder overlaps between the two sections (*Figure 31*). For ladders up to 36 feet (10.5 m) long, the overlap must be at least 3 feet (0.9 m). For ladders 36 to 48 feet (10.8 to 14.6 m) long, the overlap must be at least 4 feet (1.2 m). For ladders 48 to 60 feet (14.6 to 18.3 m) long, the overlap must be at least 5 feet (1.5 m).

- Never stand above the highest safe standing level on a ladder. On an extension ladder, this is the fourth rung from the top. If you stand higher, you may lose your balance and fall. Some ladders have colored rungs to show where you should not stand.

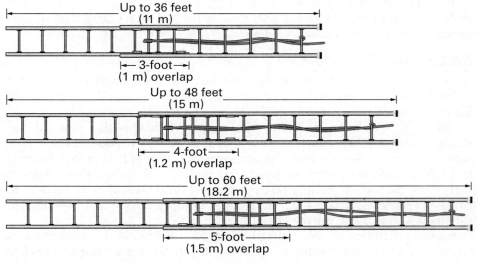

Figure 31 Overlap lengths for extension ladders.

- Avoid carrying anything on a ladder, because it will affect your balance and may cause you to fall. Haul materials up on a line instead.
- Keep yourself centered on the ladder. Do not over-reach, lean to one side, or try to move a ladder while standing on it.

2.3.3   Stepladders

Stepladders are self-supporting ladders made of two sections hinged at the top (*Figure 32*). The section of a stepladder used for climbing consists of rails and rungs like those on straight ladders. The other section consists of rails and braces. Spreaders are hinged arms between the sections that keep the ladder stable and keep it from folding while in use.

When positioning a stepladder, be sure that all four feet are on a hard, even surface. Otherwise, the ladder can rock from side to side or corner to corner when you climb it. With the ladder in position, be sure the spreaders are locked in the fully open position. The following safety precautions must be followed when using a stepladder:

- Never stand on the top step or the top of a stepladder. Putting your weight this high will make the ladder unstable. The top of the ladder is made to support the hinges, not to be used as a step.
- Although the rear braces may look like rungs, they are not designed to support your weight. Never use the braces for climbing or climb the back of a stepladder. However, there are specially designed two-person ladders available with steps on both sides.
- Check the load capacity of the ladder and do not exceed it.

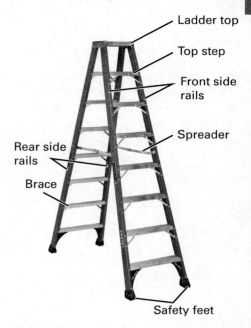

Figure 32 Typical fiberglass stepladder.
Source: Courtesy of Louisville Ladders

WARNING!

Never stand on a step with your knees higher than the top of a stepladder. You need to be able to hold on to the ladder with your hand. Keep your body centered between the side rails.

2.3.4   Inspecting Ladders

Always inspect a ladder before you use it. Check the rails and rungs for cracks or other damage. Also, check for loose rungs. If you find any damage, do not use the ladder. Check the entire ladder for loose nails, screws, brackets, or other hardware. If you find any hardware problems, tighten the loose parts or have the ladder repaired before you use it. OSHA requires regular inspections of all ladders and an inspection just before each use.

The same rules for inspecting straight ladders apply to extension ladders. In addition, the rope that is used to raise and lower the movable section of the ladder should be inspected. If the rope is frayed or has worn spots, it should be replaced before the ladder is used.

The rung locks support the entire weight of the movable section and the person climbing the ladder. Inspect them for damage before each use. If they are damaged, they should be repaired or replaced before the ladder is used.

Inspect stepladders the way you inspect straight and extension ladders. Pay special attention to the hinges and spreaders to be sure they are in good repair. Also, be sure the rungs are clean. The rungs of a stepladder are usually flat, so oil, grease, or dirt can build up on them and make them slippery.

2.3.5   Stairways

Stairways are also routinely used on construction sites where there is a break in elevation of 19 inches (46 cm) or more, and no ramp, runway, sloped

CAUTION

Wooden ladders should never be painted. The paint could hide cracks in the rungs or rails. Clear varnish, shellac, or a preservative oil finish will protect the wood without hiding defects.

embankment, or personnel hoist is provided. Observe the following regulations, based on OSHA standards, when using stairways on a job site:

- Stairways having four or more risers or rising more than 30 inches (76 cm), whichever is less, must be equipped with at least one handrail and one stair railing system along each unprotected side.
- Winding and spiral stairways must be equipped with a handrail offset sufficiently to prevent walking on those portions of the stairways where the tread width is less than 6 inches (15 cm).
- Stair railings must be not less than 36 inches (91 cm) from the upper surface of the stair railing system to the surface of the tread, in line with the face of the riser at the forward edge of the tread.

To reduce the likelihood of slips, trips, or falls, keep stairways clean and clear of debris. Do not store any tools or materials on stairways, and clean up liquid spills, rain water, or mud immediately.

Stairways must have adequate lighting. This can sometimes be a problem because permanent lighting is usually installed after stairway construction is completed. If the lighting is inadequate, temporary lighting should be installed in the stairway. Each bulb should be equipped with a protective cover, and the string should be inspected daily for burned-out or broken bulbs.

Whenever possible, avoid using stairways to transport materials between floors. Carrying small materials and tools is fine, as long as the materials do not block your vision. Going up or down a stairway while carrying large items is physically demanding and increases the chance of injuries and falls. Use the building elevator or crane service to transport large materials from one floor to another.

2.4.0   Scaffolds

Scaffolds provide safe elevated work platforms for people and materials. They are designed and built to comply with high safety standards, but normal wear and tear or incidentally putting too much weight on them can weaken them and make them unsafe. That's why it is important to inspect every part of a scaffold before each use. Personnel who assemble scaffolds must be certified to do so.

2.4.1   Types of Scaffolds

Two basic types of scaffolds—self-supporting scaffolds and suspended scaffolds—are used in the construction industry. The rules for safe use apply to both of them. Self-supporting scaffolds can be manufactured units or can be assembled at the site.

Manufactured scaffolds (*Figure 33*) are made of painted steel, stainless steel, or aluminum. They are stronger and more fire-resistant than wooden scaffolds. They are supplied in ready-made, individual units, which are assembled on site. A rolling scaffold has wheels on its legs so that it can be easily moved. The scaffold wheels have brakes so the scaffold will not move while workers are standing on it.

> **CAUTION**
>
> Only a competent person has the authority to supervise setting up, moving, and taking down scaffolds. Only a competent person can approve the use of scaffolds on the job site after inspecting the scaffolds.

> **WARNING!**
>
> Never unlock the wheel brakes of a rolling scaffold while anyone is on it. People on a moving scaffold can lose their balance and fall.

Built-up scaffolds (*Figure 34*) are built from the ground up at a job site using steel framework sections and lumber. Swing (suspended) scaffolds (*Figure 35A*) are suspended by ropes or cables in a manner that allows them to be raised or lowered as needed. Another type of suspended scaffolding is a work cage (*Figure 35B*). A work cage is typically suspended with rigging devices that attach to I-beams with various sizes of clamps and rollers.

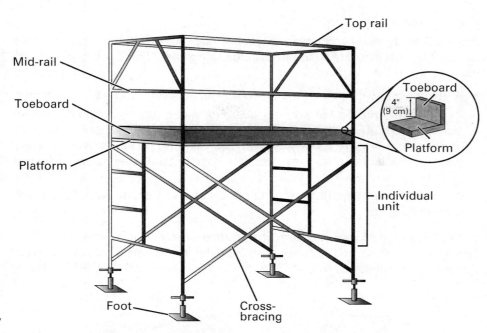

Figure 33 Typical manufactured scaffold.

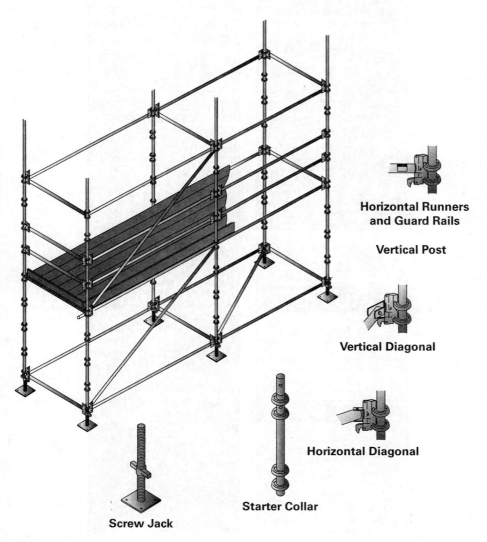

Figure 34 Built-up scaffold.
Source: Courtesy of Safway Services LLC

(A) Suspended Scaffold

(B) Work Cage

Figure 35 Suspended scaffold and work cage.
Source: Spider Staging

2.4.2   Inspecting Scaffolds

Any scaffold that is assembled on the job site must be tagged to indicate whether the scaffold meets OSHA standards and is safe to use. Three colors of tags are used: green, yellow, and red (*Figure 36*).

- A green tag means the scaffold meets all OSHA standards and is safe to use.
- A yellow tag means the scaffold does not meet all OSHA standards. An example is a scaffold on which a railing cannot be installed because of equipment interference. To use a yellow-tagged scaffold, you must wear a safety harness attached to a lanyard. You may have to take other safety measures as well.
- A red tag means a scaffold is being put up or taken down. Never use a red-tagged scaffold.

Don't rely on the tags alone; inspect all scaffolds before you use them. Check for bent, broken, or badly rusted tubes. Also check for loose joints where the tubes are connected. Any of these problems must be corrected before the scaffold is used.

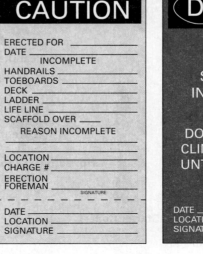

Figure 36 Typical scaffold tags.

Make sure you know the weight limit of any scaffold you will be using. Compare this weight limit to the total weight of the people, tools, equipment, and material you expect to put on the scaffold. Scaffold weight limits must never be exceeded.

If a scaffold is more than 10 feet (3.1 m) high, check to see that it is equipped with **top rails**, midrails, and toeboards; otherwise, use a PFAS. All connections must be pinned. That means they must have a piece of metal inserted through a hole to prevent connections from slipping. **Cross-bracing** must be used. A handrail is not the same as cross-bracing. The walking area must be completely **planked**.

If it is possible for people to walk under a scaffold, the space between the toeboard and the top rail must be screened. This prevents objects from falling off the work platform and injuring those below.

When you examine a rolling scaffold, check the condition of the wheels and brakes. Be sure the brakes are working properly and can stop the scaffold from moving while work is in progress. Be sure all brakes are locked before you use the scaffold.

Top rails: Top-level, horizontal boards required on all open sides of scaffolds and platforms that are more than 14 inches (36 cm) from the face of the structure and more than 10 feet (3 m) above the ground.

Cross-bracing: Braces (metal or wood) placed diagonally from the bottom of one rail to the top of another rail that add support to a structure.

Planked: Having pieces of material 2 inches (5 cm) thick or greater and 6 inches (15 cm) wide or greater used as flooring, decking, or scaffold decks.

2.4.3   Using Scaffolds

Be sure that a competent person inspects the scaffold before you use it. There should be firm footing under each leg of a scaffold before putting any weight on it. If you are working on loose or soft soil, you can put planks or matting under the scaffold's legs or wheels, as shown in *Figure 34*. When moving a rolling scaffold, first unlock the brakes and then move the scaffold. Once the scaffold is repositioned, don't forget to relock the brakes.

WARNING!

Keep scaffolds a minimum of 3 feet (91 cm) from power lines up to 300 volts and a minimum of 10 feet (3.1 m) from power lines above 300 volts. Never move a scaffold while someone is on it.

2.0.0 Section Review

1. A barrier with yellow and purple markings indicates a _____.
   a. fire hazard
   b. fall hazard
   c. radiation hazard
   d. confined space hazard

2. Positioning lanyards are used only as fall restraint devices and may *not* be used as fall arrest devices.
   a. True
   b. False

3. When positioning a straight ladder against a wall, how far from the wall should the base of the ladder be?
   a. 4 feet (1.2 m)
   b. One-fourth the distance from the ground to the point where the ladder touches the wall
   c. The height of the wall minus 4 feet (1.2 m)
   d. One-half the distance from the ground to the point where the ladder touches the wall

4. If a scaffold has a yellow tag, it means _____.
   a. the scaffold may only be used by one person at a time
   b. the scaffold is condemned and cannot be used
   c. a safety harness must be worn when using it
   d. the scaffold is under assembly and cannot be used

3.0.0 Struck-By and Caught-in-Between Hazards

Performance Tasks

There are no performance tasks in this section.

Objective

Recognize and avoid struck-by and caught-in-between hazards.

a. Describe struck-by hazards and how to avoid them.

b. Describe common caught-in/caught-between hazards and steps that can prevent them.

There is an apparent overlap in struck-by and caught-in-between hazards that needs to be clarified, as they can involve the same kinds of equipment. *Struck-by* means being hit by a moving object, while *caught-in-between* means being trapped between a moving object and a solid surface. Assume a worker is working near a crane that is swinging a load and the load hits the worker. That is an example of a struck-by incident. However, if the worker is behind the crane and the crane backs up, pinning the worker against a wall, it becomes a caught-in-between incident.

3.1.0   Struck-By Hazards

On any job site, there is a risk of being struck by falling objects, such as tools dropped from above. Flying objects such as debris from grinding and chipping metal are another struck-by hazard. On any site where there is moving equipment, or where workers are near roadways, there is also the danger of being struck by a vehicle.

3.1.1   Falling Objects

Workers are at risk from falling objects when they are beneath machinery and equipment such as cranes and scaffolds; where overhead work is being performed; or when working around stacked materials. To protect against struck-by injuries from falling objects, always wear an approved hard hat. Employers generally require workers to wear hard hats at all times on construction sites.

When working near machinery and equipment such as cranes, never stand or work beneath the load or in the fall zone. Barricade hazard areas where rigging equipment is in use, and post warning signs to inform other workers of falling object hazards. Inspect cranes and rigging components before use and do not exceed the rated load capacity.

When performing overhead work, be sure all tools, materials, and equipment are secured to prevent them from falling on people below. Use protective measures such as toeboards, debris nets, catch platforms, or canopies to catch or deflect falling objects. Use tool lanyards to prevent tools from falling.

Many workers are hurt or killed by falling stacks of material. Do not stack materials higher than 4:1 height-to-base ratio. Secure all loads by blocking and interlocking them. Interlocking means placing alternate layers at right angles to each other. Be aware of changing weather conditions, such as wind, that may lift and shift loads.

3.1.2   Flying Objects

There is a danger from flying objects when power tools or activities such as pushing, pulling, or prying causes objects to become airborne. Chipping, grinding, brushing, or hammering are all examples of job tasks that may cause flying objects. Tools that move at very high speed, like pneumatic and powder-actuated tools, can be very dangerous. Injuries from flying objects can range from minor abrasions to concussions, blindness, or death. The workers shown in *Figure 37* are using a pneumatic chipping hammer and a pneumatic grinder, and are properly using protective face shields.

Figure 37 Protection from flying particles.

To protect against flying object hazards, follow these guidelines:

- Use eye protection such as safety glasses, goggles, or face shields where machines or tools may cause flying particles.

- Inspect tools and machines to ensure that protective guards are in place and in good condition.

- Make sure you are trained in the proper operation of pneumatic and powder-actuated tools.

- Use shielding devices such as a welding screen or similar equipment to block flying debris.

3.1.3 Vehicle and Roadway Hazards

A common cause of incidents for workers on large job sites and highway projects are vehicle-related hazards such as being run over by vehicles or equipment (especially backing equipment) or by equipment tip-over. The most common cause of death for equipment operators is equipment rollover. If vehicle safety practices are not observed at your site, you risk being struck by swinging backhoes, moving vehicles, or swinging crane loads. If you work near public roadways (*Figure 38*), you risk being struck by passenger or commercial vehicles. When working near moving vehicles and equipment, follow these guidelines:

- Stay alert at all times and keep a safe distance from vehicles and equipment.

- Maintain eye contact with vehicle or equipment operators to ensure that they see you.

- Never get into blind spots of equipment operators.

- Keep off equipment unless authorized.

- Wear reflective or high-visibility vests or other suitable garments.

- Never stand between pieces of equipment unless they are secured.

- Never stand under loads handled by lifting or digging equipment, or near vehicles being loaded or unloaded.

Operators must also use caution when driving vehicles. The operator of any vehicle is responsible for the safety of passengers and the protection of the load. Follow these safety guidelines when you operate a vehicle on a job site:

- Always wear a seat belt.

- Be sure that each person in the vehicle has a firmly secured seat and seat belt.

- Obey all speed limits. Reduce speed in crowded areas.

Did You Know?

Struck-by Fatalities

Nearly one in four struck-by vehicle deaths involve construction workers—more than any other occupation.

Figure 38 A busy job site.
Source: Carolina Bridge Co.

Signaler: A person who is responsible for directing a vehicle when the driver's vision is blocked in any way.

- Look to the rear and sound the horn before backing up. If your rear vision is blocked, get a **signaler** to direct you.
- Every vehicle must have a backup alarm. Make sure the backup alarm works.
- Always turn off the engine when fueling.
- Turn off the engine and set the brakes before leaving the vehicle.
- Never stay on or in a truck that is being loaded by excavating equipment.
- Keep windshields, rearview mirrors, and lights clean and functional.
- Carry road flares, fire extinguishers, and other standard safety equipment at all times.
- Never use a cell phone while operating a motor vehicle.

> ### WARNING!
>
> When a vehicle is operated indoors, the carbon monoxide in its exhaust can kill or sicken anyone in the vicinity unless there is good ventilation. Carbon monoxide is especially dangerous because you cannot see, smell, or taste it. It is never a good idea to operate a vehicle, a generator, or any other machine with an internal combustion engine indoors. Make sure there is good ventilation before you operate a motorized vehicle or equipment such as a generator indoors.

Tool Lanyards

When working at elevations, a dropped tool can become a serious hazard. If the tool has moving parts, like battery-powered drills, the fall will likely destroy it.

Tool lanyards specifically designed for work in the elevated environment have been introduced by Snap-on®. Tethering tools to the tool belt or wrist can prevent injuries, save tools, prevent component damage, and prevent lost time in retrieval.

Source: Courtesy of Snap-on Industrial – Tools at Height Program

3.2.0 Caught-In and Caught-Between Hazards

Congested work sites, heavy equipment, and the presence of multiple trades can contribute to caught-in-between hazards. The primary causes of caught-in-between fatalities include trench/excavation collapse, rotating equipment, and unguarded parts.

One of the most disastrous caught-in hazards on a construction job site is being trapped by a cave-in of the walls of a trench or excavation. Other caught-in or caught-between hazards involve protective guards on power tools and machines, as well as the risk of being crushed by heavy equipment.

3.2.1 Trenches and Excavations

Trenches and excavations are common hazards, especially in construction and pipeline work. Anyone working in or around a trench or excavation must know and follow safety procedures aimed at protecting workers from cave-ins.

An excavation is any man-made cut, cavity, trench, or depression formed by removal of earth or soil. Sometimes the terms *excavation* and *trench* are used interchangeably, but there is a difference. A trench is an excavation that is deeper than it is wide, and usually not wider than 15 feet (4.6 m). Nearly all trenches are dangerous if not protected. Because trenches are narrow, workers can easily become trapped.

Hazards involved with trench and excavation work include the following:

- Cave-ins
- Water accumulation
- Falling objects
- Collapse of nearby structures
- Hazardous atmospheres produced by toxic gases in the soil

WARNING!

Just 2 to 3 feet (0.61 to 0.91 m) of soil can put enough pressure on your lungs to prevent you from breathing. In as little as four to six minutes without oxygen, you can sustain considerable brain damage.

Cave-ins are the most common and deadly hazard in excavation work. When dirt is removed from an excavation, the surrounding soil becomes unstable and gravity can force it to collapse. Cave-ins occur when soil or rock falls, or slides, into an excavation. Most cave-ins occur in trenches 5 to 15 feet (1.53 to 4.6 m) deep, and happen suddenly with little or no warning. On average, about 1,000 trench collapses occur each year in the United States.

Soil conditions can change, so they must be constantly evaluated. There are certain factors that could change the surroundings of the site, making a cave-in more likely. These factors include the following:

- Changes in weather conditions, such as freezing, thawing, or sudden heavy rain
- An excavation dug in unstable or previously disturbed soil
- Excessive vibration around the excavation
- Water accumulation in an excavation

Sometimes there are visible warning signs around the excavation that can be spotted before a cave-in occurs. Being aware of the warning signs increases your chances of getting out before a collapse occurs. Visible warning signs of a potential cave-in include the following:

- Ground settlement or narrow cracks in the sidewalls, slopes, or surface next to the excavation
- Flakes, pebbles, or clumps of soil separating and falling into the excavation
- Changes or bulges in the wall slope

If you notice any of these signs, get out of the excavation immediately and alert your co-workers as well.

Soil type is a key factor in determining the type of protective system needed to ensure that the trench will be safe. Solid rock is the most stable, while sandy soil is the least stable. The four types of soil are shown in *Table 3*.

Did You Know?

Excavation Fatalities

The fatality rate for excavation work is 112 percent higher than the rate for general construction.

TABLE 3 Soil Types

Name	Type/Characteristics
Solid Rock	Excavation walls stay vertical as long as the excavation is open.
Type A Soil	Fine-grained, cohesive: clay, hardpan, and caliche. Particles too small to see with the naked eye.
Type B Soil	Angular rock, silt, and similar soil.
Type C Soil	Coarse-grained, granular: sand, gravel, and loamy sand. Particles are visible to the naked eye.

To be safe, treat soil as if it is Type C soil, per *Table 3*, unless proven otherwise. It is better to over-prepare for a stronger soil than to not prepare enough for a weaker one.

A competent person must inspect excavations daily and decide whether cave-ins or failures of protective systems could occur, and whether there are any other hazardous conditions present. The competent person must conduct inspections before any work begins, as needed throughout the shift, and after every rainstorm or other hazard-increasing incident.

If the inspection reveals indications of protective system failure, hazardous atmospheres, or a possible cave-in, workers must be removed from the hazardous area and may not return until corrective action has been taken. Always ask the competent person on site or your immediate supervisor if you have questions about proper safety practices.

Once visual and manual tests are performed and the soil type is determined, a protective system must be chosen. Protective systems are required in nearly all excavations. There are various types of trench protective systems to meet each type of soil condition. Selecting a protective system for an excavation depends on soil conditions, the depth of the trench, and the environmental conditions surrounding the site.

Regardless of what type of system is used, if the excavation is more than 20 feet (6.1 m) deep, the entire excavation protective system has to be designed by a registered professional engineer. There are two basic systems of trench protection: sloping and benching systems and support systems. The method to be used is determined by the engineer and is based on the types of soil and the site conditions.

Sloping and benching are forms of trench protection that cut away and slant the excavation face. A sloping system is a method in which the sides of an excavation are cut back to a safe angle using relatively smooth inclines (*Figure 39A*). A benching system is similar to a sloping system, but instead of smooth inclines, the sides of the trench wall are cut back using a series of steps (*Figure 39B*). Benching systems cannot be used with Type C soil.

Shoring: A support system designed to prevent a trench or excavation cave-in.

Shielding: A structure used to protect workers in trenches.

Trenches are often located in narrow places, so sloping and benching are not options. In these situations, support systems like **shoring** or **shielding** must be used. Shoring structures are typically made of metal or wood and are used to support the sides of a trench and prevent soil from caving in. They consist of plating held firmly in place with expandable braces (*Figure 40*). There are many types of shoring systems. Some of them are easy to install, and others require experience and engineering.

Shielding structures, also known as trench boxes, are placed inside trenches or excavations, and are strong enough to protect workers in the event of a cave-in, so long as the workers are within the confines of the box (*Figure 41*). Trench

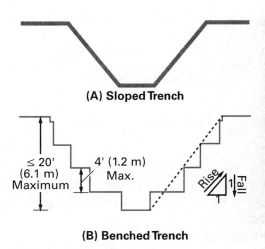

(A) Sloped Trench

≤ 20' (6.1 m) Maximum   4' (1.2 m) Max.   Rise 1 / Fall 1

(B) Benched Trench

Figure 39 Sloped and benched trenches for Type B soil.

Figure 40 Shoring structure.
Source: Kundel Industries

shields are used only to provide a protected space for workers. Shoring not only protects workers, but also prevents the trench walls from collapsing.

Spoil piles, comprised of the material removed from an excavation, and other materials represent a hazard if not handled properly. Loose rock, soil, materials, and equipment on the face or near the excavation can fall or roll into the excavation, or overload and possibly collapse excavation walls. Keep all spoil, materials, and heavy equipment at least 2 feet (0.6 m) away from the edge of an excavation, or set up barricades to contain falling material. Scale the excavation face to remove loose material, and place spoil so that rainwater runs away from the excavation. Use a retaining device strong enough and high enough to resist expected loads. If the spoil cannot be safely stored on site, remove it to a temporary site.

When working in a trench, there must be a safe means of entry and exit for workers, such as a stairway, ladder, or ramp. There must be an exit every 25 feet (7.6 m) for every trench over 4 feet (1.2 m) deep. Lifting equipment such as loader buckets and backhoe shovels are not safe means for entering or exiting a trench. Once you are in the trench and before you begin work, take a moment to look around and find the nearest ladder so that you can plan your exit, if necessary.

Your company is required to have an emergency action plan that must be communicated to every worker. If you are not sure what the emergency action plan is for your site, don't be afraid to ask questions. Your knowledge could help prevent serious injury or even death, and your supervisor wants you and everybody working with you to be as safe as possible.

Most importantly, try to prevent emergencies before they happen. When in doubt, get out! If you notice potentially dangerous conditions while working in an excavation, get yourself and your co-workers out of danger immediately, and inform your supervisor or the competent person of your concerns.

Spoil: Material such as earth removed while digging a trench or excavation.

Figure 41 Shielding structure.
Source: Kundel Industries

Call before You Dig: Dial 811 to Keep it Safe for Yourself and Others

Front   Back

Digging and excavation are likely to be part of all construction projects. It is essential that the site supervisor or project manager gets approval to dig before work begins; most states have laws that require this. Millions of lines of cable, piping, and utility fixtures are buried throughout the United States, and breaking into one of them could lead to power outages, electrocution, gas leaks, explosions, and the loss of life.

In addition to the physical dangers of digging, contractors are at risk for fines, penalties, and legal action if they start digging without having called 811 to get the okay. Local utility representatives need to come out and mark the ground to indicate where their lines may be buried. Contractors cannot assume that because a work site is in a rural location, that the risk is low for hitting a line. Utility lines are everywhere. Calling 811 a week or two in advance of the project allows local utility operators to respond and offer guidance as to where a worker may safely dig.

(continued)

All states have a One-Call Notification System center that makes the process easy. Sometimes this can be done online. Regardless of state availability, however, 811 works nationwide. This could save your life.

The Common Ground Alliance Initiative, who leads the nation's campaign to Call before You Dig, has five rules for contractors. These include:

1. Making the call to either a local center or the national 811 line and placing a request to dig.
2. Waiting to hear back from all known and potential utilities. This can take several days to a couple of weeks, so be sure to call a few weeks ahead.
3. Confirming that everything is marked on the property before getting started. Utilities will place flags or other markers to indicate where it is or is not safe to dig.
4. Respecting the process. This includes maintaining the markings made by utility operators and keeping current with the approval to dig. Where questions arise, call 811.
5. Proceeding with caution. When digging, be aware of the tolerance zone, which is 1.5 to 2 feet away from a marked line. If the dig must happen closer to the line than 1.5 to 2 feet, it must be done carefully by hand or with a vacuum excavation tool.

A Dig Safely card and other materials are readily available from the various state Dig Safely notification centers, as shown. This card has contact information and the American Public Works Association (APWA) universal color codes used for the temporary marking of underground facilities. All states have cards similar to this one, which is from New York.

Additional information on safe digging practices may be found at **call811.com**.

3.2.2   Tool, Machine, and Equipment Guards

Almost all tools and machines used in construction and industrial work are equipped with guards that protect workers from rotating parts. *Figure 42* shows the guards on a grinding machine. All tools and machines that could harm workers must have a guard shielding the hazard. The following types of tools and machines must have guards:

- Grinding tools
- Shearing tools
- Presses
- Punches
- Cutting tools
- Rolling machines
- Tools or machines with pinch points
- Tools or machines with sharp edges

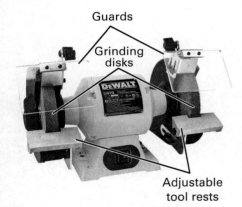

Figure 42 Bench grinder with guards.
Source: Image property of Stanley Black & Decker. Used with permission

Machine guards should prevent moving parts of the machine from coming into contact with your arms, hands, or any other part of the body, while allowing you to use the machine comfortably and efficiently. Some workers find machine guards to be aggravating and try to remove them from machines. Guards should be secure and should not be easily removed. They should be maintained in good condition, made of durable material, and bolted or screwed to the machine so that tools are needed for their removal. *Figure 43* shows a coupling guard installed over the coupling that connects a motor shaft to a pump shaft. The guard is bolted in place and should only be removed when the motor is not running. The guard is there to prevent someone from being caught in the coupling, which rotates at high speed. Its highly visible markings indicate that it is a hazardous location.

Follow these guidelines for using and caring for tool and machine guards:

- Do not remove a guard from a tool or machine except for cleaning purposes or to change a blade or perform other services. Make sure the machine is turned off and tagged out.
- When you are finished with cleaning or maintenance, replace the guard immediately.

Coupling guard

Figure 43 Coupling guard.

- Do not use any material to wedge a guard open.
- Guards and attachments are designed for the specific tool or machine you are using. Use only attachments that are specifically designed for that tool or machine.

3.2.3 Cranes and Heavy Equipment

Motorized equipment is used in many different jobs, including construction, mining, plant maintenance and operations, road maintenance, equipment transportation, and snow removal. Working with motorized equipment can be dangerous. Workers can be crushed by falling loads, fall from equipment, be electrocuted by power lines, or be struck or trapped by vehicles. The swing radius of equipment can also be a hazard if the job is not carefully planned and properly barricaded. Most heavy equipment has pinch points. *Figure 44* shows some of the pinch point hazards on an excavator.

Dangers exist for both equipment operators and other workers on the site. In one example, a contractor was operating a backhoe when another

Figure 44 Pinch and crush points.
Source: Courtesy of Atlas Copco

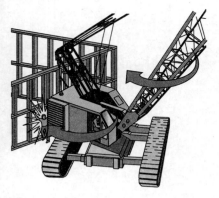

Figure 45 Example of a caught-between hazard.

employee attempted to walk between the swinging back end of the backhoe and a concrete wall. As the employee approached the backhoe from the operator's blind side, the back end hit the victim, crushing him against the wall. A similar problem can occur with a crane or excavator, as shown in *Figure 45*. As the cab swings, the rear end extends beyond the base of the machine. For that reason, barricades must be placed around the working perimeter of the machine.

Because working with or near motorized equipment is dangerous, you must understand that your first responsibility on a job is safety. This includes your own safety, the safety of others on the site, and the safe use of equipment on the site. You must know the hazards and safety procedures of every job you are on, regardless of the work you are doing.

3.0.0 Section Review

1. Which of these activities is most likely to produce flying objects?
   a. Using a chipping hammer
   b. Using a screwdriver
   c. Painting a wall
   d. Climbing a ladder

2. The type of soil most likely to result in a trench cave-in is _____.
   a. solid rock
   b. Type A
   c. Type B
   d. Type C

4.0.0 Energy Release Hazards

Performance Task

4. Inspect a typical power cord and GFCI to ensure their serviceability.

Objective

Identify common electrical hazards and how to avoid them.

a. Summarize basic job-site electrical safety guidelines.

b. Explain the importance of disabling equipment as well as basic lockout/tagout procedures.

Lockout/tagout (LOTO): A formal procedure for taking equipment out of service and ensuring that it cannot be operated until an authorized person has removed the lock and/or warning tag.

Whenever possible, you will de-energize equipment before you begin working on it. Your employer will have written guidelines that you must follow to de-energize the equipment. It is important to follow all guidelines because some circuits receive power from multiple sources. To place equipment in a safe work condition, all energy sources must be removed. Once equipment is de-energized, **lockout/tagout (LOTO)** devices must be attached to all sources to prevent someone from unknowingly restoring the energy before the work has been completed.

4.1.0 Electrical Safety Guidelines

Electrical safety is a concern for all workers, not just for electricians. On many jobs, no matter what your trade, you will use or work around electrical equipment. Extension cords, power tools, portable lights, and many other pieces of equipment use electricity.

Not all electrical incidents result in death. There are different types of electrical incidents. Any of the following can happen:

TABLE 4 Effects of Electrical Current on the Human Body

Current	Common Item/Tool	Reaction to Current
0.001 amps	Watch battery	Faint tingle.
0.005 amps	9-volt battery	Slight shock.
0.006–0.025 amps (women) 0.009–0.030 amps (men)	Christmas tree bulb	Painful shock. Muscular control is lost.
0.050–0.9 amps	Small electric radio	Extreme pain. Breathing stops; severe muscular contractions occur. Death may result.
1.0–9.9 amps	Jigsaw (4 amps); Sawsall® or Port-a-Band® saw (6 amps); portable drill (3–8 amps)	Ventricular fibrillation and nerve damage occur. Death may result.
10 amps and above	ShopVac® (15-gallon); circular saw	Heart stops beating; severe burns occur. Death may result.

- Burns
- Electric shock
- Explosions
- Falls caused by electric shock
- Fires

If the human body comes in contact with an electrically energized conductor and is also in contact with a **ground** at the same time, the body becomes an additional path of resistance for the electrical current to flow through. The electricity flows through the body in less than the blink of an eye without warning. That's why safety precautions are so important when working with and around electrical circuits. When a body conducts electrical current, and the current is high enough, the person can be electrocuted (killed by electric shock). *Table 4* shows the effects of different amounts of electrical current on the human body and lists some common tools that operate using those currents.

Here's an example: A craftworker is operating a portable power drill while standing on damp ground. The power cord inside the drill has become frayed, and the electric wire inside the cord touches the metal drill frame. Three amps of current pass from the wire through the frame, then through the worker's body and into the ground. *Table 4* shows that this worker will probably die.

There are specific policies and procedures to keep the workplace safe from electrical hazards. You can do many things to reduce the chance of an electrical incident. If you ever have any questions about electrical safety on the job site, ask your supervisor.

Ground: The conducting connection between electrical equipment or an electrical circuit and the earth.

WARNING!

Less than one amp of electrical current can kill. Always take precautions when working around electricity.

4.1.1 Grounding

Grounding is a method of protecting humans from electric shock; however, it is normally a secondary protective measure. The term *ground* refers to a conductive body, usually the earth. A ground is a conductive connection, whether intentional or incidental, by which an electric circuit or equipment is connected to earth or to an engineered grounding system. By grounding a tool or electrical system, a low-resistance path to the earth is intentionally created. When properly done, this path offers low resistance and has enough current-carrying capacity to prevent the buildup of voltages that could create a personnel hazard. This does not guarantee that no one will receive a shock. It will, however, greatly reduce the possibility of such incidents, especially when used in combination with your company's safety program.

Did You Know?

Electrocution

Electric shocks or burns are a major cause of incidents in the construction industry. According to the Bureau of Labor Statistics, electrocution is the fourth leading cause of death among construction workers.

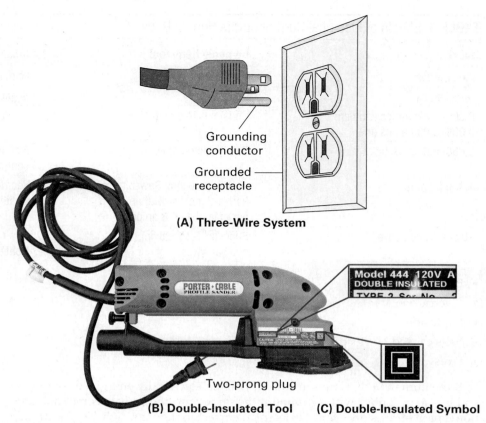

(A) Three-Wire System

(B) Double-Insulated Tool   **(C) Double-Insulated Symbol**

Figure 46 Three-wire system, double-insulated tool, and double-insulated symbol.

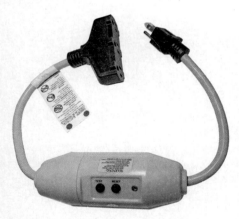

Figure 47 Extension cord with a GFCI.

Ground fault circuit interrupter (GFCI): A device that interrupts and de-energizes an electrical circuit to protect a person from electrocution.

Use three-wire extension cords for portable power tools and make sure they are properly connected (*Figure 46A*). The three-wire system is one of the most common safety grounding systems used to protect you from incidental electrical shock. The third wire is connected to a ground system. If the insulation in a tool fails, the current will pass to ground through the third wire—not through your body. Double-insulated tools (*Figure 46B*) are also very effective in preventing shocks. In fact, it has become more common to use double-insulated tools because they are safer than relying on a three-wire cord alone. *Figure 46C* also shows the double-insulated symbol that can be found on double-insulated tools. Double-insulated tools use a two-wire power cord with no ground pin. One prong of the plug is larger than the other so it can only be connected to a polarized receptacle.

4.1.2 Ground Fault Circuit Interrupters

A **ground fault circuit interrupter (GFCI)** is a fast-acting circuit breaker that senses small imbalances in the circuit caused by current leakage to ground. A GFCI continually matches the amount of current going to an electrical device against the amount of current returning from the device. Whenever the two values differ by more than five milliamps, the GFCI interrupts the electric power within $\frac{1}{40}$ of a second. *Figure 47* shows an extension cord with a GFCI.

A GFCI must be used on all receptacles that are not part of the building's permanent wiring, such as temporary power and extension cords. A GFCI provides protection against a ground fault, which is the most common form of electrical shock. It also provides protection from fires, overheating, and wiring insulation deterioration.

Tripping of GFCIs—interruption of circuit flow—is sometimes caused by wet connectors and tools. Limit the amount of water that tools and connectors come into contact with by using watertight or sealable connectors. Tripping may also be caused by cumulative leakage from several tools or from extremely long circuits. GFCIs should be periodically inspected in accordance with the manufacturer's recommendations or site safety practices. GFCIs have a TEST button that can be pressed to verify that the GFCI is working. The GFCI should be tested before any use.

4.1.3 Summary of Electrical Safety Guidelines

Here are some basic job-site electrical safety guidelines:

- All power tools used in construction should be ground-fault protected.
- Make sure that panels, switches, outlets, and plugs are grounded.
- Never use bare electrical wire.
- Never use metal ladders near any source of electricity.
- Inspect electrical power tools before you use them.
- Never operate any piece of electrical equipment that has a danger tag or lock-out device attached to it.

WARNING!

All work on electrical equipment should be done with circuits de-energized, locked out, and confirmed. All conductors, buses, and connections should be considered energized unless proven otherwise.

- Never use worn or frayed cables (*Figure 48*). If the cord is frayed or worn, disconnect the power and dispose of the cord.
- Make sure lightbulbs have protective guards to prevent incidental contact (*Figure 49*).

4.1.4 Working Near Energized Electrical Equipment

No matter what your trade, your job may include working near exposed electrical equipment or conductors. This is one example of **proximity work**. Often, electrical distribution panels, switch enclosures, and other equipment must be left open during construction. This leaves the wires and components in them exposed. Some or all of the wires and components may be energized. Working near exposed electrical equipment can be safe, but only if you keep a safe working distance.

Regulations and company policies tell you the minimum safe working distances from exposed conductors. The safe working distance can be very small or span a significant distance, depending on the voltage. The higher the voltage, the greater the required safe working distance.

You must learn the safe working distance for each situation. Make sure you never get any part of your body or any tool you are using closer to exposed conductors than that distance. You can get information on safe working distances from your instructor, your supervisor, company safety policies, and regulatory documents. This subject is also covered in greater detail in other craft curricula where it is relevant.

Proximity work: Work done near a hazard but not actually in contact with it.

Figure 48 Never use damaged cords.

Figure 49 Work light with protective guard.

One of the common causes of electrical shock is coming into contact with overhead wires with metal ladders, cranes, or excavating equipment. A distance of at least 10 feet (3 m) must be maintained from any conductor carrying 50,000 volts or less. Greater distances are required for higher voltages.

4.1.5   If Someone Is Shocked

If you are there when someone gets an electrical shock, you can save a life by taking immediate action. The best thing to do is to immediately shut off the power to the circuit. Do not touch the victim while he or she remains in contact with the power source. Once the circuit is disconnected, call an ambulance or the emergency number at your job site. Render first aid only if you are trained to do so.

WARNING!

Do not touch the victim or the electrical source with your hand, foot, or any other part of your body or with any object or material. You could become part of the circuit—and another victim.

4.2.0   Lockout/Tagout Requirements

Failure to disable machinery before working on it is a major cause of injury and death on job sites. Lockout/tagout procedures safeguard workers against unexpected releases from various energy sources. An energy source can be electrical, mechanical, hydraulic, pneumatic, chemical, or thermal. After the power has been turned off, energy can still be stored in a device or component. For example, even when a hydraulic system has been turned off, hydraulic pressure may remain in all or part of the system. This high-pressure energy can be released if the system is opened during service or repairs, creating an extremely hazardous situation. Additional dangers exist if the device or system contains chemicals, flammable liquids, high-temperature liquids, or gases.

When anyone is working on or around active systems, all equipment that could release energy must be shut down, drained, de-energized, or otherwise rendered harmless whenever possible. Switches, circuit breakers, valves and other components are switched off or closed, and then locks or tags (*Figure 50*) are applied so they cannot be re-energized while work is ongoing.

(A) Electrical Lockout

(B) Valve Lock

(C) Pneumatic Lockout

Figure 50 Lockout/tagout devices.
Sources: Honeywell Safety Products (50B); Panduit Corp. (50C)

Electrical Cord Safety

Electrical cords are frequently seen on construction sites, yet they are often overlooked. Use the following safety guidelines to ensure your safety and the safety of other workers.

- Every electrical cord should have an Underwriters Laboratory (UL) label attached to it. Check the UL label for specific wattage. Do not plug more than the specified number of watts into an electrical cord.
- A cord set not marked for outdoor use is to be used indoors only. Check the UL label on the cord for an outdoor marking.
- Do not remove, bend, or modify any metal prongs or pins of an electrical cord.
- Do not run a cord through doorways or through holes in ceilings, walls, and floors that might pinch the cord. Also, check to see that there are no sharp corners along the cord's path. Any of these situations will lead to cord damage.
- Extension cords are a tripping hazard. They should never be left unattended and should always be put away when not in use.

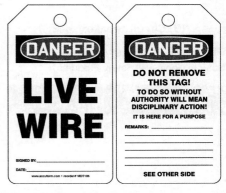

Figure 51 Typical safety tags.
Source: Courtesy of Accuform.com LLC

Generally, each lock has its own key, and the individual who applies the lock keeps the key. A variety of tags are used, depending on the circumstances and the organization. Tags (*Figure 51*) typically have the word *DANGER* on them, along with other printed information, writing space for comments, and a line for the installer's signature and date. The authorized person who applied the lock must be the one to remove it. In an emergency, a supervisor or other person authorized by the employer may remove a lockout/tagout.

In a lockout, an energy-isolating device such as a disconnect switch or circuit breaker is placed in the Off position and a lock is applied. Padlocks are popular and versatile, but other styles (*Figure 52*) are often specifically suited for the equipment being serviced. Multiple lockout devices (*Figure 53*) are used when more than one person is accessing the equipment. When a worker needs to service a system and sees that another person's lock has been applied, it is tempting to accept this as a safe situation. However, the lock may be removed by the first worker and the circuit re-energized while the second worker is still vulnerable. In these cases, the multiple lockout devices allow several workers to apply locks, and each is ensured that the system cannot be restored until theirs is removed.

In a tagout, components that control power to equipment and machinery are set to a safe position and a written warning is attached. This method may be more appropriate for valves and other energy-controlling devices that are difficult or impossible to lock, since tags alone cannot prevent a control device or switch from being repositioned.

It is important to emphasize that LOTO procedures are not solely related to electrical energy sources. Technicians cannot forget that other forms of energy may need to be rendered safe as well.

The exact procedures for LOTO will vary by organization and site. Check with your instructor or site supervisor regarding the detailed LOTO procedure at your site and be sure you are familiar with the proper steps.

4.2.1 Pressurized or High-Temperature Systems

Many jobs require that workers be close to tanks, piping systems, and pumps that contain pressurized or high-temperature fluids. Be aware that touching a container of high-temperature fluid can cause burns. Many industrial processes involve fluids that are as hot as several thousand degrees. Also, if a container holding pressurized fluids is damaged, it may leak and spray dangerous fluids. Any work around pressurized or high-temperature systems is considered proximity work. Barricades, a monitor, or both may be needed to ensure the safety of those working nearby.

(A) Electrical Plug Lockout

(C) Ball Valve Lockout

(B) Circuit Breaker Lock

(D) Electrical Switch Lockout

Figure 52 Lockout devices.
Source: Honeywell Safety Products

Figure 53 Multiple lockout/tagout device.

4.0.0 Section Review

1. The purpose of double insulating an electrically powered tool is to _____.

   a. prevent the tool from overheating
   b. protect users from electrical shock
   c. keep the tool from getting too cold
   d. allow it to use a three-prong receptacle

2. When a lockout is performed, the key to the lock is _____.

   a. left in the lock
   b. kept by the person who applied the tag
   c. given to the foreman for safekeeping
   d. hidden in a convenient place

5.0.0 Personal Protective Equipment

Performance Tasks

2. Inspect the following PPE items and determine if they are safe to use:
   - Eye protection
   - Hearing protection
   - Hard hat
   - Gloves
   - Approved footwear

3. Properly don, fit, and remove the following PPE items:
   - Eye protection
   - Hearing protection
   - Hard hat
   - Gloves

Objective

Associate personal protective equipment (PPE) with the hazards they reduce or eliminate.

a. Explain how PPE is used to protect craftworkers from different types of injuries.

b. Explain how respirators protect craftworkers from respiratory dangers.

Every worker is responsible for wearing appropriate PPE on the job. When worn correctly, PPE is designed to protect you from injury. You must keep it in good condition and use it when you need to. When workers are injured on the job, it is often because they are not using required PPE.

You will not see all the potentially dangerous conditions just by looking around a job site. It is important to stop and consider what type of incidents could happen on any job that you are about to do. Knowing how to use PPE will greatly reduce your chance of getting hurt. Keep in mind that PPE requirements are usually established by the job site. Some job sites will have different requirements than others, so always make sure you check. For example, some sites require high-top safety boots instead of the more common standard 6-inch tops.

5.1.0　PPE Items

Remember, while PPE is the last line of defense against personal injury, using it properly and taking care of it are the first steps toward protecting yourself on the job.

The first line of defense is represented by the engineering and administrative controls used by employers to eliminate or reduce the need for some PPE. Engineering controls are implemented by making physical changes, such as reducing noise by the use of mufflers on equipment engines; installing guards on moving equipment; and making sure equipment is properly maintained. An example of an administrative control is operating noisy equipment on a shift when fewer workers are present.

The best protective equipment is of no use unless you follow these rules:

- Regularly inspect it.
- Properly care for it.
- Use it properly when it is needed.
- Never alter or modify it in any way.

The sections that follow describe protective equipment commonly used on construction sites, and tell how to use and care for each piece of equipment. Be sure to wear the equipment according to the manufacturer's specifications.

WARNING!

Your clothing must comply with good general work and safety practices. Do not wear clothing or jewelry that could get caught in machinery or otherwise cause an incident, such as loose clothing, baggy shirts, or dragging pants. You must wear a shirt at all times; some tasks will require long-sleeved shirts. Your shirt should always be tucked in unless you are performing welding.

Protective equipment commonly worn by craftworkers on any job site usually includes the following:

- Hard hats
- Eye and face protection
- Gloves
- Foot and leg protection
- Hearing protection
- Respiratory protection
- High-visibility clothing

Sharing PPE

It is not a good idea for workers to share PPE. If one person has an infection, it can be passed to another through the shared equipment.

Figure 54 Typical hard hat.
Source: Photo courtesy of Bullard

5.1.1 Hard Hats

Figure 54 shows a typical hard hat. The outer shell of the hat can protect your head from a hard blow. The webbing inside the hat keeps space between the shell and your head. Adjust the headband so that the webbing fits your head and there is at least one inch of space between your head and the shell.

Do not alter your hard hat in any way. Inspect your hard hat every time you use it. If there are any cracks or dents in the shell, or if the webbing straps are worn or torn, get a new hard hat. Wash the webbing and headband with soapy water as often as needed to keep them clean. Wear the hard hat only as the manufacturer recommends. Never wear anything but approved clothing under your hard hat.

WARNING!

No articles should be worn under the hard hat that could interfere with fit and visibility. That includes ball caps or hoodies that obscure peripheral vision. Only employer-approved gear is to be worn under the hard hat.

5.1.2 Eye and Face Protection

Wear eye protection (*Figure 55*) wherever there is even the slightest chance of an eye injury. In general, eye protection is required any time you are on a job site. Face shields (*Figure 56*) are added over safety glasses for certain tasks, such as grinding. Appropriate eye or face protection is required when there is a possibility of exposure to hazards from flying particles, molten metal, liquid chemicals, acids or caustic liquids, chemical gases or vapors, or potentially injurious light radiation. The following are examples of tasks requiring eye protection:

- Grinding and chipping metal
- Using power saws and other tools/equipment that can throw out solid material
- Working with molten lead, tar pots, and other molten materials
- Working with chemicals, acids, and corrosive liquids
- **Arc welding**

On the job site, potential eye hazard areas are usually identified, but always be on the lookout for possible hazards.

Regular safety glasses will protect you from falling objects or from objects flying toward your face. Side shields provide added protection and are required whenever there is risk of flying debris. Safety goggles provide the best protection from all directions. A face shield is required, in addition to safety glasses or goggles, when there is likely to be flying debris. Grinding and chipping activities are good examples of such activities.

Handle your safety glasses and goggles with care. If they get scratched, replace them; the scratches will interfere with your vision. Clean the lenses regularly with lens tissues or a soft cloth.

Arc welding: The joining of metal parts by fusion, in which the necessary heat is produced by means of an electric arc.

CAUTION

Standard eyeglasses are not adequate protection in high-hazard work zones. Shatterproof lenses and side shields are required for eye protection in those areas.

(A) Safety Glasses

(B) Tinted Safety Glasses

(C) Safety Goggles

Figure 55 Typical safety glasses and goggles.
Source: MSA The Safety Company

(A) Clear Face Shield

(B) Tinted Face Shield

Figure 56 Full-face shields.
Source: MSA The Safety Company

Case History

A training specialist traveled nearly three hours to reach a remote job site in order to observe and photograph the installation of high-voltage transmission lines. On arriving, he parked his car, put on his work boots, hard hat, reflective vest, and safety glasses, and walked over to the location where the work was being done. He was introduced to the project manager, who shocked him by saying: "Sir, I have to ask you to leave my job site because you are not wearing electrically insulated boots." Given no choice, the training specialist headed toward his car for the long trip home. Fortunately, the project manager needed to go back to the office for a meeting and offered the loan of his boots, thus preventing six hours of wasted time, as well as the loss of a rare opportunity.

The Bottom Line: Project managers and superintendents in construction and industrial work are serious about safety. If you are not fully prepared, you may find yourself sitting on the sidelines. A worker who shows up at a job site without the required PPE could lose a day's pay and possibly face disciplinary action.

Welders must use tinted goggles or welding hoods. The tinted lenses protect the eyes from the bright welding arc or flame. Welders must use filter lenses as specified by job site requirements. Welder's helpers and all employees working in the vicinity of arc welding should not look directly at the welding process and are also required to use eye protection with the prescribed level of shading. Oxyacetylene welding and burning also require the use of a filter lens.

Follow these general precautions for eye care:

- Always report all eye injuries and suspected foreign material in your eye to your supervisor immediately. Do not try to remove foreign matter yourself.

- Keep your hands away from your eyes.

- Keep materials out of your eyes by regularly clearing debris from your hard hat brim, the top of your goggles, and your face shield. When removing a cap, hard hat, or face shield, be careful to remove it in a way that prevents accumulated dirt and debris from falling into your eyes.

- Flood your eyes with water if you feel something in them. Never rub them, as this can make the problem worse.

- Know the location of eyewash stations and how to use them.

5.1.3 Hand Protection

On many construction jobs, you must wear heavy-duty gloves to protect your hands (*Figure 57*). Construction work gloves are usually made of cloth, canvas, or leather. Make sure your work gloves are a good fit. Wearing gloves that are

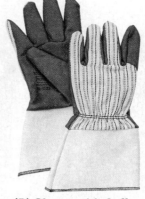

(A) Fitted Work Gloves   **(B) Gloves with Cuffs**   **(C) Rubberized Cotton**

Figure 57 Work gloves.
Sources: North Safety Products USA (57B); Honeywell Safety Products (57C)

too big for your hands can lead to injury. Never wear gloves around rotating or moving equipment. They can easily get caught in the equipment. Replace gloves when they become worn, torn, or soaked with oil or chemicals.

Gloves help prevent cuts and scrapes when you handle sharp or rough materials. Heat-resistant gloves are needed for handling hot materials. Craftworkers should be familiar with standards governing hand protection, such as ANSI/ SEA Standard 105-2011, *Hand Protection Selection Criteria* and European Standard EN388, *Protective Gloves Against Mechanical Risks*. These standards establish a rating system for gloves tested for their resistance to various hazards, including chemical burns, abrasions, cuts, punctures, resistance to heat and flame, and resistance to cold. Additional European standards govern protection from chemicals and microorganisms (EN374); thermal hazards (EN407 and EN511); and ionizing radiation and radioactive contamination (EN421).

Electricians use special rubber-insulated gloves when they work on or around live circuits. When working with solvents or other chemicals, it may be necessary to wear chemical-resistant gloves.

WARNING!

Only specially trained employees are allowed to use dielectrically tested rubber gloves to work on energized equipment. Never attempt this work without proper training and authorization.

5.1.4   Foot and Leg Protection

Approved safety footwear must be worn on all job sites. The best shoes to wear on a construction site are shoes with a safety toe to protect toes from falling objects (*Figure 58*). This type of shoe is generally required on every construction-related job site.

Some specialized work will require different footwear or gear. For example, safety-toed rubber boots are needed on job sites that are subject to chemically hazardous conditions or standing water. When climbing a ladder, your shoes or boots must have a well-defined heel to prevent your feet from slipping off the rungs. You will need foot guards when using jackhammers and similar equipment.

Never wear canvas shoes or sandals on a construction site. They do not provide adequate protection. Always replace boots or shoes when the sole tread

Figure 58 Safety shoe.

becomes worn or the shoes have holes, even if the holes are on top. Because of the risk of fire, do not wear oil-soaked shoes when you are welding or cutting metal.

Remember these general guidelines relating to leg protection:

- Never carry pointed tools, such as scissors or screwdrivers, in your pants pockets. Use a canvas or leather tool sheath with all sharp ends pointing down.
- When using certain special equipment, such as chainsaws and brush hooks, use shin guards.
- Consider stability when stepping into or onto locations where materials are stored. Materials may shift and pinch your legs and/or feet.

5.1.5 Hearing Protection

Unlike damage to most parts of the body, ear damage does not always cause pain. Exposure to loud noise over a long period of time can cause hearing loss, even if the noise is not loud enough to cause pain. Vibrations caused by sound waves enter the ear and are received by the cochlea, which consists of tiny hair cells inside the ear that convert the vibrations into sound signals. Exposure to intense sound waves over time can damage the cochlea and reduce the ability to hear. Damage to the cochlea can also cause tinnitus, which is a permanent condition characterized by ringing or buzzing in the ears that never stops. Hearing loss reduces your quality of life and makes simple, daily tasks more complicated. Save your hearing by using hearing protection whenever you are working in a noisy environment.

Here is a good rule to follow: If the noise level is so great that you have to raise your voice to be heard by someone who is less than 2 feet (61 cm) away, you need to wear hearing protection.

Most construction companies follow rules defined in official safety standards in deciding when hearing protection must be used. One type of hearing protection is specially designed earplugs that fit into your ears and filter out noise (*Figure 59A*). Another type of hearing protection is earmuffs, which are large padded covers for the entire ear (*Figure 59B*). The headband on earmuffs must be adjusted for a snug fit. If the noise level is very high, you may need to wear both earplugs and earmuffs.

Noise-induced hearing loss can be prevented by using noise control measures and personal protective devices. *Table 5* shows the recommended maximum length of exposure to sound levels rated 90 decibels and higher. When noise levels exceed those outlined in *Table 5*, an effective hearing conservation program is required. A company-appointed program administrator will oversee this program. If you have questions about the hearing conservation program on your site, see your supervisor or the program administrator.

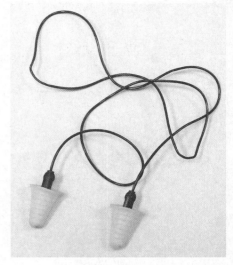

(A) Earplugs

(B) Earmuffs

Figure 59 Hearing protection.
Sources: MSA The Safety Company (59B)

TABLE 5 Maximum Noise Levels

Sound Level (decibels)	Maximum Hours of Continuous Exposure per Day	Example
90	8	Power lawn mower
92	6	Belt sander
95	4	Tractor
97	3	Hand drill
100	2	Chain saw
102	1.5	Impact wrench
105	1	Spray painter
110	0.5	Power shovel
115	0.25 or less	Hammer drill

5.2.0   Respiratory Hazards and Protection

Wherever there is danger of an inhalation hazard, you must use a respirator. There are a number of job-site conditions that require workers to wear respirators, including the following:

- Dust from metal grinding
- Drilling or cutting concrete
- Toxic fumes from welding or flame cutting of some metals
- Working with cleaning solvents
- Working in low-oxygen environments such as confined spaces
- Spray painting
- Sand blasting
- Working with chemicals such as chlorine or ammonia

5.2.1   Silica Standard

Silica is a mineral found in concrete, masonry, and rock. During construction, silica may be found in a dust form. Silica dust is created when cutting, drilling, or performing similar tasks on materials containing crystalline silica. Prolonged exposure to silica dust can cause **silicosis**, which is an incurable, and sometimes fatal, lung disease. The time it takes for silicosis to develop varies based on the length of time and the amount of silica someone is exposed to. In addition to silicosis, the inhalation of silica dust increases the risk of lung cancer and **chronic obstructive pulmonary disease (COPD)**. Individuals working closely with materials containing silica, such as workers operating abrasive blasting (sandblasting) equipment, also risk the development of kidney disease.

Silicosis: A serious lung disease resulting from the inhalation of crystalline silica particles.

Chronic obstructive pulmonary disease (COPD): Lung disease(s) that results in the obstruction of lung airflow and interferes with normal breathing. Chronic bronchitis and emphysema generally fall under the diagnosis of COPD.

> **WARNING!**
>
> When you are working in an area where silica dust is present, you must use the appropriate respiratory protection. Never work around silica dust without proper training, authorization, and PPE.

OSHA actually holds employers responsible for protecting workers from crystalline silica. *29 CFR, Part 1926.1153*, known in the workplace as the silica standard, outlines the steps for controlling exposure when working with sand, stone, concrete, and similar materials. A number of options are available for keeping workers safe, as well as methods to monitor exposure.

The basic requirements for employers include the establishment and implementation of a written silica exposure control plan that identifies tasks leading to exposure and the approaches to protecting workers. In addition, employers must do the following:

- Designate a competent person to ensure the exposure control plan is effectively implemented.
- Establish procedures to restrict access to areas where silica dust is actively generated.
- Train workers on tasks that create silica dust and ways to limit exposure.
- Limit common housekeeping practices that expose workers to silica dust when an alternative is available.
- Provide medical exams that include lung-function tests and x-rays every three years for workers that wear a respirator for 30 or more days each year.
- Keep records of all the above.

Figure 60 Concrete saw with a water-delivery system (wet sawing).
Source: Robert I. Carr, Ph. D., P.E.

Table 1 of *29 CFR, Part 1926.1153* is an extensive table and subject to change; therefore, its contents have not been included herein. All trainees are encouraged to access the standard at **https://www.osha.gov/laws-regs/regulations/standardnumber/1926/1926.1153** and become familiar with the requirements for various activities related to silica dust.

High-efficiency particulate arrestance (HEPA) filters: Special filters designed to capture very small particles. HEPA filters have a minimum efficiency of 99.97 percent, capturing particles that are 0.3 microns (0.0000117 inch) or larger in size.

Clean rooms: A room or area that is used for scientific research or specialized processes, such as the manufacture of pharmaceuticals and semiconductors. The cleanliness level of a clean room is generally based on the number of particles measured per cubic meter of area.

The OSHA silica standard guides employers and workers in managing silica dust exposure. *Table 1* of the standard provides a great deal of the information needed by craftworkers. All construction craftworkers are encouraged to review the table and become familiar with the requirements. Note that the requirements are also subject to change, and this topic is one that is consistently monitored to ensure the guidelines remain effective.

The primary means of management is to use saws and similar power tools that are equipped with a built-in water-delivery system (*Figure 60*). Water prevents the dust from becoming airborne. Handheld and stand-mounted concrete drills must have a dust collection system connected to a vacuum (*Figure 61*). Vacuums used to capture silica dust must be equipped with **high-efficiency particulate arrestance (HEPA) filters** (*Figure 62*). These filters are extremely effective for capturing the tiniest particles and are also used in critical environments such as hospital surgical areas and **clean rooms**.

There are some conditions that require the use of breathing filtration. For silica dust applications, OSHA requires filtration to meet an assigned protection factor (APF) of 10. This is a level of performance that is typical of many respirators. However, not all disposable filters meet the criteria.

It is important to your long-term health and the health of others to be aware of the sources of silica dust and how they are managed. Epoxy grout, for example, also contains silica. Any time concrete, brick, stone, grout, and similar materials are being ground or sawed, it is best to assume silica is present and act accordingly.

5.2.2 Other Respiratory Hazards

Asbestos is another hazardous material that can be harmful to your lungs. Prolonged exposure can cause lung cancer, asbestosis (scarring of the lung tissue), and a cancer called mesothelioma. It may take more than 20 years for these diseases to develop. Smoking significantly increases the risk of lung

Figure 61 Rotary Hammer Drill with Osha Table 1 Compliant On Board Dust Extractor
Source: Image property of Stanley Black & Decker. Used with permission.

Figure 62 HEPA vacuum cleaner.
Source: Photo used with permission of
Husqvarna Group

disease. If you smoke, tar from cigarettes will stick to the asbestos fibers in the lungs, making it more difficult for your body to get rid of the asbestos.

Sources of asbestos include older insulation and other building materials, such as floor tiles, pipe insulation, drywall compounds, and reinforced plaster, along with some roofing and siding materials. The United States banned production of most asbestos products in the 1970s, meaning that asbestos-containing materials (ACMs) are generally found in structures built before 1980. During renovation or maintenance operations, asbestos may be dislodged and the fibers become airborne. Asbestos fibers are particularly hazardous and must not be disturbed unless special techniques and procedures are used to remove it safely. Although you will not feel the effects immediately, this exposure can be a serious health hazard. Never handle asbestos without proper training, authorization, and PPE.

Only workers who have been trained and licensed are authorized to perform work related to asbestos including, but not limited to, the following tasks:

- Demolition
- Removal
- Alteration
- Repair
- Maintenance
- Installation
- Cleanup
- Transportation
- Disposal
- Storage

If asbestos is encountered or suspected on the job site, a supervisor must be notified and all work must stop until the asbestos is either sealed off or removed by licensed professionals.

Around the World

Asbestos Use

Asbestos was identified as a health risk many years ago. However, countries have reacted to the hazard in different ways. While some countries banned its use in products and construction decades ago, other countries, such as India, continue to use asbestos without regulation of any form. Indeed, asbestos use in India has risen over 80 percent in the last decade.

The World Health Organization (WHO) estimates that 125 million people worldwide are exposed to asbestos each year in the workplace. Even in the United States, asbestos may still be encountered in old buildings, flooring, and insulation components. Be aware of the regulations concerning the use of asbestos in your region and stay alert to sources that might do you harm if not avoided.

5.2.3   Types of Respirators

Federal law specifies which type of respirator to use for different types of hazards. There are four general types of respirators (*Figure 63*):

- Self-contained breathing apparatus (SCBA)
- Supplied air mask
- Full facepiece mask with chemical canister (gas mask)
- Half mask or mouthpiece with mechanical filter

(A) Self–Contained Breathing Apparatus (SCBA)

(B) Supplied Air Mask

(C) Full Facepiece Mask

(D) Half Mask

Figure 63 Examples of respirators.

Source: MSA The Safety Company (63A–63B) (63D); North Safety Products USA (63C)

A SCBA carries its own air supply in a compressed air tank. It is used where there is not enough oxygen or where there are dangerous gases or fumes in the air.

A supplied air mask uses a remote compressor or air tank to provide breathable air in oxygen-deficient atmospheres. Supplied-air masks can generally be used under the same conditions as SCBAs.

A full facepiece mask with chemical canisters is used to protect against brief exposure to dangerous gases or fumes. A half mask or mouthpiece with a mechanical filter is used in areas where you might inhale dust or other solid particles.

5.2.4 Wearing a Respirator

If you need to use a respirator, your employer must provide you with the appropriate training to select, test, wear, and maintain this equipment. Your employer is also responsible for having you medically evaluated to ensure you are fit enough to wear respiratory protection equipment without being harmed. You must also be tested to ensure a proper fit and a good seal.

Wearing a respirator generally places a burden on the employee. A self-contained breathing apparatus is heavy and may be difficult for some workers to carry; negative-pressure respirators restrict breathing; and other respirators may cause claustrophobia. For these and other reasons, workers must be medically evaluated by a physician or other licensed health care professional to determine under what conditions they may safely wear respirators.

Respirators are ineffective unless properly fit-tested to the user. To obtain the best protection from a respirator, perform positive (breathing out) and negative (breathing in) fit checks each time it is worn. These fit checks must be repeated until a good face seal is obtained. A respirator must be clean, in good condition, and all of its components must be in place in order to provide adequate protection.

When a respirator is required, a personal monitoring device is usually also required. This device samples the air to measure the concentration of hazardous chemicals.

5.2.5 Selecting a Respirator

Follow company and government procedures when choosing the type of respirator for a particular job. Also be sure that it is safe for you to wear a respirator. Under current regulations, workers must fill out a questionnaire to identify potential problems in wearing a respirator. Depending on the answers, a medical exam may be required. When a respirator is not required, workers may voluntarily use a dust or particle mask for general protection. These masks do not require fit testing or a medical examination.

Always use the appropriate respiratory protective device for the hazardous material involved and the extent and nature of the work to be performed. Before using a respirator, you must determine the type and concentration level of the contaminant, and whether the respirator can be properly fitted on your face. If you wear contact lenses, practice wearing a respirator with your contact lenses to see whether you have any problems. That way, you will identify any problems before you use the respirator under hazardous conditions.

CAUTION

All respirator instructions, warnings, and use limitations contained on each package must be read and understood by the wearer before use.

5.0.0 Section Review

1. Personnel working in the vicinity of arc welding work must wear eye protection with tinted lenses.
   a. True
   b. False

2. Drilling concrete requires use of a respirator because it produces _____.
   a. asbestos
   b. lead
   c. toxic fumes
   d. silica dust

6.0.0 Job-Site Hazards

Objective

Describe safety practices used with other common job-site hazards.

   a. List other types of hazards craftworkers may encounter.

   b. Describe common environmental hazards and how craftworkers should respond to them.

   c. Summarize hazards associated with hot work.

   d. Identify fire hazards and describe basic fire fighting procedures.

   e. Name different types of confined spaces and how to avoid related hazards.

Performance Tasks

There are no performance tasks in this section.

It is impossible to list all the hazards that can exist on a construction or industrial job site. This section describes some of the more common hazards and explains how to deal with them. For your safety, you must know the specific hazards where you are working and how to prevent incidents and injuries. If you have questions specific to a job site, ask your supervisor.

6.1.0 Job-Site Exposure Hazards

The term *exposure* refers to contact with a chemical, biological, or physical hazard. Exposure can be chronic or acute. Chronic exposure is long-term and repeated, and may be mild or severe. Acute exposure is short-term and intense.

According to the HAZCOM standard, your employer must inform you of any hazards to which you might be exposed. In order to better protect you, your employer may require pre-employment and periodic medical examinations to ensure that your health is not being negatively affected by chemical exposure during work. Exams are usually conducted annually, and generally involve targeted organ testing. Because most medical tests cannot directly check for specific chemical exposure, blood tests are used to check the status of the particular organs (called target organs) known to be most affected by the chemicals to which a person has been exposed.

The results of these tests are compared to pre-employment test results and any other annual or periodic exams you may have had. Employers also frequently request a final screening when a worker leaves the company.

Workers can be exposed to chemicals and other hazardous materials in a variety of ways. Routes of exposure include breathing (inhalation), eating or drinking (ingestion), and skin contact (absorption). Chemical exposure is subject to permissible exposure limits (PELs). In other words, some level of exposure to certain chemicals and fumes is considered acceptable and does not represent a hazard. A PEL is the maximum concentration of a substance that a worker can be exposed to in an eight-hour shift. When a worker approaches or reaches the PEL, he or she must be removed from that environment.

There are various types of hazardous materials you may come in contact with during various types of construction work, including the following:

- Lead
- Bloodborne pathogens
- Chemicals

Taking the time to understand each of these hazards will help you stay healthy and safe. Working around these hazards requires special training and PPE. Never handle any of these materials without proper training, authorization, and PPE.

6.1.1 Lead

Lead occurs naturally in the Earth's crust and is spread through human activities such as mining and the burning of fossil fuels. As an element, lead is indestructible—it does not break down. Once it is released into the environment, it can move from one medium to another. For example, lead in dust can be carried long distances, dissolve in slightly acidic water, and find its way into soil where it can remain for years. Lead is difficult to detect because it has no distinctive taste or smell.

Lead has many useful properties. It is soft and easily shaped, durable, resistant to some chemicals, and fairly common. Because lead is so versatile, it is used in the production of piping, batteries, and casting metals. However, lead is a toxic metal that can cause serious health problems. You can be exposed to lead by breathing air, drinking water, eating food, and swallowing or touching dust or dirt that contains lead.

You may encounter lead-based paints during demolition or renovation of structures built before 1978, when lead-based paint was banned. Dust created from sanding lead-based paints is hazardous. Protective clothing and equipment must always be used when lead levels are above the PEL. If you are unsure whether lead is present, consult your supervisor.

All waste contaminated with lead is considered hazardous. Those who handle hazardous waste must have special training. Never handle hazardous materials or waste without proper training, authorization, and PPE.

6.1.2   Bloodborne Pathogens

Bloodborne pathogens are another health hazard you may encounter on the job site. Sometimes you will hear them referred to as bloodborne infectious diseases. They are diseases that can be transmitted by contact with an infectious person's blood or other bodily fluids, the most common being HIV (the virus that causes AIDS) and hepatitis B and C.

You could be exposed to another person's blood on the job site when administering first aid to an injured person or being involved in a multi-victim incident. In order to safeguard yourself, you must know the universal precautions to prevent exposure and follow them closely:

- Always use appropriate gloves, eye protection, and a mask when administering first aid.
- If you come in contact with someone's blood, immediately wash the affected areas with soap and water. Otherwise, use antiseptic hand cleanser or antiseptic towelettes until washing with soap and running water is possible.
- Notify your supervisor of the contact right away.

You may need to seek medical attention for precautionary screening, particularly if that person has any bloodborne infectious diseases. Your employer should have a written exposure control plan and procedure for such occurrences.

6.1.3   Chemical Splashes

Chemicals such as acids and solvents can pose physical hazards, including acid reactions or burns. Acids can create toxic vapors or react violently when mixed with other chemicals. Other chemicals are flammable, combustible, or explosive. For example, solvents and compressed gases are often flammable. These materials can catch fire in their liquid or gaseous state. Some materials, such as blasting caps, are dangerous solids. These items should be kept away from open flames, sparks, intense heat, or other ignition sources to prevent fires or explosions. Know the physical hazards of the chemicals being used in order to prevent fires and chemical reactions.

Workers should always check the SDS for a chemical before working with or around it. That way, you will be prepared to take appropriate corrective action if you or another worker is exposed.

WARNING!

Chemical splashes can become medical emergencies. Before working with any chemicals, ensure that you know where the nearest shower and eyewash station are, and verify that they are functioning properly.

Remember to wear prescribed clothing and PPE on the job site at all times. This will help guard against your skin and eyes being splashed with chemicals. If you are not sure of the PPE needed on your job site or for a certain task, consult your supervisor before beginning work.

WARNING!

Never handle any chemicals without proper training and PPE. Review the SDS for each chemical, and strictly follow SDS first aid instructions if you suspect exposure.

6.1.4   Container Labeling

On a construction site, any material in a container must have a label. Labels describe what is in a container and warn of chemical hazards. The HAZCOM standard states that hazardous material containers must be labeled, tagged, or marked. The label must include the name of the material, the appropriate hazard warnings, and the name and address of the manufacturer. On December 1, 2015, OSHA required that all shipments be labeled as outlined by the Global Harmonization System (GHS), which had been adopted as a worldwide standard to enhance awareness and improve the safety and health of workers exposed to these substances.

In today's global marketplace, chemicals and similar hazardous materials are in constant movement between countries. With many countries having their own hazardous material labeling system, products must often be labeled multiple times to support the systems of both the shipping and receiving countries. The GHS will solve this problem by providing some standardized labeling for various materials. There are nine pictograms that workers must become familiar with. Those that are related to flammable, combustible, and explosive materials are shown in *Figure 64*, and those that represent other hazards are shown in *Figure 65*.

Pictograms have been in use for a number of years to depict a variety of hazards or to communicate information. The advantage of pictograms is that they cross any language barrier. Regardless of language, pictograms such as those used in the GHS tell the user what type of hazard to be concerned about. Although other systems of container labeling will be encountered for years to come, the pictograms all share common features and visually represent the hazard of concern.

If a material is transferred from a labeled container to a new container, the new container must be labeled with all of the information from the original label. Make sure that any materials being worked with are labeled. Be sure that you understand your company's labels.

Flame	**Flame Over Circle**	**Exploding Bomb**

- Flammables
- Pyrophorics
- Self-heating
- Emits flammable gas
- Self-reactives
- Organic peroxides

- Oxidizers

- Explosives
- Self-reactives
- Organic peroxides

Figure 64 GHS labels for flammable, combustible, and explosive materials.

Process Safety Management

One way to avoid the hazards associated with chemical spills is to evaluate those hazards and plan ahead for dealing with unwanted releases. Forward-thinking companies adopt a method known as process safety management for this purpose. Process safety management involves training employees and contractors; making information available; establishing formal procedures; analyzing potential hazards; and taking steps to mitigate the hazards.

Health Hazard

- Carcinogen
- Mutagenicity
- Reproductive toxicity
- Respiratory sensitizer
- Target organ toxicity
- Aspiration toxicity

Exclamation Mark

- Irritant (skin and eye)
- Skin sensitizer
- Acute toxicity (harmful)
- Narcotic effects
- Respiratory tract irritant
- Hazardous to ozone layer (non-mandatory)

Gas Cylinder

- Gases under pressure

Corrosion

- Skin corrosion/burns
- Eye damage
- Corrosive to metals

Environment (Non-Mandatory)

- Aquatic toxicity

Skull and Crossbones

- Acute toxicity (fatal or toxic)

Figure 65 GHS labels for various materials.

WARNING!

Never use chemicals from an unlabeled container.

6.1.5   Radiation Hazards

Radioactive materials are used in construction during radiographic testing of welds in piping, vessels, and pumps. These hazards are also found at nuclear power plants. Radioactive materials must be properly labeled and warning signs must be posted in the work area. Only trained workers are allowed in these areas. If you see signs containing symbols like those shown in *Figure 66*, stay away from that area unless properly trained.

Radioactive materials are a special type of physical hazard. Excessive radiation exposure can cause skin burns, nausea, vomiting, infertility, and cancer. Employees can minimize radiation exposure by limiting the amount of time they are exposed and/or increasing their distance from the source. Always use the proper shielding or PPE. Check with your supervisor to find out if there are any radioactive hazards and how to avoid them.

Figure 66 Radioactivity warning symbols.

6.1.6   Biological Hazards

Biological hazard signs are used to warn workers of the actual or potential presence of a biological hazard (*Figure 67*). Biological hazards, or biohazards, can be any infectious agents that create a real or potential health risk. Biological hazard signs are commonly used to identify contaminated equipment, containers, rooms, materials, and areas housing experimental animals.

Background colors on signs may vary, but must contrast enough for the symbol to be easily identified. Wording is used on the sign to indicate the nature of the hazard or to identify the specific hazard.

6.1.7   Evacuation

In many work environments, specific evacuation procedures are needed. These procedures go into effect when dangerous situations arise, such as fires, chemical spills, and gas leaks. In an emergency, you must know the evacuation procedures. You must also know the signal (usually a horn or siren) that tells workers to evacuate.

When the evacuation signal sounds, follow the evacuation procedures precisely. That usually means taking a certain route to a designated assembly area and telling the person in charge that you are there. If hazardous materials are released into the air, you may have to determine which way the wind is blowing. Some sites may have a **wind sock** that indicates the wind direction. Different evacuation routes are planned for different wind directions. Taking the right route will keep you from being exposed to the hazardous material.

Figure 67 Biological hazard symbol.

Wind sock: A cloth cone open at both ends mounted in a high place to show which direction the wind is blowing.

6.2.0   Environmental Extremes

Workers on construction and industrial sites are often required to work outdoors in extremes of heat and cold. It is important to know how to protect yourself in such environments and to recognize the symptoms of conditions that could be caused by environmental extremes.

6.2.1   Heat Stress

Heat stress occurs when abnormally hot air and/or high humidity, or extremely heavy exertion, prevents your body from cooling itself fast enough. When this happens, you may suffer heat cramps, heat exhaustion, or a heat stroke. To prevent heat stress, take the following precautions:

- Drink plenty of water.
- Avoid alcoholic or caffeinated drinks.
- When possible, perform the most strenuous work during cooler parts of the day.
- Wear lightweight, light-colored clothing.
- Wear loose-fitting cotton clothing if it does not create a hazard.
- Keep your head covered and your face shaded.
- Take frequent, short breaks.
- Rest in the shade whenever possible.

Workers are encouraged to drink 4 cups (≈1 L) of water per hour when the heat index is 103°F (39.5°C) or higher when working in hot conditions. Start by having one or two glasses of water before beginning work, and then drink 1 cup (237 ml) every 15 to 20 minutes when working in hot conditions. It is possible to drink too much water as well. Water intake should not exceed 6 cups (1.4 L) per hour or 3 gallons (11.4 L) per day. Drinking water should be cool but not cold.

6.2.2   Heat Cramps

Heat cramps are muscular pains and spasms caused by heavy exertion. Any muscles can be affected, but most often it is the muscles that have been used the most. Loss of water and electrolytes from heavy sweating causes these cramps. Symptoms of heat cramps include the following:

- Painful muscle spasms and cramping
- Pale, sweaty skin
- Normal body temperature
- Abdominal pain
- Nausea

As noted above, body temperature is not normally elevated during the onset of heat cramps. An increase in body temperature usually indicates that the problem is escalating to a more serious stage. If you experience heat cramps, take the following steps: move to a cool area, drink some water, and gently stretch and massage cramped muscles.

6.2.3   Heat Exhaustion

Heat exhaustion typically occurs when people exercise heavily or work in a warm, humid place where body fluids are lost through heavy sweating. When it is humid, sweat does not evaporate fast enough to cool the body properly.

Symptoms of heat exhaustion include the following:

- Cool, pale, and moist skin
- Heavy sweating
- Headache, nausea, vomiting
- Dilated pupils
- Dizziness
- Possible fainting
- Fast, weak pulse
- Slight elevation in body temperature

If someone is showing the symptoms of heat exhaustion, get the victim to a shaded area and have him or her lie down. Try to cool the victim by applying cold wet cloths, fanning, removing heavy clothing, and giving small amounts of cool water if the victim is conscious. If the victim's condition does not improve within a few minutes, call emergency medical services.

6.2.4   Heat Stroke

Heat stroke is life threatening. The body's temperature-control system, which produces sweat to cool the body, stops working. Body temperature can rise so high that brain damage and death may result if the body is not cooled quickly.

WARNING!

If you suspect someone has heat stroke, call emergency medical services immediately.

The symptoms of heat stroke include the following:

- Hot, dry, or spotted skin
- Extremely high body temperature
- Very small pupils
- Mental confusion
- Headache
- Vision impairment
- Convulsions
- Loss of consciousness

While waiting for help to arrive, try to cool the victim using the methods described for heat exhaustion. Ice packs placed in the armpits and the groin should be used, if possible.

6.2.5   Cold Stress

When your body temperature drops even a few degrees below normal, which is about 98.6°F (37°C), you can begin to shiver uncontrollably and become weak, drowsy, disoriented, unconscious, or even fatally ill. This loss of body heat is known as cold stress, or hypothermia.

People who work outdoors during the winter need to learn how to protect against loss of body heat. The following guidelines can help you keep your body warm and avoid the dangerous consequences of hypothermia, frostbite, and overexposure to the cold.

WARNING!

Always seek immediate medical attention if you suspect hypothermia or frostbite.

Outdoors, indoors, in mild weather, or in cold, it is a good idea to dress in layers. Layering your clothes allows you to adjust what you are wearing to suit changing temperature conditions. In cold weather, wear cotton, polypropylene, or lightweight wool next to your skin, and wool layers over your undergarments. For outdoor activities, choose clothing made of waterproof, wind-resistant fabrics such as nylon. Since a great deal of body heat is lost through the head, always wear a hat for added protection.

Water chills the body far more rapidly than air or wind. Always take along an extra set of clothing whenever working outdoors. If clothes become damp, change to dry clothes to prevent your body temperature from dropping in cold weather. Wear waterproof boots in damp or snowy weather, and always pack raingear even if the forecast calls for sunny skies.

6.2.6   Frostbite

Frostbite is a dangerous condition that can have lifelong effects on your body. It usually affects the hands, fingers, feet, toes, ears, and nose. Symptoms of frostbite include a pale, waxy-white skin color and hard, numb skin.

Remember the following when providing first aid for frostbite:

- Never rub the affected area. This can damage the skin and tissue.
- After the affected area has been warmed, it may become puffy and blister.
- The affected area(s) may have a burning feeling or numbness. When normal feeling, movement, and skin color have returned, the affected area(s) should be dried and wrapped to keep it warm.
- If there is a chance the affected area may get cold again, do not warm the skin. If the skin is warmed and then becomes cold again, it will cause severe tissue damage. Seek medical attention as soon as possible. First aid for frostbite includes moving the victim to a warm area, removing wet clothing, and applying warm water of about 105°F (41°C) to the affected area. Do not pour the warm water directly onto the area, however.

6.2.7   Hypothermia

Hypothermia is a serious, potentially fatal condition caused by loss of body heat. You do not have to be in below-freezing temperatures to be at risk for hypothermia. Hypothermia can happen on land or in water.

The effects of hypothermia can be gradual and often go unnoticed until it is too late. If you know you'll be working outdoors for an extended period of time, work with a buddy. At the very least, let someone know where you'll be and what time you expect to return. Ask your buddy to check you for overexposure

Around the World

Temperature Extremes

Some places in the world reach incredibly hot and cold temperatures. Although work may come to a halt in the worst conditions, the job goes on between the extremes. In Chita, Russia, for example, the average low temperature in January is roughly $-30°C$ ($\approx -22°F$). The average high during this period is around $-17°C$ ($\approx 1°F$). Since these are average temperatures, people in Chita must adapt and continue working. The admiration of the world's workers should go to those who must work in temperature extremes like this without hesitation.

to the cold, and do the same for your buddy. Check for shivering, slurred speech, mental confusion, drowsiness, and weakness. If anyone shows any of these signs, call for emergency medical assistance and see that the victim is moved indoors as soon as possible to warm up.

WARNING!

If a co-worker exhibits uncontrolled shivering, slurred speech, clumsy movements, fatigue, and confused behavior, immediately call for medical assistance.

Symptoms of hypothermia include the following:

- A drop in body temperature
- Fatigue or drowsiness
- Uncontrollable shivering
- Slurred speech
- Clumsy movements
- Irritable, irrational, confused behavior

If a person is showing symptoms of hypothermia from exposure to cold air, immediately call for medical assistance. The victim should be moved to a warm, dry area. Any wet clothing should be removed and replaced with dry clothing or blankets. If possible, give the victim warm, sweet drinks; avoid drinks with caffeine, as well as alcohol. Have the victim move his arms and legs to create muscle heat. If he is unable to do this, place warm bottles or hot packs around the armpits, groin, neck, and head. Do not rub the victim's body or place him in a warm bath. This could cause additional harm.

Body heat is lost up to 25 times faster in water than on land, and the methods used to help victims while waiting for medical help to arrive are different. If you are a victim, first of all, do not remove any clothing. Close up and tighten all clothing to create a layer of trapped water close to the body. This provides insulation that slows heat loss. Keep your head out of the water and put on a hat or hood if possible. Get out of the water as quickly as possible, or climb onto a floating object. Do not attempt to swim unless you can reach a floating object or another person. Swimming or other physical activity uses the body's heat and reduces survival time by about 50 percent.

If it is not possible to get out of the water, wait quietly, and conserve your body heat by folding your arms across your chest, keeping your thighs together, bending your knees and crossing your ankles. If another person is in the water, huddle together with your chests pressed tightly together.

6.3.0   Hot Work Hazards

Welding and torch cutting of metals is done using heat that is high enough to melt steel. In addition, most flame cutting and some welding processes are done with flammable gases. For those reasons, such processes are classified as hot work and are subject to special safety precautions. Grinding with a pneumatic or electric grinding machine (*Figure 68*) is also classified as hot work, as it gives off sparks that could cause a fire or explosion in an environment containing flammable gases or concentrated dust. Hot work always carries a risk of severe burns as well as fire hazards. Being aware of these risks is of great importance and will help keep you safe.

6.3.1 Arc Welding Hazards

Arc welding is a process in which metals are joined using a high-intensity electric arc (*Figure 69*) at temperatures in the range of 1,500 to 3,000°F (815 to 1,648°C). The arc melts and fuses the base metals. In most cases, a filler metal is melted along with the base metal to strengthen the joint. Welders are required to wear specialized PPE that protects their eyes from damage and protects their bodies from burns caused by sparks and molten metal. Such eye protection includes tinted goggles as well as a full face shield with a tinted viewing window. Anyone working in the vicinity of arc welding must also be protected, especially from damage to their eyes caused by looking at the welding arc. Never look at an arc welding operation without wearing the proper tinted eye protection, because the ultraviolet light from the arc will burn your eyes. Even a reflected arc can harm your eyes. It is extremely important to follow proper safety procedures at and around all welding operations. Serious eye injury or even blindness, as well as burns, can result from unsafe conditions.

When people are working near a welding operation, **welding curtains** (*Figure 70*) must be set up and everyone in the vicinity must wear tinted protective eyewear.

Even a brief exposure to the ultraviolet light from arc welding can cause a **flash burn** and damage your eyes badly. You may not notice the symptoms until sometime after the exposure. Here are some symptoms of flash burns to the eye:

- Headache
- Feeling of sand in your eyes
- Red or weeping eyes
- Trouble opening your eyes
- Impaired vision
- Swollen eyes

Welded material is dangerously hot. Mark it with a sign and stay clear for a while after the welding has been completed.

6.3.2 Oxyfuel Cutting, Welding, and Brazing

Oxygen and acetylene (oxyfuel) gases are combined to produce a flame with a temperature high enough to melt steel. So-called oxyfuel is commonly used to cut and weld steel and to join copper pipe by **brazing**.

Figure 71 shows a worker using an oxyfuel torch to cut steel. Persons working with or around welding, brazing, or metal cutting equipment can be severely

Figure 68 Grinding creates sparks.

Welding curtains: Protective screens set up around a welding operation designed to safeguard workers not directly involved in that operation.

Figure 69 Arc welding.
Source: The Lincoln Electric Company, Cleveland, OH, USA

Flash burn: The damage that can be done to eyes after even brief exposure to ultraviolet light from arc welding. A flash burn requires medical attention.

Brazing: A process using heat in excess of 800°F (427°C) to melt a filler metal that is drawn into a connection. Brazing is commonly used to join copper pipe.

Figure 70 Welding curtain.
Source: Sellstrom Manufacturing

Figure 71 Oxyfuel cutting.
Source: The Lincoln Electric Company, Cleveland, OH, USA

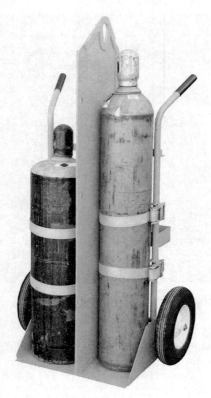

Figure 72 Compressed-gas cylinders used for oxyfuel cutting.
Source: Vestil Manufacturing

burned from the intense heat and flames generated in these processes. In addition, the gases used in these processes pose an additional danger because they are stored under pressure in metal cylinders (*Figure 72*).

Many hazards are involved in the handling, storage, and use of compressed gases. It takes energy to compress and confine the gas. That energy is stored until purposely released to perform useful work or until incidentally released by container failure or other causes.

Some gases, such as acetylene, are highly flammable. Oxygen is an explosion hazard if it comes into contact with grease or oil. Flammable compressed gases have additional stored energy besides simple compression-released energy. In case of a fire, for example, escaping oxygen will make the fire more intense. Other compressed gases, such as nitrogen, can cause asphyxiation simply because they displace oxygen.

The cylinders containing oxygen and acetylene must be transported, stored, and handled very carefully. Always follow these safety guidelines:

- Keep the work area clean and free from potentially hazardous items such as combustible materials and petroleum products.

WARNING!

Keep oxygen away from sources of flame and combustible materials, especially substances containing oil, grease, or other petroleum products. Compressed oxygen mixed with oil or grease will explode. Never use petroleum-based products around fittings that serve compressed oxygen lines.

- Use great caution when you handle compressed-gas cylinders.
- Store cylinders in an upright position where they will not be struck, where they will be away from corrosives, and where they cannot tip over or fall. Cylinders must be stored in the upright position in a secure container or secured to a wall with chains. When in storage, oxygen and fuel cylinders must be separated by 20 feet (6.1 m) or by a 5-foot (1.5 m) high ½-hour fire-rated barrier (*Figure 73*).

WARNING!

Do not remove the protective cap unless a cylinder is secured. If the cylinder falls over and the nozzle breaks off, the cylinder could shoot off like a rocket, injuring or killing anyone in its path.

6.3.3   Transporting and Securing Cylinders

Always handle cylinders with care. They are under high pressure and should never be dropped, knocked over, rolled, or exposed to heat in excess of 140°F (60°C). When moving cylinders, always be certain that the valve caps are in place. Cylinders must be transported to the workstation in the upright position on a hand truck or bottle cart such as the one shown in *Figure 72*. Oxygen and cutting gases should not be transported on the same cart unless the cart has a divider, as shown. Be aware that some job sites will not permit oxygen and cutting gases to be transported on the same cart under any circumstances.

Never attempt to lift a cylinder using the holes in a safety cap; use an approved lifting cage (*Figure 74*). Make sure that the cylinder is secured in the cage. Cages of various sizes are available for high-pressure cylinders and cylinders containing liquids.

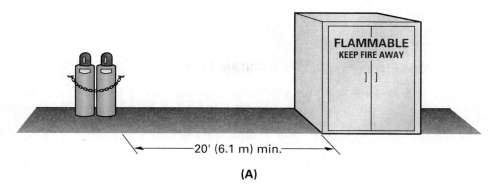

20' (6.1 m) min.

(A)

Minimum: 5' (1.5 m) high
½-hour fire rating

Fuel gas   Oxygen

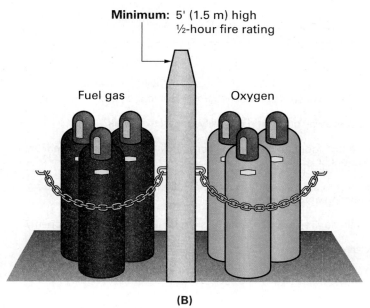

(B)

Figure 73 Cylinder storage.

Figure 74 Cylinder lifting cage.
Source: Courtesy of Saf-T-Cart

6.3.4 Hot Work Permits

A hot work permit (*Figure 75*) is an official authorization from the site manager to perform work that may pose a fire hazard. The permit includes information such as the time, location, and type of work being done. The hot work permit system promotes the development of standard fire safety guidelines. Permits also help managers keep records of who is working where and at what time. This information is essential in the event of an emergency at times when personnel need to be evacuated.

Most sites require the use of hot work permits and fire watches. When these requirements are violated, severe penalties may be imposed because of the risk it represents. Before a hot work permit is issued, a competent person must inspect the work area to ensure that no flammable or combustible materials are present. This inspection includes the areas above and below where the work will be performed, as well as the surrounding area.

A fire watch is posted when welding or cutting work is being done in many areas. One person other than the welding or cutting operator must constantly scan the work area for fires. Fire watch personnel should have ready access to fire extinguishers and alarms and know how to use them. Welding and cutting operations should never be performed without a fire watch. The area where welding is done must be monitored afterwards until there is no longer a risk of fire. The person on fire watch must be dedicated to that activity and may not perform other work during the time when hot work is being performed.

HOT WORK PERMIT

For Cutting, Welding, or Soldering with Portable Gas or ARC Equipment

(References: 2021 Uniform Fire Code Article 49 & National Fire Protection Association Standard NFPA 51B.)

Job Date_____ StartTime_____ Expiration_____ WO #_____

Name of Applicant_____ Company_____ Phone_____

Supervisor_____ Phone_____

Location / Description of work _____

_____

IS FIRE WATCH REQUIRED?

1. _____ (yes or no)   Are combustible materials in building construction closer than 35 feet to the point of operation?

2. _____ (yes or no)   Are combustibles more than 35 feet away but would be easily ignited by sparks?

3. _____ (yes or no)   Are wall or floor openings within a 35 foot radius exposing combustible material in adjacent areas, including concealed spaces in floors or walls?

4. _____ (yes or no)   Are combustible materials adjacent to the other side of metal partitions, walls, ceilings, or roofs, which could be ignited by conduction or radiation?

5. _____ (yes or no)   Does the work necessitate disabling a fire detection, suppression, or alarm system component?

YES to any of the above indicates that a qualified fire watch is required.

Fire Watcher Name(s) _____ Phone_____

NOTIFICATIONS

Notify the following groups at least 72 hours prior to work and 30 minutes after work is completed.
Write in names of persons contacted.

Notify in person OR by phone ONLY if question #5 above is answered "yes":

- Facilities Management Fire Alarm Supervisor

Notify by phone or in person: (If by phone, write down name of person and send them a completed copy of this permit.)

- Facilities Management Fire Protection Group
- Environmental Health & Safety Industrial Hygiene Group

SIGNATURES REQUIRED

University Project Manager_____ Date _____ Phone_____

I understand and will abide by the conditions described in this permit. I will implement the necessary precautions which are outlined on both sides of this permit form. Thirty minutes after each hot work session, I will reinspect work areas and adjacent areas to which spark and heat might have spread to verify that they are fire safe, and contact Facilities Management Alarm Technicians to have any disabled fire protection systems reactivated.

_____   _____ Date _____ Phone_____
  Permit Applicant   Company or Department

1/17/20

Figure 75 Hot work permit.

6.4.0   Fire Hazards and Fire Fighting

Fire is always a hazard on construction job sites. Many of the materials used in construction are flammable. In addition, welding, grinding, and many other construction activities create heat or sparks that can cause a fire. Fire safety involves two elements: fire prevention and fire fighting.

6.4.1   How Fires Start

For a fire to start, three things are needed in the same place at the same time: fuel, heat, and oxygen. If one of these three is missing, a fire will not start.

Fuel is anything that will combine with oxygen and heat to burn. Oxygen is always present in the air. When pure oxygen is present, such as near a leaking oxygen hose or fitting, material that would not normally be considered fuel (including some metals) will burn.

Heat refers to a source of ignition. It may be as simple as a single spark. Heat is anything that will raise a fuel's temperature to the **flash point**. The flash point is the temperature at which a fuel is encouraged to burn. The flash points of some fuels are quite low—room temperature or less. When the burning gases raise the temperature of a fuel to the point at which it ignites, the fuel itself will burn—and keep burning—even if the original source of heat is removed.

What is needed for a fire to start can be shown as a fire triangle (*Figure 76*). If one element of the triangle is missing, a fire cannot start. If a fire has started, removing any one element from the triangle will put it out.

6.4.2   Combustibles

Combustibles are categorized as liquid, gas, or ordinary combustibles. The term *ordinary combustibles* means paper, wood, cloth, and similar fuels. Liquids can be flammable or combustible. Flammable liquids have a flash point below 100°F (38°C). Combustible liquids have a flash point at or above 100°F (38°C). Flammable gases used on construction sites include acetylene, hydrogen, ethane, and propane. To save space, these gases are compressed so that a large amount is stored in a small cylinder or bottle. As long as the gas is kept in the cylinder, oxygen cannot get to it and start a fire. The cylinders should be stored away from sources of heat. If oxygen is allowed to escape and mix with a flammable gas, the resulting mixture will explode under certain conditions.

The easiest way to prevent fire in ordinary combustibles is to keep a neat, clean work area. If there are no scraps of paper, cloth, or wood lying around, there will be no fuel to start a fire. Establish and maintain good housekeeping habits. Use approved storage cabinets and containers for all waste and other ordinary combustibles.

6.4.3   Fire Prevention

The best way to ensure fire safety is to prevent a fire from starting. Fire can be prevented by the following actions:

- *Removing the fuel* — Liquid does not usually burn. What generally burns are the gases (vapors) given off as the liquid evaporates. Keeping liquids in an approved, sealed container prevents evaporation. If there is no evaporation, there is no fuel to burn.

- *Removing the heat* — If the liquid is stored or used away from a heat source, it will not be able to ignite.

- *Removing the oxygen* — The vapor from a liquid will not burn if oxygen is not present. Keeping safety containers tightly sealed prevents oxygen from coming into contact with the fuel.

The best way to prevent a fire is to make sure that fuel, oxygen, and heat are never present in the same place at the same time.

Flash point: The temperature at which fuel gives off enough gases (vapors) to burn.

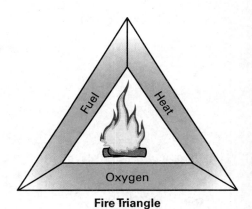

Fire Triangle

Figure 76 The fire triangle.

The following are some basic safety guidelines for fire prevention:

- Always work in a well-ventilated area, especially when using flammable materials such as shellac, lacquer, paint stripper, or construction adhesives.
- Never smoke or light matches when working with or near flammable materials.
- Keep oily rags in approved, self-closing metal containers.
- Store combustible materials only in approved containers.

6.4.4   Basic Fire Fighting

You are not expected to be an expert firefighter, but you may have to deal with a fire to protect your safety and the safety of others. You need to know the locations of fire fighting equipment on your job site as well as which equipment to use on different types of fires. However, only a **qualified person** is placed in a position where fire fighting skills are a required activity.

Most companies tell new employees where fire extinguishers are kept. If you have not been told, be sure to ask. Also, ask how to report fires. The telephone number of the nearest fire department should be clearly posted in your work area. If your company has a fire brigade, learn how to contact them. Learn your company's fire safety procedures. Know what type of extinguisher to use for different kinds of fires and how to use them. Make sure all extinguishers are fully charged. Never remove the tag from an extinguisher—it shows the date the extinguisher was last serviced and inspected.

A fire watch is required to have a fire extinguisher while on duty. A portable extinguisher such as the one shown in *Figure 77* is commonly provided for that purpose.

The function of a fire extinguisher is to remove one of the three elements (oxygen, heat, fuel) needed to sustain a fire. A fire extinguisher cannot remove fuel, so it is designed to remove either heat or oxygen. Heat is removed by using a coolant such as water; oxygen is removed by smothering the fire with dry chemicals or CO_2.

Fire extinguishers are rated for specific types of fires. Class A extinguishers, for example, are intended for fires involving ordinary combustibles such as paper, wood, and fabric. An extinguisher that is rated only for Class A fires might contain water, which could not be used on electrical or grease fires. Using water on a Class B fire (flammable liquids, grease, or gases) could spread the fire or splash burning fuel. Using water on a Class C (electrical) fire could cause you to be electrocuted because water conducts electricity. Class D fire extinguishers are required for fires involving reactive metals such as sodium, potassium, magnesium, titanium, zirconium, and the metal hydrides. A Class D fire extinguisher contains a powder designed to coat the metal to either smother the fire or keep oxygen from reaching the fire. If the powder coat fails to extinguish the fire, the best solution is to reapply the powder to keep the fire from spreading.

A pictogram label (*Figure 78*) on the extinguisher denotes the type(s) of fire for which the extinguisher is intended. The extinguisher shown in the figure is a dry-chemical extinguisher, which can be used on a Class A, B, or C fire. This type is the most common because it can be used to fight most fires. Some CO_2 extinguishers can also be used on Class A, B, or C fires, but others are designed for use on only Class B and C fires. This can be seen in *Figure 79*.

CO_2 is heavier than oxygen, so it smothers the fire by displacing the available oxygen. It also cools as it expands, so it can serve as a coolant. Some companies permit only CO_2 extinguishers to be used because the dry chemical in a dry-chemical extinguisher is a corrosive that can damage electrical systems. In addition, the fine powder emitted by the material in a dry-chemical extinguisher can cause eye and respiratory irritation when it becomes airborne. One advantage of the dry-chemical extinguisher is that its working range is 10 to 20 feet (3.1 to 6.1 m), while the range of the CO_2 extinguisher is 3 to 8 feet (0.9 to 2.4 m). A pressurized-water extinguisher, which is used only for Class A fires, has a range of about 50 feet (15 m).

Qualified person: A person who, by possession of a recognized degree, certificate, or professional standing, or by extensive knowledge, training, and experience, has demonstrated the ability to solve or prevent problems relating to a certain subject, work, or project.

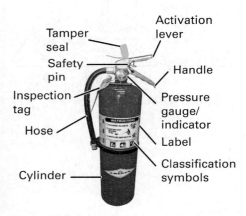

Figure 77 A portable fire extinguisher.
Source: Courtesy of Amerex Corporation.

Figure 78 Fire extinguisher pictograms.

EXTINGUISHER TYPE	TYPE OF FIRE

Ordinary Combustibles

Fires in paper, cloth, wood, rubber, and many plastics require a water-type extinguisher labeled A.

CO₂

OR

Dry Chemical

Flammable Liquids

Fires in oils, gasoline, some paints, lacquers, grease, solvents, and other flammable liquids require an extinguisher labeled B.

Electrical Equipment

Fires in liquids, fuse boxes, energized electrical equipment, computers, and other electrical sources require an extinguisher labeled C.

Ordinary Combustibles, Flammable Liquids, or Electrical Equipment

Multi-purpose dry chemical extinguishers are suitable for use on Class A, B, and C fires.

Metals

Combustible metals such as magnesium and sodium require special extinguishers labeled D.

Figure 79 Fire extinguisher applications.
Source: Courtesy of Amerex Corporation (Photos)

Prevention and Preparation Are the Keys to Fire Safety

Any fire in the workplace can cause serious injury or property damage. When chemicals are involved, the risks are even greater. Prevention is the key to eliminating the hazards of fire in the workplace. Preparation is the key to controlling any fires that do start. Take the following precautions to ensure safety from fire in the workplace:

- Keep work areas clean and clutter-free.
- Know how to handle and store chemicals.
- Know what you are expected to do in case of a fire emergency.
- Should a fire start, call for professional help immediately. Don't let a fire get out of control.
- Know what chemicals you work with. You might have to tell firefighters at a chemical fire what kinds of hazardous substances are involved.
- Make sure you are familiar with your company's emergency action plan for fires.
- Use caution when using power tools near flammable substances.

WARNING!

Do not use a CO_2 extinguisher on a Class D fire, as it can cause the fire to spread. Also note that CO_2 is heavier than air, so it will concentrate in low areas, displacing oxygen. For that reason, a CO_2 extinguisher must never be taken into, or used in, a confined space. Before it can be used in a confined space, all personnel in the space must first be evacuated. After the extinguisher has been used in the confined space, the space must be cleared by a competent person before it can be re-entered.

Finally, CO_2 is discharged from the extinguishers at a temperature of $-109°F$ $(-78°C)$, so it can cause frostbite if it contacts the skin. Never discharge it while it is pointed at someone and do not hold it by the activation lever unless the activation lever is insulated.

WARNING!

When using a CO_2 extinguisher do not hold the activation lever with your bare hand, as it will become extremely cold.

Fire extinguishers must be periodically inspected to make sure they are properly charged. The needle on the extinguisher's pressure gauge must register in the green area (*Figure 80*). The inspection must be documented on the tag attached to the extinguisher. Whenever you check out an extinguisher, make sure its inspection tag (*Figure 81*) is up to date and that the pressure gauge registers the correct charge. Not all extinguishers have gauges, however. Extinguishers without gauges must be weighed to determine if they are fully charged.

6.4.5 Using a Fire Extinguisher

Use the PASS method (*Figure 82*) when attacking a fire:

- **P**ull the pin from the handle, breaking the tamper seal in the process (*Figure 83*).
- **A**im the nozzle at the base of the fire while 8 to 10 feet away.
- **S**queeze the discharge handle.
- **S**weep the nozzle back and forth at the base of the fire.

Note that the method of attacking the fire can vary from one extinguisher to another. The instructions for a specific fire extinguisher are printed on the label,

Figure 80 Fire extinguisher pressure gauge.

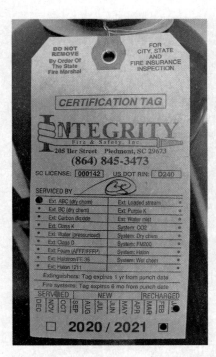

Figure 81 Fire extinguisher inspection tag.

Figure 82 PASS method.
Source: Courtesy of Amerex Corporation

Figure 83 Pin and tamper seal.

Figure 84 Example of fire extinguisher instructions.
Source: Badger Fire Protection Inc.

as shown in *Figure 84*. Be sure you are familiar with the type of extinguishers that are commonly provided on your current job site.

As the fire is extinguished, move in closer until it is completely out. Stop shooting every three or four sweeps to check progress. Try not to use all of the contents on the initial attack, in case the fire flares up. Once the extinguisher has been used, it must be re-charged.

6.5.0   Confined Spaces

Construction and maintenance work is not always done outdoors; a lot of it is done in confined spaces. A confined space is a space that is large enough to work in but that has limited means of entry or exit. A confined space is not designed for human occupancy and it has limited ventilation. Examples of confined spaces are tanks, vessels, silos, storage bins, hoppers, vaults, pits, and certain compartments on ships and barges (*Figure 85*).

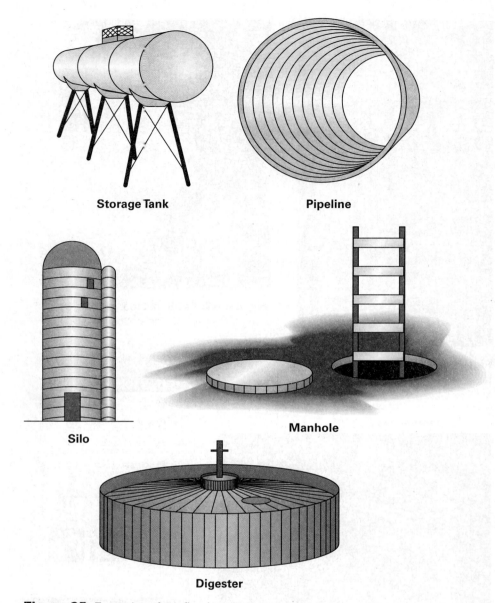

Storage Tank

Pipeline

Silo

Manhole

Digester

Figure 85 Examples of confined spaces.

OSHA defines a confined space as a space that has the following characteristics:

- Is large enough and so configured that employees can bodily enter and perform their assigned work.
- Has a limited or restricted means of entry or exit.
- Is not designed for continuous employee occupancy.

OSHA further defines a **permit-required confined space** as a space that has one or more of the following characteristics:

- Contains or has the potential to contain a hazardous atmosphere.
- Contains a material that has the potential for engulfing an entrant.
- Has an internal configuration such that an entrant could be trapped or asphyxiated by inwardly converging walls or by a floor that slopes downward and tapers to a smaller cross-section.
- Contains any other recognized serious safety or health hazard.

Permit-required confined space: A confined space that has been evaluated and found to have actual or potential hazards, such as a toxic atmosphere or other serious safety or health hazard. Workers need written authorization to enter a permit-required confined space. Also see *confined space*.

Atmospheric hazards are a concern in a confined space. The air in the space can contain flammable or explosive vapors or toxic gases. The space can also contain too much or too little oxygen. For safe working conditions, the oxygen level in a confined space atmosphere must range between 19.5 and 23.5 percent by volume as measured with an oxygen analyzer, with 21 percent being considered the normal level. Oxygen concentrations below 19.5 percent by volume are considered deficient; those above 23.5 percent by volume are considered enriched. Special meters are available to test atmospheric hazards. Forced ventilation can be used to overcome these hazards, but that does not eliminate the need to satisfy other confined space safety requirements, including inspection and permitting.

WARNING!

If too much oxygen is introduced into a confined space, it can be absorbed by a worker's clothing and ignite. If too little oxygen is present, it can lead to death in minutes. For this reason, the following precautions apply:

- Make sure confined spaces are ventilated properly for cutting or welding purposes.
- Never use oxygen in confined spaces for ventilation purposes.
- Always remain aware of the work going on around you, as conditions can change.

A permit-required confined space is a type of confined space that has been evaluated by a qualified person and found to have actual or potential hazards. Written authorization, commonly known as an entry permit, is required in order to enter a permit-required confined space. Once the entry permit is issued, it must be posted at the entry to the confined space.

When equipment is operating, confined spaces may contain hazardous gases or fluids, or may be oxygen-deficient. In addition, the work you are doing may introduce hazardous fumes into the space. Welding and metal cutting are examples of such work. For safety, you must take special precautions both before you enter and leave a confined space, and while you work there.

Until you have been trained to work in permit-required confined spaces and have taken the needed precautions, you must stay out of them. If you are not sure whether a confined space requires a permit, ask your supervisor. You must always follow your employer's procedures and your supervisor's instructions. Confined space procedures may include getting clearance from a safety representative before starting the work. You will be told what kinds of hazards are involved and what precautions you need to take. You will also be shown how to use the required PPE. Remember, it is better to be safe than sorry, so ask!

WARNING!

Without proper training, no employee is allowed to enter a permit-required or non-permit-required confined space. Employers are required to have programs to control entry to, and hazards in, both types of confined spaces.

Never work in a confined space without an attendant. An attendant must remain outside a permit-required confined space. The attendant monitors entry, work, and exit (*Figure 86*).

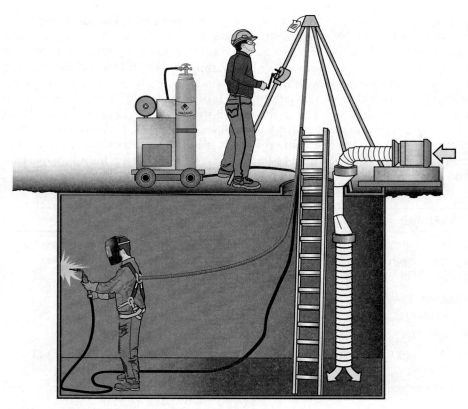

Figure 86 Permit-required confined space.

6.0.0 Section Review

1. The term PEL refers to a worker's _____.

   a. personal energy level
   b. maximum exposure limit
   c. amount of lead in the bloodstream
   d. product efficiency level

2. A condition in which body temperature drops a few degrees below normal is known as _____.

   a. heat stress
   b. cold stroke
   c. cold stress
   d. heat exhaustion

3. Swollen eyes and a sandy feeling in the eyes are symptoms of _____.

   a. a flash burn
   b. heat stroke
   c. asbestos exposure
   d. frostbite

4. The three elements needed for a fire to start are fuel, combustible liquid, and oxygen.

   a. True
   b. False

5. A confined space is an enclosed space that has _____.

   a. only one door
   b. limited access
   c. hazardous gases
   d. locked doors

1. The four leading causes of death in the construction industry include electrical hazards, struck-by hazards, caught-in or caught-between hazards, and _____.

   a. vehicular incidents
   b. falls
   c. radiation exposure
   d. chemical burns

2. A sign that has a white background with a green panel with white lettering is a _____.

   a. general information sign
   b. safety instruction sign
   c. caution sign
   d. danger sign

3. To properly dispose of oily rags, they must be _____.

   a. stored in a container designed for the purpose
   b. washed thoroughly and returned to use
   c. taken outdoors and thrown into a dumpster
   d. burned at the end of the shift

4. Keeping your work area clean and free of scraps or spills is referred to as _____.

   a. managing
   b. organizing
   c. housekeeping
   d. stacking and storing

5. HAZCOM classifies all paint, concrete, and wood dust as _____.

   a. hazardous materials
   b. common materials
   c. inexpensive materials
   d. nonhazardous materials

6. Under HAZCOM, if you spot a hazard on your job site, you must _____.

   a. report it to your supervisor
   b. leave immediately
   c. notify your co-workers
   d. correct the problem

7. Which of the following must be reported to your supervisor?

   a. Only major injuries
   b. Only incidents and major injuries
   c. All injuries, incidents, and accidents
   d. Only incidents in which a death occurred

8. Metal ladders should *NOT* be used near _____.

   a. stairways
   b. scaffolds
   c. electrical equipment
   d. windows

9. If you lean a straight ladder against the top of a 16-foot (4.9 m) wall, the base of the ladder should be _____.

    a. 3 feet (0.9 m) from the base of the wall
    b. 4 feet (1.2 m) from the base of the wall
    c. 5 feet (1.5 m) from the base of the wall
    d. 6 feet (1.8 m) from the base of the wall

10. The two basic types of scaffolds are _____.

    a. self-supporting and suspended scaffolds
    b. fixed and portable scaffolds
    c. metal and wooden scaffolds
    d. assembled and deliverable scaffolds

11. Interlocking stacked material is done by _____.

    a. applying a chain and padlock to it
    b. driving stakes around the stack
    c. securing the objects with rope
    d. placing the objects at right angles

12. To reduce the risk of workers being hurt or killed by falling materials, the maximum height-to-base ratio of a stack of materials should be _____.

    a. 2:1
    b. 4:1
    c. 8:1
    d. 10:1

13. The most common cause of death for equipment operators is _____.

    a. hit and run
    b. equipment rollover
    c. brake malfunction
    d. head-on collisions

14. Most cave-ins happen suddenly with little or no warning and occur in trenches 5 feet (1.5 m) to _____ .

    a. 10 feet (3.05 m) deep
    b. 15 feet (4.6 m) deep
    c. 20 feet (6.1 m) deep
    d. 25 feet (7.6 m) deep

15. In every trench over 4 feet (1.2 m) deep, there must be an exit every _____.

    a. 10 feet (3.05 m)
    b. 12 feet (3.7 m)
    c. 18 feet (5.5 m)
    d. 25 feet (7.6 m)

16. The minimum distance that a spoil pile must be located from the edge of an excavation is _____.

    a. 6 inches (15 cm)
    b. 2 feet (0.6 m)
    c. 5 feet (1.5 m)
    d. 100 yards (91.4 m)

17. The type of trench protection designed to prevent trench wall cave-ins is
_____.

    a. a trench shield
    b. a trench box
    c. a spoil pile
    d. shoring

18. Protective guards are provided on power tools and machines in order to
keep _____.

    a. dirt out of the tool or machine
    b. workers from being caught in rotating or moving parts
    c. the tool or machine from being struck by moving equipment
    d. unauthorized personnel from using the tool or machine

19. The minimum safe working distance from exposed electrical conductors
_____.

    a. depends on the voltage
    b. is 6 inches (15 cm)
    c. is 1 foot (30 cm)
    d. is unlimited

20. Work that is performed near a hazard but not in direct contact with it is
called _____.

    a. close-call work
    b. near-miss work
    c. proximity work
    d. barricade work

21. Circuit breakers and disconnect switches are examples of _____.

    a. energy-isolating devices
    b. energy-removal devices
    c. lockout/tagout devices
    d. multiple lockout devices

22. Which of the following provide the *best* eye protection from all
directions?

    a. Welding hoods
    b. Face shields
    c. Safety goggles
    d. Strap-on glasses

23. The type of respirator that has its own clean air supply is the _____.

    a. half mask
    b. mouthpiece with mechanical filter
    c. self-contained breathing apparatus
    d. full facepiece mask

24. A worker can be exposed to hazardous materials by inhalation, ingestion,
and _____.

    a. osmosis
    b. absorption
    c. radiation
    d. proximity

25. Hypothermia is a condition brought on by _____.
    a. excessive alcohol consumption
    b. prolonged exposure to cold
    c. excessive sweating
    d. prolonged exposure to heat

26. When in storage, oxygen and fuel cylinders used in oxyfuel cutting must be separated by _____.
    a. 20 feet (6.1 m)
    b. a metal barrier
    c. a 20-foot (6.1 m) wall
    d. 10 feet (3.05 m)

27. Grease and oil must be kept away from oxygen tanks because _____.
    a. grease or oil will contaminate the oxygen
    b. oxygen causes oil to freeze
    c. grease or oil can cause oxygen to explode
    d. oxygen will contaminate the oil or grease

28. Which of these must be present in the same place at the same time for a fire to occur?
    a. Oxygen, carbon dioxide, and heat
    b. Oxygen, heat, and fuel
    c. Hydrogen, oxygen, and wood
    d. Grease, liquid, and heat

29. A fire extinguisher labeled C would be used to fight a(n) _____.
    a. electrical fire
    b. magnesium fire
    c. paper fire
    d. gasoline fire

30. Which of the following characteristics is typical of a confined space?
    a. It has a limited amount of ventilation.
    b. There is no means of escape.
    c. It is too small to work in.
    d. It may be entered by untrained employees.

Answers to Section Review Questions and Module Review Questions are found in *Appendix A*.

Introduction to Construction Math

Source: by McLittle Stock/Shutterstock

Objectives

Successful completion of this module prepares you to do the following:

1. Identify whole numbers and solve basic arithmetic problems with them.
   a. List the key qualities of whole numbers and summarize their place values.
   b. Add and subtract whole numbers.
   c. Multiply and divide whole numbers.
2. Name fraction types and calculate with fractions.
   a. Define equivalent fractions and calculate their lowest common denominators.
   b. Define improper fractions and convert them into mixed numbers.
   c. Add and subtract fractions.
   d. Multiply and divide fractions.
3. Identify decimal numbers and calculate with them.
   a. List the key qualities of decimal numbers and summarize their place values.
   b. Add, subtract, multiply, and divide decimal numbers.
   c. Convert between decimals, fractions, and percentages.
4. Name the common length-measuring tools and use them to measure lengths accurately.
   a. Describe English and metric rulers, using them correctly to measure lengths.
   b. Describe English and metric measuring tapes, using them correctly to measure lengths.
5. Name common length, weight, volume, and temperature units in both the inch-pound and metric systems and convert them into other comparable units.
   a. List and convert between common inch-pound and metric length units.
   b. List and convert between common inch-pound and metric weight units.
   c. List and convert between common inch-pound and metric volume units.
   d. List and convert between common inch-pound and metric temperature units.
6. Classify angles and geometric shapes, as well as calculating their areas or volumes.
   a. List each angle type.
   b. Name common geometric shapes and summarize their qualities.
   c. Calculate the area of two-dimensional shapes.
   d. Calculate the volume of three-dimensional shapes.

Performance Tasks

Under supervision, you should be able to do the following:

1. Using a measuring tape, measure lumber pieces in both English and metric units.

2. Using a measuring tape, measure a room-sized space in both English and metric units.

3. Using a measuring tape, determine a short inside measurement in both English and metric units.

4. Add English measurements that include fractions.

Overview

Craft professionals rely on math to do their jobs accurately and efficiently. Plumbers calculate pipe lengths, plan drain slopes, and interpret dimensioned plans. Carpenters meet code requirements by using math to frame walls and ceilings properly. HVAC professionals develop ductwork and calculate airflow with practical geometry. Whichever craft lies in your future, math will play a role in it. This module reviews the math that you will need and sharpens the skills that you will be using in the exciting modules ahead.

NOTE

Unless the text indicates otherwise, attempt all math problems without a calculator. However, you may use a calculator to check your answers.

Industry Recognized Credentials

If you are training through an NCCER-accredited sponsor, you may be eligible for credentials from NCCER's Registry. The ID number for this module is 00102. Note that this module may have been used in other NCCER curricula and may apply to other level completions. Contact NCCER's Registry at 888.622.3720 or go to **www.nccer.org** for more information.

1.0.0 Whole Numbers

Performance Tasks	**Objective**
There are no performance tasks in this section.	Identify whole numbers and solve basic arithmetic problems with them. a. List the key qualities of whole numbers and summarize their place values. b. Add and subtract whole numbers. c. Multiply and divide whole numbers.

Whole numbers: Complete number units without fractions or decimals.

Fraction: A portion of a whole number represented by two numbers. The upper number of a fraction is called the numerator, while the bottom number is called the denominator.

Decimal: The part of a number represented by digits to the right of a point, called a decimal point. For example, in the number 1.25, .25 is the decimal portion of the number. In this case, it represents the fraction $^{25}/_{100}$.

No matter which construction craft you enter, you can be sure that you will be working with whole numbers. Accurately reading, writing, and communicating whole numbers to others is extremely important on any job site. Carpenters often use whole numbers during the planning phase to estimate square feet of drywall or linear feet of baseboard. Sheet metal workers use whole numbers for almost every process they do.

Never forget that mistakes with numbers waste time, effort, and materials. Mistakes hurt the bottom line. Even more importantly, however, your skill with numbers directly reflects on you. Your reputation as a craft professional heavily depends upon your developing a friendly relationship with numbers and math.

In this section, you will learn how to work with whole numbers. **Whole numbers** are complete number units without a **fraction** or **decimal**. This section examines whole numbers only. Later sections cover working with fractions and decimals.

1.1.0 Place Values of Whole Numbers

Each **digit** making up a whole number has a specific **place value**. A digit is any of the numerical symbols from 0 to 9. *Figure 1* shows an example of a whole number with seven digits. To read this seven-digit whole number out loud, say "five million, three hundred sixteen thousand, two hundred forty-seven."

Each of this whole number's seven digits represents a specific place value. The digit's value depends on its place, or location, within the whole number. For example, in this whole number, the place value of the 5 is five million, while the place value of the 2 is two hundred.

Keep in mind the following points about whole numbers:

- Numbers larger than zero are called **positive numbers** (such as 1, 2, 3 . . .). Except for zero, all numbers without a minus sign in front of them are positive.
- Numbers less than zero are called **negative numbers** (such as −1, −2, −3 . . .). Negative numbers have a negative (−) sign in front of them.
- Zero (0) is neither positive nor negative.
- Whole numbers can be positive, negative, or zero.

Some whole numbers may contain the digit zero. For example, the whole number 7,093 has a zero in the hundreds place. When reading that number out loud, say "seven thousand ninety-three."

1.1.1 Study Problems: Place Values of Whole Numbers

1. Look at the following description of a number. This number would be written as _____.

   Digit in the hundreds place: 9
   Digit in the ones place: 4
   Digit in the thousands place: 3
   Digit in the tens place: 6

   a. 3,964
   b. 4,693
   c. 30,964
   d. 39,064

2. In the number 25,718, the numeral 5 is in the _____.

   a. tens place
   b. thousands place
   c. ones place
   d. hundreds place

3. An estimate for a commercial flooring job requires one thousand, six hundred ninety-three square meters of carpet to complete the first floor. Written as a whole number, this value is _____.

   a. 163
   b. 1,693
   c. 10,693
   d. 16,093

4. A project supervisor estimates that a commercial building will require sixteen thousand, five hundred feet of copper piping to complete its restroom facilities. Written as a whole number, this value is _____.

   a. 1,650
   b. 16,500
   c. 160,500
   d. 16,000,500

Digit: Any of the numerical symbols 0 to 9.

Place value: The value that a digit represents in a whole number, determined by its place within the whole number or by its position relative to the decimal point. In the number 124, the number 2 represents 20, since it is in the tens position.

Positive numbers: Numbers greater than zero. For example, 1, 2, and 3 are positive numbers. Except for zero, any number without a negative (−) sign in front of it is treated as a positive number.

Negative numbers: Numbers less than zero. For example, −1, −2, and −3 are negative numbers.

Did You Know?

Whole Numbers

The following are whole numbers...

   1 5 67 335 2,654

... but the following are not whole numbers:

   ½ ¾ 7⅛ 0.45 4.25

Whole numbers never have a fractional or decimal part.

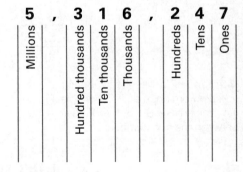

Figure 1 Place values.

5. An engineer estimates that the total cost to install a building's HVAC system will be three hundred twenty-two thousand nine hundred seven dollars. Written as a whole number, this value is _____.
   a. $3,297
   b. $300,297
   c. $322,907
   d. $322,000,907

1.2.0   Adding and Subtracting Whole Numbers

To add means to combine the values of two or more numbers together into one **sum** or total. To add whole numbers, perform the following steps:

Sum: The result of an addition problem. For example, in the problem $7 + 8 = 15$, 15 is the sum.

$$\begin{array}{r} 723 \\ + 84 \\ \hline \end{array}$$

Step 1 Line up the digits in the top and bottom numbers by place value columns.

$$\begin{array}{r} 723 \\ + 84 \\ \hline 7 \end{array}$$

Step 2 Beginning at the right side, add the numbers in the ones column (3 and 4) together first.

$$\begin{array}{r} {}^{1}723 \\ + 84 \\ \hline 07 \end{array}$$

Step 3 Continue adding the digits in each column, moving from right to left, one column at a time. In this example, adding the 2 and 8 in the tens column gives 10. This requires carrying the 1 from the tens column over to the next column on the left. To do so, place the 0 in the tens column and carry the 1 over to the top of the hundreds column as shown. Always add the carried-over number to the rest of the digits in that column.

$$\begin{array}{r} {}^{1}723 \\ + 84 \\ \hline 807 \end{array}$$

Step 4 Add the 7 already in the hundreds column to the 1 carried over. The resulting sum is 807.

$$\begin{array}{r} {}^{1\ 1\ 1}66,723 \\ + 5,784 \\ \hline 72,507 \end{array}$$

Some problems may require several carries. See example on the left.

To subtract means to take away the quantity of one number from the quantity of a second number. The result is the **difference**. To subtract whole numbers, perform the following steps:

Difference: The result of subtracting one number from another. For example, in the problem $8 - 3 = 5$, 5 is the difference between the two numbers.

$$\begin{array}{r} 12,766 \\ - 1,483 \\ \hline \end{array}$$

Step 1 Line up the digits in the top and bottom numbers by place value columns. Generally, position the larger number on the top. If not, the result will be a negative number. Negative numbers are uncommon in job site math.

$$\begin{array}{r} {}^{6\ 1}12,\cancel{7}66 \\ - 1,483 \\ \hline 83 \end{array}$$

Step 2 As with addition, start with the right column—the ones column. Subtract the 3 from the 6 to get 3 and record it under the ones column. Move to the tens column, noting that 8 cannot be subtracted from 6. Solve this problem by borrowing a 1 from the hundreds column. To borrow, cross out the 7 in the hundreds column, change it to a 6, and bring a 1 over to the 6 in the tens column. This action makes it 16 instead of 6. Subtract 8 from 16 to get 8 and record it in the tens column.

$$\begin{array}{r} {}^{6\ 1}12,\cancel{7}66 \\ - 1,483 \\ \hline 11,283 \end{array}$$

Step 3 Continue working towards the left, column by column, completing each subtraction for the remaining place values.

Some problems may require several borrowing operations. See example on the right.

$$
\begin{array}{r}
^{3\,12}\;^{10\,15}\\
9\cancel{4}\cancel{3},\cancel{1}\cancel{5}3\\
-436,372\\
\hline
506,781
\end{array}
$$

1.2.1 Study Problems: Adding and Subtracting Whole Numbers

Use addition and subtraction to solve the following problems. Read each question carefully to determine the appropriate procedure. Be sure to show all your work.

1. In calculating a bid for a roof restoration, a contractor estimates $850 for lumber, $460 for roofing shingles, and $170 for hardware. What is the total cost for the materials portion of the bid?

2. Brazil's currency is called the *real* and is identified with the R$ symbol. A plumbing contractor allots R$10,236 to complete three residential jobs. If Job 1 will cost R$2,477, Job 2 will cost R$2,263, and Job 3 will cost R$3,218, how much money remains to handle unexpected costs?

3. An HVAC contracting company sent out three work crews to complete three installations over the past week. If Crew One worked 10-, 9-, 11-, 12-, and 9-hour days, Crew Two worked 9-, 12-, 12-, 9-, and 9-hour days, and Crew Three worked 12-, 12-, 10-, 9-, and 11-hour days, how many total hours did the three crews work for the week?

4. A general contractor ordered three different sized windows to complete a job on a residential home. She ordered a bow window that cost $874; one 36"×36" double-hung window that cost $67; and one 36"×54" double-hung window that cost $93. If she budgeted $1,250 for windows, how much will she have left after buying them?

5. Russia's currency is called the *ruble* and is identified with the Rub symbol. An electrical contractor has Rub850,360 in the business's bank account. After depositing Rub119,980 in payments made by clients and paying out Rub79,205 in wages, how much remains in the account?

NOTE

The dimensional units *feet* and *inches* often appear as the symbols ' and " in construction drawings and calculations. This module explains dimensional units and symbols in a later subsection.

Did You Know?

Numeral Systems

People in ancient Egypt, Babylon, Greece, and Rome developed different ways of writing numbers. Some of these early systems were complex and difficult to use. A new system slowly emerged from India over several centuries. Arab traders discovered and then popularized it, leading to Europeans adopting it during the Middle Ages. The Hindu-Arabic system uses only ten symbols—0, 1, 2, 3, 4, 5, 6, 7, 8, 9— and continues as the standard today. Combining these ten symbols (also called numerals or digits) produces any desired number.

1.3.0 Multiplying and Dividing Whole Numbers

Multiplication quickly adds a number to itself numerous times. For example, assume six different people gave their crew leader $8 to pick up lunch. Adding $8 together six times gives the total amount. However, multiplying 6 times $8 to get a **product** of $48 is easier and faster. With larger numbers, repeatedly adding takes a long time. Multiply whole numbers with the following steps:

Product: The result of a multiplication problem. For example, the product of 6×6 is 36.

Step 1	Line up the digits of the top and bottom numbers by place value columns. Position the larger number on top. This arrangement makes the task easier but has no effect on accuracy.	$\begin{array}{r}374\\ \times\,26\end{array}$
Step 2	Start with the ones column (the right column) and multiply the 6 by the 4 ($6 \times 4 = 24$). Write the 4 in the ones column and carry the 2 up to the tens column.	$\begin{array}{r}^{2}\\ 374\\ \times\,26\\ \hline 4\end{array}$
Step 3	Multiply the 6 by the 7 next, and then add the carried-over 2 ($6 \times 7 = 42$, $42 + 2 = 44$). Write the 4 in the tens column and carry the 4 up to the hundreds column.	$\begin{array}{r}^{4\,2}\\ 374\\ \times\,26\\ \hline 44\end{array}$
Step 4	Next, multiply the 6 by the 3, and then add the carried-over 4 ($6 \times 3 = 18$, $18 + 4 = 22$). Write a 2 in the hundreds column and place the remaining 2 in the thousands column (since there are no places left in the upper number).	$\begin{array}{r}^{4\,2}\\ 374\\ \times\,26\\ \hline 2,244\end{array}$

$$374$$
$$\times\ 26$$
$$\overline{2{,}244}$$
$$80$$

Step 5 Now, multiply each digit in 374 by the 2 in 26. Multiply the 2 by the 4 in the ones column ($2 \times 4 = 8$). Write the 8 in the tens column, directly under the 2,244. Do not write the 8 in the ones column, because the 2 in the number 26 is in the tens column. To help keep things aligned, write a zero (0) in the ones column.

$$1$$
$$374$$
$$\times\ 26$$
$$\overline{2{,}244}$$
$$480$$

Step 6 Multiply the 2 by the 7 in the tens column ($2 \times 7 = 14$). Write down a 4 in the hundreds column and carry the 1.

$$1$$
$$374$$
$$\times\ 26$$
$$\overline{2{,}244}$$
$$+7{,}480$$

Step 7 Multiply the 2 by the 3, and then add the carried-over 1 ($2 \times 3 = 6$, $6 + 1 = 7$). Write a 7 in the thousands column.

$$1$$
$$374$$
$$\times\ 26$$
$$\overline{2{,}244}\ \text{product of } 6 \times 374$$
$$+7{,}480\ \text{product of } 20 \times 374$$
$$\overline{9{,}724}\ \text{final product}$$

Step 8 Complete the operation by adding the two products to get the final product of 9,724. There will be as many products to add together as there are digits in the bottom number. In this case, there are two digits in the bottom number (the 2 and the 6). So, there are two products to add.

Division is the opposite of multiplication. Multiplication adds a number to itself several times ($5 + 5 + 5 = 15$, or $5 \times 3 = 15$). Division subtracts one number from another several times. The number being divided is the **dividend**. The number being divided into the dividend is the **divisor**. The result is the **quotient**. For example, in $15 \div 3 = 5$, 15 is the dividend, 3 is the divisor, and 5 is the quotient.

Adding, subtracting, or multiplying whole numbers always gives a whole number result. With division, however, the outcome may not be so neat. If the divisor does not divide perfectly into the dividend (such as $10 \div 3$), the answer will be a whole number plus a leftover value. The leftover, called the **remainder**, completes the answer. For example, dividing 10 by 3 gives a quotient of 3 and a remainder of 1.

Dividend: In a division problem, the number being divided is the dividend. For example, in the problem $16 \div 8 = 2$, 16 is the dividend.

Divisor: In a division problem, the number that is divided into another number is called the divisor. For example, in the problem $16 \div 8 = 2$, 8 is the divisor.

Quotient: The result of a division problem. For example, when dividing 6 by 2, the quotient is 3.

Remainder: The amount left over in a division problem. For example, in the problem $34 \div 8$, 8 goes into 34 four times ($8 \times 4 = 32$) with 2 left over, so the remainder is 2.

Construction Estimates

Contractors consider creating accurate estimates an essential job skill. Many projects have a budget that the contractor must stay within. For example, granite cannot be specified for a countertop unless the budget contains enough money. Once the design is ready, contractors estimate the costs, including materials, labor, and overhead. Overhead refers to the costs of putting the labor on the job, covering office expenses, and paying company staff. Finally, the contractor adds the desired profit to the estimate. The lowest bid usually gets the job.

Long division requires a sequence of operations. Division, multiplication, and subtraction all contribute to the process.

Divide whole numbers with the following steps:

$$24\overline{)2{,}638}$$

Step 1 Set up the division problem by using the division bar as shown. The $\div$ symbol means the same thing, but the division bar keeps things organized on paper. Position the divisor on the left side of the division bar and place the dividend on the right side. In this example, 24 is the divisor and 2,638 is the dividend.

Step 2	Addition, subtraction, and multiplication operations begin in the ones column. By contrast, division starts on the dividend's left side. In this example, try to divide 24 into 2. This operation fails because 24 is larger than 2. Next, try using the two leftmost digits together (26). Divide 24 into the 26. This operation works because 24 divides into 26 once. Write a 1 above the 6 and write a 24 under the 26. The 24 equals 1×24. Subtract 24 from 26 to get 2. Using zeros (0) helps keep everything aligned.

$$\begin{array}{r} 01?? \\ 24\overline{)2{,}638} \\ -24 \\ \hline 02 \end{array}$$

Step 3	Bring down the next number in 2,638 (the 3) and place it next to the 2. Determine if 24 divides into 23. The answer is no, so place a 0 on the answer line above the 3. Write zeros in the appropriate place value columns since $24 \times 0 = 0$. Subtract the two numbers to get 23.

$$\begin{array}{r} 010? \\ 24\overline{)2{,}638} \\ -24\downarrow \\ \hline 023 \\ -000 \\ \hline 023 \end{array}$$

Step 4	Bring down the next number in 2,638 (the 8) and place it next to the 3. Determine if 24 can go into 238. Yes, it can, but how many times? Knowing the multiplication tables helps here. Think about numbers that are easy to work with. Multiplying 10 by 24 is 240. It is close to 238 but still too large. However, 24 does go into 238 nine times ($24 \times 9 = 216$). Write a 9 on the answer line next to the 0. Write 216 below the 238 and subtract to get a remainder of 22. The final answer indicates that 24 goes into 2,638 a total of 109 times with a remainder of 22. Write the complete answer as 109 r22.

$$\begin{array}{r} 010? \\ 24\overline{)2{,}638} \\ -24\downarrow| \\ \hline 023 \\ -000\downarrow \\ \hline 0238 \end{array} \qquad \begin{array}{r} 0109 \text{ r22} \\ 24\overline{)2{,}638} \\ -24\downarrow| \\ \hline 023 \\ -000\downarrow \\ \hline 0238 \\ -0216 \\ \hline 22 \end{array}$$

Refer to Step 4. Notice the discussion about how many times the number 24 divides into 238. Always remember that the answer to each individual division step should be a number from 0 to 9. If the answer is 10 or more, something went wrong. Either a multiplication error has happened, or the divisor really could be divided into the digits at least one full time.

1.3.1 The Order of Operations

Many math problems require several operations to solve them. Problems with multiple operations often appear on one or both sides of an **equation**. Although this module contains no complex equations, even simple equations require a specific operation order.

If the answer is to be accurate, operations must be done in the proper order. For simple equations, the correct order is multiplication, division, addition, and subtraction. Use a simple memory aid to remember this order: MDAS ("My Dear Aunt Sally").

The following equation is a good example:

$$6 + 3 \times 5 = A$$

Adding 6 to 3 and then multiplying the sum by 5 results in an answer of 45, which is wrong. However, multiplying 3 by 5 first and then adding 6 results in an answer of 21, which is correct. Following the proper order makes a big difference.

Use a second memory aid—PEMDAS—for more complicated equations. Remember the memory aid with the phrase, "Please Excuse My Dear Aunt Sally." The P represents parentheses, and the E represents exponents. If an equation has an operation in parentheses, such as (6×3), do that operation first.

Handle any exponents next. This module examines them only briefly, so do not get too worried if they are a new idea. An exponent is a superscript number next to a normal-sized number—for example, 6^2. The exponent, 2, specifies that two sixes get multiplied together ($6 \times 6 = 36$). If the exponent is 3 (6^3), three sixes get multiplied together ($6 \times 6 \times 6 = 216$).

Equation: A mathematical statement indicating that the value of two mathematical expressions, such as 2×2 and 1×4, are equal. An equation is written using the equal sign in this manner: $2 \times 2 = 1 \times 4$.

Numerators and Denominators

This module introduces working with fractions. Think of a fraction as another way to write a division problem. The number on top of a fraction is called the numerator. It is identical to the dividend in a long division problem. The lower number, called the denominator, is the same as the divisor. The biggest difference between fractions and long division is that most fractions encountered on the job site result in a quotient that is less than one. In the average long division problem, the quotient is usually greater than one. But, in the end, fractions simply represent another way to write a division problem.

Number of Courses

Masons often use the term *courses* when erecting block walls. This refers to the number of block rows stacked up to make the wall. A 15-course wall made from $8'' \times 8'' \times 16''$ concrete blocks is 10' high.

Block height = 8"
Number of courses = 15
12 inches = 1 foot
Therefore:
$8'' \times 15 = 120''$
$120'' \div 12'' = 10'$

1.3.2 Study Problems: Multiplying and Dividing Whole Numbers

Use multiplication and division to solve the following practical problems. Read each question carefully to determine the appropriate procedure. Addition or subtraction may also be required. Be sure to show all your work.

1. A supervisor sends a trainee to the truck for 180 special fasteners. The fasteners come in bags of 15. How many bags will the trainee bring back?

2. If the following amounts of lumber need to be delivered to each of 2 different staging areas at 4 different job sites, how many total boards of each size are needed?
    a. (65) 2 × 4s
    b. (45) 2 × 8s
    c. (25) 2 × 10s

3. One plumbing job requires 45 meters of PVC pipe, and a second job requires 30 meters. The pipe comes in 6-meter lengths only. How many sections should the plumber buy? How much pipe will be left over, assuming that there are no errors?

4. If a crane rental company charges $800 per day, $3,400 per week (5 days), and $10,500 per month,
    a. How much would it cost to rent the crane for 3 days?
    b. How much would it cost to rent the crane for 12 days?
    c. How much would it cost for one month and 11 days?

1.0.0 Section Review

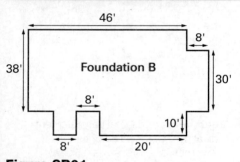

Figure SR01

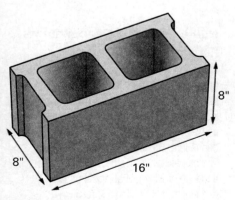

Figure SR02

1. Given the number 92,475, which digit is in the ten-thousands place value column?
    a. 2
    b. 4
    c. 7
    d. 9

2. What is the distance around the foundation shown in *Figure SR01*?
    a. 206 feet
    b. 214 feet
    c. 224 feet
    d. 234 feet

3. How many concrete blocks like the one shown in *Figure SR02* are required to erect a 2' high wall for the foundation shown in *Figure SR01*? Assume there is no waste. Round the answer up to the nearest whole number. (Hint: Multiply the distance around the foundation by 12 to change it from feet to inches first.)
    a. 336 blocks
    b. 504 blocks
    c. 514 blocks
    d. 528 blocks

2.0.0 Fractions

Objective

Name fraction types and calculate with fractions.

   a. Define equivalent fractions and calculate their lowest common denominators.

   b. Define improper fractions and convert them into mixed numbers.

   c. Add and subtract fractions.

   d. Multiply and divide fractions.

Performance Tasks

There are no performance tasks in this section.

A fraction divides whole numbers into parts. Fractions often appear as two numbers separated by a slash or a horizontal line, such as ½.

The slash or horizontal line means the same thing as the ÷ sign. Think of a fraction as another way to write a division problem. The fraction ½ means 1 divided by 2, or one divided into two equal parts. This fraction is spoken as "one-half."

The lower number of the fraction, the **denominator**, indicates the number of parts into which the upper number is being divided. In a typical division problem, it is the divisor. The upper number, the **numerator**, is the whole number being divided into parts. It is the dividend of the division problem. In the fraction ½, the 1 is the upper number, or numerator, and the 2 is the lower number, or denominator. A fraction in which the numerator and denominator are identical (such as ²/₂, ⁸/₈, ¹⁶/₁₆) always equals 1.

Denominator: The part of a fraction below the dividing line. For example, the 2 in ½ is the denominator. It is equivalent to the divisor in a long division problem.

Numerator: The part of a fraction above the dividing line. For example, the 1 in ½ is the numerator. It is equivalent to the dividend in a long division problem.

2.1.0 Equivalent Fractions and Lowest Common Denominators

Equivalent fractions, such as ½, ²/₄, and ⁴/₈, have the same value. *Figure 2* shows the individual units making up ½, ¼, and ⅛. It is often easier and more practical to work with fractions that have been reduced to the lowest common denominator. The lowest common denominator is the smallest denominator that all the fractions can share. For the three equivalent fractions ½, ²/₄, and ⁴/₈, the fraction ½ is the version with the lowest common denominator. This section introduces equivalent fractions and the way to find the lowest common denominator.

Equivalent fractions: Fractions having different numerators and denominators but still having equal values, such as the fractions ½ and ²/₄.

2.1.1 Finding Equivalent Fractions

To someone used to working with fractions, ½, ²/₄, and ⁴/₈ are obviously equivalent fractions. However, there is a mathematical way to confirm this. Determine each fraction's value by dividing the denominator by the numerator. So, 2 divided by 1 equals 2; 4 divided by 2 equals 2; and 8 divided by 4 also equals 2. Since the quotients (2) match, the fractions are equivalent. A physical example proves equivalence too. Cut pieces of wood ½" long, ²/₄" long, and ⁴/₈" long. All three pieces will match.

When measuring objects, try to record all measurements using the same fractions—in sixteenths of an inch, for example. In other words, use the same denominator for each measurement. Doing this makes comparing, adding, and subtracting fractional measurements easier.

Sometimes it is necessary to convert a fraction to an equivalent one. For example, to find out how many sixteenths of an inch are equal to ½ inch, multiply both the numerator and the denominator by the same number. Multiplying the numerator and denominator by the same number creates an equivalent fraction.

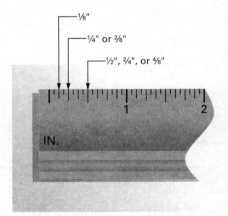

Figure 2 Equivalent fractions.

To change ½ into sixteenths, work out what number times 2 gives 16. The answer is 8, so multiply both the numerator and the denominator by 8:

$$\frac{1 \times 8}{2 \times 8} = \frac{8}{16}$$

This calculation shows that ⁸⁄₁₆" equals ½".

2.1.2 Reducing Fractions to Their Lowest Terms

If something's measurement is ⁴⁄₁₆", it may be easier to work with it by reducing the number to its lowest terms. Calling out a measurement of ⁴⁄₁₆" on a construction job site will probably cause a few puzzled looks! To find the lowest terms of ⁴⁄₁₆", use division as follows:

Step 1 Determine the largest number that divides perfectly into both the numerator and the denominator. With a little practice, this is an easy mental task. If there is no number (other than 1) that divides perfectly into both numbers, the fraction is already in its lowest term form. For this example, however, the largest number is 4.

$$\frac{4 \div 4}{16 \div 4} = \frac{1}{4}$$

Step 2 Divide both the numerator and the denominator by the number determined in Step 1. For this example, divide both the numerator and the denominator by 4.

This operation shows that the lowest term of ⁴⁄₁₆" is ¼. Always remember to divide the numerator and the denominator by the same number. Be sure that both numbers divide perfectly, with a whole number answer as the result.

Around the World
Fractions on the Job

In the real world, most measurements are not whole numbers. Typically, craftworkers measure and cut pipe, lumber, and other materials to fractional lengths, such as ⅜, ⁵⁄₁₆, or ¾ of an inch or a foot. All craftworkers must be comfortable with fractions, especially those who use the inch-pound measuring system. The metric system almost eliminates the need for fractions in most day-to-day measurements.

2.1.3 Comparing Fractions and Finding Lowest Common Denominators

Which measurement is larger: ¾" or ⅝"?

Tackle this question by asking a different question: Which is longer, three sections of a board cut into four equal pieces or five sections of a board cut into eight equal pieces (*Figure 3*)?

Comparing fractions with different denominators is challenging. While the visual example makes it obvious that ¾ is more than ⅝, this method is inappropriate for the job site. A mathematical approach makes the task easier. If the fractions have the same denominator, just compare the numerators. Fractions with different denominators require another method:

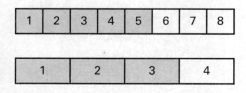

Figure 3 Lumber divided into equal sections.

$$\frac{3}{4} \text{ or } \frac{5}{8}$$

To compare them, find a common denominator. The common denominator is a number that both denominators can change into cleanly. The following method calculates a common denominator:

Step 1 Multiply the two denominators together ($4 \times 8 = 32$). The result (32) is a common denominator, because each denominator divides into it perfectly, giving a whole number result.

Step 2 Convert the two fractions so both have the same denominator (32):

$$\frac{3 \times 8}{4 \times 8} = \frac{24}{32}$$

$$\frac{5 \times 4}{8 \times 4} = \frac{20}{32}$$

Now, comparing the two fractions is trivial: ¾ is longer than ⅝.

Working with fractions like ²⁴⁄₃₂ or ²⁰⁄₃₂ can be challenging. To make the task easier, find the lowest common denominator possible between them, which means reducing the fractions to their lowest terms. Although 32 is a common denominator for the two fractions, it is not the lowest common denominator.

To find the lowest common denominator, follow these steps:

Step 1 Reduce each fraction to its lowest terms.

Step 2 Find the lowest common multiple of the denominators. Sometimes one denominator is already a multiple of the other. If this is the case, find the equivalent fraction for the term with the smaller denominator.

Step 3 If no common multiples exist, multiply the denominators together to get a common denominator.

Consider the lumber example again. The fractions ¾ and ⅝ are already in their lowest terms. Comparing the denominators shows that 8 is a multiple of 4. Calculate the equivalent fraction with a denominator of 8 for ¾:

$$\frac{3 \times 2}{4 \times 2} = \frac{6}{8}$$

Comparing ⁶⁄₈ to ⅝ reveals that ⁶⁄₈ is the larger fraction. This method gives the same result as comparing ²⁴⁄₃₂ to ²⁰⁄₃₂.

The job determines the method that craft professionals use. For example, carpenters often work in sixteenths, so they favor this fraction in calculations.

2.1.4 Study Problems: Finding Equivalent Fractions

Find the equivalent for the following fractions:

1. ¼ equals how many sixteenths?

   a. ²⁄₁₆
   b. ⁴⁄₁₆
   c. ⁶⁄₁₆
   d. ⁸⁄₁₆

2. ²⁄₁₆ equals how many thirty-seconds?

   a. ¹⁄₃₂
   b. ²⁄₃₂
   c. ⁴⁄₃₂
   d. ⁸⁄₃₂

3. ¾ equals how many eighths?

   a. ²⁄₈
   b. ⁴⁄₈
   c. ⁵⁄₈
   d. ⁶⁄₈

4. ¾ equals how many sixty-fourths?

   a. ⁴⁸⁄₆₄
   b. ⁵⁰⁄₆₄
   c. ⁵²⁄₆₄
   d. ⁵⁴⁄₆₄

Did You Know?

Nominal Measurements

A so-called 8" × 8" × 16" concrete block actually measures 7⅝" × 7⅝" × 15⅝". The masonry industry refers to it by its nominal dimensions of 8" × 8" × 16" to allow for a ⅜" mortar joint between blocks. Adding ⅜" to the block's actual dimensions also makes the block's measurements come out in whole number inches, rather than fractions. Other block types and sizes use this same practice.

5. $\frac{3}{16}$ equals how many thirty-seconds?

    a. $\frac{2}{32}$

    b. $\frac{4}{32}$

    c. $\frac{6}{32}$

    d. $\frac{8}{32}$

Without using a calculator, find the lowest term for each of the following fractions:

6. $\frac{2}{16} =$ _____

7. $\frac{2}{8} =$ _____

8. $\frac{12}{32} =$ _____

9. $\frac{4}{8} =$ _____

10. $\frac{4}{64} =$ _____

Find the lowest common denominator for the following pairs of fractions:

11. $\frac{2}{6}$ and $\frac{3}{4}$.

    a. 6

    b. 8

    c. 12

    d. 16

12. $\frac{1}{4}$ and $\frac{3}{8}$.

    a. 4

    b. 8

    c. 12

    d. 18

13. $\frac{1}{8}$ and $\frac{1}{2}$.

    a. 3

    b. 5

    c. 7

    d. 8

14. $\frac{1}{4}$ and $\frac{3}{16}$.

    a. 8

    b. 16

    c. 18

    d. 20

15. $\frac{4}{32}$ and $\frac{5}{8}$.

    a. 64

    b. 32

    c. 16

    d. 8

2.2.0   Improper Fractions and Mixed Numbers

Improper fraction: A fraction whose numerator is larger than its denominator. For example, $\frac{8}{4}$ and $\frac{6}{3}$ are improper fractions.

Mixed number: A combination of a whole number with a fraction or decimal. Examples of mixed numbers are $3\frac{7}{16}$, 5.75, and $1\frac{1}{4}$.

An **improper fraction** has a larger numerator than its denominator. The fractions $\frac{43}{6}$, $\frac{19}{16}$, and $\frac{55}{8}$ are all improper fractions.

Converting an improper fraction to a **mixed number** makes calculations easier. A mixed number consists of a whole number combined with a fraction. The mixed numbers $7\frac{1}{6}$, $1\frac{3}{16}$, and $6\frac{7}{8}$ are equivalent to the improper fractions $\frac{43}{6}$, $\frac{19}{16}$, and $\frac{55}{8}$. Some improper fractions convert to a whole number without a fraction. For example, the improper fraction $\frac{42}{6}$ converts to 7.

Remember, fractions are just another way of writing a division problem. To convert an improper fraction to a mixed number, perform the division.

The following steps demonstrate converting the improper fraction $^{67}/_{32}$ to a mixed number:

Step 1	Divide the numerator (dividend) by the denominator (divisor). The denominator must be able to be divided into the numerator at least once, or the fraction is not an improper fraction. Set up the problem on paper if necessary.	$32\overline{)67}$
Step 2	Perform the long division.	$\begin{array}{r} 02\text{ r}3 \\ 32\overline{)67} \\ -64 \\ \hline 3 \end{array}$
Step 3	The quotient of the problem (2) is the whole number part of the mixed number. Any remainder (3 in this example) becomes the fractional part's numerator. Retain the original denominator.	$2\dfrac{3}{32}$

2.2.1 Study Problems: Changing Improper Fractions to Mixed Numbers

Change these improper fractions to mixed numbers:

1. $^{35}/_{8} = $ _____.

2. $^{97}/_{16} = $ _____.

3. $^{13}/_{4} = $ _____.

4. $^{14}/_{5} = $ _____.

5. $^{48}/_{16} = $ _____.

2.3.0   Adding and Subtracting Fractions

To add fractions, first find a common denominator and convert them to equivalent fractions. Once the denominators are the same, add the numerators. The following steps show the process for adding $^{3}/_{4}$ and $^{5}/_{8}$:

Step 1	Find the fractions' common denominator. One common denominator is 32. The lowest common denominator is 8. Both choices work fine and give equally valid results.	
Step 2	Convert the fractions to equivalent fractions with the same denominator.	$\dfrac{3 \times 8}{4 \times 8} = \dfrac{24}{32}$ $\dfrac{5 \times 4}{8 \times 4} = \dfrac{20}{32}$
Step 3	Add the numerators. Place this sum over the common denominator (32).	$\dfrac{24}{32} + \dfrac{20}{32} = \dfrac{44}{32}$
Step 4	Reduce the fraction to its lowest terms. For this example, both the numerator and denominator divide by 4. Doing this reduces the fraction to $^{11}/_{8}$, an improper fraction. Finally, change the improper fraction, $^{11}/_{8}$, to the mixed number $1^{3}/_{8}$.	

Any common denominator will work, but the lowest common denominator gives a reduced result. Of course, the result may still be an improper fraction that must be changed to a mixed number.

Subtracting fractions is like adding them. Begin by finding a common denominator. For example, a painter has $^{7}/_{8}$ of a liter of paint. Another worker asks for $^{1}/_{4}$ of a liter. How much will be left?

$\frac{7}{8}$ $\frac{1}{4}$	**Step 1**	Find a common denominator. In this case, 8 is a common denominator and is also the lowest common denominator.
$\frac{1 \times 2}{4 \times 2} = \frac{2}{8}$	**Step 2**	Since one fraction is already in eighths, only convert ¼ to eighths. Multiply the numerator and denominator by 2 to get a fraction with a denominator of 8.
$\frac{7}{8} - \frac{2}{8} = \frac{5}{8}$	**Step 3**	Subtract the numerators. Place the difference over the common denominator (8). The result is ⅝.

Sometimes a project requires subtracting a fraction from a whole number. To do this, change the whole number into a fraction. Then, find a common denominator and subtract the numerators.

Changing a whole number into a fraction is easy. Make the whole number the numerator and use 1 for the denominator. For example, the whole number 5 becomes the fraction $5/1$. To subtract $2/7$, from 5, find the lowest common denominator (7). Converting $5/1$ into sevenths gives the fraction $35/7$. Subtract $2/7$ from $35/7$, giving a difference of $33/7$. Change the improper fraction to the mixed number $4\,5/7$.

Around the World

The Roots of Fractions

The word *fraction* comes from a Latin word that means "to break." Other English words, such as *fracture*, also come from the same source. Ancient cultures, including the Egyptians, the Babylonians, and the Indians devised their own ways of writing and using fractions. The Arabs added the line between the numerator and denominator.

2.3.1 Study Problems: Adding and Subtracting Fractions

Add the following fractions. Reduce the results to the lowest terms and change any improper fractions to mixed numbers.

1. $1/8 + 4/16 =$ ________
2. $4/8 + 6/16 =$ ________
3. $2/4 + 3/4 =$ ________
4. $3/4 + 2/8 =$ ________
5. $14/16 + 3/8 =$ ________

Subtract the following fractions. Reduce the results to the lowest terms.

6. $3/8 - 5/16 =$ ________
7. $11/16 - 5/8 =$ ________
8. $3/4 - 2/6 =$ ________
9. $11/12 - 4/8 =$ ________
10. $11/16 - 1/2 =$ ________

Subtract the following fractions and whole numbers. Reduce the results to the lowest terms and change any improper fractions to mixed numbers.

11. $8 - 3/4 =$ ________
12. $12 - 5/8 =$ ________
13. Two punches are made from steel bar stock $9\,7/16$" long. If one punch is $4\,1/64$" long and the other is $4\,3/32$" long, how many inches of stock are wasted?
    a. $11/16$"
    b. $15/16$"
    c. $1\,15/64$"
    d. $1\,21/64$"

14. After sawing 12$\frac{1}{16}$" off a board that is 20$\frac{3}{4}$" long, the length of the remaining board will be _____.

   a. 8$\frac{1}{4}$"

   b. 8$\frac{11}{16}$"

   c. 11$\frac{1}{4}$"

   d. 17$\frac{3}{8}$"

15. A rough opening for a window measures 36$\frac{3}{8}$". The window itself measures 35$\frac{15}{16}$". The total clearance that will exist between the window and the rough opening is _____.

   a. $\frac{7}{16}$"

   b. 1"

   c. 1$\frac{7}{16}$"

   d. 1$\frac{3}{4}$"

2.4.0 Multiplying and Dividing Fractions

Multiplying and dividing fractions differs from adding and subtracting fractions. A common denominator is not required. Follow these steps to multiply $\frac{4}{8}$ and $\frac{5}{6}$:

Step 1	Multiply the numerators together to get a new numerator. Multiply the denominators together to get a new denominator.	$\dfrac{4 \times 5 = 20}{8 \times 6 = 48}$
Step 2	If possible, reduce the result to its lowest terms. In this example, $\frac{20}{48}$ reduces to $\frac{5}{12}$, since both the numerator and denominator are divisible by 4.	

Fractions do not need to be reduced before multiplying. But reducing them first makes the multiplication easier, since the numbers are smaller. Reducing the product to the lowest terms becomes easier too. What may seem like an extra step can save time in the long run. Regardless of the method, however, the final answer will be the same.

Dividing fractions is like multiplying fractions but with one difference. Before multiplying, **invert**, or flip, the divisor fraction. Follow these steps to divide $\frac{1}{2}$ and $\frac{3}{4}$:

Invert: To reverse the order or position of numbers. In fractions, inverting means reversing the positions of the numerator and denominator, so $\frac{3}{4}$ becomes $\frac{4}{3}$. When dividing by fractions, one fraction is inverted.

Step 1	Invert the divisor fraction ($\frac{3}{4}$).	$\dfrac{3}{4}$ becomes $\dfrac{4}{3}$
Step 2	Change the division sign ($\div$) to a multiplication sign ($\times$).	$\dfrac{1}{2} \times \dfrac{4}{3}$
Step 3	Multiply the numerators and denominators in the normal way.	$\dfrac{1 \times 4 = 4}{2 \times 3 = 6}$
Step 4	Reduce to the lowest terms when possible.	$\dfrac{4}{6}$ reduces to $\dfrac{2}{3}$

Check a division result for correctness by using multiplication. Multiplying $\frac{3}{4}$ times $\frac{2}{3}$ gives a product of $\frac{1}{2}$—the problem's original dividend.

To divide a fraction by a whole number, convert the whole number to a fraction by putting it over 1. Remember that $\frac{4}{1}$ is the same as 4. Perform the following steps to divide $\frac{1}{2}$ by 4:

Step 1	Change the whole number (4) into a fraction.	$4 = \dfrac{4}{1}$

$$\frac{4}{1} \text{ becomes } \frac{1}{4}$$

Step 2 Invert the fraction.

$$\frac{1}{2} \times \frac{1}{4} = \frac{1}{8}$$

Step 3 Multiply the fractions.

Convert mixed numbers, such as $2\frac{1}{3}$, into an improper fraction before inverting it. Do this by multiplying the fraction's denominator (3) by the whole number (2). Then, add the fraction's numerator (1):

$$(3 \times 2) + 1 = 7$$

Place the result (7) over the fraction's denominator (3). The following shows the complete process to convert a mixed number into an improper fraction:

$$2\frac{1}{3} = \frac{(3 \times 2) + 1}{3} = \frac{7}{3}$$

2.4.1 Study Problems: Multiplying and Dividing Fractions

Without using a calculator, find the answers to the following multiplication problems. Reduce results to their lowest terms and change any improper fractions to mixed numbers.

1. $\frac{4}{16} \times \frac{5}{8} = $ _____

2. $\frac{3}{4} \times \frac{7}{8} = $ _____

3. $\frac{2}{8} \times 15 = $ _____

4. $\frac{3}{7} \times 49 = $ _____

5. $\frac{8}{16} \times \frac{32}{64} = $ _____

Without using a calculator, find the answers to the following division problems. Reduce results to their lowest terms and change improper fractions to mixed numbers.

6. $\frac{3}{8} \div 3 = $ _____

7. $\frac{5}{8} \div \frac{1}{2} = $ _____

8. $\frac{3}{4} \div \frac{3}{8} = $ _____

9. On construction drawings, smaller dimensions often represent larger ones. This technique allows large objects or structures to fit on the paper. If $\frac{1}{4}$" represents 1', then a line on the drawing measuring $8\frac{1}{2}$" represents _____.

   a. 34'
   b. 36'
   c. 38'
   d. 40'

10. How many $\frac{7}{8}$" long strips can be cut from a single 7" long strip of material?

    a. 5
    b. 6
    c. 7
    d. 8

2.0.0 Section Review

1. Which of the following fraction pairs are equivalent fractions?

   a. $\frac{1}{2}$ and $\frac{17}{32}$
   b. $\frac{1}{2}$ and $\frac{32}{64}$
   c. $\frac{1}{2}$ and $\frac{3}{4}$
   d. $\frac{1}{2}$ and $\frac{63}{128}$

2. The improper fraction $^{76}\!/_{64}$ converts to the mixed number $1^{9}\!/_{32}$.
   a. True
   b. False

3. $^{11}\!/_{16} + \frac{3}{8} =$ _____.
   a. $^{9}\!/_{16}$
   b. $\frac{7}{8}$
   c. $^{11}\!/_{16}$
   d. $1^{1}\!/_{16}$

4. $1\frac{1}{4} - ^{7}\!/_{16} =$ _____.
   a. $^{5}\!/_{16}$
   b. $\frac{5}{8}$
   c. $\frac{3}{4}$
   d. $^{13}\!/_{16}$

5. $\frac{1}{8} \times ^{1}\!/_{10} =$ _____.
   a. $^{1}\!/_{80}$
   b. $^{1}\!/_{40}$
   c. $\frac{4}{5}$
   d. $1\frac{1}{4}$

6. $^{7}\!/_{16} \div \frac{7}{8} =$ _____.
   a. $^{49}\!/_{128}$
   b. $\frac{3}{8}$
   c. $\frac{1}{2}$
   d. 2

3.0.0 The Decimal System

Objective

Identify decimal numbers and calculate with them.
a. List the key qualities of decimal numbers and summarize their place values.
b. Add, subtract, multiply, and divide decimal numbers.
c. Convert between decimals, fractions, and percentages.

Performance Tasks

There are no performance tasks in this section.

The number 10 lies at the heart of the decimal system. Learning the place values of digits earlier in this module established the groundwork for the decimal system. Decimal numbers offer numerous advantages, including simpler calculations and easier fractions.

3.1.0 Decimals

Decimal values often include a dot (or period), called a decimal point, within the number. Digits to the decimal point's left represent a whole number. Digits to the decimal point's right represent a fraction of the number 1. In fact, decimals often go by the name of decimal fractions, since they represent part of a whole number. Even if a decimal point is not visible, assume one to the right of the ones place. For example, with the number 5,316,247, the decimal point is on the right side of the number 7.

Figure 4 shows the place values of digits to the left and to the right of the decimal point. For applications requiring extreme precision, extending a number five places to the right of the decimal point usually provides the

Millions		Hundred thousands	Ten thousands	Thousands		Hundreds	Tens	Ones		Tenths	Hundredths	Thousandths	Ten-thousandths	Hundred-thousandths	Millionths
5	,	3	1	6	,	2	4	7	.	4	2	9	6	3	5

Figure 4 Place values on both sides of the decimal point.

required precision. For more common applications, extending the number two places to the right of the decimal point is usually adequate. The application determines the requirements, however. In some cases, a whole number alone may be sufficient.

To read a decimal fraction aloud, say the number as it is written. Then say the name of the place value for the rightmost digit. For example, 0.56 is spoken as "fifty-six hundredths." The rightmost digit (6) is in the hundredths place.

Note that the number 0.56 begins with a zero to the left of the decimal point. It adds clarity when the decimal represents a value less than one but usually is not spoken.

Mixed numbers also appear in decimal form. The number 15.7 is an example. Speak it aloud as "fifteen and seven-tenths." Notice that the word "and" separates the whole number from the decimal fraction.

3.1.1 Rounding Decimals

Sometimes an answer can be too precise. For example, assume tubing costs €3.76 (the € symbol represents the *euro*, the currency of the European Union) per meter. If the budget specifies €800 for the tubing, how much tubing can be purchased?

The answer, approximately 212.7659574 meters, goes on for many decimal places. However, a supplier will not sell 0.7659574 meters of tubing. An answer to the nearest tenth suffices for this application. Therefore, round the number to a single digit to the decimal point's right. The following steps show the process:

212.<u>7</u>659574	**Step 1**	Underline the rounding place column.
212.<u>7</u>659574	**Step 2**	Look at the digit one place to its right (6).
212.8	**Step 3**	If the digit to the right is 5 or more, round up by adding 1 to the underlined digit. If the digit is 4 or less, leave the underlined digit alone. In this example, the digit to the right is 6, which is more than 5, so round up by adding 1 to the underlined digit. Finally, drop the remaining digits to the right of it.

NOTE

In this case, following the rounding rules causes a slight budget overrun (€800.13). Budgets sometimes follow their own rounding rules to avoid financial problems. Unless instructed otherwise, however, always follow standard rounding rules.

3.1.2 Comparing Decimals with Decimals

Deciding if one decimal value is larger or smaller than another is simple. The process resembles comparing whole numbers. Examine the following numbers:

4,214 and 4,217

Obviously 4,217 is larger. Adding a decimal point does not change anything:

0.4214 and 0.4217

The result is the same. The number with the digits 4217 is still larger.

3.1.3 Study Problems: Working with Decimals

For the following problems, identify the words that represent the proper way to read aloud the decimal values shown.

1. 0.4 = _____.

   a. four
   b. four-tenths
   c. four-hundredths
   d. four-thousandths

2. 0.05 = _____.

   a. five
   b. five-tenths
   c. five-hundredths
   d. five-thousandths

3. 2.5 = _____.

   a. two and five-tenths
   b. two and five-hundredths
   c. two and five-thousandths
   d. twenty-five-hundredths

For the following problems, identify the written numerical value of the spoken number.

4. Eighteen-hundredths = _____.

   a. 0.0018
   b. 0.018
   c. 0.18
   d. 1.8

5. Five and eight-tenths = _____.

   a. 5.0
   b. 5.008
   c. 5.08
   d. 5.8

For the following problems, select the answer that arranges the decimals from smallest to largest.

6. 0.400, 0.004, 0.044, and 0.404

   a. 0.400, 0.004, 0.044, 0.404
   b. 0.004, 0.044, 0.404, 0.400
   c. 0.004, 0.044, 0.400, 0.404
   d. 0.404, 0.044, 0.400, 0.004

7. 0.567, 0.059, 0.56, and 0.508

   a. 0.508, 0.56, 0.567, 0.059
   b. 0.059, 0.56, 0.508, 0.567
   c. 0.567, 0.059, 0.56, 0.508
   d. 0.059, 0.508, 0.56, 0.567

8. 0.320, 0.032, 0.302, and 0.003

   a. 0.003, 0.032, 0.302, 0.320
   b. 0.320, 0.302, 0.032, 0.003
   c. 0.302, 0.320, 0.003, 0.032
   d. 0.003, 0.032, 0.320, 0.302

9. 0.867, 0.086, 0.008, and 0.870

   a. 0.870, 0.867, 0.086, 0.008
   b. 0.008, 0.086, 0.867, 0.870
   c. 0.086, 0.008, 0.867, 0.870
   d. 0.008, 0.870, 0.867, 0.086

10. 0.626, 0.630, 0.616, and 0.641
    a. 0.616, 0.641, 0.630, 0.626
    b. 0.616, 0.626, 0.630, 0.641
    c. 0.641, 0.616, 0.626, 0.630
    d. 0.630, 0.616, 0.626, 0.641

3.2.0 Adding, Subtracting, Multiplying, and Dividing Decimals

Calculating with decimals requires the same skills as working with whole numbers. However, the correct answer depends on placing the decimal point correctly.

3.2.1 Adding and Subtracting Decimals

When adding or subtracting decimals, always keep the decimal points aligned.

To add 4.76 and 0.834, line up the problem as shown. Adding a 0 keeps the numbers aligned correctly.

$$
\begin{array}{r}
\overset{1}{4}.760 \\
+\ 0.834 \\
\hline
5.594
\end{array}
$$

Remember, all the decimal points, including the answer's, must align vertically. The same rule applies when subtracting decimals. To subtract 2.724 from 5.6, line up the decimal points vertically. Again, add zeros to keep the numbers aligned and to make borrowing easier.

$$
\begin{array}{r}
\overset{4\ \ 15\ 9\ 10}{5.600} \\
-\ 2.724 \\
\hline
2.876
\end{array}
$$

3.2.2 Multiplying Decimals

Multiplying decimal numbers closely resembles whole number multiplications. For example, an office cubicle partition panel is 4.5' wide. How wide a wall will seven side-by-side panels make?

$\begin{array}{r} 4.5 \\ \times\ 7 \end{array}$	**Step 1** Set up the problem just like a whole number multiplication.
$\begin{array}{r} \overset{3}{4}.5 \\ \times\ 7 \\ \hline 315 \end{array}$	**Step 2** Multiply the two numbers.
	Step 3 To place the decimal point, count the number of digits to the right of the decimal point of both numbers being multiplied. In this example, only 4.5 has a decimal point. There is just one digit to its right.
$\begin{array}{r} 4.5 \\ \times\ 7 \\ \hline 31.5 \end{array}$	**Step 4** Place the answer's decimal point by counting over the same number of digits from the right.

If there are more digits to the right of the decimal points than there are in the answer, add extra zeros to the left of the digits.

$$
\begin{array}{r}
0.507 \\
\times\ 0.022 \\
\hline
1014 \\
10140 \\
000000 \\
+\ 0000000 \\
\hline
11154 = 0.011154
\end{array}
$$

Total up the digits to the right of the decimal points in the two multiplied numbers (6). Place the answer's decimal point six digits from the right. In this case, add an extra zero since there are only five digits in the answer.

3.2.3 Dividing with Decimals

Dividing decimal numbers can take one of three forms:

Type 1: Those with a decimal point in the number being divided (the dividend):

$$22 \overline{)44.5}$$

Type 2: Those with a decimal point in the number being divided by (the divisor):

$$0.22 \overline{)4,450}$$

Type 3: Those with decimal points in both numbers (the dividend and the divisor):

$$0.22 \overline{)44.5}$$

The following examples demonstrate each type.
Divide Type 1 numbers using the following steps (dividend = 44.5 and divisor = 22):

Step 1 Place the answer's decimal point directly above the dividend's.	$22 \overline{)44.5}$
Step 2 Divide as usual.	$\begin{array}{r} 02.0 \\ 22 \overline{)44.5} \\ -44 \\ \hline 00.5r \end{array}$

The answer is 2, with a remainder of 0.5. Note that the quotient's decimal point aligns with the dividend's decimal point. This alignment is critical!
Divide Type 2 numbers using the following steps (dividend = 4,450 and divisor = 0.22):

Step 1 Move the divisor's decimal point all the way to the right. Count the number of places that it moved (2).	0.22 becomes 22
Step 2 Next, move the dividend's decimal point to the right by the same number of places (2). Since the dividend is a whole number, add extra zeros. Place the answer's decimal point directly above the dividend's.	
Step 3 Divide as usual.	$\begin{array}{r} 20227.2 \\ 22 \overline{)4450.00.0} \\ -44 \\ \hline 0050 \\ -0044 \\ \hline 00060 \\ -00044 \\ \hline 000160 \\ -000154 \\ \hline 0000060 \\ -0000044 \\ \hline 000001.6r \end{array}$

The answer is 20,227.2 with a remainder of 1.6.

> **Did You Know?**
>
> **Scientific Notation**
>
> Scientific notation (sometimes called exponential notation) is a system that manages very large or very small decimal numbers. Numbers in this form have two parts: the base and the exponent. The base is a regular whole number or decimal number. The exponent represents the number of times to multiply 10 by itself ($10^2 = 10 \times 10 = 100$). A negative exponent represents the number of times to multiply $\frac{1}{10}$ by itself ($10^{-2} = \frac{1}{10} \times \frac{1}{10} = \frac{1}{100}$). Scientists, engineers, and mathematicians often use scientific notation. The following are some examples:
>
> | Decimal Notation | Scientific Notation |
> |---|---|
> | 1 | 1×10^0 |
> | −50 | -5×10^1 |
> | 7,530,000,000 | 7.53×10^9 |
> | 0.0000000082 | 8.2×10^{-9} |

Divide Type 3 numbers using the following steps (dividend = 44.5 and divisor = 0.22):

0.22 becomes 22	**Step 1** Move divisor's decimal point all the way to the right. Count the number of places it moved (2).
44.5 becomes 4,450	**Step 2** Move the dividend's decimal point the same number of places to the right (2). Add extra zeros if required.
$22\overline{)44.50.}$	**Step 3** Place the answer's decimal point directly above the dividend's.

$$
\begin{array}{r}
202 \\
22\overline{)4450} \\
-44 \\
\hline
005 \\
-000 \\
\hline
0050 \\
-0044 \\
\hline
0006r
\end{array}
$$

Step 4 Divide as usual.

The answer is 202 with a remainder of 6.

Double-check any division answer by multiplying the quotient by the divisor and adding the remainder. The result should be the original dividend.

3.2.4 Using the Calculator to Add, Subtract, Multiply, and Divide Decimals

Doing decimal math with a calculator resembles calculating with whole numbers. Follow these steps using the problem 45.6 + 5.7 as an example.

Step 1 Turn the calculator on.

Step 2 Press 4, 5, . (decimal point), and 6. The number 45.6 appears.

> **NOTE**
>
> For the next step, press whichever operation key the problem specifies: + to add, − to subtract, × to multiply, ÷ to divide.

Step 3 Press the + key. The 45.6 should stay on the display.

Step 4 Press 5, . (decimal point), and 7. The number 5.7 appears.

Step 5 Press the = key. The answer should appear.

Step 6 Press the On or C key to clear the calculator fully. The CE key clears the most recent entry only. A 0 should appear.

Step 7 Repeat the previous steps, but select a different operation ($-$, $\times$, or $\div$). Continue repeating the sequence to perform all four operations.

$$45.6 + 5.7 = 51.3$$
$$45.6 - 5.7 = 39.9$$
$$45.6 \times 5.7 = 259.92$$
$$45.6 \div 5.7 = 8$$

3.2.5 Study Problems: Decimals

Without using a calculator, find the answers to these addition and subtraction problems. Be sure to keep the decimal points aligned.

1. $2.50 + 4.20 + 5.00 = $ _____

2. $1.82 + 3.41 + 5.25 = $ _____

3. $6.43 + 86.4 = $ _____

4. The combined thickness of a piece of sheet metal 0.078 cm thick and a piece of band iron 0.25 cm thick is _____.

   a. 0.308 cm

   b. 0.328 cm

   c. 3.08 cm

   d. 32.8 cm

5. Yesterday, a lumber yard contained 6.7 tons of wood. Since then, 2.3 tons were removed. How many tons of wood remain?

   a. 3.4 tons

   b. 4.4 tons

   c. 5.4 tons

   d. 6.4 tons

Use the following information to answer Questions 6 and 7:
A machinist mills a part. The starting thickness is 6.18 inches. The machinist makes three cuts three-tenths of an inch each.

6. How many inches of material have been removed at this point?

   a. 0.6 inches

   b. 0.8 inches

   c. 0.9 inches

   d. 1.09 inches

7. The remaining part thickness is _____.

   a. 5.28 inches

   b. 6.08 inches

   c. 6.10 inches

   d. 6.15 inches

8. Ceramic tile weighs 4.75 pounds per square foot. Therefore, 128 square feet of ceramic tile weighs _____.

   a. 598 pounds

   b. 608 pounds

   c. 908 pounds

   d. 1,108 pounds

Without using a calculator, find the answers to the following division problems. Round answers to the nearest hundredth.

9. $45.36 \div 18 = $ _____

10. $4.536 \div 18 = $ _____

11. $0.4536 \div 18 = $ _____

12. $25\overline{)10.20}$

13. $6\overline{)31.2}$

Did You Know?

Other Counting Systems

Ancient Babylonians developed one of the earliest place value systems—called base-60. It used special symbols combined in 59 different ways to represent numbers. Interestingly, the system had no zero symbol. Without the zero, calculations were harder and required remembering the location of empty place values. While nobody uses base-60 today, it survives in one important place. We measure time in units of 60 (60 seconds per minute and 60 minutes per hour).

Computer programmers sometimes use a special system called base-16, or hexadecimal. It uses sixteen numerals instead of ten: 0–9, plus A, B, C, D, E, and F. In the hexadecimal system, each place value is 16 times greater than the one to the right. For example, the hexadecimal number 1E is the same as the decimal number 30 (1 = 16, E = 14, so 16 + 14 = 30).

Without using a calculator, find the answers to the following division problems. Round answers to the nearest hundredth.

14. $282 \div 14.1 =$ _____

15. $694 \div 3.2 =$ _____

16. $99 \div 0.45 =$ _____

17. $2.5\overline{)102}$

18. $0.6\overline{)312}$

Without using a calculator, find the answers to the following division problems. Round answers to the nearest hundredth.

19. $20.82 \div 4.24 =$ _____

20. $38.9 \div 3.7 =$ _____

21. $9.9 \div 0.45 =$ _____

22. $0.25\overline{)10.20}$

23. $0.6\overline{)31.2}$

Use a calculator to find the answers to the following problems. Round answers to the nearest hundredth.

24. $\begin{array}{r} 45.89 \\ +\ 7.85 \\ \hline \end{array}$

25. $\begin{array}{r} 7.6 \\ \times\ 0.12 \\ \hline \end{array}$

26. $\begin{array}{r} 685.79 \\ -\ 562.66 \\ \hline \end{array}$

27. $6.45 \div 3.25 =$

28. $\begin{array}{r} 34.76 \\ +\ 3.64 \\ \hline \end{array}$

Solve these problems to practice rounding decimals.

29. A 90.5" pipe must be cut into 3.75" pieces. How many complete 3.75" pieces are possible?
    a. 14
    b. 24
    c. 34
    d. 44

30. If a car traveled 1,001 kilometers on 151.8 liters of gas, to the nearest tenth of a kilometer, how many kilometers per liter did it achieve?
    a. 0.1
    b. 0.2
    c. 6.5
    d. 6.6

31. A supplier sells wire for $4.30 per pound. If the total sale is $120.95, to the nearest hundredth of a pound, how much did the customer buy?
    a. 0.28
    b. 2.8
    c. 28.1
    d. 28.13

Measuring Coating Thickness

Coating thickness is important on many parts and surfaces because too little or too much can cause problems. A paint coating must be thick enough to prevent corrosion, withstand abrasion, and look nice. But too thick a coating might not cure properly or could crack, flake, or blister.

Customers often specify the desired coating thickness. A painter can use a wet-film thickness gauge to check a coat of paint. Typically, these gauges measure in mils or microns. The gauge shown here uses the mil, which equals one one-thousandth of an inch (0.001"). This value can also be written as 10^{-3}. A mil equals 0.0254 millimeters.

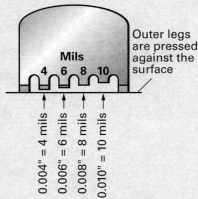

Thickness is determined by the longest tooth with paint and the shortest tooth without paint. The coating thickness here is 6–8 mils.

32. XYZ Supply Company sells a particular vent pipe at a special price of $0.37 per linear foot. To the nearest tenth of a foot, how many feet of pipe can be purchased with $115.38?

    a. 308.1
    b. 310.8
    c. 311.8
    d. 311.9

33. The same vent pipe's regular price is $0.48 per linear foot. To the nearest tenth of a foot, how many feet can be purchased for the same price ($115.38)?

    a. 240.38
    b. 240.4
    c. 241
    d. 241.4

3.3.0 Converting Decimals, Fractions, and Percentages

Sometimes craft professionals must convert one number form into another to make work easier. Some numbers may appear as decimals, some as percentages, and others as fractions. Decimals, percentages, and fractions are all different ways of expressing the same thing. The decimal 0.25, the percentage 25%, and the fraction ¼ all represent the same quantity. Each form can change into one of the others.

3.3.1 Converting Decimals to Percentages and Percentages to Decimals

To understand percentages, think of a whole number divided into 100 parts. A $1.00 bill is a good example, since it equals 100 equal parts—pennies. Each penny represents 1% of a dollar. Any part of a whole can be expressed as a percentage. *Figure 5* shows a good craft example. The tank holds 100 gallons. At present, it contains 50 gallons. To what percentage is the tank filled?

The correct answer is 50%. Percentage reflects a quantity based on there being 100 parts of something. How many gallons out of 100 does the tank contain? It contains 50 out of 100, or 50 percent. Percentages are an easy way to express parts of a whole.

Decimals and fractions also express parts of a whole. The tank in *Figure 5* is 50 percent full. Expressed as a fraction, it is ½ full. The decimal equivalent is 0.50 full.

Percentages can include decimals too. Adding a half-gallon of liquid to the tank in *Figure 5*, increases it to 50.5% full.

Sometimes, workers must convert decimals to percentages. Suppose a maintenance worker needs to prepare one gallon of cleaning solution. The mixture must contain 10% to 15% of the cleaning concentrate. The rest is water. There is 0.12 gallon of concentrate available. Will this be enough to make one gallon? To answer the question, convert the decimal value of the concentrate (0.12) to a percentage. Follow these steps to convert to a percentage:

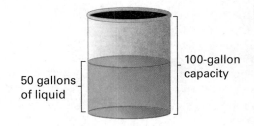

Figure 5 100-gallon tank.

Step 1 Multiply the decimal value by 100. (Tip: To multiply by 100, simply move the number's decimal point two places to the right.)	$0.12 \times 100 = 12$
Step 2 Add a % sign.	12%

The mixture must contain 10% to 15% of the concentrate. Since there is 12% of a gallon available, there is enough concentrate to make one gallon of working solution.

Converting percentages to decimals is simple too. Suppose a one-pound mixture must contain 22% of a certain chemical by weight. A digital scale, which reads in decimal values, weighs the ingredients. To determine the correct amount of chemical in decimal form, convert the percentage (22%) to a decimal by following these steps:

22	**Step 1** Drop the % sign.
22 ÷ 100 = 0.22	**Step 2** Divide the number by 100. (Tip: To divide by 100, simply move the number's decimal point two places to the left.)

Make the mixture by weighing out 0.22 pound of the chemical. Add it to 0.78 pound of the other ingredients, giving a total of one pound.

Decimals at Work

Several crafts professions, including millwrights, use decimal measurements daily. Millwrights often align pump and drive motor shafts. Extremely accurate alignment eliminates vibration and prevents damage to couplings and seals. Laser technology helps align this pump and motor. The pump manufacturer specifies the alignment, often to within 0.002 or 0.003 of an inch. The millwright places shims of various thicknesses under the pump and/or motor feet to tweak the alignment.

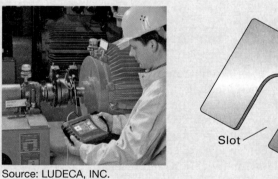

Source: LUDECA, INC.

3.3.2 Converting Fractions to Decimals

Changing a fraction into a decimal requires nothing more than division. Remember, a fraction is already a kind of division problem. Changing a fraction into a decimal value requires dividing the numerator by the denominator. The following steps show how to convert ¾ of a liter into a decimal value:

4)3.0	**Step 1** Divide the fraction's numerator by the denominator.
	For this example, place a decimal point and a 0 after the number 3. Since 4 will not divide into 3, the extra digit is necessary.
4)3.0	**Step 2** Place the answer's decimal point directly above the dividend's.
0.75 4)3.00 − 2.8 0.20 − 0.20 0.00	**Step 3** Divide as usual.

The decimal equivalent of ¾ is 0.75. If required, change this value into a percentage by moving the decimal point two positions to the right and adding a % sign. So, ¾ of a liter is also 0.75 liters or 75% of a liter.

3.3.3 Converting Decimals to Fractions

Converting a decimal to a fraction is also simple. Remember that both decimals and fractions express the same thing in different ways. Saying a decimal value aloud sounds like a fraction. Writing the spoken form automatically translates it.

For example, drill bits come in fractional sizes, decimal sizes, and even sets identified by special codes. A carpenter must drill a 0.125 inch hole. However, the available drill bits have fractional size markings on them. One of those sizes might be equivalent to 0.125 inch. To find out, the carpenter must follow these steps to convert the decimal to a fraction:

Step 1 Say the decimal value aloud.	0.125 is one hundred twenty-five thousandths
Step 2 Write the words down as a fraction.	0.125 written as a fraction is $\dfrac{125}{1000}$
Step 3 Reduce the fraction to its lowest terms.	$\dfrac{125}{1000} = \dfrac{125 \div 125}{1000 \div 125} = \dfrac{1}{8}$

Clearly, a ⅛ inch drill bit will drill a 0.125 inch hole.

3.3.4 Converting Inches to Decimal Equivalents in Feet

Unlike the United States, most other countries use the metric system of measurement. The metric system perfectly matches the decimal system since both rely on units of 10. However, for those working with the inch-pound system, measurements may require conversion to decimal. For example, what decimal part of a foot does 3 inches represent?

First, express the inches as a fraction with 12 as the denominator. Remember, one inch equals ¹⁄₁₂ of a foot. The fraction for 3 inches is ³⁄₁₂.

Reducing the fraction gives ¼. Convert this to a decimal by dividing the 4 into 1.00:

$$
\begin{array}{r}
0.25 \\
4\overline{)1.00} \\
-\,0.8 \\
\hline
0.20 \\
-\,0.20 \\
\hline
0.00
\end{array}
$$

So, 3 inches equals 0.25 foot.

Practical Math Application

When contractors calculate the expenses for building a house, they must pay close attention to the percentages they choose for various factors. Of course, they want to maximize profit as well. For example, if a contractor is building a house that will sell for $175,000, what will the profit be with the following percentage breakdown?

Profit = ?%
Overhead = 27%
Materials = 35%
Labor = 24%

Solution

Since $175,000 represents 100% of the revenue, the total cost and profit together must equal 100%. Adding the known percentages reveals that the total job cost is 86% of the

(Continued)

revenue. The profit is therefore 14% of the revenue (100% − 86% = 14%). Find the profit in dollars with this method:

$$\text{Profit} = \$175{,}000 \times 14\% = \$175{,}000 \times 0.14 = \$24{,}500$$

Calculators make working with percentages easy. In fact, some calculators have a % key to make the task even easier.

For calculators with a % button, do the following:

Step 1   Enter 175000 into the calculator.
Step 2   Press the × key.
Step 3   Enter 14 into the calculator.
Step 4   Press the % key.
Step 5   Press the = key.

For calculators without a % button, do the following:

Step 1   Enter 175000 into the calculator.
Step 2   Press the × key.
Step 3   Enter the decimal equivalent of 14% (0.14) into the calculator.
Step 4   Press the = key.

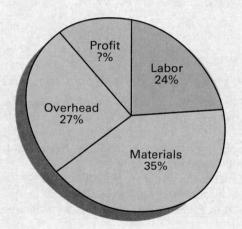

NOTE

Do not combine these two methods when calculating percentages. Either enter the percentage directly and press the % button or convert it to a decimal first. Doing both will give an incorrect result.

3.3.5   Study Problems: Converting Different Values

Convert these decimals to percentages.

1. 0.62 = _____

2. 0.475 = _____

3. 0.7 = _____

Convert these percentages to decimals.

4. 72% = _____

5. 12.5% = _____

Without using a calculator, convert the following fractions to their decimal equivalents.

6. ¼ = _____

7. ¾ = _____

8. $\frac{1}{8} =$ _____

9. $\frac{5}{16} =$ _____

10. $\frac{20}{64} =$ _____

Without using a calculator, convert the following decimals to fractions. Reduce them to their lowest terms.

11. $0.5 =$ _____

12. $0.12 =$ _____

13. $0.125 =$ _____

14. $0.8 =$ _____

15. $0.45 =$ _____

Convert the following measurements to a decimal value in feet. Round to the nearest hundredth.

16. 9 inches = _____ feet

17. 10 inches = _____ feet

18. 2 inches = _____ feet

19. 4 inches = _____ feet

20. 17 inches = _____ feet

Going Green

City Governments Thinking Green

Governing bodies around the world come up with innovative ways to cut energy costs and lower greenhouse gas emissions. Ideas include creating energy by harnessing wave power, air conditioning a building using rooftop gardens or lake water, and integrating wind turbines into a building's construction to generate energy.

New York

In 2006, New York City Mayor Michael Bloomberg announced a plan to cut the city's greenhouse-gas emissions 30% by the year 2030. The plan would generate enough new clean energy to provide 640,000 homes with electricity. Bladed turbines submerged in New York's East River would generate power using the tidal currents. In 2007, the first turbines went into operation 30 feet (9.14 meters) below the river's surface.

The pilot project succeeded so well that the city officials expanded the project. As of 2020, upgraded fifth-generation turbines supply power to thousands of New York City residents.

Aspen, CO

Many years ago, Aspen became the first municipality located west of the Mississippi River to make use of hydroelectric power—a renewable energy resource. After years of work, Aspen achieved its goal of powering the city with nothing but renewable energy. Today, the city's power comes from hydroelectric (46%), wind (53%), and landfill gas (1%) sources.

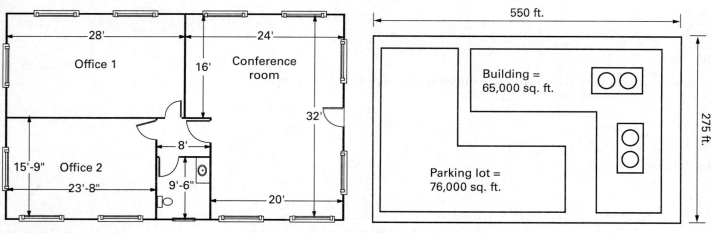

Figure 6 Office floor plan.

Figure 7 Site plan.

3.3.6 Practical Applications

The following are applications that craftworkers encounter daily. Use a calculator and the information provided to solve each conversion problem. Be sure to show all your work.

1. Find the cost of baseboard needed for the office building shown by the floor plan in *Figure 6*. While all door widths are a standard 36", do *not* subtract the door openings from the room measurements. Wise craftworkers always allow some extra material in case somebody makes a mistake. Round the number of baseboard feet up to the nearest linear foot. The lumber company charges $1.19 per linear foot of baseboard and gives a 12% discount. Add 6% tax after subtracting the discount amount.

   Costs:

   a. Total linear feet needed _____
   b. Initial baseboard cost $ _____
   c. Discount amount $ _____
   d. Baseboard cost after discount $ _____
   e. Amount added for 6% tax $ _____
   f. Total baseboard cost $ _____

2. Using the site plan shown in *Figure 7* determine the percentage of the lot that will be used for parking, the building, and for walkways. Round percentages off to the nearest tenth of a percent.

   a. Percentage for parking _____ %
   b. Percentage for building _____ %
   c. Percentage for walkways _____ %

Did You Know?

Direct and Indirect Costs

When contractors determine a percentage of profit, they must consider many different costs. These divide into two broad categories: direct costs and indirect costs.

Direct costs are expenses directly associated with the job. They include materials (paint, insulation, lumber) and the crew's labor hours. Large-scale jobs may also include subcontracted demolition and cleanup crews, cranes and earth-moving machinery, and job-site security.

Indirect costs (or overhead costs), on the other hand, are expenses required to keep the contracting company running. They include insurance, administrative staff payroll, marketing, office supplies, and property taxes. Think of some other indirect costs!

3.0.0 Section Review

1. In the number 135.792, what value does the 9 represent?

    a. 90

    b. nine tenths

    c. nine one-hundredths

    d. nine one-thousandths

2. The number 0.960 is larger than the number 0.0962.

    a. True

    b. False

3. $3.625 + 4.9 =$ _____.

    a. 7.525

    b. 8.525

    c. 41.15

    d. 52.63

4. $42.58 - 7.577 =$ _____.

    a. 35.003

    b. 35.523

    c. 36.523

    d. 50.157

5. $9.64 \times 12 =$ _____.

    a. 11.568

    b. 21.64

    c. 115.2

    d. 115.68

6. $123.82 \div 6.5 =$ _____.

    a. 19.04923

    b. 18.76

    c. 1.905

    d. 0.190

7. Express the number 0.479 as a percentage.

    a. 0.00479%

    b. 0.479%

    c. 47.9%

    d. 479%

8. The decimal equivalent of the fraction ⅞ is _____.

    a. 0.0875

    b. 0.75

    c. 0.875

    d. 8.75

9. If a fuel tank is ⅔ full, then it is more than 65% full.

    a. True

    b. False

4.0.0 Measuring Length

Performance Tasks

Under supervision, you should be able to do the following:

1. Using a measuring tape, measure lumber pieces in both English and metric units.

2. Using a measuring tape, measure a room-sized space in both English and metric units.

3. Using a measuring tape, determine a short inside measurement in both English and metric units.

4. Add English measurements that include fractions.

Objective

Name the common length-measuring tools and use them to measure lengths accurately.

a. Describe English and metric rulers, using them correctly to measure lengths.

b. Describe English and metric measuring tapes, using them correctly to measure lengths.

CAUTION

Most rulers, especially wooden ones (like the metric ruler shown in *Figure 8*) have extra material at the ends to maintain accuracy if the ends get damaged. Always start a measurement at the first marked line rather than at the ruler's physical end.

All construction workers use measuring tools to determine the dimensions of various objects. Tape measures and rulers, both English and metric, are the most common (*Figure 8*).

Standard English Tape Measure

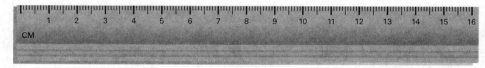

Metric Ruler

Figure 8 Various measuring tools.

4.1.0 Reading English and Metric Rulers

Reading a ruler is easy. But those who use them only occasionally may struggle to remember the values of the small marks between major units. Misreading a ruler leads to errors, expensive mistakes, and wasted material.

4.1.1   The English Ruler

NOTE

In a measurement setting, the terms *inch-pound*, *English*, and *Imperial* mean the same thing. They are names for the measuring system used in the United States, Canada, and a few other countries. This module usually prefers *inch-pound*. However, this section uses *English* when discussing rulers and measuring tapes since an "inch-pound ruler" sounds a little strange!

People using the inch-pound system measure with English rulers. These provide measurements in inches. Marks divide English rulers into whole inches and then halves, fourths, eighths, and sixteenths. Some rulers may have marks for thirty-seconds or even sixty-fourths. Except for the whole inch marks, everything else represents fractions of an inch. This subsection explains the standard English ruler and standard fractions. In *Figure 8*, the English ruler has two sets of marks. The top row's marks are ⅛" apart. The bottom row uses marks ¹⁄₁₆" apart. Other rulers may have different marking systems. For example, *Figure 9* shows a ruler with just ¹⁄₁₆" marks.

In *Figure 9*, the distance between the ⅛" and ¼" positions is two-sixteenths of an inch (²⁄₁₆"). Similarly, the mark immediately after the ¾" position represents thirteen-sixteenths of an inch (¹³⁄₁₆").

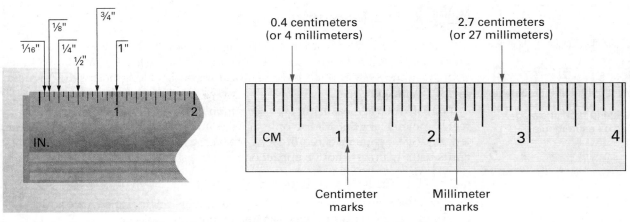

Figure 9 Standard ruler showing ¹⁄₁₆" marks and larger values.

Figure 10 Marks on a metric ruler.

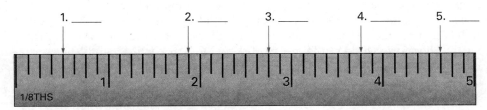

Figure 11 Practice with a ⅛" mark ruler.

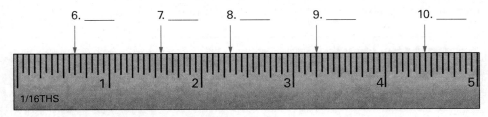

Figure 12 Practice with a ¹⁄₁₆" mark ruler.

4.1.2   The Metric Ruler

People using the metric system use metric rulers, which provide measurements in centimeters and millimeters. Marks on metric tape measures and rulers (*Figure 10*) represent centimeters and millimeters. The larger numbered lines are centimeters. The smaller lines represent millimeters. The metric system is base-10, since all measurements use some form of the number 10. For example, each millimeter is ¹⁄₁₀ of a centimeter.

A measurement 7 marks after 2 centimeters (*Figure 10*), is 2.7 (two point seven) centimeters or 27 millimeters. Both numbers represent the same distance but use different units.

The measurement six marks behind the 1-centimeter mark (*Figure 10*) is 0.4 (point four) centimeters or 4 millimeters.

4.1.3   Study Problems: Reading Rulers

Use the information from *Figure 9* to help identify the marked lengths numbered 1 through 10 in *Figure 11* and *Figure 12*. Label each length in the space provided or on a separate sheet of paper.

Identify the marked lengths in *Figure 13*. Record the correct answers for 11, 12, and 13 in centimeters. Record the correct answers for 14 and 15 in millimeters.

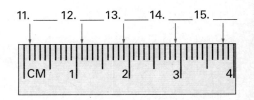

Figure 13 Practice with a metric ruler.

Figure 14 Measuring tape showing English and metric measurements.

Loadbearing: Carrying a significant amount of weight and/or providing necessary structural support. A loadbearing wall typically carries some portion of the roof's weight and cannot be removed without risking structural failure or collapse.

Stud: A vertical support inside a wall to which the wall finish material attaches. The base of a stud rests on a horizontal baseplate, and a horizontal cap plate rests on top of a series of studs.

Joists: Lengths of wood or steel that usually support floors, ceilings, or a roof. Roof joists will be at the same angle as the roof itself, while floor and ceiling joists are usually horizontal.

4.2.0   The Measuring Tape

Measuring tapes (also known as tape measures) may have English or metric markings along the blade. Some, such as the one in *Figure 14*, have both. Tapes using just one system often have identical markings along both sides to make measuring easier.

English/metric measuring tapes commonly come in 16 feet (5 meters) and 25 feet (8 meters) lengths. Metric-only tapes commonly come in 3.5, 5, and 8 meter lengths. Site layout tasks require longer tapes. Carpentry, pipefitting, and other crafts usually prefer short, compact ones.

4.2.1   The English Measuring Tape

English measuring tapes have markings like rulers but may also have additional markings. These marks make tasks like wall framing easier.

As *Figure 15A* shows, every 24 inches has a black background. Carpenters usually use 24 inch spacing for wall studs that do not carry a heavy load from above. Marks 16 inches apart, *Figure 15B*, have a red background. These help carpenters space studs 16 inches apart for **loadbearing** walls.

Figure 15 shows common **stud** spacing. With the measuring tape hook pulled tightly against the left side of one stud, the left side of the next stud is 16 inches away. Therefore, these studs are placed on 16-inch centers. Small black diamonds, like those in *Figure 15C*, identify 19.2 inch spacings for specialized studs and **joists**.

Figure 16 shows other markings that may appear on a standard tape measure. The red foot number (1F) works with the red inch number (2) to provide quick measurements. In this case, the numbers indicate 1 foot, 2 inches. The number below it (the black 14) indicates 14 inches. Adding the top scale numbers together gives the bottom scale number (1 foot or 12 inches + 2 inches = 14 inches). The red numbers appear along the length of the blade.

Around the World

Nominal and True Dimensions

A 2 × 4 board used in the United States is not really 2" by 4". Only the rough board cut from the log had those dimensions. After drying and planing, the finished board is 1½" by 3½". The following are examples of the true dimensions of common U.S. board sizes:

Nominal Size	Actual Measure	Actual Metric Measure
1" × 10"	¾" × 9¼"	19 × 235 mm
2" × 6"	1½" × 5½"	38 × 140 mm
2" × 8"	1½" × 7¼"	38 × 184 mm
2" × 10"	1½" × 9¼"	38 × 235 mm
4" × 4"	3½" × 3½"	89 × 89 mm

To convert to equivalent metric lumber sizes, use 25 mm per inch as a conversion. This method is not very precise, but it works. A 2" × 4" is 50 mm × 100 mm in the metric world. However, the actual dimensions are typically 40 mm × 90 mm. Over the years, actual lumber size has steadily decreased compared to nominal size. This trend may provoke changes in building codes to keep structures safe.

4.2.2   The Metric Measuring Tape

Metric measuring tapes look like standard measuring tapes. In the United States, metric-only tapes are rare. In most cases, a combination tape will be the only option. Outside the United States, however, metric-only tapes are common.

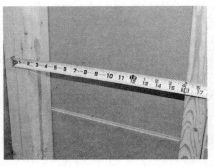

A   Every 24 inches is marked with a contrasting black background. 24-inch spacing on center is used most commonly for nonbearing walls.

B   Every 16 inches is marked with a red background. 16-inch spacing on center is used most commonly for loadbearing walls.

C   Every 19.2 inches is marked with a small black diamond. 19.2-inch spacing on center is an alternate, less-commonly used spacing scheme for loadbearing walls and special joists.

Figure 15 Stud spacing and wall framing markings.

Figure 16 Other markings on a standard tape measure.

Metric tapes have markings to help with framing tasks too. But they will be spaced differently to accommodate other building codes and lumber dimensions.

4.2.3 Using a Tape Measure

Most people learn to use a measuring tape easily. Measuring tapes curve slightly to make them stiffer. This feature helps them stay straight when extended. Some tapes have thicker blades and a greater curve for longer extension. Manufacturers call the amount of extension the tape's *standout*. The following steps outline the correct way to measure with a tape:

Step 1 Pull the tape straight out of its case with one hand.

Step 2 Hook the metal tape end to the measured object. Another worker may have to hold the tape end if it cannot be hooked onto the object. The tape hook should slide slightly to compensate for its thickness. It slides in for making inside measurements and slides out for outside ones. If the hook binds instead of sliding, measurements could be off by as much as $\frac{1}{16}$" (1.6 mm). A bent or damaged hook can have the same effect.

Step 3 Extend the tape by pulling the case. Go a bit beyond the measurement point.

Step 4 After extending the tape the correct amount, slide the case's thumb lock to hold the tape in place. Not all tape measures have a thumb lock, and some tape measures lock automatically. A thumb release retracts the tape on automatically locking tape measures. Never let the tape snap back into the case, or the hook may break.

Step 5 Read the measurement from the correct location on the measured object.

Measuring between two objects can be challenging. For example, installing a shelf between two walls requires this kind of measurement. Measuring tapes usually have the case's exact length stamped on the side. Take the measurement by firmly bumping the hook against one wall and the back of the case against the other. Read the length where the tape enters the case. Then, add the case length to the value.

> **NOTE**
>
> Always double-check measurements before cutting a piece of material. Remember the old saying, "Measure twice, cut once."

4.2.4 Study Problems: Reading Measuring Tapes

Identify the marked lengths numbered 1 through 5 in *Figure 17*. Record the answers in inches, using fractions as necessary. Identify the marked lengths numbered 6 through 10 in *Figure 18*. Record the answers in centimeters, using tenths of a centimeter as necessary.

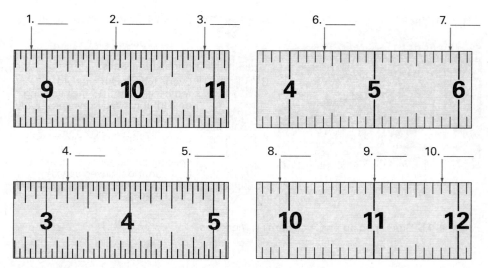

Figure 17 English tape measure practice.   **Figure 18** Metric tape measure practice.

Around the World
Visualizing Metric Units

The following examples make it easier for inch-pound users to visualize metric measurements.
Metric Examples Visualized (approximation only):

- 1 millimeter = the thickness of a dime
- 1 centimeter = the width of a standard paperclip
- 1 decimeter = the length of a crayon
- 1 meter = the distance from a door handle to the floor (about 1.1 yards)
- 1 kilometer = the length of 6 city blocks (about 0.6 miles)
- 1 gram = the weight of a paperclip
- 1 kilogram = the weight of a brick (about 2.2 pounds)

Did You Know?
Standard Measure

Determining the distance to a neighbor's farm may have been the earliest reason for measuring. Imagine that you wanted to measure the distance to a favorite fishing pond. You could pace off the distance and count the number of steps. You might find that it was 570 paces between your house and the pond. You could then pace off the distance to another pond. Comparing paces would tell you which distance was longer, so the pace became a measurement unit. Since everyone has a different pace length, however, this method is not fully accurate.

Early humans used feet, arms, hands, and fingers for measuring. A few of these units survive today. We measure length in feet and horse height in hands. An inch comes close to a thumb's width. But body part sizes differ too. A person with small hands could sell a horse as taller than it really is.

These problems led to standardized measurement units. Manufactured physical objects, such as special sticks or metal rods, became the basis for consistent units. Craft associations made careful copies of these so-called standards and used them to produce their measuring tools. Crafts guilds insisted on their apprentices, journeymen, and masters following the standards. A craft's reputation depended on quality and accuracy, just as it does today. Guilds often penalized those who were careless or who deliberately falsified measurements.

4.0.0 Section Review

1. The metric system can be referred to as a(n) _____.

   a. base-10 system
   b. base-100 system
   c. geometric progression
   d. open-ended system

2. A measuring tape's hook is slightly loose. What is the correct action?

   a. Using a small hammer, carefully pound the attaching rivets enough to tighten them and prevent movement.
   b. Nothing; the hook end should be slightly loose.
   c. Replace the tape measure.
   d. Return the tape measure to the manufacturer for repair.

5.0.0 Metric and Inch-Pound Measurement Systems

Objective

Name common length, weight, volume, and temperature units in both the inch-pound and metric systems and convert them into other comparable units.

   a. List and convert between common inch-pound and metric length units.

   b. List and convert between common inch-pound and metric weight units.
   c. List and convert between common inch-pound and metric volume units.
   d. List and convert between common inch-pound and metric temperature units.

Performance Tasks

There are no performance tasks in this section.

Most of the world uses the metric system of measurement. Companies manufacture and ship goods to global destinations daily. They may provide dimensions, weights, temperatures, and pressures in metric values, inch-pound values, or both.

Companies within the United States have started using the metric system. Many manufacturers use both inch-pound and metric values so they can support customers from any country. Many crafts require its workers to use both systems and convert between them. *Figure 19* shows some familiar metric and inch-pound system examples.

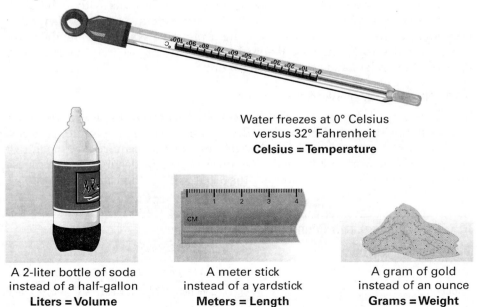

Water freezes at 0° Celsius versus 32° Fahrenheit
Celsius = Temperature

A 2-liter bottle of soda instead of a half-gallon
Liters = Volume

A meter stick instead of a yardstick
Meters = Length

A gram of gold instead of an ounce
Grams = Weight

Figure 19 Common inch-pound and metric measured values.

TABLE 1 Metric System Prefixes.

Unit Prefix	Expressed Quantity as Powers of 10			Decimal System Expression
Micro- (μ)	1/1,000,000	0.000001	10^{-6}	One-millionth
Milli- (m)	1/1,000	0.001	10^{-3}	One-thousandth
Centi- (c)	1/100	0.01	10^{-2}	One-hundredth
Deci- (d)	1/10	0.1	10^{-1}	One-tenth
Deka- (da)	10	10.0	10^{1}	Tens
Hecto- (h)	100	100.0	10^{2}	Hundreds
Kilo- (k)	1,000	1,000.0	10^{3}	Thousands
Mega- (M)	1,000,000	1,000,000.0	10^{6}	Millions
Giga- (G)	1,000,000,000	1,000,000,000.0	10^{9}	Billions

Unit: A standardized quantity used for a particular kind of measurement.

Science, engineering, and many crafts require exact measurements of physical quantities. Measurements involve comparing whatever is being measured to a standard dimension called a **unit**. Measurements must always include the appropriate unit (such as foot, liter, or pound). A number alone is not a measurement.

Measuring with the metric system is simpler than measuring with the inch-pound system. Scaling metric units always involves multiplying or dividing by 10's. On the other hand, the inch-pound system requires many conversion factors. For example, one mile is 5,280 feet, and 1 inch is ¹⁄₁₂ of a foot. On the other hand, a centimeter is ¹⁄₁₀₀ of a meter and a kilometer is 1,000 meters. Metric calculations usually take place in decimal form, eliminating having to add and subtract fractions.

Scaling metric units up or down involves attaching a prefix to the unit. *Table 1* shows the common metric scaling prefixes. For example, the prefix *milli-* scales a unit by 0.001 (one one-thousandth). One millimeter is therefore 0.001 meter. The prefix *kilo-* scales a unit by 1,000. One kilometer is therefore 1,000 meters. While the table can look intimidating, it is actually quite simple. Just find the prefix and read across to find the scaling factor.

5.1.0 Units of Length Measurement

Converting measurement units may involve changing units within the same system—for example, inches to yards or centimeters to meters. On other occasions, conversion requires changing from one system to another—for example, miles to kilometers. Before looking into conversion, however, it is important to understand common length units.

5.1.1 Inch-Pound System Units of Length

Table 2 shows the inch-pound system's common length units, along with their relationships. Many applications require breaking inches down into fractions. For everyday applications, ¹⁄₆₄" is the smallest fraction required. A machinist doing precision work might need to express inches in decimal values. These numbers may be as small as thousandths (0.001") or ten-thousandths (0.0001").

TABLE 2 Common Inch-Pound Units of Length.

Inch-Pound Length Units		
1 inch	=	¹⁄₁₂th of a foot; 0.0833 feet
1 foot	=	12 inches; ¹⁄₃rd of a yard
1 yard	=	36 inches; 3 feet
1 mile	=	5,280 feet; 1,760 yards

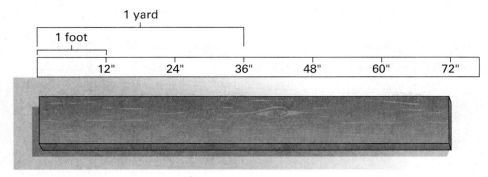

Figure 20 Converting inches to yards.

Abbreviate the following common inch-pound units as shown:

- An inch or inches may be abbreviated as *in* or identified by the symbol ".
- A foot or feet may be abbreviated as *ft* or identified by the symbol '.
- A yard or yards may be abbreviated as *yd*.
- A mile or miles is rarely abbreviated, but the most common abbreviation is *mi*.

Converting one unit to another requires multiplying or dividing by a conversion factor. A calculator simplifies unit conversions. For example, to change from inches to yards, divide the number of inches by 12 (the number of inches in a foot) to convert to feet. Then, divide that number by 3 (the number of feet in a yard) to convert to yards. Alternatively, just divide the number of inches by 36 (*Figure 20*) to convert directly to yards.

Many in the craft professions convert between units with specialized smartphone apps. These contain conversion factors for many inch-pound and metric units. Search engines can also convert between units or suggest online calculators. Just type in the two units as a search term, such as "inches to meters."

5.1.2 Metric System Units of Length

Table 3 shows the metric system's common length units, along with their relationships. Notice that all the units relate to each other by 10's.

Abbreviate the common metric length units as shown:

- A millimeter is abbreviated as *mm*.
- The centimeter is abbreviated as *cm*.
- The meter is abbreviated as *m*.
- The kilometer is abbreviated as *km*.

Converting metric units requires multiplying or dividing by some value based on 10. The easiest way to do this operation is to move the decimal point to the left or right. Which way to move the decimal point depends on size. To convert a smaller value to a larger one, move the decimal point to the left. To convert a larger value to a smaller one, move the decimal point to the right. *Table 1* helps you determine how many places to move.

For example, to convert 72 centimeters to meters (*Figure 21*), remember that meters are bigger than centimeters. The third row of *Table 1* indicates that the

TABLE 3 Common Metric Units of Length.

Metric Length Units		
1 millimeter	=	0.1 centimeters; 0.001 meters
1 centimeter	=	10 millimeters; 0.01 meters
1 meter	=	100 centimeters; 0.001 kilometers
1 kilometer	=	1,000 meters

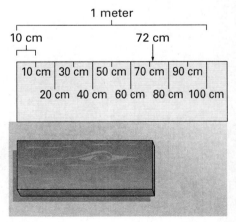

Figure 21 Converting centimeters to meters.

TABLE 4 Length Conversion Factors.

Unit	Millimeter	Centimeter	Inch	Foot	Yard	Meter	Kilometer	Mile
1 millimeter	1	0.1	0.03937	0.003281	0.001094	0.001	0.000001	0.0000006214
1 centimeter	10	1	0.3937	0.03281	0.01094	0.01	0.00001	0.000006214
1 inch	25.4	2.54	1	0.08333	0.02778	0.0254	0.0000254	0.00001578
1 foot	304.8	30.48	12	1	0.3333	0.3048	0.0003048	0.0001894
1 yard	914.4	91.44	36	3	1	0.9144	0.0009144	0.0005681
1 meter	1,000	100	39.37	3.281	1.094	1	0.001	0.0006214
1 kilometer	1,000,000	100,000	39,370	3,281	1,094	1,000	1	0.6214
1 mile	1,609,340	160,934	63,360	5,280	1,760	1,609.3	1.6093	1

centi- prefix stands for 0.01. Therefore, move the decimal point two places to the left. As a result, 72 centimeters becomes 0.72 meters.

Remember that moving decimal points is identical to multiplying or dividing. Dividing 72 centimeters by 100 also gives 0.72 meters.

5.1.3 Converting Length Units Between Systems

Converting measurements between the inch-pound and metric systems always involves multiplying or dividing by a conversion factor. Reference books, dictionaries, and textbooks all contain conversion factor charts. *Appendix C* contains an example. Online tools and smartphone apps perform conversions automatically using the same conversion factors.

Converting between units requires accurate math and the correct conversion factor. Choose the wrong factor or make a calculation mistake and the result is worthless. *Table 4* provides common factors for converting length measurements from one system to another. After converting, the answer may require rounding.

According to *Table 4*, 1 centimeter is equal to 0.3937 inches. To convert 13 centimeters to inches, multiply 13 by this factor. The result is 5.1181 inches. For most applications, round to 5.12 inches, or even to 5.1 inches. Rounding decisions depend on the application. Locating the center of a steel plate to drill a hole requires high precision. Placing a mailbox post the correct distance from the end of a driveway is less demanding.

5.1.4 Study Problems: Converting Measurements

Without using a calculator, find the answers to the following conversion problems. Round your answers to the nearest hundredth.

1. 0.45 meter = _____ centimeters

2. 3 yards = _____ inches

3. 36 feet = _____ yards

4. 90 inches = _____ yards

5. 1 centimeter = _____ meters

6. 66 inches = _____ centimeters

7. 47 feet = _____ meters

8. 54.5 centimeters = _____ feet

9. 19 yards = _____ meters

10. 4.7 meters = _____ inches

5.2.0 Units of Weight Measurement

This section focuses upon inch-pound and metric units used to measure weight.

5.2.1 Inch-Pound Units of Weight

Weight is the **force** that an object exerts on a surface due to its **mass** and the Earth's gravity. In the inch-pound system, common units for weight include the ounce, the pound, and the ton. *Table 5* shows the relationship between these three units.

Abbreviate the common inch-pound weight units as shown:

- An ounce or ounces may be abbreviated as *oz* or *ozs*.
- A pound or pounds may be abbreviated as *lb* or *lbs*.
- The ton may be abbreviated as *t*.

Converting from one inch-pound weight unit to another requires multiplying or dividing by an appropriate conversion factor. For example, to convert 134 ounces to pounds, divide by 16—the number of ounces in one pound. The result is 8.375 pounds. Round the number to the appropriate level.

Force: A push or pull on a surface. In this module, a force is the weight of an object or a fluid.

Mass: The quantity of matter present.

TABLE 5 Common Inch-Pound Units of Weight.

Inch-Pound Weight Units		
1 ounce	=	¹⁄₁₆th of a pound; 0.0625 pounds
1 pound	=	16 ounces; 0.0005 tons
1 ton	=	2,000 pounds

5.2.2 Metric Units of Weight

The metric system uses the milligram, the gram, and the kilogram as its weight units. As with length, these metric units relate to each other by 10's. *Table 6* shows their relationships.

Abbreviate each metric weight unit as shown:

- A milligram is abbreviated as *mg*.
- The gram is abbreviated as *g*.
- The kilogram is abbreviated as *kg*.

Converting metric weight units requires multiplying or dividing by 10's. Alternatively, just shift the decimal point to the left or right. *Table 6* shows that 1 gram is equal to 1,000 milligrams. To convert 6,439 milligrams to grams, divide 6,439 grams by 1,000—the number of milligrams in a single gram. A milligram is 0.001 grams, so an easier method is to move the decimal point three places to the left. Regardless of the method chosen, the result is the same—6.439 grams.

TABLE 6 Common Metric Units of Weight.

Metric Weight Units		
1 milligram	=	0.001 grams; 0.000001 kilograms
1 gram	=	1,000 milligrams; 0.001 kilograms
1 kilogram	=	1,000 grams; 0.001 metric tons
1 metric ton	=	1,000 kilograms

Around the World

Metric System as a Modern Standard

As modern scientific thinking started to develop during the 1500s, scientists realized that they needed to agree on measurement standards. Every country had its own units, so scientists had difficulty communicating their measurements to each other.

In the 1790s, French scientists started developing what became the metric system. They began with a standard length, called the meter. Over the next century, the world scientific community adopted the metric system. Eventually, most of the world successfully made the change. However, not until the 1970s did the United States and Canada truly acknowledge the metric system. Increasing global trade stimulated this shift, but many people resisted learning a new measurement system. Even today, many in the United States do not understand metric units.

5.2.3   Converting Weight Units Between Systems

Table 7 shows the relationship between weight units in the inch-pound and metric systems. The conversion factors may be more precise than required. Always round answers to the values appropriate to the application.

An HVAC technician might have to convert air-conditioning refrigerant weight from one system to the other. Or a crane operator might need to verify that his inch-pound rigging could handle a particular metric-specified load. Imagine that the load weighs 1,200 kilograms (kg). How many pounds is this?

Table 7 shows that 1 kg = 2.205 pounds, so multiplying 1,200 kg by 2.205 yields 2,646 pounds. The slings or lifting equipment must have a rated capacity greater than 2,646 pounds.

When working with smaller weights, ounces or grams may be more appropriate. One ounce is equal to 28.35 grams, so the gram is the smaller unit. Convert ounces to grams by multiplying the number of ounces by 28.35. Convert grams to ounces by dividing by 28.35.

Water and the Gram

The milliliter is a popular metric unit for measuring liquids. It equals $1/1000$ (0.001) of a liter. One milliliter of pure water weighs exactly one gram. This is yet another example of the metric system's practicality. Remember, however, that other liquids have different weights.

TABLE 7 Inch-Pound and Metric Weight Conversion Chart.

Unit	Milligram	Gram	Ounce	Pound	Kilogram	Ton	Metric Ton
1 milligram	1	0.001	0.00003527	0.000002205	0.000001	0.000000001102	0.000000001
1 gram	1,000	1	0.03527	0.002205	0.001	0.000001102	0.000001
1 ounce	28,349.5	28.35	1	0.0625	0.02835	0.00003125	0.00002835
1 pound	453,592	453.6	16	1	0.4536	0.0005	0.000454
1 kilogram	1,000,000	1,000	35.27	2.205	1	0.0011	0.001
1 ton	907,200,000	907,185	32,000	2,000	907.185	1	0.9072
1 metric ton	1,000,000,000	1,000,000	35,274	2,204.62	1,000	1.102	1

5.2.4   Study Problems: Converting Weight Units

Convert these weights from inch-pound to metric weight units, or vice versa. Round answers to the nearest hundredth.

1. 50 pounds = _____ kilograms

2. 50 kilograms = _____ pounds

3. 15.9 ounces = _____ grams

4. 94 grams = _____ ounces

5.3.0   Units of Volume Measurement

Volume: The amount of space contained within a three-dimensional space.

Cube: A three-dimensional object whose sides are all the same length.

This section examines units of **volume** measurement in the inch-pound and metric systems. Volume applies to three-dimensional objects such as a **cube** or a cylinder. This module does not examine liquid measurement units like the gallon or liter.

An object's volume refers to the amount of space it occupies. For example, a barrel might take up 7.35 cubic feet of space. An engine's cylinders might have a total volume of 250 cubic centimeters. Calculating volume depends on the object. The appropriate calculations appear later in this module.

A cube is the simplest object since its sides are all the same length. For example, if a cube measures $1" \times 1" \times 1"$, it has a volume of 1 cubic inch. Notice that volume units use the same units as length measurements. Just add the word "cubic" to the unit.

5.3.1 Inch-Pound Units of Volume

Table 8 shows the three most common volume units in the inch-pound system. Note that these units use the familiar length measurement units.

Abbreviate each of the inch-pound volume units as shown:

- The cubic inch is *cu in* or in^3. It sometimes appears as *CI*, but this is not a standard abbreviation.
- The cubic foot is *cu ft* or ft^3.
- The cubic yard is *cu yd* or yd^3.

Convert between these units in the usual way. For example, to convert 864 cubic inches to cubic feet, multiply 864 by the conversion factor 0.0005787. The result is 0.4999, or 0.5, cubic feet.

5.3.2 Metric Units of Volume

The cubic centimeter and the cubic meter are the most common metric volume units. The crafts use these two most often. *Table 9* shows their relationship to each other. As with all metric system units, these, and other volume units, relate to each other by 10's. Convert between these in the usual way for metric units. The cubic meter represents a three-dimensional object equivalent to a cube 100 centimeters $\times$ 100 centimeters $\times$ 100 centimeters in size. Therefore, 1 cubic meter contains 1,000,000 cubic centimeters.

Abbreviate cubic centimeters as *cu cm* or cm^3. Abbreviate cubic meters as *cu m* or m^3.

5.3.3 Converting Volume Units Between Systems

Table 10 shows the relationship between inch-pound and metric volume units. A worker excavating soil to prepare a foundation might have to convert between systems. In most cases, the units in each system should be similar in size (example: cubic yards to cubic meters). Rarely would the worker have to convert a large unit to a small one (example: cubic yards to cubic centimeters).

TABLE 8 Common Inch-Pound Units of Volume.

Inch-Pound Volume Units		
1 cubic inch	=	0.0005787 cubic feet
1 cubic foot	=	1,728 cubic inches; 0.037 cubic yards
1 cubic yard	=	27 cubic feet

TABLE 9 Common Metric Units of Volume.

Metric Volume Units		
1 cubic centimeter	=	0.000001 cubic meters
1 cubic meter	=	1,000,000 cubic centimeters

TABLE 10 Inch-Pound and Metric Volume Conversion Chart.

Unit	Cubic Centimeter	Cubic Inch	Cubic Foot	Cubic Yard	Cubic Meter
1 cubic centimeter	1	0.061	0.00003531	0.000001308	0.000001
1 cubic inch	16.387	1	0.0005787	0.000021433	0.000016387
1 cubic foot	28,316.85	1,728	1	0.037	0.0283
1 cubic yard	764,554.86	46,656	27	1	0.765
1 cubic meter	1,000,000	61,023.7	35.315	1.308	1

A welder fabricates a box to hold a battery. The battery's specification sheet says that the battery is cube-shaped, with a volume of 1,600 cubic centimeters. The space allowed for the battery box is cube-shaped and contains 112 cubic inches. Will the battery fit?

To convert cubic centimeters to cubic inches, multiply the number of cubic centimeters by the conversion factor. Since the conversion factor is 0.0610, multiply 1,600 by 0.061. The result is 97.6 cubic inches. The battery box will fit the available space with some room to spare.

A concrete craftworker calculates that a slab requires 59 cubic yards of cement. However, the supplier sells cement in cubic meters. How much should the worker order?

According to *Table 10*, the conversion factor is 0.765. Multiply 59 cubic yards by 0.765, which gives 45.1 cubic meters.

5.3.4   Study Problems: Converting Volume Units

Convert these volumes from the inch-pound system to the metric system, or vice versa. Round to the nearest tenth.

1. 11,600 cubic inches = _____ cubic feet

2. 1.9 cubic meters = _____ cubic centimeters

3. 512 cubic meters = _____ cubic yards

4. 7 cubic feet = _____ cubic centimeters

5. 0.2 cubic meters = _____ cubic feet

5.4.0   Temperature Units

Many crafts regularly work with temperatures. Temperature is a measurement of how fast the atoms or molecules in a solid, liquid, or gas are moving. If they move rapidly, the temperature is high. When they slow down, the temperature is low. People often use the terms *hot* and *cold* to describe the way they sense temperature. A welder running an arc welding machine makes the metal atoms in two pieces of steel move rapidly. The metal gets hot, softens, and joins. A refrigerator chills the water in a water bottle by slowing down the water molecules.

Expressing temperature can be tricky since there are many different temperature units. Converting between these happens often, not only in the crafts, but in science, engineering, and industry. This section examines common temperature units and the proper ways to convert them.

Temperature units come from a measuring scale based on one or more standard temperature points. These points usually come from water's behavior. For example, at sea level water freezes and boils at specific temperatures. Several temperature scales use these points as references.

Most temperature measurements are in degrees, represented by the degree symbol (°). However, a degree's size depends on the specific temperature scale. A complete temperature unit is the degree symbol followed by a single letter

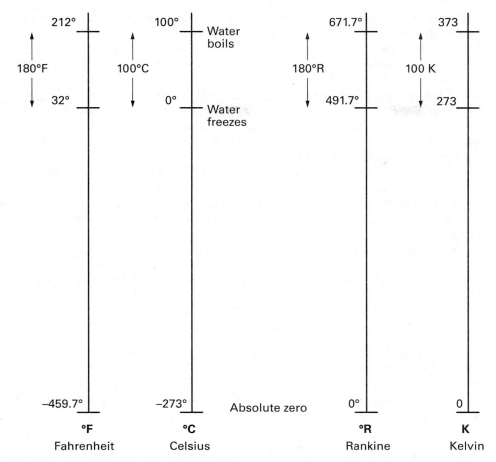

Figure 22 Comparing temperature scales.

identifying the scale. For example, °F and °C are popular temperature units. One major temperature scale does not use degrees, however.

The four most common temperature scales are the Fahrenheit, Celsius, Rankine, and Kelvin scales (*Figure 22*). Everyday temperature measurements are usually in Fahrenheit or Celsius. On the Fahrenheit scale, water freezes at 32°F and boils at 212°F. On the Celsius scale, water freezes at 0°C and boils at 100°C. *Figure 22* shows the abbreviation for each unit.

In the 1700s, scientists speculated that a minimum temperature must exist. Nothing could be colder than this value. Today, we call this point *absolute zero*. At this temperature, atoms and molecules barely move. Two temperature scales, the Rankine scale and the Kelvin scale, use absolute zero as one endpoint.

On the Rankine scale, water freezes at 491.7°R and boils at 671.7°R—a 180° range. This range is identical to that of the Fahrenheit scale (212°F − 32°F = 180°). Degrees on the Rankine scale are therefore the same size as those on the Fahrenheit scale. For this reason, some call the Rankine scale the *absolute Fahrenheit scale*.

Around the World

Coldest Recorded Temperature

Scientists routinely produce extremely low temperatures using a variety of exotic laboratory techniques. To date, the coldest temperature ever achieved came to within less than a billionth of a degree of absolute zero (0 K).

In 1983, Russian scientists at the Vostok Station in Antarctica measured the lowest natural temperature on Earth—a chilly −128.6°F (−89.2°C)!

Digital Thermometers

Thermometers with digital (numeric) readouts can provide temperature readings in both Celsius and Fahrenheit. Many analog thermometers have dual scales as well. The unit on the top is a digital temperature probe. The one on the bottom is a non-contact infrared thermometer. It determines temperature by measuring how much invisible infrared radiation an object emits. Both provide their readings directly on a numeric display.

Source: Courtesy of Extech Instruments, a FLIR Company

On the Kelvin scale, water freezes at 273 K and boils at 373 K—a 100-unit range. This range is identical to that of the Celsius scale ($100°C - 0°C = 100°$). Units on the Kelvin scale are therefore the same size as those on the Celsius scale. For this reason, some call the Kelvin scale the *absolute Celsius scale*. Note that the unit identifier for the Kelvin scale is the letter K without a degree symbol in front of it. In fact, the term *degree* is not used when referring to temperatures on the Kelvin scale. For example, to express room temperature on the Celsius and Kelvin scales, say something like, "Room temperature is around 22 degrees Celsius or 295 kelvins." Both the Kelvin and Celsius scales are part of the metric system of measurement.

Most crafts use the Fahrenheit and Celsius scales. Scientific and some engineering applications require the Rankine or Kelvin scale.

Charts, tables, and smartphone apps make converting between temperature scales simple. However, the craftworker benefits from understanding the underlying math.

The Fahrenheit scale has 180° between water's boiling and freezing points ($212° - 32° = 180°$). The Celsius scale has 100° between these same points ($100° - 0° = 100°$). The following fraction shows the relationship between the scales:

$$\frac{\text{Fahrenheit range}}{\text{Celsius range}} = \frac{180°}{100°} = \frac{9}{5}$$

One degree Fahrenheit is $\frac{5}{9}$ of one degree Celsius. One degree Celsius is $\frac{9}{5}$ of one degree Fahrenheit. To convert a Fahrenheit temperature to a Celsius temperature, subtract 32° (since 32° corresponds to 0° on the Celsius scale) and then multiply by $\frac{5}{9}$. To convert a Celsius temperature to a Fahrenheit temperature, multiply by $\frac{9}{5}$ and then add 32°. The following equations show this process:

$$°C = \frac{5}{9} \times (°F - 32°)$$

$$°F = \left(\frac{9}{5} \times °C\right) + 32°$$

Practice these calculations to become more comfortable with them. *Figure 23* shows two examples.

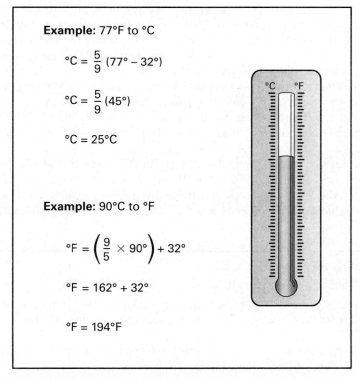

Example: 77°F to °C

$$°C = \frac{5}{9}(77° - 32°)$$

$$°C = \frac{5}{9}(45°)$$

$$°C = 25°C$$

Example: 90°C to °F

$$°F = \left(\frac{9}{5} \times 90°\right) + 32°$$

$$°F = 162° + 32°$$

$$°F = 194°F$$

Figure 23 Sample temperature conversions.

5.4.1　Study Problems: Converting Temperatures

Convert these temperatures from Fahrenheit to Celsius, or Celsius to Fahrenheit. Round to the nearest tenth of a degree.

1. $180°F =$ _____ $°C$

2. $66°F =$ _____ $°C$

3. $-26°C =$ _____ $°F$

4. $71°C =$ _____ $°F$

5.0.0　Section Review

1. When converting from a larger metric unit to a smaller one, the decimal point must move _____.
   a. to the left
   b. to the right
   c. all the way to the left end of the number
   d. all the way to the right end of the number

2. Weight is the force that an object exerts on a surface due to its mass and _____.
   a. density
   b. volume
   c. the current temperature
   d. the pull of the Earth's gravity

3. The cubic meter represents a three-dimensional object equivalent to a cube that measures _____.
   a. 10 cm × 10 cm × 10 cm
   b. 100 mm × 100 mm × 100 mm
   c. 100 cm × 100 cm × 100 cm
   d. 10 m × 10 m × 10 m

4. On the Celsius temperature scale, the difference between the freezing and boiling points of water is _____.
   a. 32°C
   b. 100°C
   c. 180°C
   d. 212°C

6.0.0　Introduction to Geometry

Objective

Classify angles and geometric shapes, as well as calculating their areas or volumes.
  a. List each angle type.
  b. Name common geometric shapes and summarize their qualities.
  c. Calculate the area of two-dimensional shapes.
  d. Calculate the volume of three-dimensional shapes.

Performance Tasks

There are no performance tasks in this section.

Geometry might sound complicated, but it is a very practical kind of math. It deals with many familiar things—circles, triangles, squares, and rectangles. The construction industry uses geometry constantly as it works with measurements and shapes. Understanding practical geometry helps every craftworker to carry out daily tasks.

The first part of this section focuses on **plane geometry**. In plane geometry, shapes are two-dimensional (2D). They have length and width only. In contrast, **solid geometry**, covered in the second part, works with three-dimensional (3D) shapes. These have three dimensions—length, width, and height.

Circles: A closed curved line drawn at a constant distance around a central point. A circle measures 360°.

Plane geometry: The mathematical study of two-dimensional (flat) shapes.

Solid geometry: The mathematical study of three-dimensional shapes.

6.1.0 Angles

The construction crafts use the term **angle** frequently. It describes the shape made by two straight lines that meet at a point, called the **vertex**. Angle measurements use the **degree** as the unit, often represented by the ° symbol. Degrees describe the relationship between the two lines. A protractor measures angles. *Figure 24* shows typical angles used in construction.

- **Acute angle** — An angle that measures more than 0° and less than 90°. The most common acute angles are 30°, 45°, and 60°.
- **Right angle** — An angle that measures exactly 90°. The two lines that form the right angle are perpendicular to each other. Imagine the shape of a capital letter L. The sides of the L are perpendicular and form a right angle. The construction industry uses this angle frequently. Plans and drawings indicate right angles with a square symbol at the vertex, as *Figure 24* shows.
- **Obtuse angle** — An angle that measures more than 90° and less than 180°.
- **Straight angle** — An angle measuring exactly 180° (a flat line).
- **Adjacent angles** — These angles have the same vertex and one side in common. Adjacent refers to objects that are next to each other.
- **Opposite angles** — Angles formed by two straight lines that cross. Opposite angles are always equal.

Angle: The shape made by two straight lines coming together at a point. The space between those two lines is measured in degrees.

Vertex: The point at which two or more lines or curves come together.

Degree: A unit of measurement for angles. For example, a right angle measures 90°.

Acute angle: Any angle greater than 0° and less than 90°.

Right angle: An angle that measures 90°. The two lines that form a right angle are perpendicular to each other. This is the angle most used in the crafts.

Obtuse angle: Any angle greater than 90° and less than 180°.

Straight angle: A 180° angle or flat line.

Adjacent angles: Angles that have the same vertex and one side in common.

Opposite angles: Two angles that are formed by two straight lines crossing. They are always equal.

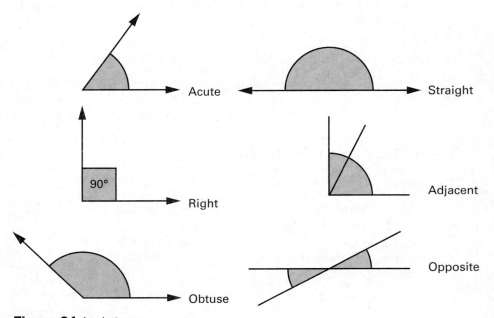

Figure 24 Angle types.

6.2.0 Shapes

Figure 25 shows shapes common to many crafts: rectangles, squares, triangles, and circles.

6.2.1 Rectangle

A **rectangle** is a four-sided shape with four 90° angles. The four angles always add up to 360°. A rectangle has two pairs of equal sides that are parallel to each other. The **diagonal** lines of a rectangle are always equal too. Diagonals are lines connecting opposite corners. Cutting a rectangle on the diagonal gives two identical right triangles, as *Figure 26* shows. All right triangles have one 90° angle.

6.2.2 Square

A **square** is a special rectangle with four equal sides. It too has four 90° angles that add up to 360°. Cutting a square on the diagonal between opposite corners gives two right triangles. Each **right triangle** will have two 45° angles and one 90° angle, as *Figure 27* shows.

Measuring the outside lines of a rectangle or a square determines the **perimeter**. The perimeter is the distance around any two-dimensional figure. The term can also apply to some three-dimensional figures. Craftworkers often measure or calculate a shape's perimeter when estimating material or planning a cut. For example, before installing molding along all four walls of a room, measure the perimeter first. If the room is 14' by 12', purchase 52' of molding (14' + 12' + 14' + 12' = 52'). Another way to calculate it is

$$(2 \times 14') + (2 \times 12') = 52'$$

Since a square has four equal sides, calculate its perimeter by multiplying one side's length by four.

Rectangle: A four-sided shape with four 90° angles. Opposite sides of a rectangle are always parallel and the same length. Adjacent sides are perpendicular and may or may not be equal in length.

Diagonal: Line drawn from one corner of a rectangle or square to the farthest opposite corner.

Square: (1) A special type of rectangle with four equal sides and four 90° angles. (2) The product of a number multiplied by itself. For example, 25 is the square of 5; 16 is the square of 4.

Right triangle: A triangle that includes one 90° angle.

Perimeter: The distance around the outside of a closed shape, such as a rectangle, circle, square, or any irregular shape.

Did You Know?

Rope Stretchers

The word *geometry* comes from two Greek words meaning "earth measuring." One of geometry's earliest applications was surveying—measuring land.

Ancient Egyptians located their farms on the banks of the Nile River. Every year, the Nile overflowed its banks, depositing rich silt on the farmland. But the annual flood destroyed property markers too. So professional surveyors, called rope stretchers, had to measure the land and replace the markers in the correct locations.

The rope stretchers calculated distances and directions using ropes with equally spaced knots tied along their length. Measuring flat land was easy, but dealing with hills and other obstacles was more difficult. Solving these challenges laid the foundations of geometry. Today, geometry remains important to many crafts and professions.

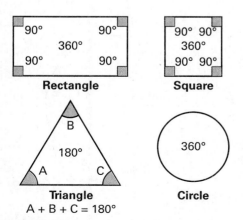

Figure 25 Common plane geometry shapes.

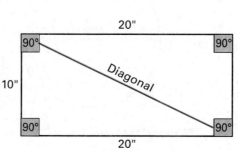

Figure 26 Cutting a rectangle on the diagonal produces two right triangles.

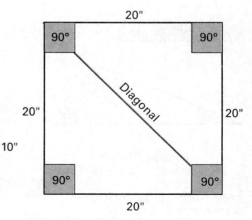

Figure 27 Cutting a square on the diagonal produces two right triangles.

6.2.3 Triangle

A **triangle** is a closed shape that has three sides and three angles. Although the angles in a triangle can vary, they always add up to 180° (*Figure 28*). The following are different types of triangles used in construction:

- *Right triangle* — A right triangle has one 90° angle.
- **Equilateral triangle** — An equilateral triangle has three equal angles and three equal sides.
- **Isosceles triangle** — An isosceles triangle has two equal angles and two equal sides. A line used to **bisect** the triangle (cut it in half from the top to the center of the **base**) creates two adjacent right angles.
- **Scalene triangle** — A scalene triangle has three sides of unequal lengths.

Triangle: A closed shape that has three sides and three angles.

Equilateral triangle: A triangle that has three equal sides and three equal angles.

Isosceles triangle: A triangle that has two equal sides and two equal angles.

Bisect: To divide into two equal parts. For example, when an angle is bisected, the two resulting angles are equal.

Base: As it relates to triangles, the base is the line forming the bottom of the triangle.

Scalene triangle: A triangle with sides of unequal lengths.

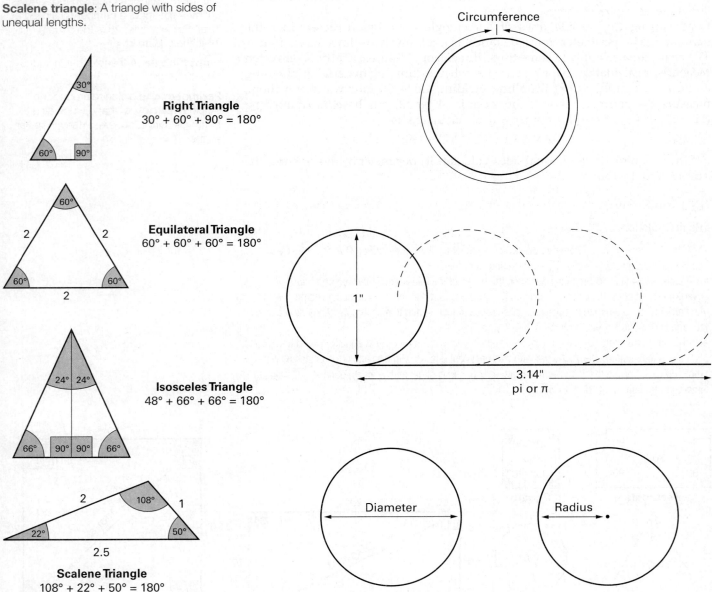

Right Triangle
30° + 60° + 90° = 180°

Equilateral Triangle
60° + 60° + 60° = 180°

Isosceles Triangle
48° + 66° + 66° = 180°

Scalene Triangle
108° + 22° + 50° = 180°

Figure 28 The sum of a triangle's three angles always equals 180°.

Figure 29 Measurements that apply to circles.

6.2.4 Circle

A circle is a closed curved line around a center point. Every point on the curved line is the same distance from the center point. A circle measures 360°. The following measurement terms apply to circles (*Figure 29*):

- **Circumference** — The circumference is the length of the curved line that forms the circle. The **formula** for finding circumference is **pi** (3.14) × **diameter**.

- *Diameter* — The diameter is the length of a straight line that crosses from one side of the circle through the center point to a point on the opposite side. The diameter is the longest straight line possible inside a circle.

- *Pi* or π — Pi is a mathematical constant value of approximately 3.14 (or $^{22}/_7$) used to determine the area and circumference of circles.

- **Radius** — The radius is the length of a straight line from the center point of the circle to any point on the curved line that forms the circle. The radius equals half the diameter.

Circumference: The distance around the curved line that forms the circle.

Formula: A mathematical process used to solve a problem. For example, the formula for finding the area of a rectangle is Side A times Side B = Area, or A × B = Area.

Pi: A mathematical constant value approximately equal to 3.14 (or $^{22}/_7$) used to determine the area and circumference of circles. It is sometimes symbolized by the Greek letter π.

Diameter: The length of a straight line that crosses from one side of a circle, through the center point, to a point on the opposite side. The diameter is the longest straight line that can be drawn inside a circle.

Radius: The distance from a circle's center point to any point on the curved line, or half the width (diameter) of the circle.

6.3.0   Calculating the Area of Shapes

Area refers to the space contained within a two-dimensional object's perimeter. To carpet an office floor or paint its walls, carpet layers and painters must calculate areas before ordering material. Squared measurement units specify the surface area. Inch-pound area measurement units include square inches (*sq in* or in^2), square feet (*sq ft* or ft^2), and square yards (*sq yd* or yd^2). Metric area measurements include square centimeters (*sq cm* or cm^2) and square meters (*sq m* or m^2). For larger areas, such as land measurement, square miles or square kilometers are common.

Area: The amount of space contained in a two-dimensional object, such as a rectangle, circle, or square.

- 1 square inch = 1 in × 1 in = 1 in^2
- 1 square foot = 1 ft × 1 ft = 1 ft^2
- 1 square yard = 1 yd × 1 yd = 1 yd^2
- 1 square centimeter = 1 cm × 1 cm = 1 cm^2
- 1 square meter = 1 m × 1 m = 1 m^2

Craftworkers must be able to calculate the areas of basic shapes. Mathematical formulas make this task simple. *Appendix C* lists the important ones. Become familiar with these now since they appear again and again in future modules. The following are a few significant examples:

- *The area of a rectangle* = length × width. For example, a painter measures a wall 20 feet long and 8 feet high. To calculate its area, multiply 20 ft × 8 ft = 160 ft^2.

- *The area of a square* = length × width. A worker tiles a 12 meter square room. Since all sides of a square are equal, the area is 12m × 12m = 144 m^2.

- *The area of a circle* = pi × radius². This formula includes the mathematical constant pi, which is approximately 3.14. Multiply pi by the radius of the circle squared. To square a number, multiply it by itself. For example, a worker wishing to seal a circular driveway must measure its radius first. If the radius is 20 feet, the area is 3.14 × (20 ft × 20 ft) = 1,256 ft^2.

- *The area of a triangle* = ½ × base × height. The base is the side that the triangle sits on. The height is the length of a line from the base to the triangle's highest point. A carpenter is installing pieces of siding on a wall section. The triangle's base is 2 feet, and its height is 4 feet. The area is ½ × 2 ft × 4 ft = 4 ft^2.

6.3.1   Study Problems: Calculating Area

1. The area of a rectangle that is 8 feet long and 4 feet wide is _____.
   a. 12 ft^2
   b. 22 ft^2
   c. 32 ft^2
   d. 36 ft^2

2. The area of a 16 centimeter square is _____.
   a. 256 cm^2
   b. 265 cm^2
   c. 276 cm^2
   d. 278 cm^2

3. The area of a circle with a 14 foot diameter is _____.
   a. 15.44 ft^2
   b. 43.96 ft^2
   c. 153.86 ft^2
   d. 196 ft^2

4. The area of a triangle with a base of 4 centimeters and a height of 6 centimeters is _____.
   a. 12 cm^2
   b. 24 cm^2
   c. 32 cm^2
   d. 36 cm^2

5. The area of a rectangle that is 14 meters long and 5 meters wide is _____.
   a. 60 m^2
   b. 65 m^2
   c. 70 m^2
   d. 75 m^2

6.4.0   Volume of Three-Dimensional Shapes

Volume is the amount of space contained within three dimensional objects. Calculating volume may require measuring as many as three dimensions. For example, calculating a cube's volume requires its length, width, and height. Some people refer to height as *depth* or *thickness*. Inch-pound volume units include cubic inches (*cu in* or *in³*), cubic feet (*cu ft* or *ft³*), and cubic yards (*cu yd* or *yd³*). Metric measurements include cubic centimeters (*cu cm* or *cm³*) and cubic meters (*cu m* or *m³*).

- 1 cubic inch $= 1 \text{ in} \times 1 \text{ in} \times 1 \text{ in} = 1 \text{ in}^3$
- 1 cubic foot $= 1 \text{ ft} \times 1 \text{ ft} \times 1 \text{ ft} = 1 \text{ ft}^3$
- 1 cubic yard $= 1 \text{ yd} \times 1 \text{ yd} \times 1 \text{ yd} = 1 \text{ yd}^3$
- 1 cubic centimeter $= 1 \text{ cm} \times 1 \text{ cm} \times 1 \text{ cm} = 1 \text{ cm}^3$
- 1 cubic meter $= 1 \text{ m} \times 1 \text{ m} \times 1 \text{ m} = 1 \text{ m}^3$

When calculating volumes, be sure that all measurements are in the same units. For example, a concrete slab has dimensions of 4 ft × 6 ft × 6 in. But multiplying these values to calculate volume will give an incorrect result. Convert feet to inches or inches to feet first. For example, 6 inches equals 0.5 feet. $4 \text{ ft} \times 6 \text{ ft} \times 0.5 \text{ ft} = 12 \text{ ft}^3$.

People have many names for geometric objects like the concrete slab. Some are quite technical and hard to pronounce. To make things easier, this module refers to all such objects as *three-dimensional rectangles*.

Geometry Practical Application

Before erecting a metal building, concrete workers must pour a slab and install anchor bolts. The structure attaches to the slab by the anchor bolts. Since cement comes in cubic yards, a volume measure, the worker must calculate how much cement to use. To determine the slab's volume, multiply its length times its width times its depth (thickness). Alternatively, multiply the slab's area by its depth.

Solution

$$\text{Volume} = \text{area} \times \text{depth}$$
$$\text{Area of slab} = 50' \times 40' = 2{,}000 \ (\text{ft}^2)$$
$$\text{Slab depth} = 6''$$
$$(\text{convert inches to feet:}$$
$$6'' = {}^{6}\!/_{12} = {}^{1}\!/_{2} \ \text{foot})$$
$$\text{Volume} = 2{,}000 \ \text{ft}^2 \times {}^{1}\!/_{2} \ \text{foot} = 1{,}000 \ \text{ft}^3$$
$$\text{Volume of slab} = 1{,}000 \ \text{ft}^3$$

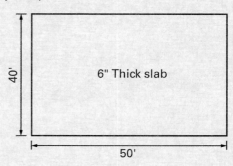

6" Thick slab

40'

50'

Craft professionals often calculate the volumes of three-dimensional spaces. *Appendix C* contains a list of formulas for many objects. The following examples illustrate the more common ones. Do not forget that all units must be the same!

6.4.1 Three-Dimensional Rectangles

Calculate a three-dimensional rectangle's volume by multiplying length×width×depth. Perform the following steps to calculate the volume of *Figure 30*:

Step 1 If necessary, convert all units to a single type.

$$4 \text{ in} = \frac{4}{12} = 0.33333 \text{ ft}$$

Step 2 Determine the volume.

$$20 \text{ ft} \times 8 \text{ ft} \times 0.33333 \text{ ft} = 53.3328 \text{ ft}^3$$

Step 3 If necessary, convert cubic feet to cubic yards.

$$53.3328 \text{ ft}^3 \div 27 \text{ (cu ft per cu yd)} = 1.98 \text{ yd}^3$$

6.4.2 Cubes

A cube is a special three-dimensional rectangle. Calculate its volume in the same way: length × width × depth. However, since all sides are equal in length (*Figure 31*), just multiply the side length by itself three times. Perform the following steps to calculate the volume of *Figure 31*:

Step 1 Determine the volume.

$$8 \text{ ft} \times 8 \text{ ft} \times 8 \text{ ft} = 512 \text{ ft}^3$$

Step 2 If necessary, convert cubic feet to cubic yards.

$$512 \text{ ft}^3 \div 27 \text{ (cu ft per cu yd)} = 18.96 \text{ yd}^3$$

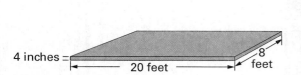

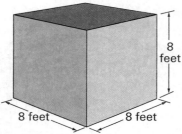

4 inches

20 feet

8 feet

Figure 30 A three-dimensional rectangle.

8 feet   8 feet   8 feet

8 feet

Figure 31 A cube.

Around the World

Pi

Pi is a mathematical constant that describes the relationship between a circle's circumference and its diameter. The value of pi goes on forever without any repeating patterns. In most cases, a few digits suffice for accurate calculations. Craftworkers often round pi to 3.14 or 3.14159. Mathematicians and computer scientists have developed software that calculates pi to trillions of digits. These calculations have little practical value beyond breaking records or testing new software techniques.

6.4.3 Cylinders

Calculate a cylinder's volume using the following formula: $pi \times radius^2 \times height$. This formula uses the area of a circle formula multiplied by the cylinder's height. Perform the following steps to calculate the volume of *Figure 32*:

Area of the circle $= 3.14 \times 11^2$ $= 379.94 \text{ ft}^2$	**Step 1** Determine the circle's area. Since the diameter is 22 feet, the radius is 11 feet (half the diameter).
$379.94 \text{ ft}^2 \times 10 \text{ ft} = 3{,}799.4 \text{ ft}^3$	**Step 2** Calculate the cylinder's volume (circle area $\times$ height).
$3{,}799.4 \text{ ft}^3 \div 27$ (cu ft per cu yd) $=$ 140.72 yd^3	**Step 3** If necessary, convert cubic feet to cubic yards.

6.4.4 Triangular Prisms

A triangular prism is a three-dimensional triangle. Calculate its volume with the following formula: $\frac{1}{2} \times base \times height \times depth$ (thickness). Perform the following steps to calculate the volume of *Figure 33*:

$\frac{1}{2} \times 6 \text{ cm} \times 12 \text{ cm} = 36 \text{ cm}^2$	**Step 1** Calculate the triangle's area.
$36 \text{ cm}^2 \times 2 \text{ cm} = 72 \text{ cm}^3$	**Step 2** Calculate the prism's volume.

6.4.5 Study Problems: Calculating Volume

1. A rectangular shape 5 feet high, 6 feet thick, and 13 feet long has a volume of _____.
   a. 24 ft^3
   b. 43 ft^3
   c. 95 ft^3
   d. 390 ft^3

2. A 3 centimeter cube has a volume of _____.
   a. 6 cm^3
   b. 9 cm^3
   c. 12 cm^3
   d. 27 cm^3

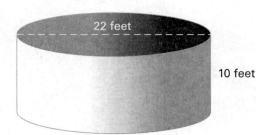

Figure 32 A cylinder.

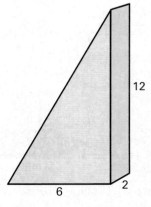

Figure 33 A triangular prism.

3. A triangular prism that has a 6 inch base, a 2 inch height, and a 4 inch depth has a volume of _____.

   a. 12 in³
   b. 24 in³
   c. 36 in³
   d. 48 in³

4. A cylinder that is 6 meters in diameter and 60 centimeters high has a volume of _____.

   a. 16.96 m³
   b. 18.23 m³
   c. 1130.4 m³
   d. 6782.4 m³

5. If a work crew must spread gravel across an area that measures 17 feet square and the gravel layer is to be 6 inches thick, the volume of gravel required is _____.

   a. 3.77 yd³
   b. 5.35 yd³
   c. 102 yd³
   d. 144.5 yd³

Did You Know?

3-4-5 Rule

The building crafts have been using the 3-4-5 rule for centuries. This simple layout method confirms that a rectangular space has true right-angle corners. It requires nothing more than a tape measure. The numbers 3-4-5 represent the triangle's dimensions in any single unit. Any right triangle whose sides are multiples of 3-4-5 will work. Examples include 9-12-15, 12-16-20, and 15-20-25.

To use this method, pick a 3-4-5 triangle that fits within the space being checked. For example, to check the area shown, a 15-20-25 triangle works fine. Measure and mark 15'-0" down the side in one direction. Next, measure and mark 20'-0" down the adjacent side. Finally, measure the diagonal between these points. If it is not exactly 25'-0", the corners are not square (90°).

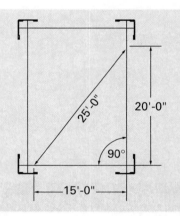

6.4.6 Practical Applications Using Volume

The following examples illustrate practical craft applications that require geometry. Use the provided information to solve each problem. Be sure to show all your work.

1. Calculate the volume of the sheet metal duct pipe shown in *Figure 34*. Round to the nearest cubic inch. _____ in³

2. How many cubic feet of topsoil must a worker remove before pouring the concrete sidewalk shown in *Figure 35*? The sidewalk must be 4" thick and level with the surrounding soil. Round to the nearest cubic foot. _____ ft³

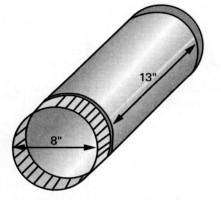

Figure 34 Finding a pipe section's volume.

Figure 35 Removing topsoil for a sidewalk.

6.0.0 Section Review

1. An angle of 66° is a(n) _____.

    a. right angle
    b. straight angle
    c. obtuse angle
    d. acute angle

2. What is the difference between circumference and perimeter?

    a. The circumference applies only to shapes with angles, while the perimeter applies only to circles.
    b. The perimeter applies to all shapes, while the circumference applies only to circles.
    c. The perimeter represents one-half of the circumference.
    d. The circumference is never larger than the perimeter.

3. Calculate the area of a rectangle that is 27.3 meters × 9.3 meters.

    a. 245.7 m^3
    b. 251.1 m^2
    c. 253.89 m^2
    d. 253.89 m^3

4. Calculate the volume of a shipping container that is 26' long × 13'8" wide × 3' deep.

    a. 355.32 ft^3
    b. 355.32 ft^2
    c. 1,065.48 ft^3
    d. 1,065.48 ft^2

Module 00102 Review Questions

1. The number matching the words "two thousand, six hundred eighty-nine" is _____.

    a. 2,286
    b. 2,689
    c. 6,289
    d. 20,689

2. A bricklayer lays 649 bricks the first day, 632 the second day, and 478 the third day. Without using a calculator, determine how many bricks the worker laid during the three-day period.

    a. 1,759
    b. 1,760
    c. 1,769
    d. 1,770

3. A supplier delivered a total of 1,478 feet of electrical cable for a job. Due to a delay, the electricians installed only 489 feet. Without using a calculator, determine how many feet of cable remain.

    a. 978
    b. 980
    c. 989
    d. 1,099

4. A worker has been asked to deliver 15 scaffolds to each of 26 different sites. Without using a calculator, determine how many scaffolds the
   worker will deliver.

   a. 120
   b. 240
   c. 375
   d. 390

5. An insulating company has 400 rolls of insulation that must be equally distributed to 5 different job sites. Without using a calculator, determine how
   many rolls of insulation will go to each site?

   a. 8
   b. 20
   c. 80
   d. 208

6. Which of the following fractions is an equivalent fraction for $\frac{3}{8}$?

   a. $\frac{3}{64}$
   b. $\frac{6}{64}$
   c. $\frac{24}{64}$
   d. $\frac{36}{64}$

7. The lowest common denominator for the fractions $\frac{8}{64}$ and $\frac{8}{32}$ is _____.

   a. 8
   b. 16
   c. 24
   d. 32

8. Perform the following calculation and reduce the answer to its lowest
   term: $\frac{3}{8} + \frac{9}{16} =$ _____.

   a. $\frac{3}{4}$
   b. $\frac{12}{16}$
   c. $\frac{7}{8}$
   d. $\frac{15}{16}$

9. Perform the following calculation and reduce the answer to its lowest
   term: $\frac{11}{32} - \frac{2}{8} =$ _____.

   a. $\frac{3}{32}$
   b. $\frac{1}{8}$
   c. $\frac{9}{32}$
   d. $1\frac{9}{32}$

10. Put the following decimals in order from smallest to largest: 0.402, 0.420,
    0.042, 0.442

    a. 0.420, 0.042, 0.442, 0.402
    b. 0.042, 0.402, 0.420, 0.442
    c. 0.442, 0.420, 0.402, 0.042
    d. 0.042, 0.402, 0.442, 0.420

11. A worker applies two coatings to a pipe. The first coating is 51.5 nanome-
    ters thick, and the second coating is 89.7 nanometers thick. How thick is
    the combined coating?

    a. 141.2 nanometers
    b. 142.12 nanometers
    c. 144.2 nanometers
    d. 145.02 nanometers

12. $3.53 \times 9.75 =$ _____.

    a. 12.28

    b. 34.42

    c. 36.14

    d. 48.13

13. It costs a painting contractor $2.37 per square foot to paint a wall. What will it cost to paint an 864.5 square foot wall? Round to the nearest cent.

    a. $204.88

    b. $2,048.87

    c. $2,848.88

    d. $2,888.86

14. $89.435 \div 0.05 =$ _____.

    a. 1788.7

    b. 17.887

    c. 4.47175

    d. 447.175

15. Without using a calculator, determine the decimal equivalent of 13.9% = _____.

    a. 0.009

    b. 0.013

    c. 0.139

    d. 1.39

16. Without using a calculator, convert 14.75 to its equivalent fraction expressed in lowest terms.

    a. $14^{75}/_{100}$

    b. $29^{5}/_{20}$

    c. $^{59}/_{4}$

    d. $14^{7}/_{4}$

17. The arrow in *Figure RQ01* is pointing to _____.

Figure RQ01

    a. $4^{3}/_{8}$"

    b. $4^{1}/_{4}$"

    c. $4^{1}/_{2}$"

    d. $4^{5}/_{8}$"

18. The arrow in *Figure RQ02* is pointing to _____.

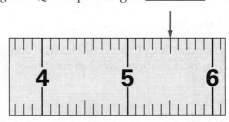

Figure RQ02

a. 5.2 cm
b. 5.3 cm
c. 5.4 cm
d. 5.5 cm

19. The red background around the number 32 in *Figure RQ03* identifies
_____.

Figure RQ03

a. that the tape measure is metric
b. the inch-pound measurement equivalent to one meter
c. the marking for studs placed on 16-inch centers in a wall structure
d. the marking for studs placed on 24-inch centers

20. A centimeter is ¹⁄₁₀₀ of a meter, while a kilometer is equal to 1000 meters.

a. True
b. False

21. To convert a measurement in centimeters to meters, move the decimal
point _____.

a. two places to the left
b. three places to the right
c. three places to the left
d. four places to the right

22. Convert 67 inches to centimeters.

a. 17.01 cm
b. 26.38 cm
c. 170.18 cm
d. 263.80 cm

23. A pound of water weighs more than a kilogram of water.

a. True
b. False

24. Convert the weight of 49 ounces to grams.

a. 1.73 grams
b. 3.06 grams
c. 747.42 grams
d. 1389.15 grams

25. Convert 15°F to Celsius.

a. −9.4°C
b. 9.4°C
c. −59°C
d. 59°C

26. The two angles shown in *Figure RQ04* are _____.

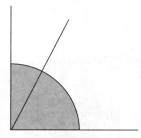

Figure RQ04

   a. obtuse and adjacent
   b. obtuse and opposite
   c. right and opposite
   d. acute and adjacent

27. The sum of the four angles in a rectangle is always 180°.

   a. True
   b. False

28. The mathematical constant pi is needed for calculating the area of a _____.

   a. circle
   b. square
   c. rectangle
   d. cube

29. Using a calculator, calculate the volume of a cylindrical tank that is 23 feet high with a radius of 6.25 feet. Round the answer to the nearest cubic foot. _____

   a. 144 ft^3
   b. 898 ft^3
   c. 903 ft^3
   d. 2,821 ft^3

30. Using a calculator, calculate the volume of a three-dimensional triangular prism that has a 7 centimeter base, a 4 centimeter height, and a 30 millimeter depth. Remember to convert all measurements to the same unit before applying the formula. _____

   a. 42 cm^3
   b. 42 mm^3
   c. 82 mm^3
   d. 82 cm^3

Answers to Section Review Questions and Module Review Questions are found in *Appendix A*.

Introduction to Hand Tools

Source: Zachary McNaughton,
River Valley Technical Center

Objectives

Successful completion of this module prepares you to do the following:

1. Name common hand tools and state how to use them.
    a. Identify various hammers and demolition tools and explain how to use them.
    b. Describe chisels and punches and how they are used.
    c. Match screwdrivers to the appropriate hardware.
    d. Differentiate between non-adjustable, adjustable, and socket wrenches.
    e. Describe various types of pliers and explain how they are used.
2. Identify common measurement and layout tools and describe how to use them.
    a. Explain how to use a variety of measuring tools.
    b. Define various types of levels and layout tools and indicate how they are used.
3. Identify and describe other hand tools common to shops and job sites.
    a. Differentiate between various handsaws and their designated applications.
    b. Identify common clamp designs.
    c. Explain how different files and utility knives are used with various materials.
    d. Describe shovels and picks and the tasks for which each one is best suited.

Performance Task

Under supervision, you should be able to do the following:

1. Inspect and demonstrate the safe and proper use of the following hand tools:
    - Hammers
    - Demolition tools
    - Chisels and punches
    - Screwdrivers
    - Adjustable wrenches
    - Non-adjustable wrenches
    - Sockets
    - Pliers
    - Tape measures
    - Levels
    - Squares
    - Handsaws
    - Clamps
    - Files
    - Utility knives
    - Shovels

Overview

Every profession has its tools. A surgeon uses a scalpel, an instructor uses a whiteboard, and an accountant uses a calculator. The construction crafts require a broad array of hand tools. Even if you are familiar with some of the tools, all craftworkers need to learn how to select, maintain, and use them safely. A quality hand tool may cost more up front, but if it is properly used and maintained, it will last for years. A true craft professional invests wisely in hand tools, and uses, maintains, and stores them with the same wisdom.

Industry Recognized Credentials

If you are training through an NCCER-accredited sponsor, you may be eligible for credentials from NCCER's Registry. The ID number for this module is 00103. Note that this module may have been used in other NCCER curricula and may apply to other level completions. Contact NCCER's Registry at 888.622.3720 or go to **www.nccer.org** for more information.

1.0.0 Common Hand Tools

Performance Task

1. Inspect and demonstrate the safe and proper use of the following hand tools:
   - Hammers
   - Demolition tools
   - Chisels and punches
   - Screwdrivers
   - Adjustable wrenches
   - Non-adjustable wrenches
   - Sockets
   - Pliers

Objective

Name common hand tools and state how to use them.

a. Identify various hammers and demolition tools and explain how to use them.
b. Describe chisels and punches and how they are used.
c. Match screwdrivers to the appropriate hardware.
d. Differentiate between non-adjustable, adjustable, and socket wrenches.
e. Describe various types of pliers and explain how they are used.

Construction craftworkers require a combination of knowledge and practical skills to be successful. While there are many things that can be done with hands alone, every craft must use hand tools to get the job done. How craftworkers handle and care for their tools often demonstrates their level of skill and respect for the craft.

Many unique hand tools are used across the crafts—far too many to present in a single module. You will likely be exposed to more hand tools in the early stages of your craft training program. This module presents some of the most common hand tools craftworkers are likely to use. Regardless of your chosen craft, some of these hand tools may become constant companions.

Hand tools are often grouped by their purpose. In each group, there are different physical characteristics that determine how each similar tool is applied to the job at hand. To use each tool properly and safely, it is important to know the purpose and characteristics. This section presents common hand tools used in construction crafts, including:

- Hammers and demolition tools
- Chisels and punches
- Screwdrivers
- Wrenches
- Pliers and wire cutters

In the construction environment, some personal protective equipment (PPE) must be worn consistently on the job site. This includes safety glasses, safety shoes, hard hats, and gloves. This protective gear is typically required on every job site for your protection.

Also remember that, when working at height, all tools must be handled carefully so that they do not fall and injure someone below. A tool should be connected to a tool belt with a **lanyard** to prevent injury to others and damage to the work below.

Lanyard: A cord, rope, or similar material used to limit the travel of two connected objects.

1.1.0 Hammers and Demolition Tools

Hammers are made in different sizes and weights for specific types of work. The safest hammers are those with heads made of special alloys or drop-forged steel. Titanium hammers are also becoming popular due to their light weight and durability.

Some of the most common hammers are the claw hammer, the drywall hammer, and the ball peen hammer. Demolition tools include various styles of sledgehammers and nail pullers. Several examples of each tool will be presented in this section.

1.1.1 Claw Hammers

The claw hammer (*Figure 1*) typically has a steel head and a handle made of wood, steel, fiberglass, and graphite. Claw hammers are generally associated with carpentry and similar crafts. The head is used to drive a nail, wedge, or **dowel**, and the claw is used to pull nails out of wood. The nailing face of the hammer may be slightly rounded or have a milled or waffled surface. While some specialized hammers do have a flat face, such faces are rarely found on hammers designed for driving nails.

A flat hammer face would leave deeper depressions when you drive the head of a nail **flush** with the surface of the work. A slightly rounded (convex) face allows a skilled worker to drive a nail head flush without seriously damaging the surface of the work. A milled or waffled nailing face is another type of hammer face (*Figure 2*). The milled face is generally reserved for framing hammers, where the marks it leaves will be permanently concealed behind finish work.

There are several common types of claw hammers. Finish hammers (*Figure 1*) are designed for lighter work such as finish carpentry, as the name implies. They are well suited for driving slender nails without **overstriking** or bending them. Most have curved claws and are around 16–20 ounces (oz), or 454–567 grams (g), in weight. Smaller versions for hammering brads or small finish nails may be as light as 7 oz (198 g).

Framing hammers (*Figure 3*) are often heavier than common finish hammers, since heavier blows are often needed for the larger nails and thicker material used in wood framing. The additional weight is in the head. A framing hammer

Dowel: A pin or rod, usually round, that fits into a corresponding hole to fasten or align two workpieces.

Flush: As an adjective, to be even or at the same level with another object. For example, when a nail is fully seated, the top of the head is flush (even) with, or slightly beyond, the surface of the wood.

Overstriking: When a hammer misses the head of a nail, with the head going past it and allowing the hammer handle to impact the nail head instead. Overstriking is a common cause of hammer damage.

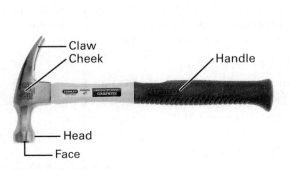

Figure 1 Claw hammer.
Source: Image property of Stanley Black & Decker. Used with permission

Figure 2 Milled and smooth nailing faces.
Source: ATU Images/The Image Bank/Getty Images

Level: Used as an adjective in this context, meaning flat or even with another surface or in relation to the Earth's horizon.

Figure 3 Framing hammer.
Source: Image property of Stanley Black & Decker. Used with permission

(A) Curved Claw Hammer

(B) Rip Claw Hammer

Figure 4 Curved claw and rip claw hammers.
Source: Image property of Stanley Black & Decker. Used with permission

may weigh as much as 28 oz (794 g). They also have longer handles to generate more power. Framing hammers may also have a milled face, as it helps prevent the face from slipping off the nail on impact.

Hammers also differ by the design of the claws (*Figure 4*). A curved claw hammer provides some additional leverage for pulling nails. Rip claw hammers have flatter claws and are best for ripping boards and woodwork apart.

Driving a nail with a claw hammer is simple using the following technique, although practice is required to develop rhythm and accuracy:

- Don the appropriate PPE, including safety glasses.
- Hold the nail in position and rest the face of the hammer on it to give you a visual starting place.
- Draw the hammer back and lightly tap the nail to start it.
- Once the nail is started, move your fingers away.
- Hold the hammer **level** with the head of the nail, using the motion of the wrist, elbow, and shoulder to strike the nail squarely in the center of the hammer face.

The hammer is designed to produce a certain amount of force on the object it strikes. If you hold the hammer incorrectly, the benefits of its design are not evident. The distance between your hand and the head of the hammer affects the force you apply to drive a nail. The closer to the head you hold the hammer, the harder you will need to swing to achieve the desired force. Hold it with the end of the handle even with the lower edge of your palm. Make it easier on yourself by holding the hammer properly; it takes less effort to drive the nail.

Pulling a nail with a claw hammer is as easy as driving one:

- Slip the claw of the hammer under the nail head.
- Use the leverage of the hammer to pry the nail up until the hammer's handle is nearly straight up (vertical), partially drawing the nail out of the wood.
- Try to pull the nail straight up from the wood. If it is too tight and more leverage is needed, place a small block of wood between the hammer head and the wood to elevate it, then repeat the prying motion.

Longer and larger nails require more effort to remove, depending on the material in which they are driven. Placing a small block of wood beneath the head when pulling a nail also protects the workpiece from being scarred by the hammer.

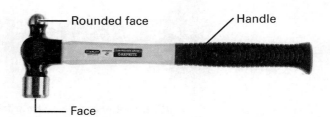

Figure 5 Drywall hammer.
Source: Image property of Stanley Black & Decker. Used with permission

Figure 6 Ball peen hammer.
Source: Image property of Stanley Black & Decker. Used with permission

1.1.2 Drywall Hammer

A **drywall** hammer (*Figure 5*) is used for driving nails in drywall, also referred to as *sheet rock*. Instead of claws, the opposite end of the hammer is shaped like a small hatchet.

Drywall hammers are lighter than normal hammers, generally weighing 12–14 oz (340–397 g). This light weight helps prevent deep indentations into the drywall. The nailing face is more rounded than other hammers as well, for the same reason. The hatchet-like tail of the hammer was originally used by plasterers to cut **lath**. However, it is equally valuable for drywall work. It can cut through small woodwork, and the U-shaped area is used to help grab and carry drywall sheets.

1.1.3 Ball Peen Hammer

A ball peen hammer (*Figure 6*) is primarily used in metalworking. It has a flat face for striking and a spherical or hemispherical head for **peening** (forming) metal or rivet heads. This hammer is also used with chisels and punches. In **welding** operations, the ball peen hammer is used to reduce stress in a fresh weld by peening, or repeatedly striking, the **joint** as it cools.

Ball peen hammers are also referred to as *engineer's hammers* or *machinist's hammers*. They are classified by weight, which ranges from 4 oz to 2 pounds (lb), or 113 g to 0.9 kilograms (kg).

It is best to strike material with a ball peen hammer in a controlled manner. Heavy blows are not often required. Never use a ball peen hammer to drive nails, because the head of the hammer is made from a milder steel. Continual pounding on nails can deform or damage the face. Ball peen hammers and hammers for woodworking are not interchangeable.

1.1.4 Sledgehammers

A sledgehammer is a heavy-duty tool that is used to drive posts or take on demolition work. You can also use it to break up cast iron or concrete. The head of the sledgehammer is made of high-carbon steel and weighs 2 to 20 lb (0.9 to 9 kg).

The shape of the head depends on the job the sledgehammer is designed to do. Sledgehammers can be either long-handled or short-handled. *Figure 7* shows three styles of sledgehammers.

Drywall: A large, flat board made of layers of fiberboard and gypsum used primarily for wall construction and finishing. Drywall, also known as *sheet rock* and *wallboard*, is typically manufactured in 4' × 8' panels and readily accepts paint.

Lath: A thin, flat strip of wood used to form latticework or a foundation for wall plaster.

Peening: The process of bending, shaping, or cutting metal by striking it with a tool.

Welding: The process of heating two or more metal workpieces to their liquid state and fusing them to create a very strong joint.

Joint: The intersection of workpiece components where a physical connection is generally made. Joints are often named by the style of connection, such as butt joints and miter joints.

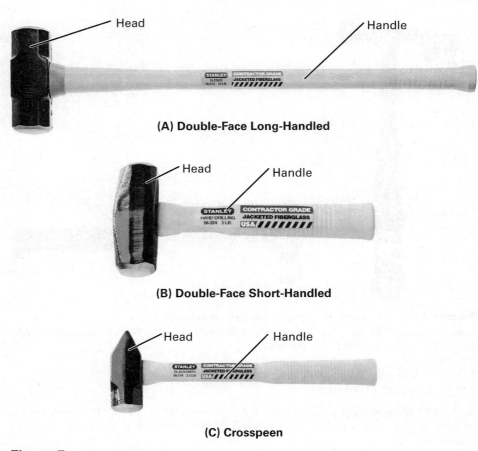

(A) Double-Face Long-Handled

(B) Double-Face Short-Handled

(C) Crosspeen

Figure 7 Sledgehammers.
Source: Image property of Stanley Black & Decker. Used with permission

A sledgehammer can cause injury to you or to anyone working near you. Follow these steps to use a sledgehammer properly and safely:

Step 1   Wear the appropriate PPE, including gloves, eye protection, and safety shoes.
Step 2   Inspect the sledgehammer to ensure that there are no defects.
Step 3   Be sure there is no one nearby who could be struck on the back swing.
Step 4   Hold the sledgehammer with both hands apart (hand over hand).
Step 5   Stand directly in front of the object to be driven.
Step 6   Lift the sledgehammer straight up above the target.
Step 7   Set the head of the sledgehammer on the target.
Step 8   Begin delivering short blows to the target and gradually increase the length and force of the stroke.

WARNING!

Do not swing a sledgehammer beyond vertical and over your head. Doing so may cause a back injury and significantly reduce control over the tool.

Hold the sledgehammer with both hands. Never use your hands to hold an object while someone else drives with a sledgehammer. Doing so could result in serious injury, such as crushed or broken bones.

1.1.5   Nail Pullers

An existing wooden structure often needs to be disassembled. Nail pullers and wrecking bars (*Figure 8*) are tools for the task.

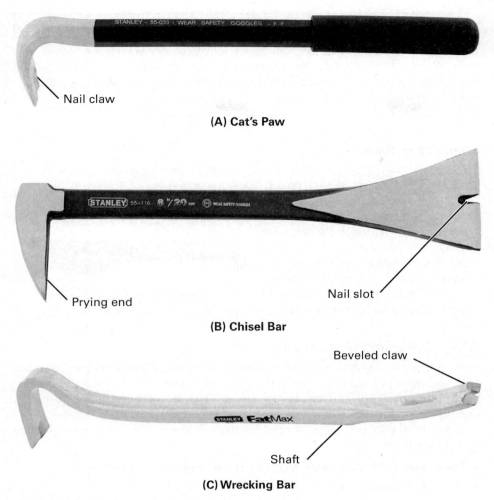

Nail claw

(A) Cat's Paw

Prying end

Nail slot

(B) Chisel Bar

Beveled claw

Shaft

(C) Wrecking Bar

Figure 8 Nail pullers and wrecking bars.
Source: Image property of Stanley Black & Decker. Used with permission

Three common nail-pulling tools are the cat's paw, the chisel bar, and the wrecking bar. There are many other names used by craftworkers to describe these tools. The cat's paw (*Figure 8A*) is a straight steel rod with a curved claw at one end. It is used to pull nails that have been driven flush with the surface of the wood or slightly below it. The cat's paw is used to pull the nail head just above the surface so another tool can completely pull it out. A cat's paw may lack the leverage needed to completely remove a long nail due to the tight curve at the head.

A chisel bar often has a claw at one end like a cat's paw, but the other end is flattened into a broad chisel-like head with a long **bevel** (*Figure 8B*). The chisel bar can be used like a claw hammer to pull a nail. The chisel-shaped end can be driven into wood joints to split and rip apart the pieces, or to get beneath a nail head.

Wrecking bars (*Figure 8C*), also called *ripping bars* or *wonder bars*, have a nail slot at the end to pull nails. An angled, wedge-shaped face at the other end is used like a chisel as well as a prying tool. They are used for dismantling woodwork, such as demolishing wooden framing, opening crates, and disassembling pallets. Use the straighter end to force wood joints apart and use the claw to pull large nails and spikes. Wrecking bars are generally 12" to 42" (31 cm to 107 cm) long. Longer bars provide more leverage and pulling power.

Before using a ripping bar or similar tool, make sure that you are wearing the appropriate PPE. Eye protection is extremely important due to the risk of flying debris. Before prying on wood or a nail, ensure that you are in a stable position and prepared for the release of resistance against your effort.

Bevel: The inclination of a surface that intersects another surface at any angle less than 90 degrees; also used as a verb, meaning to cut or trim an edge to an angled profile.

> **WARNING!**
>
> Wear a hard hat, safety glasses, and gloves for protection when using these tools. When manipulating the bar, never pull the bar toward your face. Instead, pull the bar toward your shoulder. Do not push hard away from your body, as this may cause you to lose your balance.

Demolition Tools

Many tools on the market today are designed for demolition. Small design differences often make certain tools work better for a specific task than another style.

The Stanley Fubar®, for example, combines the advantages of a pry bar with those of a hammer. Three different striking surfaces allow the tool to be driven under or between parts for disassembly. It can be used to pull nails several different ways and can be used to chop through softer materials such as drywall. Tougher materials can be chopped by first driving the chisel end through and then using the chopping edge. This type of tool has many variations.

Source: Image property of Stanley Black & Decker. Used with permission

1.1.6 Safety and Maintenance

No matter what the job is, to avoid accidents and injuries a safe work environment and safe work practices are extremely important. Don the required PPE before beginning any work and always be aware of your surroundings before and during the performance of a task.

Consider the following guidelines for the safe use and maintenance of hammers and similar tools:

- Make sure the handle is set securely in the head of a hammer.
- Replace cracked or broken handles.
- Make sure the face of the hammer is clean.
- Hold the hammer properly. Grasp the handle firmly near the end and hit the nail squarely.
- Do not hit with the cheek or side of the hammer head unless the hammer has been designed for that purpose.
- Do not use hammers with chipped or otherwise damaged heads.
- Do not use a hammer with a cast head.
- Never strike hammer heads together.

The following are additional considerations for safety and maintenance when working with sledgehammers:

- Replace cracked or broken handles before you use a sledgehammer.
- Use the right amount of force for the job.
- Keep your hands away from the object being driven.
- Never swing until you have checked behind you to make sure you have plenty of clearance from both people and objects.

The following are safety guidelines for ripping and nail pulling:

- Use two hands when ripping to enhance control.
- To prevent injury, be sure the material holding the nail is braced securely before pulling.

Many accidents with prying tools occur when a pry bar slips, and the worker falls. Be sure to maintain your balance and keep a firm grip on the tool. This technique also helps reduce damage to materials that must be reused, such as concrete forms.

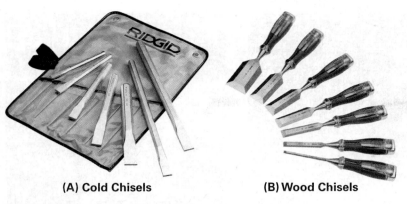

(A) Cold Chisels   **(B) Wood Chisels**

Figure 9 Cold and wood chisels.
Source: RIDGID® (9A); Image property of Stanley Black & Decker. Used with permission (9B)

Figure 10 Cape chisel.
Source: Mayhew Steel Products, Inc.

1.2.0   Chisels and Punches

Chisels are used to cut and shape wood, stone, or metal. Punches are used to indent metal, drive pins, and align holes.

1.2.1   Chisels

A chisel is a metal tool with a sharpened, beveled (sloped) edge. Wood chisels and cold chisels (*Figure 9*) are presented in this section. Both types are made from steel that is heat-treated to make it harder. Technically, a chisel can cut any material that is softer than the steel of the chisel. However, wood chisels are used exclusively on wood, while cold chisels are used on metal. The term *cold chisel* stems from the use of the tool with cold metal, as opposed to metal heated by a torch or furnace. Both cold and wood chisels are manufactured in a variety of shapes and sizes to support specific tasks.

Cold chisels are designed for hard materials, such as steel and stone. The flat cold chisel is the most common. Flat cold chisels have a tip ground to a 60-degree angle. They are often used to cut through bolts and similar hardware that is rusted or damaged and cannot be removed with wrenches.

The cape chisel (*Figure 10*) is manufactured so the cutting tip is slightly wider than the body. This feature keeps the chisel from binding in the cut when chiseling grooves. Cape chisels are used for cutting single grooves as well as cutting a series of parallel grooves when excessive material needs to be cut from a surface. The round nose (half-round) chisel is generally used for specialized work like forming flutes and channels. The diamond-pointed chisel has a square section at the tip with a single bevel. This chisel is typically used for cleaning up the internal angles of channels and chipping through thin plates.

Before using a cold chisel, it is important to wear the appropriate PPE, especially proper eye protection. The first step to using a cold chisel is to secure the object to be cut in a vise. Place the blade of the chisel at the spot where the material is to be cut. Strike the chisel with a ball peen hammer to force the chisel into and through the material. Repeat as necessary.

There are also several different styles of wood chisel. Bevel-edged chisels (*Figure 9*) are easy to push into corners because of the slightly angled edges. This chisel is commonly used for completing or cleaning up dovetail joints. Stronger wood chisels have a rectangular cross-section blade and are typically used for heavier work requirements due to their strength.

Paring chisels are longer and thinner than standard wood chisels. They can reach deeper into long joints to clean them for a precise fit. Note that most paring chisels are designed for hand-use only; they should not be struck with a hammer and must be kept very sharp to work properly. If the handle is wooden or plastic with no obvious metal striking surface on the end, it should not be struck.

Before using a wood chisel, it is important to wear the appropriate PPE, especially proper eye protection. First, outline the area to be chiseled out. Next, set the chisel at one end of the outline, with its edge on the cross-grain line and the bevel facing the recess to be made. Lightly strike the chisel head with a mallet.

Did You Know?

Stonemasons

Stone has always been an important building material. Because of its strength, stone has often been used for dams, bridges, fortresses, foundations, and monumental buildings. Today, steel and concrete have replaced stone as basic construction materials. Stone is now used primarily as sheathing for buildings, for flooring in high-traffic areas, and for decorative uses.

A stonemason's job requires precision. Stones have uneven, rough edges that must be trimmed and finished before each stone can be set. The process of trimming projections and jagged edges is called dressing the stone. This requires skill and experience using specialized hand tools, such as chisels and sledgehammers. Many craftworkers consider stonework to be an art form.

Stonemasons build stone walls, and they also set stone exteriors and floors, working with natural cut and artificial stones including marble, granite, limestone, cast concrete, marble chips, and other masonry materials.

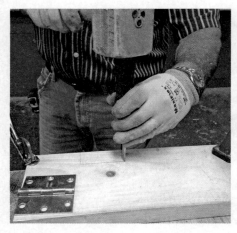

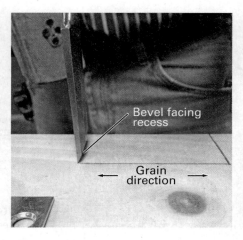

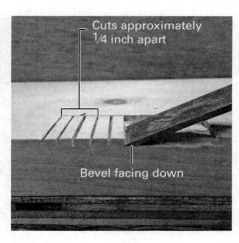

Figure 11 Proper use of a wood chisel.
Source: Cianbro Corporation

Repeat this process at the other end of the outline, again with the bevel of the chisel blade toward the recess. Then make a series of cuts approximately ¼" apart from one end of the recess to the other.

To pare (trim) away the notched wood, hold the chisel bevel-side down to slice inward from the end of the recess (*Figure 11*). Remember that most paring chisels should not be struck with a hammer. The palm of your hand can be used at times to lightly bump the chisel ahead.

1.2.2 Punches

A punch (*Figure 12*) uses the impact of a hammer to indent metal to start a drill bit, drive pins, and align holes in parts that are being mated. Punches are made of hardened and **tempered** steel.

Tempered: Treated with heat to enhance or restore steel hardness.

Three common types of punches are the center punch, the prick punch, and the tapered, or straight, punch. The center and prick punches are both used to mark points on metal. The center punch is usually heavier with a tip ground at a 60-degree angle. The prick punch tip usually has a shallower angle and makes a smaller impression. The wider depression of the center punch makes it best for marking points for drilling, while the prick punch is best for general layout work on sheet metal.

The tapered punch (*Figure 12*) has a slight taper along much of its length. Tapered punches typically have a blunt end rather than a point and are used to drive out pins or align the holes of mating parts for assembly.

To use a punch, hold it straight with one hand in the desired position and strike it squarely with a hammer.

1.2.3 Safety and Maintenance

Follow these guidelines for the safe use and proper maintenance of chisels and punches:

- Always wear the appropriate PPE, including work gloves and safety goggles.
- Make sure the wood chisel blade is beveled at a precise 25-degree angle so it will cut well.

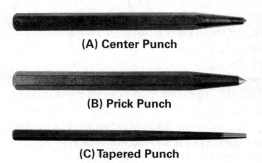

Figure 12 Punches.
Source: Image property of Stanley Black & Decker. Used with permission

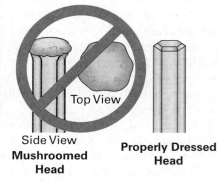

Figure 13 Chisel damage.

- Ensure a cold chisel blade is beveled at a 60-degree angle.

- Sharpen the cutting edge of a chisel on an oilstone to produce a keen edge.

- Inspect the point of the punch to ensure it is sharp and not damaged in any way. Punches can easily be sharpened with a proper bench grinder.

- Do not use a chisel, punch, or hammer with a head that has become **mushroomed** or flattened (*Figure 13*). Also shown is a properly dressed head, showing the bevel around the top. Repairing mushroomed heads is best done with a bench grinder.

- Ensure the right tool for the job is used.

Mushroomed: Refers to the head of a tool that is struck or used for striking when the head becomes wider and distorted, like the head of a mushroom.

WARNING!

Striking a chisel or punch that has a mushroom-shaped head can cause metal chips to break off. These flying chips can cause serious eye injuries. If a chisel has a mushroom-shaped head, remove the tool from service until it is repaired or replaced according to company policy.

1.3.0 Screwdrivers

Screwdrivers are used to drive or remove screws. They are identified by the type of screw they fit. The most common screwdrivers are slotted, also known as *straight, flat,* or *standard* tip, and Phillips head. Specialized screwdrivers include the clutch-drive, Torx®, Robertson®, and Allen® head, all of which are designed to fit the head of fasteners with the same name. The three latter heads offer greater grip and allow more **torque** to be applied than slotted or Phillips screws, without **stripping** the head. *Figure 14* shows six common types of screws.

The screwdrivers designed to fit common screws are described here:

- *Slotted* — A very common type of screwdriver, but it also provides the least amount of grip in the screw head, limiting the amount of torque that can be applied.

- *Phillips* — The most common type of crosshead screwdriver, offering better grip over the slotted design.

- *Clutch-drive* — Has an hourglass-shaped head that provides good grip; common where security from vandalism is desirable.

- *Torx®* — Has a star-shaped tip, widely used for mechanical assembly and repair work; also popular in residential construction.

- *Robertson® (square)* — Has a square drive that provides extra grip to apply torque. Like the Torx®, the Robertson® drive has also become popular in residential construction, for both interior and exterior applications.

- *Allen® (hex)* — Works with socket-head screws that have a female, hex-shaped (six-sided) recess. Allen® wrenches, which are typically L-shaped steel bars, are generally referred to as wrenches instead of screwdrivers.

Screwdrivers may be color-coded according to their size, to make visual identification easier.

To choose the right screwdriver and use it correctly, it helps to know something about screwdriver construction. Each section has a name, as shown in *Figure 15*. The handle is designed to give you a firm grip. The shank is the metal portion between the handle and the blade. The blade is the formed end that fits into the head of a screw.

Industrial screwdriver blades are made of tempered steel to resist wear and to prevent bending and breaking. Good-quality screwdrivers have shanks that run through the length of the handle. Some also have a tip that is hardened, resulting in the darker tips shown in *Figure 15*.

It is important to choose the right screwdriver for the screw. The blade should fit snugly into the screw head and not be too long, too short, or loose (*Figure 16*). If you use the wrong size blade, you might damage the screwdriver or the screw head.

Torque: The rotating or twisting force applied to an object such as a fastener or the shaft of a motor.

Stripping: Damaging or distorting the head or threads of a fastener.

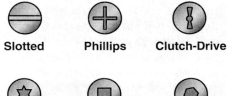

Slotted   Phillips   Clutch-Drive

Torx®   Robertson®   Allen

Figure 14 Common screw head types.

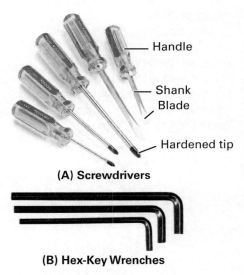

Handle

Shank

Blade

Hardened tip

(A) Screwdrivers

(B) Hex-Key Wrenches

Figure 15 Screwdrivers and hex-key wrenches.
Source: RIDGID® (15A); Klein Tools, Inc. (15B)

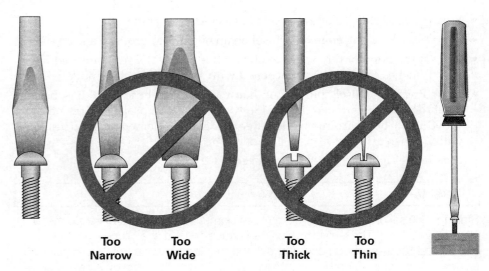

Figure 16 Proper use of a screwdriver.

Note that there are different sizes of each screw type. In addition, there are several styles of screws that are commonly called Phillips screws, but with slight differences. A Reed and Prince screw, for example, looks like a Phillips head but comes to a sharper point at the socket base. The base of a Phillips head screw slot is slightly rounded. Using the wrong screwdriver on these two styles will often result in slippage and damage to the screw.

It is very important to use a screwdriver correctly. Using one incorrectly can damage the screwdriver, screw head, or both. Wear work gloves to protect your hands. Once the right size and type of blade for the screw head is chosen, position the shank perpendicular (at a right angle) to the work. Apply firm, steady pressure to the screw head and turn clockwise to tighten, or counterclockwise to loosen. Do not allow the screwdriver bit to rotate inside the screw head, as it will damage the head.

> ### Screws vs. Nails
>
> Screws hold better than nails in most situations. The spiral ridges, called *threads*, help hold the screw tightly inside the material, unlike the smooth surface of most nails. Self-tapping screws end in a sharp point and have sharper threads. These screws cut their own threads in the material, eliminating the need to drill a starter hole. In woodworking, however, making a small starter hole with a drill helps keep the wood from splitting. This is especially important in cabinetry, trim work, and furniture construction.

1.3.1   Safety and Maintenance

Properly maintaining a screwdriver not only gives it a longer life, it also helps to prevent injuries. The following are guidelines for the proper use and maintenance of screwdrivers:

- Keep screwdrivers free of dirt, grease, and grit so the blade will not slip out of the screw-head slot, causing injury or equipment damage.
- Visually inspect a screwdriver before using it. If the handle is worn or damaged, or if the tip is not straight and smooth, the screwdriver should be repaired or replaced.
- Never use a screwdriver as a punch, chisel, or pry bar.
- Do not use or lay a screwdriver near electrically energized conductors.
- Do not expose screwdriver handles to excessive heat.
- Carry screwdrivers carefully with the tip down.

> ### Did You Know?
>
> #### Types of Screws
>
> The demand for different screw designs came about to resolve specific needs in the workplace. Phillips head screws were developed with four-point contact so that a higher torque could be applied. However, there was always room for more improvements. Torx® head screws were developed to be compatible with robotic assembly line equipment used in production applications. A pin was later added in the center to add to its tamper resistance. To drive the screw shown here, a special bit with a hole in the center to fit over the pin is required.
>
> Tamper-resistant screws, such as the Torx® and the pig nose shown here, were developed to resist vandalism to public facilities, such as school and park restrooms. Many variations of these designs have been developed over the years, and more unique designs continue to be introduced.
>
>
>
> **Tamper-Resistant Torx® Screw**
>
>
>
> **Pig Nose**

1.4.0 Wrenches

Wrenches are used to hold and turn nuts, bolts, pipes, and similar objects that require leverage to turn. There are many types of wrenches, but they fall into two main categories: non-adjustable and adjustable.

Non-adjustable wrenches fit only one size of nut or bolt. Wrenches and hardware are available in both Imperial and metric sizes. An adjustable wrench can be adjusted to fit different sizes of nuts and bolts.

1.4.1 Non-Adjustable Wrenches

Non-adjustable wrenches (*Figure 17*) include the open-end wrench, the box-end wrench, the ratcheting box-end wrench, and the combination wrench.

The open-end wrench is one of the easiest wrenches to use. The open end allows you to position the tool around the fastener easier than a box-end wrench, but with less grip. Most open-end wrenches offer two different sizes in a single wrench, such as ⁷⁄₁₆" and ½". Note that metric equivalents are not offered in this context, as metric wrenches are sized by the millimeter (mm) and there is no functional equivalency between metric and Imperial wrenches. These sizes refer to the distance between the **flats**—the straight sides of the wrench opening.

The size of the wrench must match the distance across the head of a fastener. For example, a ½" bolt head requires a ½" wrench. Note, however, that this does not refer to the size of the bolt itself. Bolts are sized according to the nominal diameter of the threaded shaft, not the head. A ⁵⁄₁₆" bolt typically has a ½" head, while a ³⁄₈" bolt has a ⁹⁄₁₆" head. You will learn more about hardware later in your training. If you work with hardware often in your chosen craft, their sizes and specifications will quickly be committed to memory.

Box-end wrenches form a continuous perimeter around the head of a fastener. The wrench opening typically has 6 or 12 flats. A hexagonal shape, typical of bolts and nuts, has 6 sides. Wrenches with 6 flats are less likely to round-off and damage the flats of a bolt or nut. Although wrenches with 12 flats offer slightly less security, the advantage is that there are 12 possible positions for the wrench to settle onto the fastener instead of just 6. In tight quarters, this is a significant advantage.

Like open-end wrenches, box-end wrenches typically fit two different sizes of hardware. Box-end wrenches offer a firmer grip than open-end wrenches. A box-end wrench is also safer because it will not slip off the fastener quite so easily. Box-end wrenches with a ratcheting feature are popular, but are generally shorter in length, reducing the leverage. Ratcheting box-end wrenches are excellent for assembly work but are not the best choice for freeing corroded or tight hardware assemblies.

Combination wrenches are, as the name implies, a combination of two types of wrenches. One end of the combination wrench is open and the other is a box-end. Each combination wrench generally fits only one size of fastener; both ends are the same size.

Flats: The flat, straight sides of a wrench opening; also, the sides on a nut or bolt head.

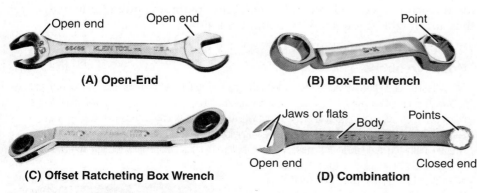

(A) Open-End

(B) Box-End Wrench

(C) Offset Ratcheting Box Wrench

(D) Combination

Figure 17 Non-adjustable wrenches.

Source: Klein Tools, Inc. (17A); Courtesy of SK Professional Tools (17B); Snap-on Incorporated (17C); Image property of Stanley Black & Decker. Used with permission (17D)

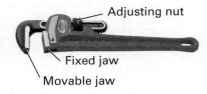

- Adjusting nut
- Fixed jaw
- Movable jaw

(A) Pipe Wrench

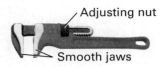

- Adjusting nut
- Smooth jaws

(B) Spud Wrench

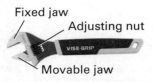

- Fixed jaw
- Adjusting nut
- Movable jaw

(C) Adjustable Wrench

Figure 18 Adjustable wrenches. Source: RIDGID® (18A–18B); Image property of Stanley Black & Decker. Used with permission (18C)

Force

Figure 19 Loosening a fastener with an adjustable wrench.

Always use the correct wrench size for a fastener and learn to pull the wrench toward your body to apply torque. Pushing the wrench away from the body is more likely to result in an injury.

Be sure that the wrench fits snug on the nut, bolt, or other fastener. If the fit is too loose, it may slip and round off the nut or bolt head. A damaged head makes a fastener much more difficult to remove.

1.4.2 Adjustable Wrenches

Adjustable wrenches, often called Crescent® wrenches due to name recognition, are used to loosen or tighten nuts and bolts, as are non-adjustable models. Although a fixed-jaw wrench is usually better and safer, an adjustable wrench offers a lot of flexibility in a single tool. In many cases, the two types are used together. For example, while a nut is being tightened with a box-end wrench, the adjustable wrench might be used to prevent the bolt head from rotating with the nut.

Adjustable wrenches have one fixed jaw and one movable jaw. The adjusting nut on the wrench interlocks with the teeth in the body of the wrench and positions the movable jaw. These wrenches come in lengths from 4" to 30" with openings several inches wide. Adjustable wrenches include the pipe wrench, spud wrench, and common adjustable wrench (*Figure 18*).

To use an adjustable wrench properly, set the jaws to the rough size for the nut, bolt, or threaded pipe. Ensure that pressure, regardless of whether the hardware is being tightened or loosened, is applied to the fixed jaw rather than the movable jaw. Then fully tighten the jaws on the hardware; it helps to lightly shake the wrench as you use the thumb and forefinger to tighten the adjusting nut. Pull on the wrench as needed to tighten or loosen the component. Pulling towards the body is safer and more power can be developed than when pushing away from the body (*Figure 19*).

WARNING!

Improperly adjusted jaws could cause the wrench to slip, resulting in a loss of balance and serious injury. Whenever possible, pull the wrench toward you. Pushing the wrench away from the body is more likely to result in a loss of balance and an injury. If you must push on the wrench, brace yourself with your free hand. Gloves should always be worn when working with wrenches.

Pipe wrenches are used to tighten and loosen threaded pipe. The upper jaw of the wrench is adjusted by turning the adjusting nut. Both jaws have serrated teeth for gripping power on smooth surfaces. The jaw is spring-loaded and slightly angled so you can release the grip and reposition the wrench without having to readjust the jaw.

The spud wrench operates in the same manner as a pipe wrench. However, the jaws are smooth, so it has no value in gripping smooth, round pipe. Instead, the wrench is used on large nuts common to plumbing and piping. Large, yet thin nuts might be found on the bottom of a sink drain, for example.

Spud Wrenches

The definition of a spud wrench varies among the crafts. Ridgid® and others refer to the example shown in *Figure 18* as a spud wrench. However, another type of spud wrench, shown here, looks like a common adjustable wrench on one end, while the opposite end comes to a blunt point, shaped like a tapered punch. This wrench was developed for structural steel workers who need a wrench to tighten large bolts and nuts. However, before the bolt can be inserted through two pieces of structural steel or pipe flanges, another tool is required to align the holes. The spud wrench handles both tasks without changing tools. In some cases, the wrench is not adjustable, but an open-end or box-end style instead.

Source: Klein Tools, Inc.

Although there are left-hand threads on some hardware, most hardware has right-hand threads. Right-hand threads are turned clockwise to tighten them, and counterclockwise to loosen them. This is very important to remember, as it is easy to turn a fastener the wrong way when attempting to remove it, which will only result in it becoming tighter.

1.4.3 Socket Wrenches

Socket wrenches typically use a ratcheting mechanism that forces the socket to rotate in one direction, while it can freely spin when turned in the opposite direction. This enables the user to quickly tighten or loosen a fastener, without removing and refitting the wrench after each turn, or when a complete revolution cannot be made because of poor accessibility.

Socket wrench sets include different combinations of sockets and ratcheting handles that are used to turn the sockets. The ratcheting mechanism on the handle attaches to a variety of socket sizes using a short square shaft. The size of the shaft is referred to as the drive size. Common drive sizes are ¼", ⅜", and ½". Note, though, that these sizes refer only to the size of the drive, not the size of the hardware they fit. Much larger sizes are available, up to 2½" for heavy industrial work. Oddly enough, ratchet drive sizes are the same globally; there are no metric drive sizes. This is reportedly because the ratchet was largely an American invention. However, metric sockets, designed to fit metric hardware, are very common today.

Socket sets are manufactured to fit many Imperial and metric hardware sizes. Most sockets (*Figure 20*) have 6 or 12 gripping flats, like box-end wrenches, and come in different lengths. Like box-end wrenches, those with 6 flats offer better grip. Deep sockets, which are longer than standard sockets, allow the socket to accommodate a bolt that protrudes through the nut. A short extension is also shown in *Figure 20*. Extensions are available in a variety of lengths, and are used to lengthen the socket and reach into tight places.

Socket sets may contain different types of handles. The ratchet handle shown in *Figure 20* has a knob at the top, used to change the direction of free rotation. Ratchet handles come in a variety of lengths. The longer the handle, the more leverage that can be applied. Larger hardware requires more torque, so handles that are used with the largest sockets are longer and stronger. In addition to ratchet handles, hinged handles are available (*Figure 21*). A hinged socket handle, also referred to as a *breaker bar*, can be used when a ratcheting action is not needed. Hinged handles are often longer and more durable, designed to break a difficult fastener loose.

To use sockets and ratchets properly, first select a socket that fits the fastener head. Then select a ratchet that matches the socket drive. Push the square end of the socket over the spring-loaded ball on the drive shaft. Place the socket over the fastener and pull the handle in the appropriate direction to turn the fastener. Reverse the direction of free rotation by using the knob or lever located on the head of the ratchet handle.

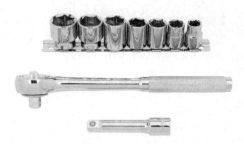

Figure 20 Socket set.
Source: Image property of Stanley Black & Decker. Used with permission

Figure 21 Hinged socket handle, or breaker bar.
Source: Image property of Stanley Black & Decker. Used with permission

Torque Wrenches

Torque wrenches are designed to apply a specific amount of torque to a fastener. Torque is measured in units of distance × force. In the Imperial system, torque values are stated in inch-pound or foot-pounds. Inches and feet represent the distance and pounds represent the force. In the metric system, torque is measured in Newton-meters. Newtons are the measure of force and meters represent the distance. One inch-pound is equal to 0.113 Newton-meter. Metric system users commonly refer to torque as *moment*.

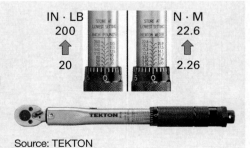

Source: TEKTON

1.4.4 Safety and Maintenance

Note the following safety and maintenance guidelines for working with wrenches:

- Pull the wrench toward your body, not toward your face or away from the body. Leaning into the wrench and pushing away can result in serious injury if the wrench slips off.

- Check the condition of the handle and jaw working surfaces before use.

- Keep adjustable wrenches clean. Do not allow mud or grease to clog the adjusting screw and slide.
- Do not use a wrench as a hammer.
- Do not use any wrench beyond its capacity. Never use a **cheater bar** or pipe on a wrench. This could snap the tool or break the head off the bolt.
- Thoroughly clean sockets and ratchets after each use. Grease and grit can collect inside the socket and on the ratchet shaft, interfering with the operation of the wrench.

Cheater bar: A length of pipe or rod used to increase the leverage on a wrench or similar tool. The increased leverage provided by a cheater bar often exceeds the tool's capacity and strength, resulting in a damaged tool as well as damaged workpieces.

Source: Image property of Stanley Black & Decker. Used with permission

Tool Lanyards

Tools are as important when working at heights as they are when working on the ground. However, when working at heights, a dropped tool becomes a serious hazard. If the tool has moving parts, such as a battery-powered drill, the fall may destroy it. Almost any dropped tool can cause a serious injury or even death, especially when the impact is unexpected.

Tool lanyards specifically designed for work in an elevated environment have been introduced by many companies. Tethering tools to the tool belt or wrist when working at height can prevent injuries, protect the tools, and eliminate the time lost to retrieve a dropped tool.

1.5.0   Pliers and Wire Cutters

Pliers are hinged tools used to hold, cut, and bend wire and soft metals. Pliers should not be used for tightening or loosening hardware such as bolts and nuts. If any significant torque is applied, they will round off the corners of the bolt or nut. Wrenches may no longer fit the damaged hardware properly, creating a larger problem.

High-quality pliers are made of hardened tool steel and come in many different styles, depending on their use (*Figure 22*). The following types of pliers are the most common:

- Slip-joint pliers
- Long-nose pliers

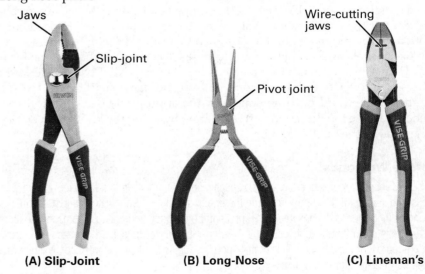

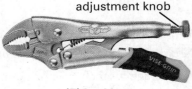

Figure 22 Common types of pliers.
Source: Image property of Stanley Black & Decker. Used with permission

- Lineman's pliers
- Tongue-and-groove pliers
- Locking pliers

It is important to note that there are many variations on these styles, and some have alternate names.

1.5.1 Slip-Joint Pliers

Slip-joint pliers are used to hold and bend wire and to grip and hold objects during assembly. They have adjustable jaws with two or three possible positions. To use slip-joint pliers properly, place the jaws on the object to be held and squeeze the handles until the pliers grip the object. Adjust the jaw position to improve the grip if necessary.

1.5.2 Long-Nose Pliers

Long-nose pliers, also known as *needle-nose pliers,* are used to get into tight places where other pliers cannot reach, or to grip parts that are too small to hold with fingers. They are also useful for bending angles or loops in electrical wire. Most have a set of wire-cutting jaws near the pivot. Long-nose pliers, like many other types of pliers, are available with spring openers—a spring located between the handles that keep them apart and the jaws open until pressure is applied to the handles. Otherwise, the tips of your fingers must push the handles apart to open the jaws.

1.5.3 Lineman's Pliers

Lineman's pliers, also known as *side cutters,* have wider, heavier jaws than slip-joint pliers. They are typically used to cut electrical wire and to hold workpieces. The wedged jaws reduce the chance that wires will slip, and the hook bend at the base of both handles allows for a better grip.

To properly cut wire with lineman's pliers, always point the loose end of the wire down. Squeeze the handles as you cut at a right angle to the wire.

1.5.4 Tongue-and-Groove Pliers

Tongue-and-groove pliers have serrated teeth that grip flat, square, round, or hexagonal objects. You can set the jaws in any of several positions by slipping the curved tongue into the desired groove. Large tongue-and-groove pliers are often used to grip pipe because the longer handles provide good leverage.

Figure 23 Straight-jaw tongue-and-groove pliers.
Source: Image property of Stanley Black & Decker. Used with permission

They are available with several basic jaw designs: straight, smooth, V-shaped, or curved. V-shaped or curved jaws work best with round objects. Straight jaws, as shown on the pliers in *Figure 23,* are best for working with flat material. Smooth jaws are chosen when serrated jaws may damage a delicate surface, but the grip on the object is compromised due to the lack of teeth.

Use the pliers by opening them to the widest position and placing the jaws on the object. Determine which jaw position provides the best and most comfortable grip, and then squeeze the handles until the pliers grip the object.

1.5.5 Locking Pliers

Locking pliers clamp firmly onto objects like a vise. Due to brand recognition, they are often called Vise-Grips®. They are especially handy for holding metal components together for welding or assembly, and for gripping a nut or bolt that has already been damaged.

A knurled knob on the handle is used to adjust jaw spacing. Simply close the handles to lock the pliers and release the lock by depressing the lever. There are many different styles of locking pliers that have been developed for unique applications. In *Figure 24,* large-jaw locking pliers are being used in a fabrication shop. Note the unusual construction of the tool.

To use locking pliers, first place the jaws on the object to be held. Turn the adjusting screw in the handle until the jaws contact the work piece and then squeeze the handles together to lock the pliers. Re-adjust the jaws if the locking tension is too loose or too tight. Squeeze the release lever to unlock and remove the pliers.

Figure 24 Using large-jaw locking pliers while operating a plasma arc cutter.
Source: The Lincoln Electric Company, Cleveland, OH, USA

1.5.6 Safety and Maintenance

Pliers may not seem to be a dangerous tool, but they can cause an injury if misused. Remember the following guidelines when using pliers:

- Wear the appropriate PPE, especially if you cut wire.
- Hold pliers close to the end of the handles to avoid pinching your fingers in the hinge.
- Do not extend the length of the handles for greater leverage; use a larger pair of pliers instead.
- Hold the scrap end of cut wire to avoid flying metal bits when you cut.
- Always cut at right angles. Do not rock the pliers from side to side or bend the wire back and forth against the cutting blades.
- Lubricate pliers regularly to prevent rust and to keep them working smoothly.
- Never use pliers around energized electrical circuits. Although the handles may be plastic-coated, they are not always rated to prevent electrical shock.
- Do not expose plier handles to excessive heat.
- Do not use pliers to turn nuts or bolts. They are poor substitutes for wrenches.
- Never use pliers as hammers.

1.0.0 Section Review

1. The type of hammer that is the best choice for pulling a nail is the _____.

   a. rip claw hammer
   b. drywall hammer
   c. curved claw hammer
   d. ball peen hammer

2. The cape chisel is made with the cutting edge wider than the body, so it does not bind when chiseling _____.

   a. deep grooves
   b. layers of wood
   c. dowels
   d. without a hammer

3. The type of screw head shown here is the _____.

   a. Torx®
   b. Robertson®
   c. Phillips
   d. clutch-drive

4. When using an adjustable wrench, pressure to tighten or loosen should always be applied _____.

   a. clockwise
   b. counterclockwise
   c. to the fixed jaw
   d. to the adjustable jaw

5. Locking pliers are also commonly referred to as _____.

   a. tongue-and-groove pliers

   b. clutch grips

   c. Robertsons®

   d. Vise-Grips®

2.0.0 Measurement and Layout Tools

Objective

Identify common measurement and layout tools and describe how to use them.

   a. Explain how to use a variety of measuring tools.

   b. Define various types of levels and layout tools and indicate how they are used.

Performance Task

1. Inspect and demonstrate the safe and proper use of the following hand tools:
   - Tape measures
   - Levels
   - Squares

Knowing how to collect precise measurements, create straight vertical lines, and level accurately are foundational skills for a craftworker. Incorrect measurements and crooked components have adverse effects on the job. It is important to ensure that measurements are precise, and components are set **plumb** and level.

Plumb: Perfectly vertical, meaning a surface is at a right angle (perpendicular) to the horizon or other surface used for comparison.

2.1.0 Rules and Other Measuring Tools

In Module 00102, *Introduction to Construction Math*, you learned how to read various types of measuring tools. This section reviews measuring tools and guidelines for their use.

Craftworkers most often use four basic types of measuring tools:

- Flat steel rule
- Tape measure
- Wooden folding rule
- Laser measuring tool

Accuracy, ease of use, durability, and readability are several factors in choosing a measuring tool. Of course, the chosen tool must also be suited for the length of the measurement.

2.1.1 Steel Rule

The flat steel rule (*Figure 25*) is a simple measuring tool. Flat steel rules are usually 6" or 12" long with metric versions available at 15 centimeters (cm) and 30 cm. Longer lengths, generally for shop use, are 1 yard (yd) or 1 meter (m) long. Steel rules can be thin and flexible, or thicker and rigid. While steel rules are accurate measuring tools, they can also be used as straightedges for layout work.

Figure 25 Steel rule.

Concave: Having a shape that curves inward, such as the interior of a circle. *Convex* is an antonym, meaning a shape that curves outward.

Figure 26 Tape measure.
Source: Image property of Stanley Black & Decker. Used with permission

Figure 27 Wooden folding rule.
Source: RIDGID®

Figure 28 Laser measuring tool.
Source: Image property of Stanley Black & Decker. Used with permission

2.1.2   Tape Measure

The blade of a tape measure (*Figure 26*), or measuring tape, is usually marked in $\frac{1}{16}$" or 1 mm increments, although some tapes offer smaller Imperial increments. A tape measure may include both Imperial and metric markings for versatility. Tape measures are also available with special markings that apply to residential and light commercial construction wall framing.

The **concave** (curved, as viewed from the top) shape of a tape measure blade is designed to strengthen the blade when it is extended, helping it to remain rigid. Once the blade tip is secure and the proper measurement is established, move the blade edge until it lies flat on the surface, then mark the material. Using this method makes it easier to read the tape and mark the material accurately.

Module 00102, *Introduction to Construction Math,* explains how to use and read a tape measure effectively. Remember that the hook on the end of the tape is designed to slide in and out slightly, allowing for accurate inside and outside measurements. The movement compensates for the thickness of the hook. If the hook becomes damaged or distorted in any way, it is best to replace the tape measure.

When taking an inside measurement, remember to check and add the dimension stamped or printed on the housing. This represents the exact width of the housing.

2.1.3   Wooden Folding Rule

A wooden folding rule (*Figure 27*) is usually marked in sixteenths of an inch on both edges of each side. Folding rules with Imperial markings come in 6' and 8' lengths, with 1-m and 2-m models available for metric users. Because it is rigid, a folding rule is better than a cloth or steel tape in some situations, such as vertical measurement. Like a tape measure, the folding rule can also have special marks at the 16", 19.2", and 24" increments to make wall framing easier. These markings will differ on metric versions due to differences in construction standards.

2.1.4   Laser Measuring Tools

A laser measuring tool (*Figure 28*) is a battery-powered, electronic version of a tape measure. The tool works by simply pointing it at a specific object and then pressing a button. When the button is pressed, a laser beam shoots toward the object and returns to the instrument. The travel time of the beam determines the distance, which is displayed on the screen. Measurements are reported very quickly. Laser measuring tools are designed to report and record in both Imperial and metric units.

These tools are precision instruments, so ensure they are handled with care and stored appropriately. The following are some of the advantages a laser measuring tool has over a traditional tape measure:

- Long vertical measurements can be taken from the ground.

- Longer measurements can be taken. Some laser tools can measure up to 600' (183 m). A special target plate may be required for long distances where more powerful lasers are required.

- Some laser measuring tools have built-in options that make length-related calculations (such as addition, subtraction, and area) easier to perform.

- Measurements can be electronically stored on the tool, so that measurements can be written down all at once after the task is completed. In addition, many are equipped with Bluetooth® communication so that measurements can be sent to a smartphone or tablet.

- Some laser measuring tools can have built-in electronic levels or spirit levels.

The disadvantages of a laser measuring tool may include accuracy, especially over long distances. This is not to say that they fail to meet their specifications for accuracy, but some craftworkers struggle to place their trust in electronic measurement, regardless of reported accuracy. In construction, absolute precision is rarely required for very long measurements. It may also be difficult to take accurate measurements in very hot or smoky environments due to the beam bouncing from particle to particle in the atmosphere, causing the light to scatter.

Note that highly reflective targets may also interfere with accurate measurement. In these cases, apply a piece of tape or similar material with a matte finish to the target.

2.1.5 Safety and Maintenance

Follow these guidelines to ensure measuring tools are safely and properly used:

- Occasionally apply a few drops of lubricant on the spring joints of a wooden folding rule and steel tape.

- Wipe moisture from steel tapes to help prevent rust in areas where the coating has become worn or damaged.

- Do not kink or twist steel tapes. Once a tape is significantly distorted, the damage is permanent.

- Do not use a steel measuring device near energized electrical components.

- Do not allow laser measuring tools to get wet or be exposed to excessive heat.

Precision Measuring Tools

Precision measuring tools, such as micrometers and calipers, make it possible to accurately measure parts that are being machined to one ten-thousandth of an inch (0.0001") or 2.54 microns. However, a standard micrometer's smallest division is 0.001" (25.4 microns). Digital micrometers are available today that can read increments as small as 0.00005", or 1.27 microns. Of course, there are micrometers designed for metric applications as well.

Source: The LS Starrett Company

2.2.0 Levels and Layout Tools

A level is a tool used to determine how close a surface is to a true horizontal plane (level) or vertical plane (plumb). Layout tools include the carpenter's square and combination square, which are also introduced here. Some squares are designed specifically for a given craft. For example, although a carpenter's square and a pipefitter's square may appear to be identical from a distance, they have different formulas and other information important to their craft. Other layout tools include chalk lines and plumb bobs.

2.2.1 Spirit Levels

The standard spirit level (*Figure 29*) is the most common leveling instrument in the construction trade. The spirit level got its name because the vials in it are often filled with colored ethanol, a type of alcohol, and alcohol is often referred to as spirits. Alcohol is used because it does not freeze like water and moves more freely inside the vial.

Most spirit levels are made of tough, lightweight metals such as magnesium or aluminum, or types of plastic that can reliably maintain their true form. They typically have three vials filled with alcohol. The center vial is used to check for level, and the two end vials are used to check for plumb. Some spirit levels, like the torpedo level shown in *Figure 29*, will also show a true 45-degree angle. Torpedo levels usually have a magnetic and/or a V-shaped base, making it useful for pipefitting work. Spirit levels are available in a variety of lengths.

The limited amount of liquid in each vial is intentional, so there is always a bubble in the vial. When the bubble is centered precisely between the lines on the vial, the surface is either level (perfectly horizontal) or plumb (perfectly vertical) as shown in *Figure 30*. Before placing a level against a surface, be sure there is no debris on the surface that will prevent an accurate measurement.

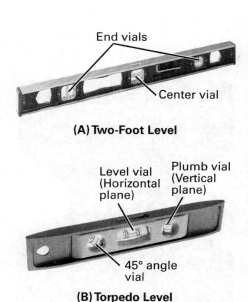

Figure 29 Spirit levels.
Source: RIDGID® (29A); Image property of Stanley Black & Decker. Used with permission (29B)

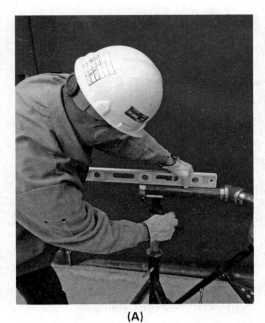

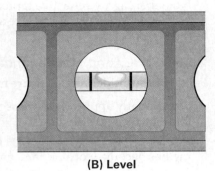

(B) Level

(A)

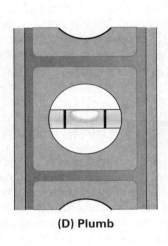

(D) Plumb

(C)

Figure 30 Levels showing level and plumb.
Source: Cianbro Corporation (27A, 27C)

When setting posts, you must be sure they are plumb in two different planes. A post level (*Figure 31*) is designed to be strapped to a post, providing a view of the condition in both planes. Many also have magnets that allow them to be used on steel posts without a strap.

2.2.2 Digital Levels

Digital levels provide a digital readout of degrees of slope, plus rate of rise to lay out stairs and roofs or determine the fall of a drain line. Some may also use a simulated bubble display for a more traditional feel. Digital levels, like the one shown in *Figure 32*, are becoming more common on construction sites due to their versatility. Always handle and store a digital level carefully to avoid damaging it, following the manufacturer's recommendations for its use and care.

2.2.3 Laser Levels

With a laser level (*Figure 33*), a single worker can accurately establish plumb, level, or square measurements. Laser levels are used to establish foundation

Figure 31 Post level.
Source: Image property of Stanley Black & Decker. Used with permission

Figure 32 Digital level display.
Source: Image property of Stanley Black & Decker. Used with permission

Figure 33 Laser level.
Source: Image property of Stanley Black & Decker. Used with permission

elevations and drainage slopes, square framing, align plumbing and electrical lines, and hang suspended ceilings. A laser level may be mounted on a tripod, fastened onto pipes or framing studs, or be suspended from a ceiling. They are used for leveling over a significant distance, well beyond the reach of a conventional level.

Laser levels come in a variety of sizes and weights. They are generally housed in sturdy casings designed to withstand the construction environment. Always handle and store laser levels carefully, following the manufacturer's recommendations for care and use. Although they are constructed for the environment, they are truly instruments that require care in their use and movement from place to place.

WARNING!

Never look directly into a laser beam. It can permanently damage your eyes and impair your vision.

2.2.4 Level Safety and Maintenance

All levels are precision instruments that must be handled with care. Unless the user looks directly into the laser beam, there is little risk of personal injury when working with this type of tool. All craftworkers must learn this valuable safety guideline. Remember these guidelines when working with levels:

- Never look directly into a laser beam or its source. Serious and permanent eye injury can result.

- Replace a level if a crack or break appears in any of the vials, or if they appear loose and insecure.

- Keep levels clean and dry.

- Do not bend or otherwise apply excessive pressure to a level.

- Do not drop or throw any type of level.

Bricklayers

Working with your hands to create buildings is a time-honored craft. People have used bricks in construction for more than 10,000 years. Archaeological records prove that bricks were one of the earliest man-made building materials.

In bricklaying, it is important that each course (row) of bricks is level and that the wall is straight. An uneven wall is weak as well as unattractive. To ensure that the work stays true, a bricklayer uses a straight level and a plumb line.

Today, bricklayers build walls, floors, partitions, fireplaces, chimneys, and other structures with brick, precast masonry panels, concrete block, and other materials. They lay brick for houses, schools, sports stadiums, office buildings, and other structures. Some bricklayers specialize in installing heat-resistant linings inside huge industrial furnaces, called refractory brick. Bricklaying is considered a segment of the masonry craft, which also includes the more complex task of stone masonry.

Square: Used as an adjective in this context, to have the shape of a square where two components are at precise 90-degree angles (perpendicular) to each other.

Rafter: One of a series of beams that extend from the perimeter of a building to the high point of a roof, providing structural support for the roof above.

2.2.5 Squares

Squares (*Figure 34*) are used for marking, measuring, and checking to see if a joint is **square**. The type of square you use depends on the type of job and your preference. Common squares are the carpenter's square, the **rafter** angle square (also called a *speed square*), the try square, and the combination square.

The carpenter's square, also called a *framing square*, is L-shaped and is used for squaring up sections of work such as baseplates and walls to ensure components are perpendicular to each other. As shown in *Figure 35*, it is also used to mark square cutting and matching lines, especially when working with wide framing lumber. The carpenter's square has a 24" blade and a 16" tongue, forming a right angle. Metric versions have a 600 mm blade with a 400 mm tongue. The blade and tongue are marked with inches and fractions of an inch, or metric centimeters and millimeters. Both the blade and tongue can be used to measure or as a straightedge. Tables and formulas relevant to carpentry are usually printed on the blade as a convenience.

Note that the carpenter's square is virtually identical to a pipefitter's square, except for one feature. On the carpenter's square, there are formulas and other information relevant to carpentry work. The pipefitter's square has formulas and other helpful information relevant to the pipefitting craft.

The rafter angle square is another type of square popular in carpentry. Frequently made of cast aluminum or tough plastics, it combines a protractor, try square, and framing square. It is marked with degree gradations for fast, easy rafter layout. The square is small, so it is easy to store and carry.

The try square has a fixed, 90-degree angle and is used mainly for woodworking. Like other squares, it can be used to lay out 90-degree cutting lines; to check a joint to ensure it is square; and to check if a piece of lumber is warped or cupped (bowed).

The combination square has a ruled blade that slides back and forth in a head. The position of the ruled blade can be changed as needed. The head is permanently formed into 45-degree and 90-degree angles. Many combination squares also contain a small spirit level and a metal scribe for marking metal. The combination square is one of the most useful squares for layout work because of its versatility. Good combination squares have all-metal parts, with the head often made from cast iron and the blade from tempered steel or stainless steel.

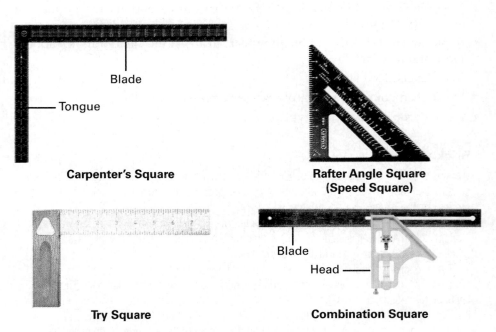

Carpenter's Square

Rafter Angle Square (Speed Square)

Try Square

Combination Square

Figure 34 Types of squares.
Source: Image property of Stanley Black & Decker. Used with permission

Figure 35 Marking a cutting line.
Source: Zachary McNaughton, River Valley Technical Center

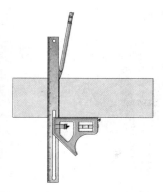

Figure 36 Using a combination square to mark a 90-degree angle.

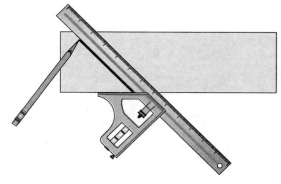

Figure 37 Using a combination square to mark a 45-degree angle.

To mark a 90-degree angle using a combination square (*Figure 36*), set the blade so that it is long enough to reach fully across the lumber or metal. Position the front edge of the head firmly against the material to be marked. Start at the edge of the material and follow the blade with a pencil to make the mark.

To mark a 45-degree angle using a combination square (*Figure 37*), again set the blade so that it is long enough to reach fully across the lumber or metal. Position the back of the head firmly against the edge of the material to be marked and use the pencil to follow the blade at a 45-degree angle to make the mark.

Did You Know?

Geometry and the Square

Some combination squares, like the one shown here, can be set at any angle. The head rotates around a precision protractor. Like other combination squares, the metal rule can slide back and forth in the head, and then be locked in the desired position. Combination square sets, like the one shown here, are used to measure and mark any angle, including the common 30-, 45-, 60-, and 90-degree angles. The centering head is used on round stock to locate the center or scribe a line across the end. A standard combination square head is also part of this set. Basic geometric principles like those exhibited by squares are used daily in many crafts.

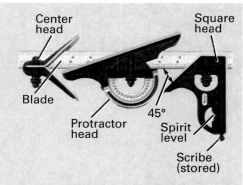

Source: The LS Starrett Company

2.2.6 Use and Maintenance

Follow these guidelines for using squares:

- Do not use a square for something it was not designed for, especially prying or hammering. It is especially important to protect the edges from damage and avoid bending the blades.

- Keep squares dry to prevent them from rusting. Use a light coat of silicone spray or lubricant on the blade for protection, and occasionally clean and lubricate the moving parts of combination squares.

- Do not drop or strike the square hard enough to change the angle between the blade and the head. A try square can be more easily damaged by a drop than a carpenter's square, because of the differences in construction.

A square can be checked to ensure that it has not become distorted. Use it on a straight and true piece of lumber to scribe a line. Then turn the square over and scribe a line in the same location. The two lines should be identical and true to each other.

Plumb
bob

Figure 38 Plumb bob.
Source: Image property of
Stanley Black & Decker. Used
with permission

Figure 39 Using a plumb bob.

2.2.7   Plumb Bob

The plumb bob consists of a pointed weight attached to a string (*Figure 38*).
It uses the force of gravity to make the line hang perfectly vertical, or plumb.
Plumb bobs come in different weights, with 20-, 16-, 12-, and 8-oz weights (566,
454, 340, and 227 g) being the most common. Although metric equivalents are
provided here, metric plumb bobs are likely found in even weights such as 570
or 450 grams, rather than weights that relate to an Imperial unit of measure.
Heavier weights are generally better for longer drops to ensure the plumb bob
can hold its position.

When the bob hangs freely and is motionless, the string is plumb (*Figure 39*).
For example, a plumb bob can be used to make sure a wall or a doorjamb is vertical. When installing a support post under a beam, a plumb bob can show the
point on the floor directly under the section of the beam that requires support.

Follow these steps to use a plumb bob properly:

Step 1   Make sure the line is attached at the exact top center of the plumb bob.
Step 2   Hang the bob from a horizontal member, such as a doorjamb, joist,
or beam. To avoid damage, be careful not to drop a plumb bob on
its point.
Step 3   When the bob hangs freely and stops swinging, the string is plumb.
When using a plumb bob outdoors, be aware that the wind may blow it
out of true vertical and cause it to swing.
Step 4   Mark the point directly below the tip of the plumb bob. This point is
precisely below the point of attachment above.

2.2.8   Chalk Lines

Chalk lines are used to mark long, straight lines on smooth surfaces. A chalk
line is a simple tool consisting of a length of string with a coating of chalk on a
reel (*Figure 40*). The line is stretched taut between two points and then snapped
to mark a temporary line.

The chalk line case is filled with colored chalk powder. The line is automatically chalked each time it is pulled out of the case. Some models can also serve
as a lightweight plumb bob.

To snap a chalk line, simply pull the line from the case and secure the end.
Stretch the line between the two points to be connected. After the line has been
pulled taut, pull the string straight up or away from the work and then release it.
This marks the surface with a straight line of chalk (*Figure 41*). If necessary, spray the
chalk line with clear lacquer to prevent it from wearing away quickly. Otherwise,
the chalk is easily washed away by rain or distorted by foot traffic. Store the chalk
line and chalk supply in a dry place, as damp or wet chalk is unusable.

Figure 40 Chalk line and chalk.
Source: Image property of Stanley
Black & Decker. Used with permission

Figure 41 Using a chalk line.
Source: Image property of Stanley Black &
Decker. Used with permission

2.0.0 Section Review

1. Folding rules with Imperial markings come in _____.
   a. 2' and 4' lengths
   b. 2' and 6' lengths
   c. 3' and 8' lengths
   d. 6' and 8' lengths

2. The carpenter's square is also called a _____.
   a. framing square
   b. speed square
   c. try square
   d. combination square

3.0.0 Other Common Hand Tools

Objective

Identify and describe other hand tools common to shops and job sites.
   a. Differentiate between various handsaws and their designated applications.
   b. Identify common clamp designs.
   c. Explain how different files and utility knives are used with various materials.
   d. Describe shovels and picks and the tasks for which each one is best suited.

Performance Task

1. Inspect and demonstrate the safe and proper use of the following hand tools:
   - Handsaws
   - Clamps
   - Files
   - Utility knives
   - Shovels

Many different saws are on the market. Some have extremely specific uses, while others are more versatile. Handsaws were in use before 1500 BC, and they remain on the job site today. However, smaller, lighter, and more powerful power tools have taken over a great deal of the work previously done by hand.

Clamps and vises have been in use for years as well, and a power tool cannot replace them. Today's clamps are convenient and can be used in a variety of situations. This section will introduce these basic tools as well as a few others you will encounter on the job site.

3.1.0   Saws

Hand sawing can be hard work. Using the right saw for the job makes sawing easier. Saws come in a variety of shapes and sizes, yet all share one important feature—cutting teeth. The shape and size of the teeth are related to the texture of cut that results, and to the speed at which the cut is made.

Differences in blade and tooth design make it possible to cut straight or curved lines through wood, metal, plastic, and many other materials. Generally, the fewer teeth per inch (tpi) of length found on a saw blade, the coarser and faster the cut will be. More teeth per inch, which means smaller teeth, results in a slower but smoother cut.

Figure 42 shows several types of saws. The following are brief descriptions of these saws.

> **WARNING!**
>
> Saw teeth are very sharp. Use gloves and do not handle saw teeth with bare hands. When cutting with a saw, ensure that your fingers always remain clear of the teeth.

Miter joints: Joints made by fastening parts together with the ends cut at a similar angle, usually 90 degrees.

- *Backsaw* — The standard blade of this saw is 8" to 14" long (20 cm to 36 cm) with 11 to 14 tpi. A backsaw has a broad, flat blade and a reinforced back edge. The reinforced back keeps the blade straight and true. It is used for cutting finish lumber and **miter joints**, as the blade is designed to produce smooth cuts.
- *Keyhole saw* — The standard blade of this saw is 12" to 14" (30 cm to 36 cm) long with 7 to 8 tpi. This saw cuts curves quickly in wood, plywood, or wallboard. It is used to cut holes for large-diameter pipes, vents, and electrical boxes. The name, however, comes from its original use of cutting holes in doors for door hardware.

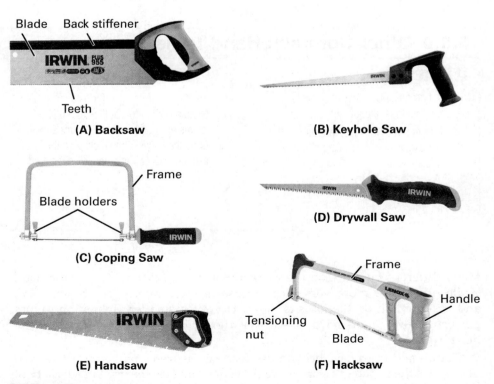

Figure 42 Common saws.
Source: Image property of Stanley Black & Decker. Used with permission (42A–42F)

- *Coping saw* — This saw has a narrow, flexible blade attached to a U-shaped frame. The blade can be rotated in the frame to any angle. The blade holders also maintain proper blade tension. Standard blades range from 10 to 20 tpi. The coping saw is used for tight curves and circles in finish carpentry. Generally, coping saw blades are mounted with the teeth pointing back toward the handle, allowing it to cut on the pull stroke. However, the blade can also be placed so the cutting action occurs on the push stroke, like most handsaws.

- *Drywall saw* — A drywall saw is a long, narrow saw used to cut softer building materials quickly. Drywall saws can have a fixed or retractable blade mounted in a wood or plastic handle. The blade on a drywall saw has a very sharp point to easily make a hole in soft material and start a cut.

- *Handsaw* — The standard blade of this saw is 26" long with 8 to 14 tpi for a crosscut saw and 5 to 9 tpi for a ripsaw. A crosscut saw, as the name implies, is designed to cut across the grain of the wood. A ripsaw is designed to cut with the grain.

- *Hacksaw* — The standard blade of this saw is 8" to 16" (20 cm to 41 cm) long with 14 to 32 tpi. The hacksaw is used to cut through metals such as bolts and pipe. For this reason, the blades have more and finer teeth than most handsaws. It has a sturdy frame and a pistol-grip handle. The blade is tightened using a tensioning nut or similar mechanism. When installing a hacksaw blade, be sure that the teeth face away from, not toward, the saw handle. Hacksaws are designed to cut on the push stroke, not on the pull stroke.

3.1.1 Handsaws

The handsaw's blade is made of tempered steel so it will stay sharp and will not buckle. When a handsaw is not handled properly and the blade does buckle, it is virtually useless. Handsaws are classified mainly by the characteristics of the teeth. Saw teeth are set or angled alternately in opposite directions to make a **kerf** slightly wider than the thickness of the saw blade body.

Kerf: A cut taken by a saw blade, with a width equal to that of the blade teeth.

Two common types of handsaws are the crosscut saw and the ripsaw. The crosscut saw, which has 8 to 14 tpi, is designed to cut across the grain of wood. It cuts slower than a ripsaw but leaves a smoother surface. Blade lengths range from 20" to 28" (51 cm to 71 cm). For general use, a 24" or 26" (61 cm or 66 cm) blade is a good choice. Longer blades allow longer cutting strokes, and longer strokes contribute to a better, more consistent cut.

Follow these steps to use a crosscut saw properly:

Step 1 Don safety glasses and work gloves before beginning. Mark the cut to be made with a square. Also place an X or similar mark on one side of the squared line to show on which side of the line the cut should be made. The kerf should always be taken from the scrap side of the mark.

Step 2 Make sure the piece to be cut is well-supported on a sawhorse or similar support. Secure the material with one or more clamps. Plan to support the scrap end as well as the main part of the lumber to keep it from breaking away as the blade nears the end of the cut. With short pieces of lumber, you can support the scrap end of the piece with a free hand. With longer pieces, additional support or assistance is required.

Step 3 Place the saw teeth on the edge of the wood farthest from you, just at the outside edge of the mark. Remember that the edge of the blade should be aligned with the line on the scrap side of the lumber, so that the kerf is removed from the scrap side of the cut.

Step 4 Start the cut with the blade touching the lumber near the handle of the saw, pulling toward your body on the first stroke. Place the saw at about a 45-degree angle to the wood and pull the length of the blade to make a small kerf.

Step 5   Push the blade in the opposite direction to cut and deepen the kerf. Start sawing slowly, increasing the length of the stroke as the kerf deepens. Remember that the cutting is done on the push stroke. Therefore, the push stroke is more forceful and faster than the pull stroke.

Step 6   Use the thumb of the hand that is not sawing to guide the saw as the cut progresses, so the blade stays perpendicular to the workpiece.

Step 7   Do not push too hard or ride the saw into the wood. Let the weight of the saw set the cutting rate. This will make it easier to control the saw and is less tiring.

Step 8   Continue to saw with the blade at a 45-degree angle to the wood. If the saw starts to wander from the line, work the blade back toward it. If the saw blade binds in the kerf, place a wooden wedge into the kerf to hold it open, or reposition the support under the scrap end to prevent binding.

Figure 43 shows the kerf that results from a crosscut. Note that the kerf is parallel with the line on one side, with the left side of the line being the scrap side. This ensures the length of the cut lumber remains accurate.

The ripsaw has 5 to 9 tpi and is designed to cut with the grain (parallel to the wood fibers), meeting less resistance than a crosscut saw. Because it has fewer teeth than the crosscut saw, it will leave a coarser surface, but cuts faster.

To use a ripsaw properly, mark and start a cut the same way you would with a crosscut saw. Once the cut is underway, saw with the blade at a steeper, 60-degree angle to the wood.

3.1.2   Safety and Maintenance

It is important to focus on the blade when sawing. Saw teeth are sharp enough to cause a serious injury. Here are some basic guidelines for caring for and working with handsaws:

Figure 43 Crosscut saw kerf.
Source: Cianbro Corporation

Emery cloth: A material made with a thin cloth backing, coated with an abrasive that is tightly bonded to the backing; typically sold in sheets or narrow rolls.

- Brace yourself when sawing so you are not thrown off balance when the scrap separates from the workpiece.

- Clamp or otherwise secure the workpiece to prevent it from moving while sawing.

- Clean saw blades occasionally with fine **emery cloth** or steel wool and apply a coat of silicone lubricant if it starts to rust.

- Always lay a saw down gently. If the blade develops a permanent bend or buckle, it cannot be repaired.

- Good quality saws can be sharpened by craftworkers with the proper equipment and skill.

- Do not allow the handsaw blade to saw into stone, concrete, or metal, as these materials can easily damage the teeth.

3.2.0   Clamps

There are many types and sizes of clamps (*Figure 44*). Clamps are sized by the maximum opening of the jaws. The depth, or throat, of the clamp determines how far from the edge of the workpiece the clamp can be placed. The following descriptions are related to the clamps shown in *Figure 44*:

- *C-clamp* — This multipurpose clamp is named for the C-shaped frame. The clamp has a metal shoe at the end of a screw. Using a sliding T-bar handle, the clamp is tightened so it holds material between the metal jaw of the frame and the shoe. C-clamps are strong and durable, usually made of cast iron or steel.

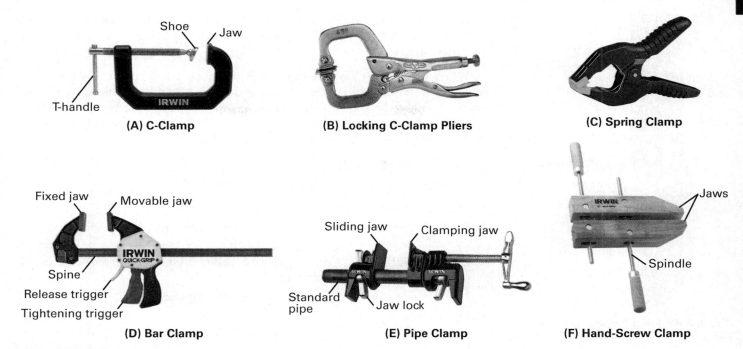

Figure 44 Types of clamps.
Source: Image property of Stanley Black & Decker. Used with permission (44A–44F)

- *Locking C-clamp pliers* — Technically, this clamp also falls into the category of pliers. The locking feature of these pliers, though, also makes them a versatile clamp. A knob in the handle controls the width and tension of the jaws. The handles are closed to lock the clamp and a lever is pressed to unlock the jaws.

- *Spring clamp* — These simple clamps are spring-operated. When the handles are released, the spring holds the clamp tightly shut. The jaws are usually cushioned with a plastic or rubbery material to protect the workpiece surface.

- *Bar clamp* — A rectangular piece of steel or aluminum forms the spine of the bar clamp. It has a fixed jaw at one end and a sliding jaw that moves along the bar. Small grooves in the spine lock the moving head into position. Pads on the jaws prevent it from damaging delicate surfaces. It is also designed to release the material quickly and easily simply by squeezing the release trigger. Another feature of this tool is that you can change the direction of the jaw to turn it into a spreader bar.

- *Pipe clamp* — The spine of a pipe clamp is a selected length of pipe. It has one fixed jaw, and a movable jaw that slides along the pipe. The movable jaw has a lever mechanism that is squeezed to slide it along the spine. The pipe connecting the two jaw assemblies can be any length required; even a full length of pipe can be used. The pipe clamp is popular for large and wide clamping tasks as a result.

- *Hand-screw clamp* — This clamp has broad wooden jaws that spread the pressure over a wide area. Each jaw works independently along two threaded rods. The jaws can be angled toward or away from each other or remain parallel.

- *Web (cargo, band) strap* — The web strap (*Figure 45*), although not technically a clamp, uses a belt-like canvas or nylon strap or band to apply even pressure around a bundle of material or to secure a load. After looping

Figure 45 Web strap.
Source: Image property of Stanley Black & Decker. Used with permission

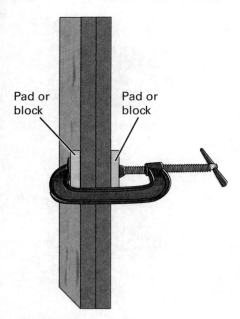

Figure 46 Placing cushions for clamp jaws.

the band around the work, the clamp head is used to tighten the band with a ratcheting action. A quick-release lever loosens the band when finished.

When clamping wood or other soft material, place rubber pads or thin blocks of wood between the workpiece and the jaws to protect the workpiece, as shown in *Figure 46*. When wrapping heavy web straps around bundles of finish lumber, it is also a good idea to protect the corners of the wood with cushions, preventing the straps from digging into the edges.

3.2.1 Safety and Maintenance

The following are guidelines for using clamps:

- Store clamps by lightly clamping them to a rack.
- Discard clamps with bent frames or spines. The spine of a pipe clamp is easily replaced.
- Clean and lubricate the moving parts occasionally, but do not overdo it, as oil tends to attract and hold dirt and dust.
- Never use any of the clamps shown for hoisting work.
- Do not use pipe or other tools to add leverage when tightening a clamp. Doing so allows you to exceed the clamp's capacity and will likely lead to permanent damage.

Did You Know?

The Romans and Their Tools

Ancient Egyptians and Greeks used a variety of tools. The Romans, however, are known as the toolmakers of the ancient world. The plane, metal-cutting saw, drawknife (pictured), frame saw, level, square, and claw hammer were all created by the Romans.

Figure 47 Files with handles installed.
Source: Image property of Stanley Black & Decker. Used with permission

3.3.0 Files and Utility Knives

Files and rasps are shaping tools. They are commonly used to remove wood and metal burrs and smooth out rough spots.

As the name implies, utility knives are general-purpose tools. Because of their versatility, they are common among all crafts. However, they are also hazardous and have been the cause of many injuries. Many companies have specific policies regarding their use. Some companies and job sites require all utility knives to have a self-retracting blade; those that do not are prohibited.

> **WARNING!**
>
> All types of files and utility knives have sharp points and edges that can cause serious injury. Handle knives and files with extreme caution and the greatest of care. Never use a file or rasp without first installing a handle.

3.3.1 Files and Rasps

Files and rasps are used to cut, smooth, or shape metal and wood. A variety of files are shown in *Figure 47*.

Files have slanting, parallel rows of teeth, while rasps and wood shavers (*Figure 48*) have individual teeth which are more aggressive (deeper). Rasps and wood shavers, which are also called *shredder rasps*, are designed for wood and similar soft materials, while files are used on metal. Rasps are much like aggressive files, made from a single piece of solid metal. Wood shavers have an open construction, much like a cheese grater, that allows the debris to pass through the cutting surface (*Figure 49*).

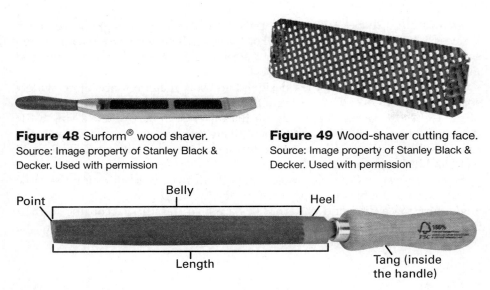

Figure 48 Surform® wood shaver.
Source: Image property of Stanley Black & Decker. Used with permission

Figure 49 Wood-shaver cutting face.
Source: Image property of Stanley Black & Decker. Used with permission

Figure 50 Parts of a file.
Source: Woodstock International, Inc.

Files and rasps are made from a single hardened piece of high-grade steel. They are sized by the length of the body (*Figure 50*). The length does not include the handle because the handle is generally a separate accessory. The sharp metal point at the end of the file, called the **tang**, fits into the handle. Handles can easily be transferred from one file to another. Files and rasps range in length from 4" to 16" (10 cm to 40 cm).

Choose a file or rasp with a profile that fits the area to be filed. Files are available in round, half-round, flat, square, mill, hand, and triangular shapes (*Figure 51*),

Tang: Tapered, pointed end of a file designed for insertion into a file handle.

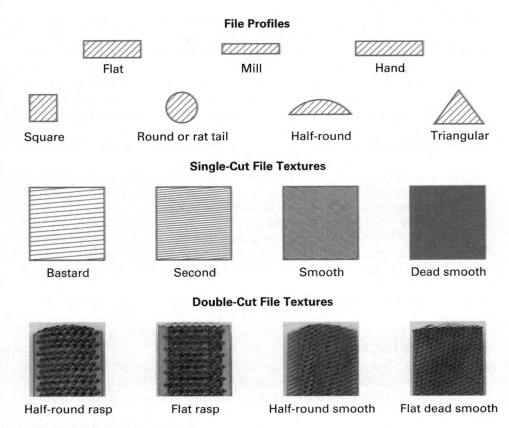

Figure 51 File profiles and textures.

as well as other shapes for unique applications. For filing large concave or flat surfaces, a half-round shape is used. For filing small curves or for enlarging and smoothing holes, a round shape with a tapered end is best. For filing angles and inside corners, a triangular file is used.

Generally, the teeth of files designed for soft materials are more widely spaced. File teeth designed for harder materials are closer together. The shape of the teeth also varies depending on the material to be worked. Using a file designed for soft material on metal will likely result in chipped, dull teeth. On the other hand, using a file designed for metal on wood will result in clogged teeth and little progress.

File classifications include single-cut, with all rows of teeth facing the same direction; and double-cut, with rows of teeth crisscrossing each other at an angle, forming a diamond pattern. Files are also classified according to their texture and depth of teeth. File textures include the following, listed from the roughest to the smoothest:

- Rough
- Coarse
- Bastard
- Second cut, or medium
- Smooth
- Dead smooth

Smooth files have more teeth per inch across the face and result in the smoothest finish but remove the least amount of material per stroke. Coarse files have fewer, deeper teeth per inch. Bastard files are relatively coarse, classified just below coarse files. The unusual name *bastard file* originated from the fact that it is an irregularity in the file line-up. Rasps are also classified by the nature of their teeth: coarse, medium, and fine.

Using a file the wrong way is inefficient and can be frustrating to the user. Follow these steps to use a file properly:

Step 1 Don work gloves and safety glasses. Select a file and ensure the file handle is securely mounted to the file tang.

Step 2 Mount the workpiece to be filed in a vise or with a clamp to a sturdy bench at approximately elbow height.

Step 3 Stand back from the vise with your feet approximately 24" (61 cm) apart, with the right foot ahead of the left. If you are left-handed, the left foot will be ahead of the right. Do not lean directly over the work.

Step 4 Hold the file handle with one hand and steady the tip of the blade with the other hand.

Step 5 For average work, hold the tip with your thumb on top of the blade and the first two fingers under it. For heavy work and larger files, place all the fingers underneath, with the thumb on top.

Step 6 Apply pressure only on the forward stroke, as the file cuts on this stroke. Placing pressure on the file during the back stroke while the file remains in contact with the workpiece only wears out the file.

Step 7 Raise the file slightly from the work on the back stroke to prepare for the next stroke.

Step 8 Keep the file flat on the work as you continue the sequence of strokes.

3.3.2 Utility Knives

A utility knife (*Figure 52*) is used to cut a variety of materials including roofing felt, fiberglass or asphalt shingles, vinyl or linoleum floor tiles, fiberboard, and drywall. Utility knives can also be used for trimming insulation and opening cartons.

The utility knife has a replaceable, thin, razorlike blade. The handle, which is approximately 6" (15 cm) long and is made of die-cast metal or plastic, holds the blade. The least expensive, simpler models are made in two halves, held together with a screw for blade replacement. Others fold to a more compact size. For increased safety, some utility knives automatically retract the blade when thumb pressure is taken off the blade control. Note that some job sites or employers may require self-retracting knives to maximize safety.

Note the different blade used in the carpet knife in *Figure 52*. Carpet is highly abrasive and hard on blades. Carpet knife blades have two sharpened edges, so they can be reversed to extend their usable life.

When using a utility knife or any sharp tool, it is important to wear the appropriate gloves. Gloves made with Kevlar® offer greater protection against injury. Kevlar® is reported to have five times the strength of steel at equal weights.

(A) Self-Retracting Knife

(B) Folding Utility Knife

(C) Carpet Knife

Figure 52 Utility knives.
Source: Image property of Stanley Black & Decker. Used with permission

---

WARNING!

Utility knife blades are extremely sharp, having a razor edge. Handle these knives with great care and ensure the blade is safely secured after each use.

To use a retractable-blade utility knife safely and properly, place a protective barrier such as a piece of scrap wood, under the object to be cut. Don an appropriate pair of gloves. Unlock and push the blade out by pushing on the lever and sliding it forward. Release pressure on the lever to lock the blade open or hold the lever if the knife automatically retracts. Always cut away from the body, or perpendicular to it. Once cutting is complete, unlock and pull the blade in by pushing on the lever and sliding it back. Folding models have a blade lock that must be depressed to fold and unfold.

3.3.3 Safety and Maintenance

The following are guidelines for the safe use and maintenance of files and rasps:

- Use the correct file for the material being worked.
- Always put a handle on a file before using it; files and rasps often come without handles when purchased in bulk.
- After using a file, brush the filings from between the teeth using a file card or wire brush, brushing parallel to the teeth. A file card (*Figure 53*) has tightly spaced, short, stiff wires on the face that dislodge material from between the file teeth.
- Store files in a dry place and keep them separated so that they cannot chip or damage each other.
- Do not allow the workpiece to vibrate in the vise as you file, as it dulls the teeth and produces an irregular surface on the workpiece.

These guidelines apply to utility knives:

- Replace blades when they stop cutting and start tearing the material. Sharp blades are safer than dull ones.
- Always keep the blade closed and locked when the knife is not in use.
- Be sure to position yourself properly, keeping your free hand out of the line of the cut. Cut perpendicular to the body (left-to-right or right-to-left) or away from the body.
- Utility knife blades may be somewhat brittle and can snap off easily, leaving a dangerous jagged edge.

Figure 53 File card.
Source: Nancy Lamm

3.4.0 Shovels and Picks

Shovels are used by many different construction trades. An electrician running underground wiring may dig a trench, and a mason may dig footers for a foundation. Carpenters, plumbers, and many other craftworkers need to use a shovel occasionally.

Three basic shapes of shovel blades are round, square, and spade. Round shovels (*Figure 54A*) are used to dig holes or move soil. A square shovel (*Figure 54B*) is used to move gravel or clean up construction debris. A traditional spade (*Figure 55*) is used to shape trenches or holes that need smooth, straight sides. It is flatter than the others and does not have curved sides. However, many variations of the traditional spade are available today.

Round shovels and spades are best for digging fresh earth. When using them, place the tip of the shovel where digging or soil removal will begin. Balance a foot on the step at the top of the blade, shift some of your weight to it, and press down to cut into the soil.

Square shovels are best for collecting debris from a hard surface, or for moving sand and gravel. When using a square shovel, place the leading edge of the shovel blade against the gravel or construction debris and push until the shovel is loaded.

All types of shovels can have wooden or fiberglass handles. They generally come in two lengths. A long handle is usually 47" to 48" (1.2 m) long; a short handle is usually 27" (66 cm) long. Short handles usually have a D-ring on the end to provide a better grip. Generally, the longer handles provide better leverage, but a shovel is not meant to be used as a lever. Roots and other obstructions should first be cleared with other tools.

A pick (*Figure 56*) consists of a wooden handle that is 36" to 45" (91 cm to 1.1 m) in length and a forged steel head weighing 2 to 3 lb (900 g to 1 kg). Depending on the size and strength of the pick, it can be used to break hardened or rocky soil, to level out stones and pavers, to loosen soil, and to break up stones and concrete. Picks with the longest handles are used when a normal amount of swing force is needed and there is clearance to do so. Short-handled picks are especially useful when working below the surface and down on your knees.

Similar to a pick, a mattock is also used for breaking hard, rocky soil and to dig trenches. Mattocks are much better than picks for clearing tree roots. One side of the mattock head has a wide cutting blade, as shown in *Figure 56*. The other end may be like a pick, or another blade turned 90 degrees from the mattock blade (as shown). Mattock handles are rarely longer than 36" (91 cm).

The first step to using a pick or mattock safely and effectively is to select the length appropriate for your size and the task at hand. Gloves should always be worn, as well as steel-toe work boots or shoes.

Place one hand at the end of the handle and place the dominant hand about two-thirds of the way up the handle. For tasks using a short-handled pick, to strike hard, raise the pick to a near vertical position. With the knees bent, plunge the tool into the ground. Raise the pick again and then swing it back toward the ground, using the weight of the tool head and the leverage of the handle to deliver a solid blow.

> **WARNING!**
>
> Do not swing a pick or mattock beyond vertical and over your head. Doing so may cause a back injury and significantly reduce control over the tool.

3.4.1 Safety and Maintenance

The following are guidelines for working safely with shovels:

- Always check to ensure that the blade is fixed firmly to the handle and no cracks or splits are present.
- Use the appropriate PPE when digging, trenching, or clearing debris, including safety glasses and gloves. Wear safety shoes or boots to protect your feet from dropped materials and tool blades.

(B) Square

(A) Round

Figure 54 Round and square shovels.
Source: RIDGID® (54A–54B)

Figure 55 Spade.
Source: Shutterstock/Sergio Schnitzler

- Do not leave dirt or debris on the blade. Rinse or brush off the shovel blade after use.

The following are guidelines for working safely with picks:

- Always check to ensure that the head is fixed firmly to the handle and no cracks or splits are present.
- Make sure that there are no other workers in the swing path before beginning the work.
- Always use a pick that is of the appropriate length and weight for your physical size.
- Only use maximum force swings when necessary. Hard swings place more strain on your back and shoulders.
- Do not swing a pick or mattock beyond vertical and over your head. Doing so may cause a back injury and significantly reduce control over the tool.
- Be sure to wear appropriate eye protection, gloves, and safety shoes or boots.

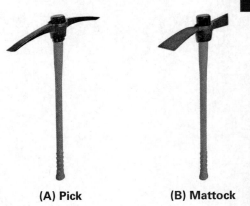

(A) Pick   (B) Mattock

Figure 56 Pick and mattock.
Source: A.M. Leonard, Inc.

3.0.0 Section Review

1. To cut through a metal pipe, the correct saw to use is a _____.
   a. coping saw
   b. crosscut saw
   c. drywall saw
   d. hacksaw

2. Which type of clamp is best suited for a very wide clamping application, such as an 8' span across the jaws?
   a. C-clamp
   b. Bar clamp
   c. Spring clamp
   d. Pipe clamp

3. Which of the following file types would leave the smoothest finish?
   a. Bastard
   b. Coarse
   c. Second cut
   d. Rough

4. Which tool is best for moving gravel?
   a. Square shovel
   b. Round shovel
   c. Mattock
   d. Spade

Module 00103 Review Questions

1. A slightly rounded face on a hammer _____.
   a. helps deliver a harder blow
   b. prevents overstriking
   c. is only found on drywall hammers
   d. helps avoid significant damage to the material when nailing

2. What type of hammer is shown in *Figure RQ01*?

Figure RQ01

Source: Image property of Stanley Black & Decker.
Used with permission

a. Curved claw hammer
b. Rip claw hammer
c. Drywall hammer
d. Ball peen hammer

3. The tool shown in *Figure RQ02* is a _____.

Figure RQ02

Source: Image property of Stanley Black & Decker.
Used with permission

a. cat's paw
b. wrecking bar
c. wood chisel
d. cape chisel

4. The cape chisel is best for _____.
a. shaving a broad area of wood
b. cutting off sections of rod
c. cutting grooves in metal
d. cutting off rusted hardware

5. Which type of chisel is typically designed to be used without a hammer?
a. Cold chisel
b. Paring chisel
c. Cape chisel
d. Diamond-pointed chisel

6. The best punch used to mark a spot for drilling a hole is the _____.
a. center punch
b. tapered punch
c. prick punch
d. angle punch

7. The most common type of crosshead screwdriver is the _____.
a. Robertson® screwdriver
b. Torx® screwdriver
c. slotted screwdriver
d. Phillips screwdriver

8. The wrench that is *most* likely to fit only one size of hardware is the _____.

    a. open-end wrench

    b. box-end wrench

    c. combination wrench

    d. adjustable wrench

9. In *Figure RQ03*, the fastener is likely being tightened.

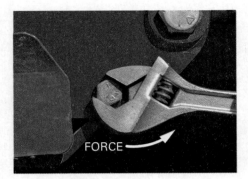

Figure RQ03

    a. True

    b. False

10. Which of the following statements about ratchets and sockets is correct?

    a. The drive size refers to the size of the hardware a ratchet handle fits.

    b. Common metric drive sizes are 1 cm, 1.5 cm, and 2 cm.

    c. Most sockets have either 6 or 12 gripping flats.

    d. Socket extensions are available in only one length for each drive size.

11. Pliers should *not* be used to _____.

    a. tighten or loosen hardware

    b. cut wire

    c. bend wire

    d. grip pipe

12. A disadvantage of a laser measuring tool is that it _____.

    a. can only measure up to about 100' (30 m)

    b. does not work well in smoky conditions

    c. must be powered with a drop cord

    d. requires several minutes to complete a measurement

13. When something is plumb, it is _____.

    a. precisely vertical

    b. level to the horizon

    c. parallel to the ground

    d. at some angle other than 90 degrees

14. The tool shown in *Figure RQ04* is called a _____.

Figure RQ04

Source: Image property of Stanley Black & Decker. Used with permission

   a. plumb level
   b. post level
   c. torpedo level
   d. laser level

15. The try square is made at a fixed _____.

   a. 45-degree angle
   b. 90-degree angle
   c. 180-degree angle
   d. 360-degree angle

16. Accurate and smooth miter joints in wood can be cut with a _____.

   a. keyhole saw
   b. hacksaw
   c. drywall saw
   d. backsaw

17. A hand-screw clamp has _____.

   a. metal jaws
   b. nylon jaws
   c. wooden jaws
   d. fiberglass jaws

18. While files have slanting parallel rows of teeth, a rasp _____.

   a. is almost smooth
   b. has deeper, individual teeth
   c. has an open cutting surface
   d. has teeth designed for cutting on the backstroke

19. How is a carpet knife different from a common utility knife?

   a. The blade has a single sharpened edge.
   b. The blade has two sharpened edges.
   c. A carpet knife is much longer than a utility knife.
   d. A carpet knife has a serrated blade.

20. A spade is used to _____.

   a. clear tree roots
   b. move gravel or clean up construction debris
   c. move large volumes of soil
   d. shape trenches and holes requiring straight sides

Answers to Section Review Questions and Module Review Questions are found in *Appendix A.*

MODULE 00104

Introduction to Power Tools

Objectives

Successful completion of this module prepares you to do the following:

1. Identify and explain how to use various types of power drills and impact wrenches.
    a. Summarize basic power tool safety guidelines.
    b. Identify common power drills and bits and explain how to use them.
    c. Describe the difference between hammer drills and impact drivers.
    d. Identify pneumatic drills and impact wrenches and explain how to use them.
2. Identify and explain how to use various types of power saws.
    a. Explain how to use a circular saw and identify different types of blades.
    b. Differentiate between jigsaws and reciprocating saws and explain how to use them.
    c. Explain how to use a portable band saw.
    d. Describe the difference between miter saws and cutoff saws.
    e. Explain how to use table saws and describe the types of jobs for which they are best suited.
3. Describe the types of jobs best suited to grinders and oscillating multi-tools.
    a. Explain how to use various types of grinders.
    b. Identify grinder accessories and the jobs for which they are used.
    c. List the type of jobs that can be performed using an oscillating multi-tool.
4. Identify and explain how to use miscellaneous power tools.
    a. Discuss the hazards of using power nailers.
    b. Describe jobs that can be performed with hydraulic jacks.

Performance Task

Under supervision, you should be able to do the following:

1. Safely and properly demonstrate the use of the following tools:
    - Electric drill (corded and cordless)
    - Hammer drill
    - Impact driver
    - Circular saw
    - Jigsaw
    - Reciprocating saw
    - Portable band saw
    - Miter or cutoff saw
    - Table saw

- Portable or bench grinder
- Oscillating multi-tool
- Power nailer

Overview

Power tools play an important role in the construction industry. Thousands of construction workers across the world use power tools every day to make holes; cut different types of materials, smooth rough surfaces, and shape a variety of products. Regardless of their specialization, all construction workers eventually use power tools on their job. This module provides an overview of the common types of power tools and how they function. It also describes the proper techniques required to ensure their safe and efficient operation.

Industry Recognized Credentials

If you are training through an NCCER-accredited sponsor, you may be eligible for credentials from NCCER's Registry. The ID number for this module is 00104. Note that this module may have been used in other NCCER curricula and may apply to other level completions. Contact NCCER's Registry at 888.622.3720 or go to **www.nccer.org** for more information.

1.0.0 Power Drills/Drivers

Performance Task

1. Safely and properly demonstrate the use of the following tools:
   - Electric drills/drivers (corded and cordless)
   - Hammer drill
   - Impact driver

Objective

Identify and explain how to use various types of power drills and impact wrenches.

a. Summarize basic power tool safety guidelines.
b. Identify common power drills and bits and explain how to use them.
c. Describe the difference between hammer drills and impact drivers.
d. Identify pneumatic drills and impact wrenches and explain how to use them.

Alternating current (AC): The common power supplied to most all wired devices, where the current reverses its direction many times per second. AC power is the type of power generated and distributed throughout settled areas.

Direct current (DC): An electric power supply where the current flows in one direction only. DC power is supplied by batteries and by transformer-rectifiers that change AC power to DC.

This module introduces four categories of power tools:

- *Electric tools (corded and cordless)* — These tools are powered by either of two types of electricity: **alternating current (AC)** or **direct current (DC)**. Tools powered by AC, plug into a wall receptacle using an integrated power cord. DC-powered tools, known as cordless tools, operate on rechargeable batteries and make up the vast majority of power tools used on today's construction sites.

- *Pneumatic tools* — These tools are powered by air. Electric or gasoline-powered compressors produce the air pressure. Air hammers and pneumatic nailers are examples of pneumatic tools.

- *Specialty tools* — These tools include powder-actuated fastening tools that use gun powder to drive pins and studs into concrete and steel. Power-actuated tools require special training and certification to use.

- *Hydraulic tools* — These tools are powered by fluid pressure. Hand pumps or electric pumps are used to produce the fluid pressure. Pipe benders, jackhammers, and wrenches are examples of hydraulic tools.

Before reviewing specific types of tools from each of these categories, it is important to address general safety guidelines that apply to all power tools, regardless of category.

1.1.0 Basic Power Tool Safety Guidelines

Always remember, if not used properly, power tools have the potential to be dangerous. Safety guidelines are covered for each power tool in this module, but general safety issues—safety in the work area, safety equipment, and working with electricity—are covered in the *Basic Safety* module. This information is vital for working with power tools, and trainees must complete the *Basic Safety* module before beginning this module.

WARNING!

Module 00101, *Basic Safety (Construction Site Safety Orientation)*, MUST be completed before studying this module. Remember that appropriate personal protective equipment (PPE) must be worn when operating any power tool or when near someone else who is operating a power tool.

Basic safety guidelines that must be followed when working with all power tools include the following:

- Always read and follow guidelines provided in the power tool's *Operating Instructions* before attempting to use it.
- Never operate power tools without proper instruction and supervision.
- Confirm you have the right power tool for the job.
- Make sure the tool has been disconnected from its source of energy when not in use and before performing maintenance or replacing parts, such as bits, blades, or discs.
- Regularly inspect power tools, including power cords and hoses, for damage. Report any damaged tools to your supervisor.
- Proper PPE, including safety glasses, are required, regardless of what power tool is in use. Some tools require the use of a face shield in addition to safety glasses. Safety shoes should also be worn, and tight-fitting gloves are required for most power tools. Loose gloves can be a safety hazard of their own.
- Keep other craftworkers who are not directly involved with the work at a safe distance.
- Do not wear loose-fitting clothing that could be caught in the mechanism of your power tool.
- Never use electric power tools in wet conditions unless they are approved for this type of application. In all cases, when a power tool has a three-pronged plug, make certain it is plugged into a grounded receptacle. Additionally, before plugging into a receptacle, confirm the power switch on the tool is set to the Off position.
- If a power tool is equipped with a **trigger lock**, do not use the lock. A trigger lock is a small lever, switch, or part that can be used to activate a locking catch or spring to hold a power tool trigger in the operating mode even when the trigger is released. Locking any power tool in the On position can be very dangerous.

Trigger lock: A small lever, switch, or part that can be used to activate a locking catch or spring to hold a power tool trigger in the operating mode without finger pressure.

WARNING!

The use of trigger locks may be prohibited by an employer or job site. If a power tool is equipped with a trigger lock, DO NOT activate it.

- Never carry an electric or pneumatic tool by its cord or hose.
- Do not allow yourself to be distracted while using a power tool. If you are distracted or interrupted while operating the tool, turn it off.

1.2.0　Types of Power Drills/Drivers

Various types of power drills, also known as *drivers*, are used in the construction industry. A power drill is most commonly used to make holes by spinning drill bits into wood, metal, plastic, or other materials. With different attachments and accessories, the power drill can be also used as a sander, polisher, screwdriver, grinder, or **countersink**.

The most common types of power drivers include the following:

- Electric (corded) drills
- Electric cordless drills
- Electromagnetic (mag drill press) drills
- Hammer drills and impact drivers
- Pneumatic drills and wrenches
- Electric screwdrivers

1.2.1　Power Drills and Bits

Most power drills have several common features. For example, most have a pistol grip with a trigger switch for controlling power (*Figure 1*).

The farther back the trigger of a variable-speed drill is pulled, the faster the drill spins. Drills also have reversing switches that enable the drill to spin backwards in order to back the drill bit out if it gets stuck in the material while drilling. Power drills use replaceable bits for different drilling tasks (*Figure 2*). On variable-speed power drills, a screwdriver bit can be used in place of a drill

Countersink: A bit or drill used to set the head of a screw at or below the surface of the material.

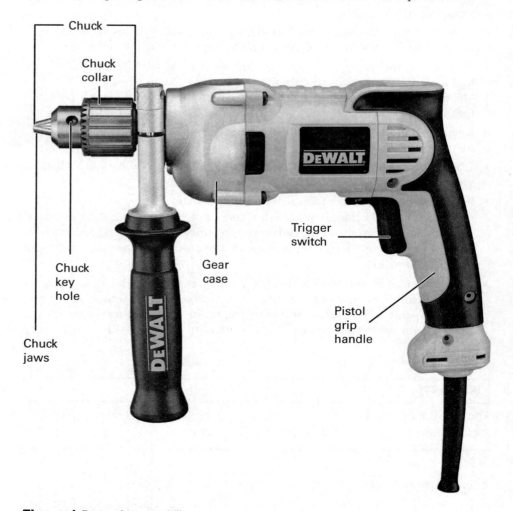

Figure 1 Parts of power drill.
Source: Image property of Stanley Black & Decker. Used with permission

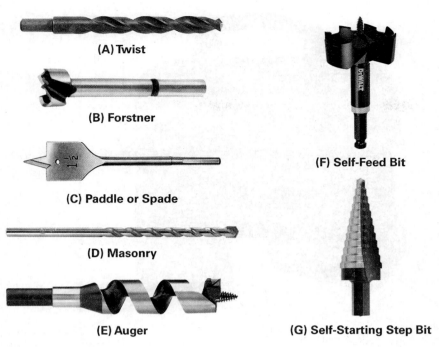

Figure 2 Drill bits.
Source: Image property of Stanley Black & Decker. Used with permission (2A, 2D–2G);
Courtesy of Milwaukee Electric Tool Corporation (2B–2C)

Figure 3 Chuck key.
Source: Image property of Stanley Black & Decker. Used with permission

bit so that the drill can be used as a screwdriver. Only use screwdriver bits designed for use in a power drill.

Twist drill bits are used to drill wood and plastics at high speeds or to drill metal at a lower speed. A **Forstner bit** is used on wood and is particularly good for boring a flat-bottom hole. A paddle bit or spade bit is also used in wood. The bit size is measured by the paddle's diameter, which generally ranges from ½ to 1½ inches (≈13 to 38 mm). A **masonry bit**, which has a carbide tip, is used in concrete, stone, slate, and ceramic material. The **auger bit**, which is designed to turn at low speeds, is used for drilling wood and other soft materials, but not for drilling metal. A **self-feed bit** uses a lead screw that pulls the larger, rear portion of the bit into the wood, making it simple to drill larger holes. The **step drill bit** is perfect for drilling holes in metal up to ¼ inch thick. As the name implies, this bit includes different 'steps' that allow the user to cut a hole to a desired diameter.

Drill bit sizes can vary a great deal. For most applications, bits are available in both fractional-inch sizes and metric sizes. For twist drill bits, common fractional-inch sizes range from ¹⁄₆₄-inch to 1 inch. Common metric sizes for twist drill bits range from 0.5 mm to 25.0 mm. It is important to note that fractional-inch bits and metric bits are not directly interchangeable. In other words, no fractional-inch bit has a corresponding metric bit, and vice versa. However, twist drill bits are also available in sets of numbered and lettered bits. Numbered sets range from #80 up to #1, which is the largest of the set. The next bit size larger is the A bit, with the Z bit being the largest of the lettered set. These sets correspond to decimal fraction and metric sizes. For example, a #8 drill bit is sized at 0.199 inches and 5.055 mm. Number and letter drills are used or specified when the precise size of the hole is important.

Bits are held in a drill by the drill **chuck**. Keyed chucks are opened and closed using a **chuck key** (*Figure 3*). Chuck keys are typically interchangeable in design, but there are several different sizes. Keyless chucks are typically found on cordless drills. Remember that the size of the chuck limits the size of the bit or tool a drill can accommodate.

1.2.2 Cordless Tools and Batteries

The vast majority of tools on today's construction sites are cordless. Cordless tools offer two important benefits: safety and mobility. They are safer, and more mobile,

Forstner bit: A bit designed for use in wood or similar soft material. The design allows it to drill a flat-bottom blind hole in material.

Masonry bit: A drill bit with a carbide tip designed to penetrate materials such as stone, brick, or concrete.

Auger bit: A drill bit with a spiral cutting edge for boring holes in wood and other materials.

Self-feed bit: A drill bit that uses a lead screw to pull the larger rear portion of the bit into the wood, making it simple to drill larger holes.

Step drill bit: A drill bit that includes steps which enable the user to cut holes in metal to a desired diameter.

Chuck: A clamping device that holds an attachment; for example, the chuck of the drill holds the drill bit.

Chuck key: A small, T-shaped steel piece used to open and close the chuck on power drills. Cordless drills typically do not have a chuck key.

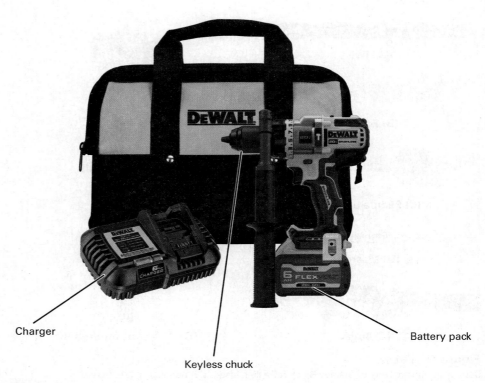

Charger

Keyless chuck

Battery pack

Figure 4 Cordless drill.
Source: Image property of Stanley Black & Decker. Used with permission

Charge cycles: Describe one complete charge and discharge cycle for a rechargeable battery.

Amp hour (Ah): A rating that describes the maximum amps a battery can provide continuously for up to 60 minutes. Drawing more amps than the battery's rating is possible, but it reduces the amount of time before the battery is drained.

Voltage rating: A measurement of the power the tool can deliver.

because they lack cumbersome power cords that can cause people to stumble and fall, and they are handy for working in areas where a power source is unavailable.

Cordless drills (see *Figure 4*) use a battery pack to power the motor. There are three battery pack types commonly used with today's cordless tools: Nickel-Cadmium (NiCd), Nickel-Metal Hydride (NiMH), and Lithium-ion (Li-ion).

The battery pack can be detached and plugged into a battery charger any time the tool is not in use. Some chargers can recharge the battery pack quickly, while others require extended periods of charging time. The quality of charge and the lifecycle of the battery must be considered when determining how to best charge the battery. Li-ion batteries have several advantages over the other NiCd and NiMH batteries, including offering the highest number of **charge cycles** as well as the ability to run the longest on a full charge. The charge cycle describes one complete charge and discharge cycle of the tool's rechargeable battery.

The term **Amp hour (Ah)** describes the maximum amps a battery can provide continuously for up to 60 minutes. Drawing more amps reduces the time before the battery is drained and needs to be recharged. For example, under ideal conditions, a 4Ah battery is capable of providing up to 4 amps continuously for one hour before the battery is drained. Drawing 8 amps from the same battery reduces the same tool's runtime to 30 minutes.

One other critical factor to consider when using cordless tools is the battery's **voltage rating**. A cordless tool's voltage rating is actually a measurement of its power. The higher the voltage, the more power the tool can deliver for the job.

Regardless of the battery type, every manufacturer provides information related to the strength and life of the battery in their product literature. Craftworkers who use cordless tools often carry extra battery packs with them.

Some cordless drills have adjustable clutches so that the drill motor can serve as a power screwdriver without applying too much power to the screw. Note that only heavy-duty cordless drills use keyed chucks, while the vast majority are keyless.

1.2.3 Power Drill Safety and Maintenance Guidelines

In addition to following basic power tool safety guidelines, familiarize yourself with these power drill safety rules before operating a power drill:

- Always wear the appropriate PPE when working with drills, especially safety glasses.
- Disconnect the power source (power cord or battery) before performing maintenance or changing drill bits.
- Keep your hands away from the drill bit and chuck.

WARNING!

Keep your hands away from a spinning drill bit. The spinning bit will cut your hands. Keep an even pressure on the drill to keep the drill from twisting or binding.

- When using a corded drill, operate only tools that are double-insulated electric power tools with proper **ground fault protection**. Using a **ground fault circuit interrupter (GFCI)** device protects equipment from continued electrical current in case of a circuit fault. The GFCI monitors the current flow and opens the circuit (which stops the flow of electricity) if it detects a difference between positive and negative flow.
- When using power drills with two-prong plugs, confirm that they are double-insulated. Tools that are not double insulated must have a third prong to provide a ground circuit. Do not use a power drill if the ground prong on the plug is broken off.
- Before connecting a drill to its power source, confirm the trigger is NOT turned on.
- To avoid hitting water lines or electrical wiring when drilling through a wall or partition, verify what is inside the wall or on the other side of the work surface beforehand.

Ground fault protection: Protection against short circuits; a safety device cuts power off as soon as it senses any imbalance between incoming and outgoing current.

Ground fault circuit interrupter (GFCI): A circuit breaker designed to protect people from electric shock and to protect equipment from damage by interrupting the flow of electricity if a circuit fault occurs.

WARNING!

Do not drill into or through a wall before confirming what is on the other side. Take steps to avoid hitting anything that would present a safety hazard or cause damage. Spaces between studs (upright pieces in the walls of a building) often contain electrical wiring, plumbing, or insulation. Care must be taken to avoid drilling directly into the wiring, pipes, or insulation.

- Prior to drilling, select the proper bit for the job and make sure it is sharp. Make sure the drill bit is tightened securely in the chuck before using it.
- When using drills with keyed chucks, attach the key to the power cord to ensure the key is available when tightening or loosening a bit.
- Hold the drill with both hands and apply steady pressure.
- Never ram the drill while drilling because the force can chip the cutting edge and damage the bearings.
- As noted earlier, never use a drill's trigger lock to hold the trigger in the operating mode.
- Do not overreach when using a power drill; maintain a stable and balanced stance.
- Keep your power drill clean at all times.

Figure 5 Right-angle drills.
Source: Courtesy of Milwaukee Electric Tool Corporation

WARNING!

If power tools are shared between workers, it is important NOT to grab the tool by the bit or any other moving part. Power tools should be disconnected from their source of power whenever they are passed between workers to avoid accidental startups. Hand off tools by their handles and/or bodies only.

1.2.4   Preparing to Use a Power Drill

Prior to using a drill, make sure the one you plan to use is designed to perform the task at hand. Some electric power drills are made to be used in tight spaces, such as between studs and joists. Drills like those shown in *Figure 5* are referred to as right-angle drills. The drill on the right is larger and develops more power for larger holes.

Preparing Drills with Keyed Chucks

Once the proper drill has been selected, it is time to prepare it for use. Follow the steps below to prepare the power drill:

Step 1   Disconnect the drill from its power source.
Step 2   Turn the chuck counterclockwise (to the left) until the chuck opening is large enough to insert the bit **shank**. The shank is the smooth part of the bit.
Step 3   Insert the bit shank into the chuck opening (*Figure 6A*). Keeping the bit centered in the opening, turn the chuck by hand until the jaws grip the bit shank.
Step 4   Insert the chuck key (*Figure 6B*) into one of the holes on the side of the chuck and turn the key clockwise (to the right) to tighten it. With larger chucks, you can tighten the individual jaws of the chuck uniformly by inserting and tightening the key in each of the holes on the chuck. Remove the key from the chuck once it is tightened. You are now ready to use the electric drill.

WARNING!

Do not forget to remove the key from the chuck after it has been tightened. Otherwise, when you start the drill, the key could fly out and injure you or a co-worker.

Shank: The smooth part of a drill bit that fits into the chuck.

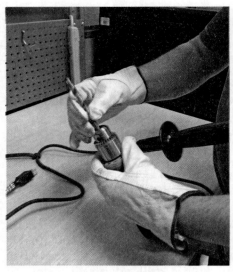
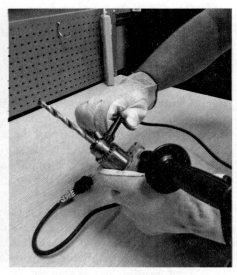

(A) Insert the Bit Shank into the Chuck Opening.

(B) Tighten with the Chuck Key.

Figure 6 Loading the bit on a keyed chuck.
Source: Cianbro Corporation

Preparing Drills with Keyless Chucks

Most cordless drills use a keyless chuck. While the steps for preparing a cordless drill are similar, there are some small differences. Follow the steps below when preparing to use drills with keyless chucks:

Step 1 Disconnect the drill from its power source by removing the battery pack before loading a bit.

Step 2 As shown in (*Figure 7A*), open the chuck by turning it counterclockwise until the jaws are wide enough to insert the bit shank.

Step 3 Insert the bit shank into the chuck opening (*Figure 7B*). Keeping the bit centered in the opening, turn the chuck by hand until the jaws grip the bit shank.

Step 4 Tighten the chuck securely with your hand so that the bit does not move (*Figure 7C*). You are now ready to use the cordless drill.

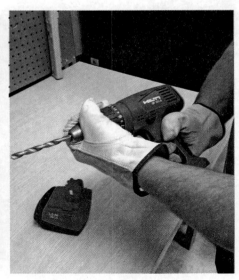

(A) Insert the Bit Shank.

(B) Keep Bit Straight and Partially Tighten the Chuck.

(C) Tighten the Chuck Securely.

Figure 7 Loading the bit on a keyless chuck.
Source: Cianbro Corporation

1.2.5   Operating Power Drills

After securing the bit, connect the drill to its power source. Use the following steps to operate the power drill:

Step 1   Put on the appropriate PPE.

Step 2   Make a small indent or mark on the material exactly where the hole needs to be drilled. In wood, use a small punch to make the indent; in metal, use a center punch.

Step 3   Firmly clamp or support the material being drilled, then hold the drill perpendicular (at a right angle) to the material surface and start the drill motor. With a variable-speed drill, start the bit slowly and verify the drill is rotating in the right direction. (When the bit is faced away, it should turn clockwise.)

Step 4   Place the tip of the bit on the indent or mark and hold the drill with both hands. Apply only moderate pressure when drilling. *Figure 8* shows the proper way to hold the drill. The drill motor should operate at approximately the same **revolutions per minute (rpm)** as it does when pressure is not exerted. If the sound of the drill indicates it is slowing considerably, apply less pressure.

Step 5   Reduce the pressure when the bit is about to emerge from the other side of the work, especially when drilling metal. If the drill bit gets stuck in the material during drilling, release the trigger, use the reversing switch to change the direction of the drill, and gently back it out of the material.

Revolutions per minute (rpm): The rotational speed of a motor or shaft, based on the number of times it rotates each minute.

Figure 8 Properly holding a drill.
Source: Cianbro Corporation

> **WARNING!**
>
> As the drill bit emerges through the opposite side of the material, be prepared for it to grab. With power still on, the tendency of the drill motor is to rotate the operator, and not the bit. Be sure to maintain firm control if this happens.

Drilling Metal

When drilling metal, it is best to lubricate the bit to help cool the cutting edges and produce a smoother finished hole. A small amount of non-combustible cutting oil or tapping fluid makes a good lubricant for drilling softer metals. Lubrication is *not* needed for drilling wood. When drilling deep holes, pull the drill bit partly out of the hole occasionally. Doing so clears the hole of shavings.

> **WARNING!**
>
> To avoid injury, adjust the drill chuck properly prior to each use. Do not use the tool motor to tighten the chuck while gripping it.

1.2.6   Electromagnetic Drill Presses

An electromagnetic drill press, commonly known as a *mag drill press*, is a portable drill mounted on an electromagnetic base (*Figure 9*). This type of tool is designed for drilling holes in thick metal. Once the drill is placed on a metal surface, the power turned on and the magnet energized, the magnetic base will hold the drill in place for drilling. Some drills can also be rotated in place while the base remains stationary.

Going Green

Recycling Rechargeable Batteries

Recycling rechargeable batteries once their power supply has become exhausted is no longer a luxury; it has become a matter of necessity. In fact, federal and state laws now regulate the disposal of some types of rechargeable batteries.

The US Environmental Protection Agency (EPA) estimates that more than 350 million rechargeable batteries are purchased annually in the United States. Likewise, hundreds of millions of rechargeable batteries and cellphones are retired each year. Rechargeable batteries are made using heavy metals and elements such as nickel, lithium, cadmium, mercury, and lead. All of these elements can be toxic to people and the environment if not disposed of properly. Most landfills are not designed to handle the toxic metals that will eventually leak out of the batteries.

Rechargeable batteries are commonly found in the following:

- Cordless power tools
- Cellular and cordless phones
- Two-way radios
- Laptop computers
- Digital cameras
- Camcorders

If a battery is rechargeable, then it is recyclable. Not only are rechargeable batteries better for the environment if they are discarded properly, but they are more cost effective. Using rechargeable batteries can help save money and protect the environment at the same time.

Recently, most cities have added hazardous waste collection centers that collect both rechargeable and regular batteries, along with paint, oil, refrigerant, and other hazardous wastes.

A switch on the junction box controls the electromagnetic base. When the switch is turned on, the magnet holds the drill in place on any surface with magnetic characteristics, such as carbon steel. The drill base must be clean and the surface must be flat. The switch on the top of the drill turns the drill motor on and off. A depth gauge can be used to set the depth of the hole being drilled. It operates like a drill press, with the operator turning a hand wheel to raise and lower the drill against the workpiece. Workers who use electromagnetic drills should be properly trained to safely operate the specific drill being used.

Figure 9 Electromagnetic drill press (mag drill press).
Source: Image property of Stanley Black & Decker. Used with permission (Photo)

Figure 10 Electromagnetic drill with safety chain.
Source: Cianbro Corporation

There are several safety rules specific to mag drill presses, including the following:

- Securely clamp the material being drilled by the press. Unsecured materials can become dangerous flying or spinning objects.
- Since mag drill presses are mounted onto an electromagnetic base, it is important to maintain electrical power to the drill's electromagnet. To prevent the electrical power supply from being interrupted, attach a *Do Not Unplug* tag (*Figure 9*) to the cord.

WARNING!

Removing power from an electromagnetic drill while it is in use can cause the drill to fall, potentially harming the operator.

- Use a safety chain to secure the electromagnetic drill in case power is lost or shut off (*Figure 10*).
- Before using a mag drill press on a construction site, confirm whether safety attachments must be used. Some sites require shields and/or safety lines to be attached before a mag drill press can be operated.

1.3.0 Hammer Drills and Impact Drivers

A hammer drill (*Figure 11*) has a light pounding action that enables it to drill into concrete, brick, or tile. The bit rotates and hammers at the same time, allowing faster drilling in these types of hard materials than would be possible with a regular drill. When the hammering action is disabled on these drills, they can also perform basic drilling tasks using other types of bits. The depth gauge on a hammer drill can be set to the depth of the hole to be drilled.

Special bits used with hammer drills are designed for pounding while they are turning. As a result, the bits used for drilling holes in masonry and rock have very hard tips, which are frequently made from **carbide**. Carbide-tipped bits are actually made by coating the tip of the drill bit with a mixture of tungsten carbide and cobalt baked at extremely high temperatures. The result is a tip that is much harder than those found on common bits.

Carbide: A very hard material made of tungsten carbide and heavy metals such as cobalt. Commonly baked on to the tips of masonry bits and to the edges of saw blades.

Did You Know

Almost as Hard as Diamonds

The hardness of a substance is often determined using the Rockwell hardness scale. To determine where a substance's hardness falls on the scale, a Rockwell hardness tester presses a pointed diamond into the substance being tested. The amount of force and depth of the indention determines the substance's hardness, which is shown as a number on the scale. Higher numbers indicate harder substances. By comparison, diamonds are one of the only substances harder than tungsten carbide. This is the reason carbide-tipped bits are harder and last much longer than regular drill bits.

Keyless chuck

Battery pack

Figure 11 Hammer drill.
Source: Image property of Stanley Black & Decker. Used with permission

The term *hammer drill* is often used to describe all tools that hammer and drill. However, rotary hammers (*Figure 12*) are designed for much heavier jobs than a typical hammer drill. They usually have slower rotational speeds than hammer drills, and strike harder and less often. Most rotary hammers require bits that fit into special chuck designs (*Figure 13*), so selected masonry bits must be compatible with the type of chuck on the tool. Chucks on rotary hammers are not keyed, but rely upon the design of the bit and chuck to hold it in place. Adapters are also available to use one bit shank design with another type of chuck when necessary.

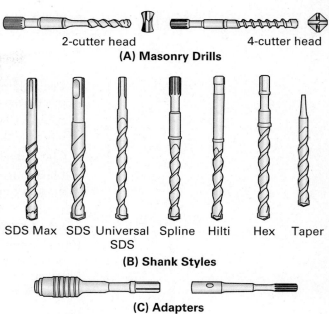

(A) Masonry Drills

2-cutter head   4-cutter head

SDS Max   SDS   Universal   Spline   Hilti   Hex   Taper
SDS

(B) Shank Styles

(C) Adapters

Figure 13 Rotary hammer bit designs.

Figure 12 Rotary hammer.
Source: Image property of Stanley Black & Decker. Used with permission

Did You Know

SDS Plus Vs SDS Max Drill Bits

SDS technology originated in the mid-1970s when Hilti & Bosch introduced a new way of fastening bits to hammer and rotary drills. In the early days of the technology, many people insisted the letters SDS stood for the German words Stecken – Drehen – Sichern, which translates to Insert, Twist, and Secure. Today, SDS is known to refer to a Slotted Drive System.

When it was introduced, SDS allowed rotary hammers to lock a bit in place while still allowing the bit to move up and down. Because the chuck was spring loaded, SDS bits could be pushed into the chuck without having to tighten them. The next evolution of the system came with the introduction of SDS Plus, which offered an improved connector yet remained backward compatible with standard SDS bits. Finally, the SDS Max system was introduced to handle the heaviest masonry and steel drilling jobs.

SDS Plus bits range in sizes from $5/32$" to $1\frac{1}{4}$", with thin wall carbide core bits up to 4" in length. SDS Max bits range from $\frac{1}{2}$"–2", with thick wall carbide core bits up to 4" in length.

SDS Plus   SDS Max

Source: Belts And Boxes

1.3.1   Hammer Drill Safety and Maintenance

In addition to following basic power tool safety guidelines, before operating a hammer drill familiarize yourself with these safety rules:

- Always wear the appropriate PPE when working with hammer drills, especially safety glasses and hearing protection.

- Never operate hammer drills or rotary hammers with one hand. Use both hands for maximum control.
- Do not operate hammer drills or rotary hammers near flammable materials. Sparks generated by the bits may ignite them.
- Use a clamp to secure smaller or lighter work pieces and never hold a work piece with any part of your body.
- Always disconnect the hammer drill from its power source before changing bits or making any adjustments.
- Check the hammer drill and bit for damage prior to using it on a job.
- Make sure you are using the correct bit for the job. Using regular, non-masonry bits on concrete can damage or destroy the bit.

1.3.2 Operating Hammer Drills

If the hammer drill is being used on concrete or similar materials that produce airborne dust, there are important silica exposure standards that must be followed. Silica exposure standards are outlined in the *Basic Safety* module.

Use the following steps to operate a hammer drill:

Step 1 Don the appropriate PPE.
Step 2 Ensure you have selected the appropriate drill bit for the job. If you need to replace the bit, disconnect the drill from its power source prior to doing so. Once the bit has been replaced, reconnect the drill to its power source.
Step 3 Make a small indent or mark on the material exactly where the hole needs to be drilled.
Step 4 Set the number of blows the bit will exert each minute.
Step 5 Set the drill to turn slowly.
Step 6 Place the tip of the bit on the indent or mark, then apply pressure on the drill against the surface.
Step 7 Drill a guide hole until it is approximately ¼" deep.
Step 8 Adjust the speed of the hammer drill to the desired speed and finish the job.
Step 9 Clean out the hole when finished.

Around the World

Power Sources

Each country provides power to its residents at a voltage and frequency that they believe best for the residents' needs. While the voltage often varies slightly from the target voltage, most power tools can accommodate these minor day-to-day differences. However, users must always ensure that the power supply for a given tool is appropriate. Although slight changes in voltage do not represent a problem, changes in power frequency and significant changes in the voltage will damage or destroy a power tool.

Here are the power characteristics for single-phase power from several different countries to show their differences. Note that the frequency, reported in hertz (Hz), refers to the cyclic nature of alternating current (AC) and how many cycles are completed per second:

- Portugal: 230 volts @ 50 Hz
- Hungary: 220 volts @ 50 Hz

- England: 240 volts @ 50 Hz
- Sweden: 230 or 400 volts @ 50 Hz
- United States: 115 or 230 volts, 60 Hz

The United States is one of very few countries that provide power at 60 Hz. Many devices can operate using power from almost all international power systems. These devices are often battery-charging devices that use a transformer to reduce the voltage and change the voltage to direct current (DC) to charge a battery. Other devices, such as computers and printers, may be equipped with external switches to allow the use of different power sources. However, few if any portable power tools have been designed with such a switch, and corded tools have no transformer. So before packing away your favorite power saw for work in another country, be sure that there is a power supply in place that can support the tool's required voltage and frequency.

1.3.3 Impact Drivers

Impact drivers (*Figure 14*) are usually lighter and more compact than common power drills and hammer drills. They also apply more torque when fastening screws or bolts. As a result, impact drivers are excellent tools for driving (or loosening) large numbers of fasteners, screws, nuts, or lag bolts. They are also a good choice when working on harder surfaces such as pressure-treated lumber, hardwoods, or even steel. Unlike most power drills, they typically do not offer variable speeds, which makes impact drivers less suited for jobs requiring precision or detailed work.

Another more subtle difference between hammer drills and impact drivers is the way they function. Hammer drills exert downward, pounding force on the bit while the chuck is turning. Impact drivers use a spring to exert perpendicular force on the bit. When the bit detects resistance as it turns, the driver exerts greater force, called **torque**, on the turning bit. Torque is the force produced by the drill as it turns the bit.

Finally, impact drivers have a hexagonal-shaped collet, or clamp, instead of a chuck. The hexagonal design is needed because of the extra torque exerted on the bit by the impact driver.

1.3.4 Impact Driver Safety and Maintenance

In addition to following basic power tool safety guidelines, familiarize yourself with these impact drill safety rules before operating the tool:

- Always wear the appropriate PPE when working with impact drivers.
- Read and understand the manufacturer's manual.
- Clean the impact driver before using it.
- Ensure the proper accessory is used for the job.
- Confirm that the fastener and material you are drilling into will not be damaged by the torque generated by the impact driver.

> **CAUTION**
>
> Impact drivers generate much more torque than regular drills. As a result, they can split, break, or rip through softer or more brittle surfaces. Be extra careful when using an impact driver to attach fasteners to wood or softer materials.

Torque: The turning force produced by the drill. As the bit's turning speed increases, torque decreases, and vice versa.

Figure 14 Impact driver.
Source: Image property of Stanley Black & Decker. Used with permission

1.3.5 Operating Impact Drivers

Use the following steps to operate an impact driver:

Step 1 Don the appropriate PPE.
Step 2 Select the appropriate bit or attachment for the job.
Step 3 Disconnect the driver from its power source.
Step 4 Pull the collet forward and insert the bit. Once the collet snaps back in place, the bit is attached.
Step 5 Reconnect the driver to its power source.
Step 6 Set the direction you want the bit to turn (tighten or loosen).
Step 7 Place the bit or socket squarely on the fastener.
Step 8 Start the tool slowly and release the trigger when the fastener is attached or removed. Driving the fastener too hard can damage it.

1.4.0   Pneumatic Drills and Impact Wrenches

Pneumatic tools are powered by compressed air. An air hose transfers the compressed air from an air compressor to the tool. Pneumatic tools tend to have more power for their weight than comparable electric tools. Two common pneumatic power tools used by construction workers are pneumatic drills and impact wrenches.

1.4.1 Pneumatic Drills

Pneumatic drills have many of the same parts, controls, and applications as electric drills. Since there is no motor, they are generally more compact in size. These drills are typically used when there is no available source of electricity, or when a high rate of production is necessary. Like electric drills, they can be used as power screwdrivers when fitted with the proper attachments. The pneumatic drill in *Figure 15* has a keyless chuck. They are also equipped with keyed chucks.

Common sizes of pneumatic drills are ¼-, ⅜-, and ½-inch. The size refers to the diameter of the largest bit shank that can be gripped in the chuck, not the drilling capacity. Some common metric sizes are 8 mm, 10 mm, and 13 mm.

1.4.2 Impact Wrenches

Impact wrenches are available in pneumatic (*Figure 16*) and electric (corded and cordless) (*Figure 17*) models. Both types are used to fasten, tighten, and loosen nuts and bolts. Like pneumatic drills, pneumatic impact wrenches must be connected with a hose to an air compressor. The speed and torque of these wrenches can be adjusted based on requirements of the job. Electric models lack the strength of pneumatic wrenches, but they are easier to maintain, especially since there is no need to service hoses and compressors.

While pneumatic impact wrenches have traditionally been more popular than their electric cousins, advances in battery-powered technology make cordless wrenches a more viable option today. Some of the most advanced cordless impact wrenches can generate up to 1,500 ft-lbs of breakaway torque.

Figure 15 Pneumatic drill.
Source: Image property of Stanley Black & Decker. Used with permission

Figure 16 Pneumatic impact wrench.
Source: Courtesy of Atlas Copco

Figure 17 Cordless impact wrench.
Source: Image property of Stanley Black & Decker. Used with permission

Although this is nowhere near the forward torque available on some larger pneumatic wrenches, it still provides a convenient and strong alternative for smaller jobs.

1.4.3 Pneumatic Tool Safety and Maintenance Guidelines

In addition to following basic power tool safety guidelines, familiarize yourself with the following pneumatic drill and impact wrench safety rules before operating either type of tool:

- Always wear the appropriate PPE, including eye, hand, and ear protection.
- Review the manufacturer's instructions provided with the pneumatic tool.
- Before changing attachments or performing any maintenance on a pneumatic drill or impact wrench, make sure the air supply is turned off and the tool is physically disconnected from the supply hose.
- Make sure that the workpiece is secure and always maintain a balanced body stance when operating the tool.
- Regularly check the air hose for damage.
- Keep your hands away from the working end of the tool.
- If you are using a pneumatic wrench to tighten a bolt-and-nut combination, use a backup wrench to keep the bolt or nut from spinning.
- Confirm the air supply is clean, dry, and at the proper pressure.
- Keep the drill or wrench lubricated.
- Use only those drill bits and impact sockets that are designed for use with the tool.

WARNING!

The amount of air pressure used to power a pneumatic drill is higher than the OSHA recommendation for cleaning a work area or similar activities. Improper use of pressurized air can lead to injuries as well as damage to the tool or materials.

1.4.4 Operating Pneumatic Drills and Impact Wrenches

Use the following steps to operate a pneumatic drill or impact wrench:

Step 1 Don the appropriate PPE.

Step 2 Prior to operating the pneumatic drill or impact wrench, ensure there is an oiler if the tool requires one, either at the air source or at the tool (*Figure 18*). Confirm the oiler is set properly.

Step 3 Attach the appropriate bit or socket to the tool.

Step 4 Make sure the pneumatic connection between the tool and the supply hose is secure, and install a whip check as required (*Figure 19*). A **whip check** is a safety attachment used to prevent whiplashing in hoses that are inadvertently uncoupled.

Whip check: A safety attachment used to prevent whiplashing in hoses when they are inadvertently uncoupled.

WARNING!

The air hose must be connected properly and securely. An unsecured air hose can come loose and whip around violently, causing serious injury. Some fittings require the use of whip checks to keep them from coming loose.

Step 5 Confirm the compressor's regulator is set to the appropriate PSI. Also, if tightening a bolt or nut, check the recommended torque specifications prior to using the tool.

WARNING!

Use only impact sockets made for pneumatic impact wrenches. Using handheld sockets can damage property and cause injury.

Figure 19 Connect the pneumatic tool and hose using the whip check.
Source: Capital Rubber Corporation. Used with permission

Figure 18 Inline pneumatic oiler reservoir.
Source: C. R. Laurence Co.

Step 6   Grip the drill or wrench firmly before operating the tool.
Step 7   When you finish using the tool, disconnect it from the air hose.

1.0.0 Section Review

1. The trigger lock on power tools should never be used to lock the trigger in the On position.
   a. True
   b. False

2. The part on a power drill that enables a user to back out a drill bit when it is stuck in the work material is called a(n) _____.
   a. chuck key
   b. reversing switch
   c. whip check
   d. auger switch

3. An important difference between hammer drills and rotary hammers is _____.
   a. rotary hammers turn faster
   b. hammer drills are designed for heavy duty jobs
   c. rotary hammers have keyed chucks
   d. rotary hammers have specially designed chucks

4. A pneumatic power tool that is best suited for fastening, tightening, and loosening nuts and bolts is a _____.
   a. cordless screwdriver
   b. hydraulic ratchet
   c. pneumatic impact wrench
   d. right-angle drill

2.0.0 Power Saws

Objective

Identify and explain how to use various types of power saws.

  a. Explain how to use a circular saw and identify different types of blades.

  b. Differentiate between jigsaws and reciprocating saws and explain how to use them.

  c. Explain how to use a portable band saw.

  d. Describe the difference between miter saws and cutoff saws.

  e. Explain how to use table saws and describe the types of jobs for which they are best suited.

Performance Task

1. Safely and properly demonstrate the use of the following tools:

- Circular saw
- Jigsaw or reciprocating saw
- Portable band saw
- Miter or cutoff saw
- Table saw

Using the right saw for the job will make your work much easier. Always make sure that the blade is right for the material being cut. This section focuses on the following types of power saws:

- Circular saws
- Jigsaw and reciprocating saws
- Portable handheld band saws
- Power miter saws
- Table saws

2.1.0   Circular Saws

Many years ago, a company named Skil® made power tool history by introducing the portable circular saw. Today, many companies make dozens of models, but a lot of people still refer to portable circular saws as Skilsaws. Other names include utility saw, electric handsaw, and builder's saw. The portable circular saw (*Figure 20*) is designed to cut lumber and boards to size and they are available in corded and cordless models.

The size of a circular saw is based on the diameter of the circular blade. Circular saws used in the United States typically use fractional-inch measurements with blade diameters that range from $3\frac{3}{8}$ to $16\frac{1}{4}$ inches. The most popular blade size for corded saws is $7\frac{1}{4}$ inches. Many smaller cordless circular saws use a $6\frac{1}{2}$-inch blade. The hole in the center of a circular saw blade fits onto the **arbor**, or shaft, of the saw. The most common arbor size for a circular saw blade is $\frac{5}{8}$ inch.

Circular saws are also available in metric sizes. Metric circular saws typically have circular blades of 165 mm, 190 mm, or 235 mm. Most metric circular saws have a 20 mm arbor, although some have a 25.4 mm arbor. It is important to note that standard saws with fractional-inch measurements are not interchangeable with metric saws. For instance, a $7\frac{1}{4}$-inch blade with a $\frac{5}{8}$-inch arbor hole is not the same as a 190 mm blade with a 20 mm arbor hole. Only metric blades can be used on metric saws and only fractional-inch blades can be used on standard saws.

Circular saw weights can vary, but most of them weigh between 7 and 14 pounds (3.175 kg and 6.35 kg). The handle of the circular saw has a trigger switch that starts the saw. The motor is protected by a rigid housing. Blade speed when the blade is not engaged in cutting is stated in rpm. The teeth of the blade point in the direction of rotation.

The blade is protected by two guards. On top, the upper blade guard protects workers from flying debris and from accidentally touching the spinning blade. The lower guard is spring-loaded; when the saw is pushed forward, it retracts up and under the top guard to allow the saw to cut.

Arbor: The end of a circular saw shaft where the blade is mounted.

Power switch   Depth adjustment

Handle   Tilt adjustment

Upper blade guard

Lower blade guard

Base

Guide slot

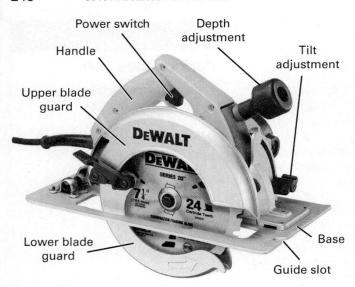

Figure 20 Portable circular saw.
Source: Image property of Stanley Black & Decker. Used with permission

Figure 21 Circular saw blades.
Source: Image property of Stanley Black & Decker. Used with permission

---

WARNING!

Never use a saw unless the lower blade guard is properly attached. The top guard protects the user from the blade and from flying particles.

Saw blades fall into two categories: standard steel and carbide-tipped. Standard steel blades must be sharpened regularly while carbide-tipped blades remain sharp for a much longer period of time. When using carbide-tipped blades, however, it is important to select the right blade for the job. *Figure 21* shows some of the most common types of saw blades, including the following:

- *Rip* — These blades are designed to cut with the grain of the wood. The square chisel teeth cut parallel with the grain and are generally larger than other types of blade teeth.

- *Crosscut* — These blades are designed to cut across the grain of the wood; that is, at 90-degree angle. Crosscut teeth cut at an angle and are finer than rip blade teeth.

- *Combination* — These blades are designed to cut hard or soft wood, either with or across the grain. The combination blade features both rip and cross-cut teeth with deep troughs (gullets) between the teeth.

- *Nail cutter* — This blade has large carbide-tipped teeth that can make rough cuts through nails that may be embedded in the work.

- *Nonferrous metal cutter* — This blade has carbide-tipped teeth for cutting aluminum, copper, lead, and brass. It should be lubricated with oil or wax before each use.

Always follow the manufacturer's instructions when using saw blades.

2.1.1 Circular Saw Safety Guidelines and Maintenance

There are numerous factors to consider in order to use a circular saw safely and effectively. In addition to following basic power tool safety guidelines, familiarize yourself with the following circular saw safety rules before operation:

- Always wear the appropriate PPE when working with saws, especially safety glasses.

- Always be aware of the blade's location as it relates to your hands or any other part of your body.

- Before connecting the saw to its source of power, ensure that the blade is tight and that the blade guard is working correctly. The chosen blade should have a maximum rpm equal to or higher than the speed of the saw.
- To avoid hitting water lines or electrical wiring, find out what is inside the wall or on the other side of a partition before cutting through a structure.
- During operation, keep both hands on the saw grips.
- Never force the saw through the work. This causes binding and overheating and may cause injury.
- Never reach underneath the work while operating the saw and never stand directly behind the work.
- Always stand to one side of the work.
- Use clamps to secure small pieces of material to be cut.
- Know where the saw's power cord is located at all times. Accidentally cutting through the power cord can cause electrocution.

WARNING!

When using a circular saw, workers should never hold material to be cut with their hands; always use a clamp instead.

The most important maintenance on a circular saw is at the lower blade guard. Sawdust builds up and causes the guard to stick. If the guard sticks and does not move quickly over the blade after it makes a cut, the bare blade may still be turning when the saw is set down and may cause damage. Remove sawdust from the blade guard area. Remember to always disconnect the power source before performing any maintenance. To avoid personal injury and damage to materials, check often to make sure the guard snaps shut quickly and smoothly. To ensure smooth operation of the guard, disconnect the saw from its power source, allow it to cool, and clean foreign material from the track. Be aware of fire hazards when using cleaning liquids such as isopropyl alcohol. Do not lubricate the guard with oil or grease. This could cause sawdust to stick in the mechanism. Always keep blades clean and sharp to reduce friction and kickback. Blades can be cleaned with hot water or mineral spirits. Be careful with mineral spirits as they are very flammable.

2.1.2 Operating Circular Saws

Use the following steps to operate a circular saw:

Step 1 Don the appropriate PPE.

Step 2 Secure and support the material you are cutting. If the work isn't heavy enough to stay in position without moving, it should be weighted or clamped down.

Step 3 Mark the cut using a pencil or marking tool.

Step 4 Unplug or remove the battery pack from the saw before adjusting the blade depth to the thickness of the wood being cut plus ¼-inch ($\approx$6 mm). This prevents the blade from protruding through the material farther than necessary.

Step 5 Place the front edge of the baseplate on the work so the guide notch and the blade are in line.

Step 6 After the saw has been started and is up to full speed, slowly move it forward to begin cutting. As the blade begins to cut, it will leave a **kerf** (*Figure 22*) that measures roughly ⅛-inch (3.2 mm) wide. It is important to consider the width of the kerf when making a cut with a circular saw. Be sure to align the blade with the waste side of the cutting line, or the finished piece will be short. When marking for the cut, mark an X on the waste side of the cut mark as a reminder of which side of the mark to cut.

CAUTION

Make sure the blade is appropriate for the material being cut.

Kerf: The channel created by a saw blade passing through the material, which is equal to the width of the blade teeth.

Figure 22 Saw kerf.
Source: Cianbro Corporation

Figure 23 Proper use of a circular saw.

Step 7 The lower blade guard will automatically rotate up and under the top guard when the saw is pushed forward. While cutting with the saw, grip the saw handles firmly with two hands, as shown in *Figure 23*. As the cut nears completion, the guide notch on the baseplate will move off the end of the work. At that point, use the blade as a guide.

Step 8 Release the trigger switch once the cut is complete. The blade will stop rotating. Make sure the blade has stopped before setting the saw down.

Source: Robert Bosch Tool Corporation

The Worm-Drive Saw

The worm-drive saw is a heavy-duty type of circular saw. Most circular saws have a direct drive. That is, the blade is mounted on a shaft that is part of the motor. With a worm-drive saw, the motor drives the blade from the rear through two gears. One gear (the worm gear) is cylindrical and threaded like a screw. The worm gear drives a wheel-shaped gear (the worm wheel) that is directly attached to the shaft to which the blade is fastened. This setup delivers much more rotational force (torque), making it easier to cut a double thickness of lumber. The worm-drive saw is almost twice as heavy as a conventional circular saw, and should be used only by an experienced craftworker.

2.2.0   Jigsaws and Reciprocating Saws

The two types of saws capable of making straight and curved cuts are jigsaws and **reciprocating** saws. Both of these saws have a blade that moves back and forth to enable the cutting action.

Reciprocating: Moving backward and forward on a straight line.

2.2.1   Jigsaws

Jigsaws have very fine blades, which makes them effective tools for doing delicate and intricate work. They are commonly used for cutting out patterns or irregular shapes from wood or even thin metals.

Today's cordless jigsaw (*Figure 24*) is an extremely useful portable power tool. It can make straight or curved cuts in wood, metal, plastic, wallboard, and other materials. The jigsaw cuts with a blade that moves up and down, unlike the spinning blade of a circular saw. This means that each cutting stroke (upward) is followed by a return stroke (downward), so the saw is cutting only half the time it is operating. This is called up-cutting or clean-cutting.

An important part of the jigsaw is the baseplate (shoeplate or footplate). Its broad surface helps to keep the blade lined up. It keeps the work from vibrating and allows the blade teeth to bite into the material.

Many models are available with tilting baseplates for cutting beveled edges. Models come with a top handle or a barrel handle. Some cordless models are available.

The jigsaw has changeable blades that enable it to cut many different materials, from wood and metal to wallboard and ceramic tile. Fine-toothed blades are used for thin materials and smoother cuts. Coarse blades are used for faster cutting of thicker materials and when smooth cuts are not a concern. Blades are rated by the number of teeth per inch or teeth per centimeter.

Most jigsaws can be operated at various blade speeds. Types of jigsaws include single-speed, two-speed, and variable-speed. The speed of a variable-speed jigsaw is controlled by how far the trigger is depressed. The low-speed range is for cutting hard materials, and the high-speed range is for soft materials.

2.2.2 Reciprocating Saws

A reciprocating saw, regardless of the manufacturer, is often referred to as a SawZall® (a trademark of the Milwaukee Electric Tool Corporation) in the United States because it was the first saw designed to serve as an electric hacksaw. Both the jigsaw and the reciprocating saw can make straight and curved cuts. They are used to cut irregular shapes and holes in plaster, plasterboard, plywood, studs, metal, and most other materials that can be cut with a saw.

Both saws have straight blades that move up and down along a straight line as they are guided along the cut. The reciprocating saw (*Figure 25*) is designed for more heavy-duty jobs than the jigsaw. It also uses longer and tougher blades than a jigsaw. In fact, the reciprocating saw is used for jobs that require brute strength, which makes it an excellent choice for general demolition work. It can saw through walls or ceilings and create openings for windows, plumbing lines, and more.

Jigsaw and reciprocating saw blades are available for a wide variety of cutting tasks and materials. To make selection easier, the blades are often labeled for the material or specific use. As a general rule, metal-cutting blades have more teeth per inch (or per centimeter) than wood-cutting blades. The teeth are also smaller.

Like the jigsaw, reciprocating saws come in single-speed, two-speed, and variable-speed models. The low-speed setting is best for metal work. The high-speed setting is for sawing wood and other soft materials.

The baseplate (shoeplate or footplate) may have a swiveling action, or it may be fixed. Whatever the design, the baseplate is there to provide a brace or support point for the sawing operation.

2.2.3 Jigsaw and Reciprocating Saw Safety and Maintenance

As with any power tool, jigsaws and reciprocating saws can be dangerous to operate if safety guidelines are ignored. In addition to following basic power tool safety guidelines, familiarize yourself with the following safety rules before operating a jigsaw or reciprocating saw:

- Always wear the appropriate PPE when working with saws, especially safety glasses.
- Before cutting through a wall or partition, find out what is inside the wall or on the other side of the partition. This will prevent the accidental cutting of water lines or electrical wiring.

Top Handle

Figure 24 Jigsaws.
Source: Image property of Stanley Black & Decker. Used with permission

---

CAUTION

Do not lift the blade out of the work while the saw is still running. If the blade is lifted out, the tip of the blade may hit the wood surface, marring the work and possibly breaking the blade.

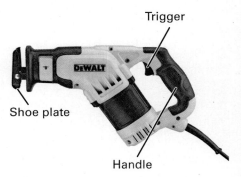

Trigger

Shoe plate

Handle

Figure 25 Reciprocating saw.
Source: Image property of Stanley Black & Decker. Used with permission

Figure 26 Proper use of a reciprocating saw.
Source: Image property of Stanley Black & Decker. Used with permission

NOTE

If cutting from the edge of a board or panel with a jigsaw, ensure the front of the baseplate is resting firmly on the surface of the work before starting the saw. The blade should not be touching the work at this stage.

NOTE

When a reciprocating saw blade first strikes a surface, it may jump back. Keep a steady hand, and the cutting action will eventually allow the blade to enter the workpiece. Forcefully plunging the blade into the surface is a common cause of broken blades.

- Make sure that the saw is disconnected from its power source before installing or changing blades, or performing any maintenance on the saw.
- When installing a blade in the saw, make sure it is in the collar as far as it will go before tightening the setscrew securely.
- When replacing a broken blade, look for any pieces of the blade that may be stuck inside the collar.
- Verify that the switch is in the Off position before it is plugged into a power source.
- Always use a sharp blade and never force the blade through the work. Forcing or leaning into the blade can cause you to lose balance, and possibly even be cut by the saw.
- When cutting metal, use a metal-cutting blade.
- Lubricate the blade with an agent such as beeswax, to help make tight turns and to reduce the chance of breaking the blade.

2.2.4 Operating Jigsaws and Reciprocating Saws

Many of the steps involved in the safe and efficient use of jigsaws and reciprocating saws are the same. Use the following steps to operate both types of saws:

Step 1 Don the appropriate PPE.
Step 2 Clamp the material being sawed to a pair of sawhorses or secure it in a vise to reduce vibration.
Step 3 Disconnect the saw from its power source.
Step 4 Confirm that the blade you have chosen is suited for the material you are cutting.
Step 5 Check the blade for dulling or damage.
Step 6 Measure the material and mark it before cutting.
Step 7 If a cut must be made from the middle of a board, drill a hole at the starting point that is large enough to allow the blade to pass through.
Step 8 Reconnect the saw to its power source.
Step 9 Squeeze the trigger to start the saw and move the blade gently but firmly into the material. Continue feeding the saw into the work at a reasonable pace without forcing it. Never force the blade into the work.
Step 10 When the cut is finished, release the trigger and let the blade come to a stop *before* removing it from the work.

Before cutting material with a reciprocating saw, set the saw to the desired speed. Use lower speeds for sawing metal; use higher speeds for sawing wood and other soft materials. Grip the saw with both hands (*Figure 26*) and place the baseplate firmly against the workpiece. Once the trigger is squeezed, the blade moves back and forth, cutting the material on the backstroke.

> **WARNING!**

Use both hands to firmly grip the reciprocating saw. Otherwise, the pull created by the blade's grip may jerk the saw out of your grasp.

2.3.0 Portable Band Saws

The portable band saw (*Figure 27*) can cut pipe, metal, plastics, wood, and irregularly shaped materials. It is especially good for cutting heavy metal, but performs equally as well on fine cutting jobs. Although it can cut wood, it is used almost exclusively to cut metal products on the job site.

The band saw has a one-piece blade that runs in one direction around guides at either end of the saw. The blade is a thin, flat piece of steel that must be of the proper length to fit the revolving pulleys that drive and support it. Proper

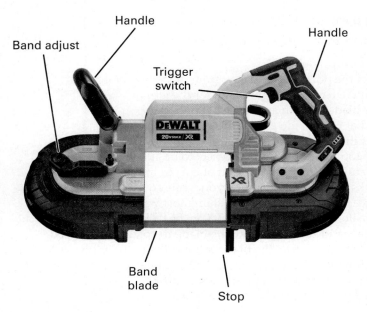

Band adjust

Handle

Trigger switch

Handle

Band blade

Stop

Figure 27 Portable handheld band saw.
Source: Image property of Stanley Black & Decker. Used with permission

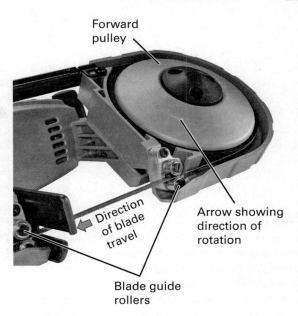

Forward pulley

Direction of blade travel

Arrow showing direction of rotation

Blade guide rollers

Figure 28 Band saw pulley and blade guide rollers.
Source: Courtesy of Milwaukee Electric Tool Corporation

blade length is determined by referencing the manufacturer's documentation. Like most blades, its coarseness is rated in teeth per centimeter or teeth per inch (TPI). As a general rule, higher TPIs produce smoother cuts. Thicker materials require coarser blades. However, if the blade is too coarse for the material, the individual teeth may begin to break off. *Figure 28* shows how the blade is routed around the pulleys and through the blade guides.

While some band saws have multiple speeds, most do not. The portable band saw generally cuts best at a low speed since higher speeds cause the blade's teeth to rub rather than cut. This can create heat through friction, which reduces the life of the blade.

WARNING!

A portable band saw always cuts in the direction of the user. For that reason, workers must be especially careful to avoid injury when using this type of saw. Always wear appropriate PPE and stay focused on the work.

2.3.1 Band Saw Safety and Maintenance

In addition to following basic power tool safety guidelines, familiarize yourself with the following band saw safety rules before operation:

- Always wear the appropriate PPE when working with saws, especially safety glasses.
- Keep your hands and fingers away from the path of the blade.
- Make sure the saw is disconnected from its power source before performing any maintenance or before adjusting the blade.
- Never use your thumbs to move wood or other materials toward the blade.
- Use only a band saw that has a stop in place.
- Before cutting through lines or pipes, confirm that they do not contain hazardous or combustible materials.
- Never force a portable band saw; let the saw do the cutting.
- Before backing out of a cut, turn off the machine, then back out the blade after it stops moving.

2.3.2 Operating a Portable Band Saw

Use the following steps to operate a portable band saw:

Step 1   Don the appropriate PPE.

Step 2   Disconnect the portable band saw from its power source.

Step 3   Verify you are using the right blade for the job.

Step 4   Check the blade tension and adjust, if necessary.

Step 5   Confirm that both blade guides are close to the blade.

Step 6   Reconnect the portable band saw to its power source.

Step 7   Place the stop firmly against the object to be cut.

Step 8   Plan saw cuts to avoid backing out of curves in the object being cut.

Step 9   Place marks on the wood to guide you while cutting. Remember to use pencil since ink will permanently stain most woods.

Step 10   Hold the material you intend to cut flat on the table before starting the cut.

Step 11   Turn on the band saw and allow the blade to come to full speed.

Step 12   Feed the material so that the blade follows the path of the marks. Little or no downward pressure is needed to make a good cut since the weight of the saw provides pressure for cutting.

Step 13   When finished cutting, turn off the bandsaw.

> **NOTE**
>
> The direction of blade rotation can pull the saw toward the stop. If the stop is not firmly against the workpiece, the saw will jump when started until the stop slams against the material being cut.

2.4.0 Miter and Cutoff Saws

Miter saws and cutoff saws are similar in that both are used to make straight or miter cuts. Both types of saws can be mounted permanently, but portable and cordless versions of the saws have the convenience of allowing users to move them from a workshop to a work site.

2.4.1 Power Miter Saws

The power miter saw combines a miter box with a circular saw, allowing it to make straight and miter cuts. There are two types of power miter saws: power miter saws and compound miter saws.

The saw blade of a standard miter saw pivots horizontally from the rear of the table and locks in position to cut angles from 0 to 45 degrees right and left. Stops are preset for common angles. The difference between the power miter saw and the compound miter saw (*Figure 29A*) is that the blade on the compound miter saw can be tilted vertically, allowing the saw to be used to make a compound cut (combined bevel and miter cut).

Similar to a miter saw and compound miter saw is the compound-slide miter saw (*Figure 29B*). A compound-slide miter saw has a rail in the table that allows the motor and blade assembly to slide forward and backward. This sliding capability enables the tool to cut wider material than a standard miter saw can cut.

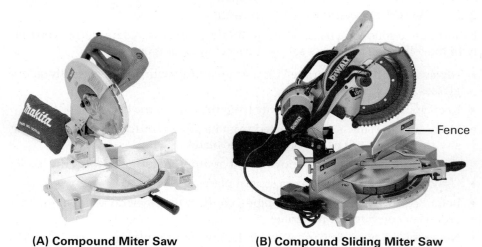

(A) Compound Miter Saw   **(B) Compound Sliding Miter Saw**

Figure 29 Miter saws.

Source: Makita USA, Inc. (29A); Image property of Stanley Black & Decker. Used with permission (29B)

2.4.2 Abrasive Cutoff Saws

An abrasive cutoff saw (*Figure 30*), also referred to as a *chop saw* or *cutoff saw*, can be used to make straight cuts or angular cuts through materials such as angle iron, flat bar, and channel. Like miter saws, cutoff saws can be either stationary or portable.

The abrasive blade on a cutoff saw can be between 10 and 18 inches in diameter. Metric blades are commonly 250 mm to 350 mm in diameter. When the saw is in operation, the blade spins at such a high speed that the resulting friction is hot enough to burn through the material. Like all rotating blades and stones, the maximum rpm of the blade must be equal to or greater than that of the saw.

2.4.3 Miter and Cutoff Saw Safety and Maintenance

Several basic safety guidelines should be followed when operating a miter saw or a cutoff saw to ensure worker safety and equipment protection. If either saw is being used on materials that produce airborne dust, there are important silica exposure standards that must be followed. Silica exposure standards are outlined in the *Basic Safety* module.

In addition to following basic power tool safety guidelines, familiarize yourself with the following safety rules before operating a miter or cutoff saw:

Figure 30 Abrasive cutoff saw.
Source: Image property of Stanley Black & Decker. Used with permission

- Always wear the appropriate PPE when working with saws, especially safety glasses.
- Never wear a watch or jewelry while operating the saw because they can get caught in the machinery.
- Securely lock the blade at the correct cutting angle.
- While operating the saw, keep your hands and fingers clear of the blade and never attempt to adjust the saw while it is running.
- Confirm the blade is in good condition and secure before using the saw.
- Do not allow other workers to stand nearby while operating a saw.
- Make sure the work area is clear of flammable materials such as chemicals and rags that could ignite from sparks. Abrasive cutoff saws tend to produce streams of sparks.
- Disconnect the saw from its power source before changing the blade or performing any sort of maintenance or setup.
- Verify that the rpm rating of the blade meets or exceeds the saw's spindle speed.
- Check all saw guards to ensure they are in place and working properly.
- Never retract a safety guard to view the material being cut while the saw is in use.
- If cutting long material, ensure the other end of the material is supported.

2.4.4 Operating Miter and Cutoff Saws

Using a power miter saw and an abrasive cutoff saw safely and efficiently involves many of the same steps used for other types of power saws. Use the following steps to operate a power miter or cutoff saw:

Step 1 Don the appropriate PPE.

Step 2 Ensure the power supply is disconnected from the saw.

Step 3 Set the angle or tilt of the blade according to the cut that needs to be made.

Step 4 Reconnect the power supply to saw.

Step 5 Place the material to be cut firmly against the fence.

Step 6 With the blade arm raised, turn on the saw and allow it to reach maximum speed.

Step 7 Firmly grasp the handle of the rotating saw arm and lower the blade on to the material. Some compound miters allow the rotating arm to slide while cutting.

Step 8 Once the cut has been made, turn off the saw and raise the rotating saw arm.

2.5.0 Table Saws

Table saws are designed with a circular saw blade that protrudes up through a slot on the bench. They are typically distinguished by the size blade they can accommodate. Table saws using 10-inch blades are excellent options for most carpentry jobs, while saws with 8-inch blades are designed for cutting smaller materials. Those with 12-inch blades are the best choice for larger jobs.

Table saws are typically used for ripping and crosscutting lumber. Ripping involves cutting lumber to a specific width while crosscutting involves cutting it to a specific length. Since the table saw has an exposed blade and a fence, users can feed lumber into the blade in a controlled manner. The result is a saw that is excellent for making precise and accurate cuts, which is one reason table saws are a favorite of furniture and cabinet makers.

While many older table saws were large, stationary tables that plugged into a receptacle for power, many of today's table saws are cordless and/or portable with folding legs (*Figure 31*).

The table part of this tool is where wood is laid before it is pushed against the rip fence and into the saw blade. The saw part of this tool is mounted under the table and can be adjusted up and down to expose more or less of the blade. Some table saws also allow the blade to be tilted up to 45 degrees.

2.5.1 Table Saw Safety and Maintenance

In addition to following basic power tool safety guidelines, familiarize yourself with the following table saw safety rules before operating it:

- Always wear the appropriate PPE when working with saws, especially safety glasses.
- Use blades designed for the job.
- Confirm the blade is sharp prior to cutting.
- Verify that the blade guard moves freely before connecting the power.
- Use a push stick or board to push any material that may require your hand to pass near the blade.
- Do not use the table saw fence during a crosscutting operation without placing a block between the material being cut and the fence. This prevents the cutoff piece from being trapped between the blade and the fence.
- Keep the table saw bench clear of all debris, especially when it is in use.
- Disconnect the table saw from its power source when you are adjusting the blade or performing any type of maintenance.
- Always anticipate kickback when ripping any type of material. This means you should not stand in line with the blade while it is cutting. Instead, stand to one side until cutting is finished.
- Set the blade to the correct height.
- Make sure any lumber you are cutting is flat on the bench and the board is pressed level against the rip fence when cutting. Otherwise, binding and kickback may occur.

2.5.2 Ripping Wood with a Table Saw

Use the following steps to rip wood or another material on a table saw:

Step 1   Don the appropriate PPE.
Step 2   Confirm the table saw is disconnected from any power source.
Step 3   Ensure the table saw is fitted with a rip blade.

Figure 31 Portable table saw.
Source: Image property of Stanley Black & Decker. Used with permission

Step 4 Adjust the blade so that it is no more than ¼-inch above the thickness of the material to be cut.

Step 5 Adjust the rip fence so that its inner edge corresponds with the desired width of the material once it is cut.

Step 6 Connect the power source and place the material you want to cut on the bench. Make sure it is aligned against the rip fence, but not touching the blade.

Step 7 Slowly push the material along the edge of the rip fence, making sure the material is held snugly against the rip fence while it is pushed.

Step 8 Use a push stick to guide the final few inches of the material past the blade. Push sticks should always be used if there is any chance that pushing the material will put your fingers within a few inches of the blade.

Step 9 When finished ripping, turn off the table saw.

CAUTION

Do not push lumber or any other material against the table saw's blade before the blade reaches full speed. Contacting the blade any earlier may cause kickback and/or serious injury.

2.0.0 Section Review

1. The proper way to start cutting material with a circular saw is to _____.
   a. power up the saw to full speed then slowly move it forward into the material
   b. hold the lower blade guard up to position the blade on the cut mark
   c. press the blade against the material being cut and set the saw rpm to Low
   d. tilt the front edge of the baseplate upward and push the saw forward

2. A jigsaw is an effective tool for _____.
   a. drilling holes in concrete or pavement
   b. making long straight cuts through thick metal
   c. cutting through walls in demolition jobs
   d. doing delicate work on thin materials

3. The blade on a portable band saw _____.
   a. moves up and down through a shoeplate
   b. spins in a circular motion on an arbor
   c. runs in one direction around guides
   d. reciprocates in and out from a guard

4. A type of miter saw in which the blade can be pivoted horizontally and vertically is called a _____.
   a. sliding jigsaw
   b. compound miter saw
   c. reciprocating saw
   d. sliding abrasive saw

5. Pushing a board against a table saw's blade before it reaches full speed _____.
   a. is the desired approach when ripping thin lumber
   b. is the best approach when using larger blades
   c. can cause kickback and serious injury
   d. can increase efficiency and speed

3.0.0 Grinders and Oscillating Multi-Tools

Performance Task

1. Safely and properly demonstrate the use of the following tools:
   - Portable or bench grinder
   - Oscillating multi-tool

Objective

Describe the types of jobs best suited to grinders and oscillating multi-tools.

a. Explain how to use various types of grinders.

b. Identify grinder accessories and the jobs for which they are used.

c. List the type of jobs that can be performed using an oscillating multi-tool.

Abrasive: A substance, such as sandpaper, that is used to wear away material.

Grit: A granular, sand-like material used to make sandpaper and similar materials abrasive. Grit is graded according to its texture. The grit number indicates the number of abrasive granules in a standard size (per in or per cm). The higher the grit number, the more particles in a given area, indicating a finer abrasive material.

Grinding tools can power all kinds of **abrasive** wheels, brushes, buffs, drums, bits, saws, and discs. These wheels come in a variety of materials and **grit**. They can drill, cut, smooth, and polish; shape or sand wood or metal; mark steel and glass; and sharpen or engrave. They can even be used on plastics.

> **WARNING!**
>
> Always wear safety goggles and a face shield when working with grinders. Make sure that the work area is free of combustible materials such as rags or flammable liquids and that a fire extinguisher is easily accessible. Clothes should be snug and comfortable and free of cuffs at the wrists and ankles. Wearing excessively loose clothing on the work site can be extremely dangerous.

3.1.0   Grinders

Grinders are available in various configurations. Common handheld grinders include angle grinders (also called side grinders or right-angle grinders), end grinders, and detail grinders. Stationary grinders, called bench grinders, are permanently mounted on a work table or bench.

Angle grinders are used to grind away hard, heavy materials and to grind surfaces such as pipes, plates, or welds (*Figure 32*). The angle grinder has a rotating grinding disc set at a right angle to the motor shaft.

End grinders are sometimes called horizontal grinders or pencil grinders. These smaller grinders are used to smooth the inside of materials, such as pipe (*Figure 33*). The grinding disc on the end grinder rotates in line with the motor shaft. Grinding is also done with the outside of the grinding disc.

Like end grinders, detail grinders (*Figure 34A*), also known as *die grinders*, have an arbor that extends from the motor shaft onto which small attachments, called points, can be mounted to smooth and polish intricate metallic work. These attachments, a sample of which is shown in *Figure 34B*, are commonly made in shaft sizes ranging from 1/16- to 1/4-inch. The shaft of these

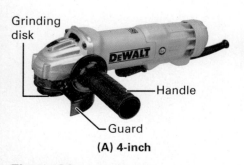

(A) 4-inch

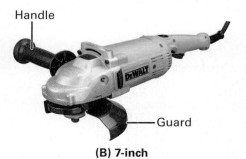

(B) 7-inch

Figure 32 Angle grinders.
Source: Image property of Stanley Black & Decker. Used with permission (32A–32B)

Figure 33 End grinder.
Source: Courtesy of Milwaukee Electric Tool Corporation

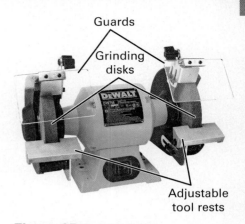

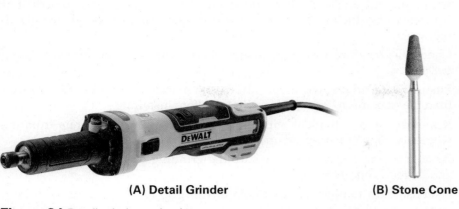

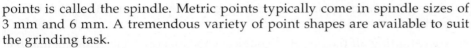

Figure 34 Detail grinder and point.
Source: Image property of Stanley Black & Decker. Used with permission (34A); Courtesy of Dremel (34B)

Figure 35 Bench grinder.
Source: Image property of Stanley Black & Decker. Used with permission

points is called the spindle. Metric points typically come in spindle sizes of 3 mm and 6 mm. A tremendous variety of point shapes are available to suit the grinding task.

The primary difference between an end grinder and a detail grinder is power. End grinders offer more power than a detail grinder.

Bench grinders (*Figure 35*) are electrically powered stationary grinding machines. They usually have two grinding wheels that are used for grinding, rust removal, and metal buffing. They are also great for renewing worn edges and maintaining the sharp edges of cutting tools. For example, the bench grinder can be used to smooth the mushroomed heads of cold chisels.

Heavy-duty grinder wheels range from $6\frac{3}{4}$ to 10 inches (150 mm to 250 mm) in diameter. Each wheel's maximum speed is given in rpm. Never use a grinding wheel above its rated maximum speed. Its rated speed must be equal to or faster than the maximum speed of the power tool.

Bench grinders come with an adjustable tool rest. This is the surface on which you position the material you are grinding, such as cold chisel heads. There should be a distance of only $\frac{1}{8}$-inch (about 3 mm) between the tool rest and the wheel.

WARNING!

Never adjust tool rests when the grinder is on or when the grinding wheels are spinning. Doing so may damage the work or cause injuries. Disconnect the power source before adjusting any type of grinder.

3.1.1 Grinder Safety and Maintenance

In addition to following basic power tool safety guidelines, familiarize yourself with these safety rules before operating any type of grinder:

For Angle, End, and Detail Grinders

- Always wear the appropriate PPE when working with grinders, especially gloves and safety glasses. Wearing PPE when using grinders is critically important since grinders generate a lot of high-speed sparks and flying debris.

- Never wear loose clothing or jewelry that can get caught in a grinder's wheels.

- Always disconnect the power source before performing maintenance.

- Keep your hands away from the grinding wheel.

- To keep sparks from flying into the path of other workers or from igniting flammable materials, use standing screens when grinding near walkways, other workers, or storage areas.

- Never use an angle grinder, end grinder, or detail grinder unless it is equipped with the manufacturer-provided guard that surrounds the grinding wheel.
- Choose a grinding disc that is appropriate for the type of work being performed.
- Ensure new grinding wheel maximum rpm markings are equal to or greater than the maximum speed of the grinder before installing.
- Remember that individual wires from a wire wheel can separate from the wheel at a high rate of speed and cause injury.
- Before starting a grinder, make sure the grinding disc is properly secured and in good condition.
- Establish firm footing and maintain a firm grip on the grinder with both hands to avoid being pulled off balance.
- Let the grinder come up to full speed before grinding.
- Direct sparks and debris (*Figure 36*) away from coworkers and any flammable materials.
- When grinding on a platform, use a flame-retardant blanket to catch falling sparks.
- After shutting off the power, do not leave the tool until the grinding disc has come to a complete stop.

> **WARNING!**
>
> Grinding discs can instantaneously disintegrate if used when they are cracked. Inspect the disc for cracks before using the grinder.

For Bench Grinders

- Always wear the appropriate PPE when working with a bench grinder, especially gloves and safety glasses. Wearing PPE when using bench grinders is critically important since bench grinders generate a lot of high-speed sparks and flying debris.
- Grinding metal on a bench grinder creates sparks, so keep the area around the grinder clean.
- Always adjust the tool rests so they are within ⅛-inch (≈3 mm) of the wheel. This reduces the chance of getting the work wedged between the rest and the wheel.

Figure 36 Be aware of sparks created by grinders.
Source: Courtesy of Atlas Copco

- Never use a grinding wheel above its rated maximum speed.
- After finishing a job with the bench grinder, shut it off.
- Before mounting a grinding wheel onto a bench grinder, inspect the wheel for chipped edges and cracks.

WARNING!

Never grind soft metals or wood on a grinding wheel. The material will wedge itself between the grit, creating stresses that may cause the wheel to shatter.

3.1.2 Operating Grinders

Use the following steps to operate a grinder:

For Angle, End, and Detail Grinders

Step 1 Don the appropriate PPE, especially a face shield.
Step 2 Secure the material in a vise or clamp it to the workbench.
Step 3 To use an angle grinder, place one hand on the handle of the grinder and one on the trigger. To use an end grinder or detail grinder, grip the grinder handle at the shaft end with one hand and cradle the opposite (motor) end of the tool in your other hand.
Step 4 Always allow the grinder to reach its maximum speed before grinding.
Step 5 Turn off the grinder and be sure that the wheel has stopped rotating before putting it down.
Step 6 Complete the work by removing any loose material with a wire brush.

For Bench Grinders

Step 1 Don the appropriate PPE, especially a face shield.
Step 2 Always use the adjustable tool rest as a support when grinding or beveling metal pieces. There should be a maximum gap of $\frac{1}{8}$-inch (≈ 3 mm) between the tool rest and the wheel and $\frac{1}{4}$-inch (≈ 6 mm) between the wheel guard (also called a spark arrestor) and wheel.
Step 3 Confirm that the bench grinder is firmly secured to a stable surface.
Step 4 Always allow the grinder to reach its maximum speed before grinding.
Step 5 Keep the metal cool as it is being ground. If the metal gets too hot, it can destroy the temper (hardness) of the material.
Step 6 Always work on the face of the wheel and not on the side. *Figure 37* shows the proper way to use a bench grinder.

CAUTION

Do not force the tool or overload the motor. If too much pressure is applied, there will be a significant reduction in the grinder's rpm.

Did You Know?

Grinding Tungsten Can Be Hazardous to Your Grinder

Workers who use power tools are almost certain to encounter objects that contain tungsten. As mentioned earlier, tungsten is an extremely hard and brittle metal with a very high melting point. Attempting to grind tungsten carbide tools with conventional grinding discs and wheels is a bad idea. Rather than actually sharpening the tool, the most likely outcome of such an attempt is a damaged and/or weakened grinding wheel.

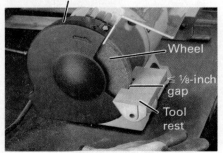

Wheel guard or spark arrestor

Wheel

$\leq \frac{1}{8}$-inch gap

Tool rest

No more than $\frac{1}{8}$" (≈ 3 mm) gap between tool rest and wheel.

Maintain a $\frac{1}{4}$" (≈ 6 mm) gap between wheel and spark arrestor.

Always work on the face of the wheel.

Figure 37 Proper use of a bench grinder.
Source: Image property of Stanley Black & Decker. Used with permission

3.2.0 Grinder Attachments and Accessories

Grinders are versatile pieces of equipment. While the primary uses of a typical grinder are to grind, cut, sand, and polish materials, numerous attachments and accessories can be used to extend a grinder's capability.

Accessories such as guards and shields (*Figure 38*) can be added to a portable grinder to provide additional protection when special grinding discs are used. For example, there are specialized guards for use with diamond discs, flap discs, and cup brushes. Side handles can be added to some grinders to improve stability and grip during use. There are also shields available to help protect against dust. Some dust shields are even equipped with a vacuum hose outlet so that dust can be vacuumed away when the grinder is in use.

Adapters and backing pads (*Figure 39*) can be installed on a grinder's arbor so that different types of sanding discs, wheels, and buffing pads can be used. The backing pad provides a stable surface for flexible discs to rest against.

There is an almost endless variety of grinder attachments available, and most of them are available in various sizes. Some of the more common accessories are shown in *Figure 40*. They include wire brushes, cup brushes and stones, flap discs, cutting wheels, and polishing and buffing wheels. Some grinders can also be equipped with a cutoff wheel. These are best described as smaller, thinner versions of the blade used on an abrasive saw. Because of their tendency to break or fly apart when too much stress is applied, many companies require workers to have specific permission and documentation before they can be used.

Wire brushes and cup brushes are good for removing rust, scale, and file marks from metal surfaces. Flap discs are used for similar jobs, but they tend to last longer than fiber discs. Cutting discs are designed for cutting through material. Cloth buffing wheels enable a grinder to be used for polishing and buffing metal surfaces. Coring bits are sometimes used for cutting holes in stone or composite materials, such as those used for countertops.

Special bits can be used on some grinders to remove rough edges and burrs from the inside and outside of pipes and other objects. These rotary bits are commonly called burr bits (*Figure 41*). As with other grinder accessories, burr bits are available in various shapes and sizes to fit almost any need. Workers should be aware that using a grinder with a burr bit requires steady hands. These bits operate at high rpm and can catch on burrs and cause the grinder to shift around quite a bit. They can also heat up the material being smoothed and possibly damage it. It is best to work in short bursts to avoid overheating the bit and the material.

(A) Diamond Blade Guard

(B) Flaring Cup Guard

Figure 38 Grinder accessories.
Source: Image property of Stanley Black & Decker. Used with permission (38A–38B)

Figure 39 Backing pad and adapter nut.
Source: Image property of Stanley Black & Decker. Used with permission

(A) Knotted Cup Brush   **(B) Cup Stone**   **(C) Cutoff Wheel**

(D) Wool Buffing Pad   **(E) Bonded Abrasive Grinding Wheel**

Figure 40 Grinder attachments.
Source: Courtesy of Milwaukee Electric Tool Corporation (40A–40D); Camel Grinding Wheels Works Sarid LTD (40E)

Figure 41 Burr bit assortment.
Source: Simpson Strong-Tie Company Inc.

No matter what type of attachment is used on a grinder, workers must always make sure that it is designed to fit the grinder and rated for the proper rpm. Always follow the attachment and accessory manufacturer's guidelines for use.

3.3.0 Oscillating Multi-Tools

An oscillating multi-tool (*Figure 42*), sometimes referred to as an *OMT*, looks like a handheld grinder, but the types of jobs it can perform are much more varied. This variety is the result of arbors that accept many different attachments and accessories. Oscillating multi-tools operate by moving these attachments back and forth at speeds of up to 21,000 oscillations per minute. Unlike saws or grinders, where you can see the blade or wheel turn, multi-tools move attachments so fast that their movement is hard to detect. This high-speed motion, combined with the smaller design of a multi-tool, makes it the tool perfect for small cutting and sanding jobs. Oscillating multi-tools can also be used for scraping grout, cutting metal and PVC piping, removing adhesives and paint, or sanding and preparing different types of surfaces.

One of the most useful characteristics of a multi-tool is its ability to fit into tight, confined places. As a result, they are an excellent choice for undercutting low baseboards, cutting small openings in drywall, or cutting pipe where there is little room for larger tools. Like most other basic power tools, oscillating multi-tools are available in corded versions, but cordless models make up the bulk of those found on today's job sites.

3.3.1 Oscillating Multi-Tool Safety and Maintenance

Follow these guidelines for the safe use and proper maintenance of oscillating multi-tools:

- Always wear the appropriate PPE when working OMTs, especially gloves and safety glasses.
- Disconnect the OMT from its power source when changing the attachment or performing maintenance.

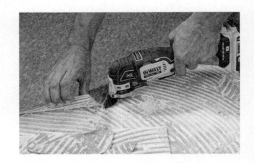

Figure 42 Oscillating multi-tool.
Source: Image property of Stanley Black & Decker. Used with permission

- Confirm that the attachment on the tool is designed for the job being performed.
- Use clamps or vices to securely hold the work piece.
- Never wear loose clothing or jewelry that can get caught on the OMTs attachment.
- Keep your hands away from the attachment when the OMT is turned on.

3.0.0 Section Review

1. A stationary grinder that has two wheels for grinding, removing rust, and buffing metal is a(n) _____.
   a. angle grinder
   b. tandem grinder
   c. bench grinder
   d. end grinder

2. Special rotary bits that can be used on grinders to remove rough edges from the inside and outside of pipes and other objects are commonly called _____.
   a. burr bits
   b. backing pads
   c. coring bits
   d. buffing wheels

3. One characteristic that makes an oscillating multi-tool so useful is its ability to _____.
   a. fit into tight places
   b. drill large holes in concrete
   c. cut through heavy steel
   d. handle large carpentry jobs

4.0.0 Miscellaneous Power Tools

Performance Task

1. Safely and properly demonstrate the use of the following tool:
   - Power nailer

Objective

Identify and explain how to use miscellaneous power tools.

a. Discuss the hazards of using power nailers.

b. Describe jobs that can be performed with hydraulic jacks.

Power tools are used for many applications on a typical construction site. In addition to drilling, cutting, and grinding, some of the more common applications include fastening and jacking. This section covers the following power tools:

- Power nailers
- Pneumatic impact wrenches
- Hydraulic jacks

4.1.0 Power Nailers

Power nailers (*Figure 43*), often referred to as *nail guns*, are common on construction sites. They greatly speed up the installation of materials such as wallboard, molding, framing members, and shingles.

Pneumatic nailers are driven by compressed air traveling through air lines connected to an air compressor. They are designed for specific purposes, such as roofing, framing, siding, flooring, sheathing, trim, and finishing. Nailers use specific types of nails, depending on the material to be fastened. Nails come in coils and in strips, and are loaded into the nail gun.

> ### WARNING!
>
> Never exceed the maximum specified operating pressure of a pneumatic nailer. Doing so will damage the pneumatic nailer and may cause injury.

Pneumatic nailers are designed to fire when the tool is pressed against the material being fastened and the trigger is pressed. An important safety feature of all pneumatic nailers is that the nailer will not fire unless it is pressed against the material.

Figure 43 Pneumatic nailer.
Source: Image property of Stanley Black & Decker. Used with permission

Power Screwdrivers

Power screwdrivers use a power source (this model uses a battery) to speed production in a variety of applications, such as drywall installation, floor sheathing and underlayment, decking, fencing, and cement board installation. A chain of screws feeds automatically into the firing chamber. Most models incorporate a back-out feature to drive out screws as well as a guide that keeps the screw feed aligned and tangle-free. Power screwdriver tools can accept Phillips or square slot screws and weigh an average of six pounds.

Source: Image property of Stanley Black & Decker. Used with permission

The use of powder-actuated anchor or fastening systems has been increasing rapidly in recent years. They are used for anchoring static loads to steel and concrete beams, walls, and so forth.

Powder-actuated fasteners are used to drive steel pins or threaded steel studs directly into masonry and steel (*Figure 44*). An advantage of these tools is that they eliminate the need for compressed air to operate the tool. However, these tools can be extremely dangerous. They look and fire like a gun and use the force of a gunpowder load (typically .22, .25, or .27 caliber) to drive the fastener into the material. The depth to which the pin or stud is driven is controlled by the density of the base material and the power level or strength of the powder load.

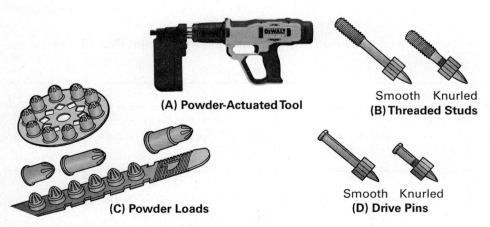

(A) Powder-Actuated Tool

Smooth   Knurled
(B) Threaded Studs

(C) Powder Loads

Smooth   Knurled
(D) Drive Pins

Figure 44 Powder-actuated fastening system.
Source: Image property of Stanley Black & Decker. Used with permission (44A)

WARNING!

Operators of powder-actuated tools must be trained and certified by the manufacturer of the tool being used.

Avoid firing a powder-actuated tool into easily penetrated materials. The fastener may pass through the material and become a flying missile on the other side.

If a powder load fails to fire, wait 10 seconds, remove it from the installation tool, and then dispose of it in a bucket of water or oil.

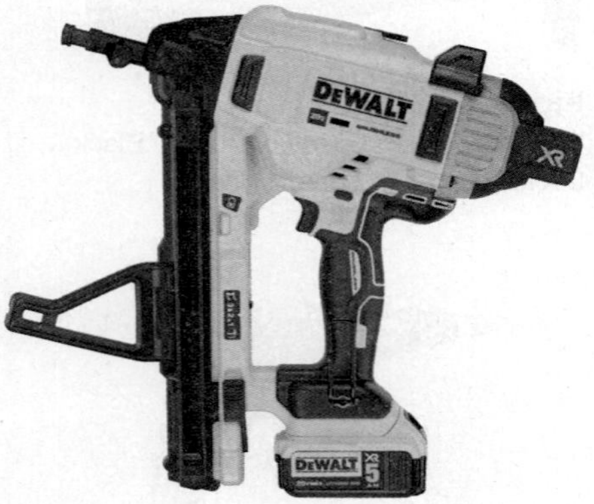

Figure 45 Cordless electric nailer.
Source: Image property of Stanley Black & Decker. Used with permission

While pneumatic and powder-actuated tools have traditionally been the choice of craftworkers nailing into concrete and steel, cordless electric nailers (*Figure 45*) are another alternative. Cordless nailers are battery operated and have enough power to fasten pins, straps, and clips to concrete and steel. Although they may not have the power that a powder-actuated tool can generate, they are strong enough to handle most concrete fastening jobs. Cordless nailers have several advantages, including the lack of a hose that must be connected to pneumatic nailers. Using cordless nailers also allows craftworkers to avoid the training and certification required to use powder-actuated tools.

4.1.1 Power Nailer Safety and Maintenance

There are numerous safety and maintenance guidelines associated with pneumatic, powder-actuated, and electric concrete fastening tools. In addition to following basic power tool safety guidelines, familiarize yourself with these safety rules before operating a power nailer:

- Always wear the appropriate PPE, especially safety glasses.
- Read and understand the manufacturer's manual.
- When using a pneumatic nailer, pay close attention to where you are pointing the nail gun. Never point the nail gun toward another worker or toward any part of your body.
- Be sure you select the correct nail gun and fastener for the job and never load the tool until you are ready to fire it.
- Never try to operate a powder-actuated tool without the proper training.
- Keep pneumatic nailers oiled according to the manufacturer's instructions.
- Disconnect the power source from the nailer before attempting to load or repair it.
- Always treat nailers as if they are loaded and ready to fire.
- Check for pipes, electrical wiring, vents, and other materials behind wallboard before nailing.
- Be careful about nailing into paneling or any other thin surface. The fastener may go through it and strike another worker.
- Never leave a pneumatic nailer connected to a compressor hose or power when it is not in use.

WARNING!

The use of powder-actuated tools may be prohibited by local codes because of safety issues, seismic concerns, or because the building is occupied. Check with your local authorities before using these tools.

- When loading the driver, install the drive pin before loading the charge.
- Never place your hands behind the material being fastened.
- Do not fire the tool close to the edge of the material, especially concrete. Pieces of concrete may chip off and strike someone.

4.1.2 Operating Power Nailers

The safe and efficient use of power fastening tools depends on careful preparation, following some commonsense guidelines, and being familiar with the tool. Follow these steps to use power nailers properly and safely:

Step 1 Don the appropriate PPE.

Step 2 If using a pneumatic nailer, disconnect it from the air hose.

Step 3 Inspect the nailer for damage and loose connections.

Step 4 Load the fastener into the nailer. When loading a powder-actuated tool, feed the pin or stud into the piston, followed by the gunpowder cartridge.

Step 5 If you are using a pneumatic nailer, connect it to the air hose. Also, check the air compressor and adjust the pressure to the recommended level. Most nailers operate at pressures of 70 to 120 pounds per square inch (psi) ($\approx$480 to 830 kPa).

Step 6 Test the nailing ability of the tool using a piece of scrap material. If the nail penetration is not deep enough, follow the manufacturer's instructions for increasing the power of the nailer.

Step 7 Hold the pneumatic or battery powered nailer firmly against the material to be fastened, then press the trigger (*Figure 46*). If you are using a powder-actuated nailer, position the tool in front of the item to be fastened and press firmly against the mounting surface (*Figure 47*).

Step 8 After finishing the job, disconnect the power source from the nailer. Turn off the air supply and disconnect the air hose from the nailer. Remove the battery from cordless nailers. Remove the powder charge(s) from powder-actuated fastening tools.

WARNING!

A nail gun is not a toy. Playing with a nail gun can cause serious injury or death. Nails can easily pierce a hand, leg, or eye. Never point a nail gun at anyone and use it only as directed by the manufacturer.

Figure 46 Proper use of a nailer.
Source: Zachary McNaughton, River Valley Technical Center

Figure 47 Using a powder-actuated fastening tool.
Source: Simpson Strong-Tie Company Inc.

4.2.0 Hydraulic Jacks

Hydraulic tools are used when an application calls for extreme force to be applied in a controlled manner. These tools do not operate at high speed, but you should exercise great care when operating them. The forces generated by hydraulic tools can easily damage equipment or cause personal injury if the manufacturer's procedures are not strictly followed.

Hydraulic jacks are portable devices used for a wide variety of purposes. They can be used to move or lift heavy equipment and other heavy material, to position heavy loads precisely, and to straighten or bend frames. Hydraulic jacks have two basic parts: a pump and a cylinder (sometimes called a ram). There are various types of hydraulic jacks including those with internal pumps and those that use a lever-operated pump. The latter type is often referred to as a Porta-Power®, the name of one common brand.

A hydraulic jack with an internal pump is a general-purpose jack that is available in many different capacities (*Figure 48*). The pump inside the jack applies pressure to the hydraulic fluid when the handle is pumped. The pressure on the hydraulic fluid applies pressure to the cylinder, which lifts or moves the load.

A typical lever-operated pump kit (*Figure 49*) includes a length of hydraulic hose, a hydraulic hand pump cylinder, and a variety of attachments made for different jobs. Lever-operated pumps are available in different capacities. Cylinders are available in many sizes; they are rated by the weight (in tons or metric tonnes) they can lift and the distance they can move it. This distance is called stroke and is measured in inches or millimeters. Hydraulic cylinders can lift more than 500 tons (≈454 metric tonnes). Strokes range from ¼-inch (≈6 mm) to more than 48 inches (≈122 cm). Different cylinder sizes and ratings are used for different jobs. Lever-operated pumps are especially useful for horizontal jacking.

4.2.1 Hydraulic Jack Safety and Maintenance

Using hydraulic jacks safely and effectively requires an awareness of the area surrounding the load, the load itself, and the jack. In addition to following basic power tool safety guidelines, familiarize yourself with these safety rules:

- Always wear the appropriate PPE, especially safety glasses.
- Operate the jack according to the manufacturer's guidelines.
- Check the area prior to jacking a load to ensure that other workers are safely out of the way and that the load will clear all obstacles.
- Make sure you have the appropriate jack for the job and never exceed its lifting capacity.

Figure 48 Portable hydraulic jack.
Source: Walter Meier Manufacturing Americas

Figure 49 Lever-operated hydraulic pump kit.
Source: Torin Jacks, Inc.

- Check the fluid level in the jack before using it and watch for any fluid leaks during use. If a Porta-Power® is being used, make sure that the hydraulic hose is not twisted or kinked. Do not move the pump if the hose is under pressure.
- Make sure the base of the jack can be placed on a solid, even, and level surface. Never place the base of the jack on bare soil or any other surface that could compact or shift under the load.
- If there is a possibility that the load could move while jacking, make sure it is chocked and restrained. Never jack metal against metal; use wood softeners as a buffer between metal surfaces.
- Place the jack under the load so that the load is centered and the weight is uniformly distributed.
- Stay clear of the object being lifted to avoid injury if the load slips off the jack.
- Do not use an extension bar, or cheater, or step on the pump handle to gain more leverage.
- Never leave a jack under a load as a support.
- Block the load up as you progress through the lift, so that it will only fall a short distance if the jack fails.
- Once at the proper height, add blocking as needed so the load is supported while the jack is removed.
- To reduce tripping hazards, remove the jack's lever any time the jack is not being pumped.

4.0.0 Section Review

1. Pneumatic nailers are designed to safely fire when the trigger is squeezed and the tool is _____.
    a. filled with a charge of compressed air
    b. pressurized with hydraulic fluid
    c. connected to its battery pack
    d. pressed against the material being fastened

2. When using a portable hydraulic jack, be sure to _____.
    a. avoid twisting or kinking the air lines
    b. leave the jack under the load as a support
    c. use an extension bar on the pump handle
    d. place the base of the jack on a solid, level surface

Module 00104 Review Questions

1. Pneumatic tools get their power from _____.
    a. air pressure
    b. fluid pressure
    c. hand pumps
    d. AC power sources

2. When operating a power tool, it is important to always _____.
    a. wear the proper PPE
    b. notify a coworker
    c. wear rubber-soled footwear
    d. ask a co-worker how to operate the tool

3. Using a trigger lock to lock a tool's trigger in an On position _____.

   a. can increase productivity
   b. ensures the tool will have enough power
   c. can be dangerous and cause injury
   d. is a great way to keep your hand from getting fatigued

4. The most common use of the power drill is to _____.

   a. cut wood, metal, and plastic
   b. drive nails into wood, metal, and plastic
   c. make holes in wood, metal, and plastic
   d. carve letters in wood, metal, and plastic

5. A masonry bit is able to drill into concrete and similar material because it has a _____.

   a. countersink shank
   b. ceramic core
   c. whip check
   d. carbide tip

6. An example of an electric power drill that is designed to be used in tight spaces is a(n) _____.

   a. electromagnetic drill
   b. right-angle drill
   c. hammer drill
   d. keyless chuck drill

7. The electromagnetic drill is a _____.

   a. handheld drill used on wood
   b. cordless drill used on masonry and tile
   c. portable drill used on thick metal
   d. pneumatic drill that has a pounding action

8. Hammer drills are designed to drill into _____.

   a. wood, metal, and plastic
   b. concrete, brick, and tile
   c. drywall, fiberglass, and wood
   d. roofing shingles, plastic, and wood

9. Compared to a common drill, an impact driver has _____.

   a. more torque
   b. more weight
   c. more variable speed options
   d. more bits for precision drilling

10. A pneumatic impact wrench requires the use of _____.

    a. impact sockets that are designed for the applicable tool
    b. an adapter so that handheld sockets will fit
    c. shear pins between the wrench and the socket
    d. a trigger lock to prevent accidental starting

11. When cutting with a circular saw, grip the saw handles _____.

    a. and pull the saw toward you
    b. firmly with two hands
    c. firmly with one hand
    d. and engage the trigger lock

12. The high speed setting on a reciprocating saw is used for _____.

    a. cutting through drywall
    b. metal work
    c. grinding surfaces
    d. sawing wood and other soft materials

13. Before using a reciprocating saw to cut through a wall or partition, always _____.

    a. find out what is on the other side
    b. remove the lower blade guard
    c. increase the revolutions per minute
    d. lubricate the guard with oil or grease

14. When using a jigsaw, avoid vibration by _____.

    a. using a low-speed setting
    b. using a clamp or vise to hold the workpiece
    c. setting a heavy object on the workpiece
    d. holding the workpiece down with your free hand

15. Use only a band saw that has a _____.

    a. stop
    b. breastplate with a broad surface
    c. battery pack
    d. thick, three-piece blade

16. A sliding compound miter saw has a rail that allows the blade to slide forward and backward, which enables the saw to _____.

    a. use much thinner blades than a standard miter saw
    b. produce much less dust than a standard miter saw
    c. cut wider material than a standard miter saw
    d. cut harder material than a standard miter saw

17. The blade of an abrasive cutoff saw spins at such a high speed that _____.

    a. the resulting friction is hot enough to burn through the material
    b. it can only be used for straight cuts
    c. the abrasive particles will melt into some metals
    d. it can never be more than eight inches in diameter

18. The end grinder is used to _____.

    a. polish intricate work
    b. grind surfaces
    c. smooth the inside of materials, such as pipe
    d. smooth the work before painting

19. A detail grinder smoothes and polishes intricate metallic work by using attachments called _____.

    a. points
    b. rollers
    c. pins
    d. studs

20. Powder-actuated fastening systems are used to _____.

    a. penetrate drywall and treated wood
    b. anchor static loads to concrete
    c. hammer nails into metal
    d. remove nails

21. Porta-Power® cylinders are rated by how much weight they can lift and by _____.

    a. their torque
    b. the amount of electromagnetic material they have
    c. the distance they can move the weight
    d. how much they weigh

22. Hydraulic jacks are used when the application calls for _____.

    a. operation at high speed
    b. quiet operation
    c. extreme force to be applied
    d. manually assisted lifting

Answers to Section Review Questions and Module Review Questions are found in *Appendix A*.

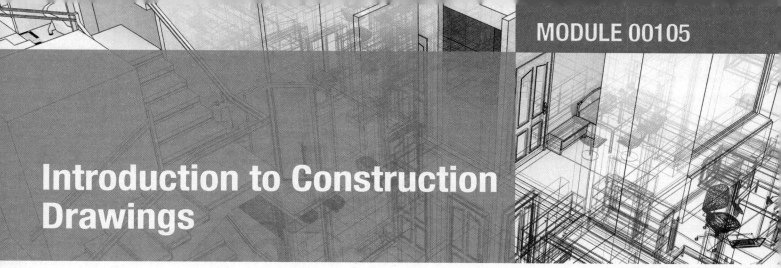

Introduction to Construction Drawings

Source: Gwycech/Shutterstock

Objective

Successful completion of this module prepares you to do the following:

1. Describe components and features used in construction drawings and identify how the drawings are different.
   a. Summarize the purpose of the six basic construction drawing components.
   b. List and explain the significance of various drawing elements, such as lines of construction, symbols, and grid lines.
   c. Explain how dimensions relate to various drawing scales.
   d. Demonstrate how to use engineer's and architect's scales.
   e. Identify the six types of construction drawings.

Performance Task

Under supervision, you should be able to do the following:

1. Using the residential floor plan and elevations supplied in the *Appendix D*:
   - Locate the wall common to bedroom #3 and the garage.
   - Determine the overall width of the home.
   - Calculate the distance from the outside northwest corner of the house to the center of the window in bedroom #2.
   - Determine the overhang of the eaves.
   - Using an architect's scale, determine the width of the garage.

Overview

Various types of construction drawings are used to represent actual components of a building project. The drawings provide specific information about the locations of the parts of a structure, the types of materials to be used, and the correct layout of the building. Knowing the purposes of the different types of drawings and interpreting the drawings correctly are important skills for anyone who works in the construction trades. This module introduces common types of construction drawings, their basic components, standard drawing elements, and measurement tools that are typically used when working with construction drawings.

Industry Recognized Credentials

If you are training through an NCCER-accredited sponsor, you may be eligible for credentials from NCCER's Registry. The ID number for this module is 00105. Note that this module may have been used in other NCCER curricula and may apply to other level completions. Contact NCCER's Registry at 888.622.3720 or go to **www.nccer.org** for more information.

1.0.0 Construction Drawings and Their Components

Performance Task

1. Using the residential floor plan and elevations supplied in the *Appendix D*:

   - Locate the wall common to bedroom #3 and the garage.
   - Determine the overall width of the home.
   - Calculate the distance from the outside northwest corner of the house to the center of the window in bedroom #2.
   - Determine the overhang of the eaves.
   - Using an architect's scale, determine the width of the garage.

Objective

Describe components and features used in construction drawings and identify how the drawings are different.

a. Summarize the purpose of the six basic construction drawing components.
b. List and explain the significance of various drawing elements, such as lines of construction, symbols, and grid lines.
c. Explain how dimensions relate to various drawing scales.
d. Demonstrate how to use engineer's and architect's scales.
e. Identify the six types of construction drawings.

Blueprints: A traditional method of reproducing technical drawings using a contact print process and light-sensitive paper that resulted in white lines on a blue background. Also, the traditional name used to describe construction drawings.

Computer-aided drafting (CAD): The making of a set of construction drawings with the aid of a computer.

Specifications: Precise written presentation of the details of a plan.

Construction drawings are architectural or working drawings used to represent a structure or system. These were referred to traditionally as **blueprints**, because years ago the process for reproducing these drawings resulted in white lines on a blue background. With the advent of **computer-aided drafting (CAD)**, and most recently building information modeling (BIM), drawings are created on a computer and reproduced using large-format plotters, printers, and copiers. Today, construction drawings are simply called drawings or prints.

Various kinds of construction drawings, including residential drawings, commercial drawings, landscaping plans, shop drawings, and industrial drawings, are used in construction. In this module, several of the most common types of drawings will be introduced.

Construction drawings, together with the set of **specifications** (often abbreviated as specs), detail what is to be built and what materials are to be used. Specifications are written statements that the architectural and engineering firm provides to the general contractors. The specs define the quality of work to be done and describe the materials to be used.

The set of construction drawings forms the basis of agreement and understanding that a building will be built as detailed in the drawings. Therefore, everyone involved in planning, supplying, and building any structure should be able to read construction drawings. For any building project, also consult the civil engineering plans for that location, including sewer, highway, and water installation plans.

Before the 1950s, construction drawings were all drawn by hand. This was a long and tedious process. Many times, if there was a mistake or a drawing change required, the drawing had to be recreated from scratch. The evolution from hand drafting to computer-aided drafting (CAD) started in 1950 when Dr. Paul J. Hanratty introduced a numerically controlled program that allowed users to draw simple lines utilizing a computer. Years later, in 1972, Dr. Hanratty invented a program called Automated Drafting and Machining. The program could draw more complex objects and included many of the commands and features still used in modern CAD programs today. By the 1980s CAD programs were becoming commonplace with designers, architects, and engineers.

Using CAD is a cost-effective way to increase drafting productivity because the computer program automates much of the repetitive work related to generating drawings. Ultimately, CAD has the following advantages over hand-drawn construction drawings:

- It is automated.
- The computer performs calculations quickly and easily.
- Changes can be made easily.
- Commonly used symbols can be retrieved without difficulty.
- CAD can produce three-dimensional modeling of the structure.

While CAD is still widely used, more and more drawings are created today using building information modeling (BIM). Traditionally, BIM was a system of paper-based documents and construction drawings designed to monitor and track specific information about a building.

In the last decade, BIM has increasingly moved toward computer technology, which allows the various trades and the building owner to track the lifecycle of the building, from the start of planning, through construction, to completion. BIM allows building owners and craftworkers to easily track and identify the various components and assemblies of individual systems (such as electrical, HVAC, and plumbing) and building materials within a specific location or locations of a building.

BIM also has the capability of simulating the lifecycles of various systems within the building to provide information on determining the wear on, and lifespan of, individual components. Additionally, BIM software allows the craftworker and the building owner to input and save information on the building as repairs, updates, or renovations are made to the building. Information such as part sizes, manufacturers, part numbers, and related drawings, or any other information ever researched in the past, can be stored in a BIM system.

Another advantage of BIM is that it allows documentation and information on a building to be easily transferred and communicated between all the members of a project.

Regardless of how drawings are created, there are almost always changes that occur during the design and construction phases. This means that the drawing needs to be updated to reflect those changes. Once there are multiple versions of a drawing, it is important to ensure that everyone is working from the most recent, up-to-date drawing. This is accomplished by using **drawing revision control**.

Drawing revision control is a system for recording, documenting, and identifying drawing revisions by using a numeric or alphabetic sequence. Drawing revisions are recorded in the revision history block of the drawing. Some revision history blocks are included as part of the title block, while some are inserted as a separate block. A revision history block typically contains columns for the letter or number designating the revision, the drafter's initials, the date of the revision, and the description of the revision. Items revised on the drawing are typically highlighted with a revision cloud around them. A numbered or lettered triangle next to the cloud identifies its sequence in the revision history. The revised drawing is saved as a new drawing file to preserve the revision history.

Drawing revision control: A system for recording, documenting, and identifying drawing revisions by using a numeric or alphabetic sequence.

1.1.0   Basic Components of Construction Drawings

Most construction drawings are laid out in a fairly standardized format. This section describes the following six parts of a construction drawing:

- **Title block**
- Border
- Drawing area
- Revision block
- Legend
- North arrow

Title block: A part of a drawing sheet that includes some general information about the project.

1.1.1 Title Block

The first thing to look at on any drawing is the title block. The title block is normally in the lower right-hand corner of the drawing or across the right edge of the paper (*Figure 1*) and serves two primary purposes. First, it provides information about the structure or assembly. Second, it is numbered so the print can be filed easily.

Different companies put different information in the title block. Generally, it contains the following:

- Company logo — Usually preprinted on the drawing.
- Sheet title — Identifies the project.
- Date — Date the drawing was checked and readied for seal, or issued for construction.
- Drawn by — Initials of the person who drafted the drawing.
- Drawing number — Code numbers assigned to a project.
- Scale — The ratio of the size of the object as drawn to the object's actual size.
- Revision blocks — Information on revisions, including (at a minimum) the date and the initials of the person making the revision. Other information may include descriptions of the revision and a revision number.

Every company has its own system for such things as project numbers and departments. It also has its own placement locations for the title and revision blocks. Your supervisor should explain the company's standards.

1.1.2 Border

The border is a clear area of approximately half an inch around the edge of the drawing area. It is there so that everything in the drawing area can be printed or reproduced on printing machines with no loss of information.

1.1.3 Drawing Area

The drawing area presents the information for constructing the project, such as the floor plan, elevations of the building, sections, and details.

1.1.4 Revision Block

A revision block is located in the drawing area, usually in the lower right corner inside the title block or near it. Different companies put the revision block in different places. This block is used to record any changes (revisions) to the drawing. It typically contains the revision number, a brief description, the date, and the initials of the person who made the revisions (*Figure 2*). All revisions must be noted in this block and dated and identified by a letter or number.

> **CAUTION**
>
> It is essential to note the revision designation on a construction drawing and to use only the latest version. Otherwise, costly mistakes may result.

PROJ	NO	REVISION	RVSD	CHKD	APPD	DATE
3483	01	RELEASED FOR CONSTRUCTION		APD	NWS	JULY 21
3483	02	DELETED PART OF LINE 12037		APD	NWS	AUG 21
3483	03	⚠ ADDED WELDING SYMBOL		APD	NWS	AUG 21

Figure 1 The title block of a construction drawing.

Figure 2 The revision block of a construction drawing.

1.1.5 Legend

Each line on a construction drawing has a specific design and thickness that identifies it. Note that some of the lines may be used to identify off-site utilities. The identification of these lines and other symbols is called the legend. Although a legend doesn't automatically appear on every construction drawing, when it does, it explains or defines symbols or special marks used in the drawing (*Figure 3*). Be aware that legends are specific only to the set of drawings in which they are contained.

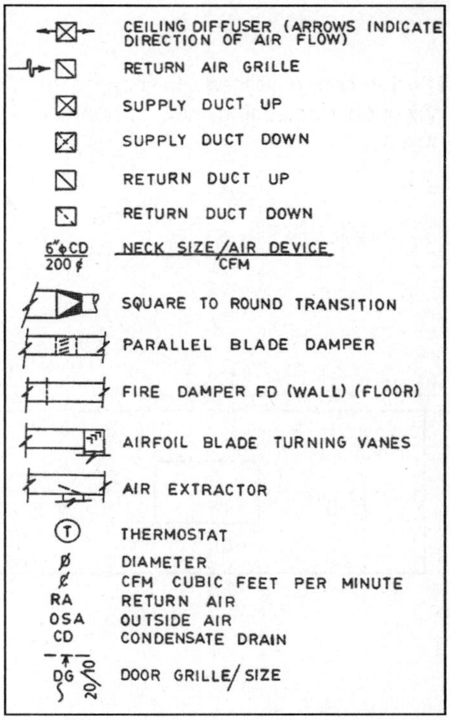

1.1.6 North Arrow

Most construction drawings have an arrow indicating north. It may be located near the title block or in some other conspicuous place. The north arrow is a reference point that helps to verify the locations of objects, walls, and building parts shown on the construction drawing.

Regional and Company Differences

Although most symbols are standard, there can be slight variations in different regions of the country. Always check the title sheet or other introductory drawing to verify the symbols on the project drawings.

Your company may also use some special symbols or terms. Your instructor or supervisor should explain the conventions that are unique to the company.

Figure 3 Sample legend.

Did You Know?
Legality of Construction Drawings
Construction drawings are incorporated into building contracts by reference, making them part of the legal documents associated with a project. When describing the project to be completed, the legal contract between the builder and the owner refers to the accompanying construction drawings for details that would be too lengthy to write out. That makes tracking changes or revisions to construction drawings over the course of a project vitally important. If an error is made along the way, either the owner or the builder must be able to find the discrepancy by reading the drawings. Taking care of construction drawings over the course of the project makes good business sense.

Did You Know?
Plan North
When true north is not noted on a drawing and cannot be verified, plan north is used to describe direction on the drawing. This helps to communicate the location of features on a drawing. Plan north is toward the top of the page, east is to the right, south to the bottom, and west to the left.

1.2.0 Drawing Elements

Various drawing elements make it easier and quicker for people to read and understand construction drawings. This section covers several of the most common drawing elements.

1.2.1 Lines of Construction

It is very important to understand the meanings of lines on a drawing. The lines commonly used on a drawing are sometimes called the alphabet of lines. Here are some of the more common types of lines (*Figure 4*):

- **Dimension line** — Establishes the dimensions (sizes) of parts of a structure. These lines end with arrows (open or closed), dots, or slashes at a termination line drawn perpendicular to the dimension line.
- **Leader** *and* *arrowhead* — Identify the location of a specific part of the drawing. They are used with words, abbreviations, symbols, or keynotes.
- *Property line* — Indicates land boundaries.
- *Cut line* — Lines around part of a drawing that is to be shown in a separate cross-sectional view.
- *Section cut* — Shows areas not included in the cutting line view.

Dimension line: A line on a drawing with a measurement indicating length.

Leader: In drafting, the line on which an arrowhead is placed and used to identify a component.

Hidden line: A dashed line showing an object obstructed from view by another object.

- *Break line* — Shows where an object has been broken off to save space on the drawing.
- **Hidden line** — Identifies part of a structure that is not visible on the drawing. You may have to look at another drawing to see the part referred to by the lines.
- *Center line* — Shows the measured center of an object, such as a column or fixture.
- *Object line* — Identifies the object of primary interest or the closest object.

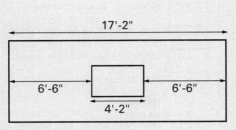

Check All the Plans

Always double-check all the plans for a project. Be familiar with other trade work that may affect your job. Determine whether an electrical conduit is close to the plumbing pipes or whether a framing member is placed too close to an HVAC duct. By knowing all the work planned for a particular area, errors can be prevented, saving time and money.

If a particular dimension is missing on the plan, check the other plans to see if it is included there, as that part of the building will also be shown on other plans.

When taking measurements from a set of plans, make sure that the dimensions of each measured section add up to the total measurement. For instance, when looking at a drawing of a wall with one window, add up the measurement from the left end of the wall to the left side of the window, the width of the window, and the measurement from the right side of the window to the right end of the wall. Make sure this total matches the measurement of the entire wall.

1.2.2 Abbreviations, Symbols, and Keynotes

Architects: Qualified, licensed people who create and design drawings for a construction project.

Engineers: People who apply scientific principles in design and construction.

Symbols: Drawings that represent a material or component on a plan.

Architects and **engineers** use systems of abbreviations, **symbols**, and keynotes to keep plans uncluttered, making them easier to read and understand. Examples of some of these items appear throughout this module. Following is additional information on these items and some variants that are commonly used.

Each trade has its own symbols, and workers should learn to recognize the symbols used by other trades. For example, an electrician should understand a carpenter's symbols, a carpenter should understand a plumber's symbols, and so on. Then, no matter what symbols are encountered on a project, the workers will understand what the symbols mean.

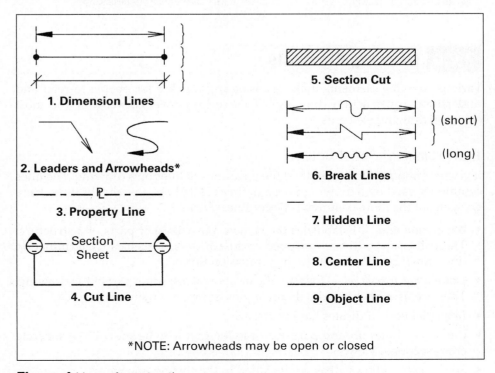

Figure 4 Lines of construction.

Abbreviations used in construction drawings are short forms of common construction terms. For example, the term *on-center* is abbreviated O.C. Some common construction abbreviations are listed in *Figure 5*.

Abbreviations should always be written in capital letters. Abbreviations for each project should be noted on the title sheet or other introductory drawing page such as the legend page. Books that list construction abbreviations and their meanings are available. There is no need to memorize these abbreviations, as they will become more familiar with time.

Symbols are used on a drawing to tell what material is required for that part of the project. A combination of these symbols, expanded and drawn to the same size, makes up the pictorial view of the plan. There are architectural symbols (*Figure 6*), civil and structural engineering symbols (*Figure 7*), mechanical symbols (*Figure 8*), plumbing symbols (*Figure 9*), and electrical symbols. Slightly different symbols may be used in different parts of the country or by different companies. The symbols used for each set of plans should be indicated on the title sheet or other introductory drawing. Many code books, manufacturers' brochures, and specifications include symbols and their meanings.

Some plans use keynotes (*Figure 10*) instead of symbols. A keynote is a number or letter (usually in a square or circle) with a leader and arrowhead that is used to identify a specific object. Part of the drawing sheet (usually on the right-hand side) lists the keynotes with their numbers or letters. The keynote descriptions normally use abbreviations.

ABBREVIATIONS

A.B.	– ANCHOR BOLT	FDN.	– FOUNDATION	RM.	– ROOM
ADD'L	– ADDITIONAL	FIN.	– FINISH	SCHED.	– SCHEDULE
ADJ.	– ADJACENT	FLR.	– FLOOR	SECT.	– SECTION
A.I.S.C.	– AMERICAN INSTITUTE OF STEEL CONSTRUCTION	F.O.B.	– FACE OF BRICK	SHT.	– SHEET
		F.O.CONC.	– FACE OF CONCRETE	SIM.	– SIMILAR
ALT.	– ALTERNATE	F.O.W.	– FACE OF WALL	S.L.V.	– SHORT LEG VERTICAL
ARCH.	– ARCHITECTURAL	FS.	– FLAT SLAB	SPC.	– SPACE
A.S.T.M.	– AMERICAN SOCIETY FOR TESTING & MATERIALS	FT.	– FOOT	SPEC.	– SPECIFICATION
		FTG.	– FOOTING	SQ.	– SQUARE
BLDG.	– BUILDING	F.W.	– FILLET WELD	STD.	– STANDARD
BM.	– BEAM	GA.	– GAUGE	STIFF.	– STIFFENER
B.O.	– BOTTOM OF	GAL.	– GALVANIZED	STL.	– STEEL
BOT.	– BOTTOM	G.L.	– GLU-LAM BEAM	STOR.	– STORAGE
BSMT.	– BASEMENT	GR.	– GRADE	SYM.	– SYMMETRICAL
BTWN.	– BETWEEN	GR. BM.	– GRADE BEAM	T.&B.	– TOP AND BOTTOM
CANT.	– CANTILEVER	H.A.S.	– HEADED ANCHOR STUD	THK.	– THICKNESS
CB.	– CARDBOARD	HORIZ.	– HORIZONTAL	T.O.	– TOP OF
CH.	– CHAMFER	H.S.B.	– HIGH STRENGTH BOLT	TYP.	– TYPICAL
C.J.	– CONTROL/CONSTRUCTION JOINT	I.D.	– INSIDE DIAMETER	U.N.O.	– UNLESS NOTED OTHERWISE
CLR.	– CLEAR, CLEARANCE	IN.	– INCH	VAR.	– VARIES
C.M.U.	– CONCRETE MASONRY UNIT	INT.	– INTERIOR	VERT.	– VERTICAL
COL.	– COLUMN	JNT.	– JOINT	V.I.F.	– VERIFY IN FIELD
CONC.	– CONCRETE	LB.	– POUND	WT.	– WEIGHT
CONN.	– CONNECTION	LIN. FT.	– LINEAL FEET		
CONST.	– CONSTRUCTION	L.L.V.	– LONG LEG VERTICAL		
CONT.	– CONTINUOUS	MAT'L.	– MATERIAL		
CONTR.	– CONTRACTOR	MAX.	– MAXIMUM	**SYMBOLS**	
CTRD.	– CENTERED	MECH.	– MECHANICAL	℄	CENTER LINE
DET.	– DETAIL	MID.	– MIDDLE		
DIAG.	– DIAGONAL	MIN.	– MINIMUM	φ	DIAMETER
DIAM.	– DIAMETER	MISC.	– MISCELLANEOUS		
DIM.	– DIMENSION	MTL.	– METAL	⊕	ELEVATION
DISCONT.	– DISCONTINUOUS	N.I.C.	– NOT IN CONTRACT		
DWG.	– DRAWING	NO.	– NUMBER	&	AND
EA.	– EACH	NOM	– NOMINAL		
E.F.	– EACH FACE	N.T.S.	– NOT TO SCALE	W/	WITH
EL.	– ELEVATION	O.C.	– ON CENTER		
ELECT.	– ELECTRICAL	O.D.	– OUTSIDE DIAMETER	℔	PLATE
ELEV.	– ELEVATOR	O.H.	– OPPOSITE HAND		
EQ.	– EQUAL	OPNG.	– OPENING	X	BY
E.W.B.	– END WALL BARS	℔	– PLATE		
E.W.	– EACH WAY	P.S.F.	– POUND PER SQUARE FOOT	#	NUMBER
EXIST.	– EXISTING	P.S.I.	– POUND PER SQUARE INCH		
EXP. JNT.	– EXPANSION JOINT	R.	– RADIUS	⊙	AT
EXT.	– EXTERIOR	REINF.	– REINFORCEMENT		
F.D.	– FLOOR DRAIN	REQ'D.	– REQUIRED	⌗	SQUARE
				L	ANGLE

Figure 5 Abbreviations.

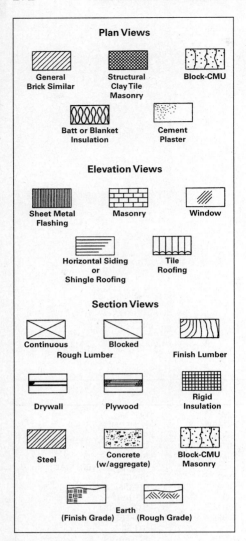

Figure 6 Architectural symbols.

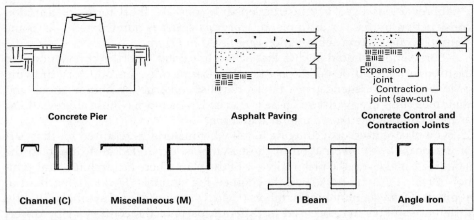

Figure 7 Civil and structural engineering symbols.

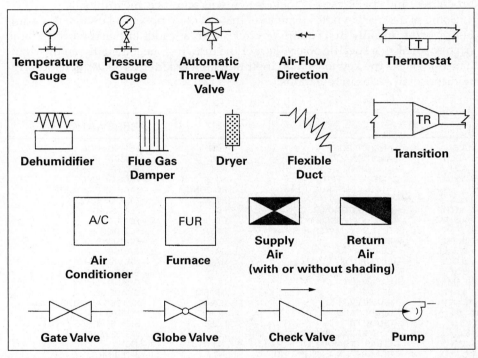

Figure 8 Mechanical symbols.

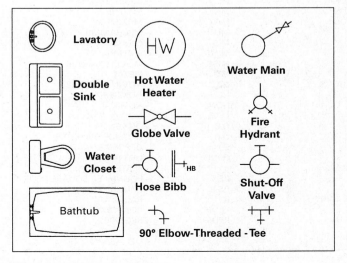

Figure 9 Plumbing symbols.

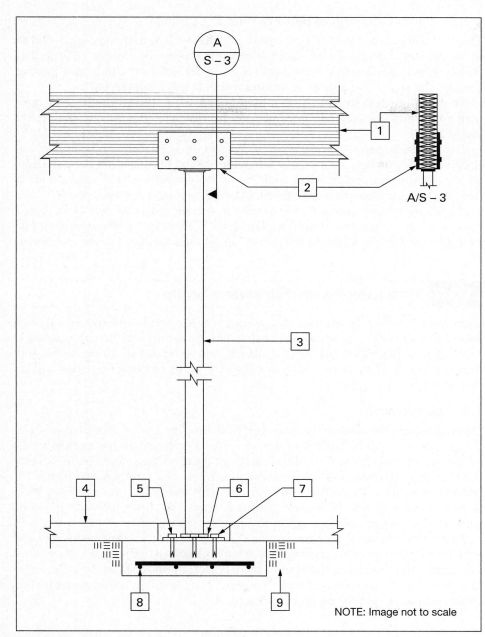

Figure 10 Keynotes.

Source: CM Fusion LLC

Request for Information (RFI)

A request for information (RFI) is used to clarify any discrepancies in the plans. If there is a discrepancy, notify the foreman. The foreman will write up an RFI, explaining the problem as specifically as possible and putting the date and time on it. The RFI is submitted to the superintendent, who passes it to the general contractor, who passes it to the architect or engineer, who then resolves the discrepancy.

Always refer to specifications and the RFI when deciding how to interpret drawings.

Care of Construction Drawings and Mobile Devices

Construction drawings and mobile devices used to store drawings and other supplemental documents are valuable records and should be protected. Follow these rules when handling construction drawings and mobile devices:

- Never write on a construction drawing without authorization.
- Keep drawings clean. Dirty drawings are hard to read and can cause errors.
- Fold drawings so that the title block is visible.
- Fold and unfold drawings carefully to avoid tearing.

- Do not lay sharp tools or pointed objects on construction drawings.
- Keep drawings away from moisture.
- Make copies for field use; don't use originals.
- Use a case when transporting or storing a mobile device.
- Keep the mobile device charged and make sure all data is backed up.
- Protect the mobile device display with a screen cover.
- Avoid prolonged exposure of the mobile device to direct sunlight and extreme heat/cold.
- Keep the mobile device away from water, food, and drink.

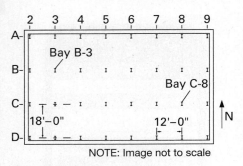

Figure 11 Grid.

1.2.3 Using Gridlines to Identify Plan Locations

Think of using a road map to find a particular street. The map may use a grid to make locating a detailed area easier. For example, the index might indicate that the street is in section B-3, referring to B along the side of the map and 3 along the top. The street was able to be located where B and 3 intersected.

The gridline system shown on a plan (*Figure 11*) is used like the grid on a map. On a drawing such as a floor plan, a grid divides the area into small parts called bays.

The numbering and lettering system begins in the upper left-hand corner of the floor plan. The numbers are normally across the top and the letters are along the side. To avoid confusion, certain letters and the symbol for zero are not used. Omitted from the gridline system are the letters I, O, and Q, and the numbers 1 and 0.

A gridline system makes it easy to refer to specific locations on a plan. A single outlet can be pointed out on a plan, even if there are a dozen shown. For example, referring to an outlet in bay C-8 helps to distinguish a particular outlet.

1.3.0 Dimensions and Drawing Scale

Dimensions and their associated drawing scales provide information that is essential for correctly translating construction drawings into actual buildings. This section describes the basic features and purpose of dimensions and drawing scales. How to read and use different types of drawing scales will be discussed later.

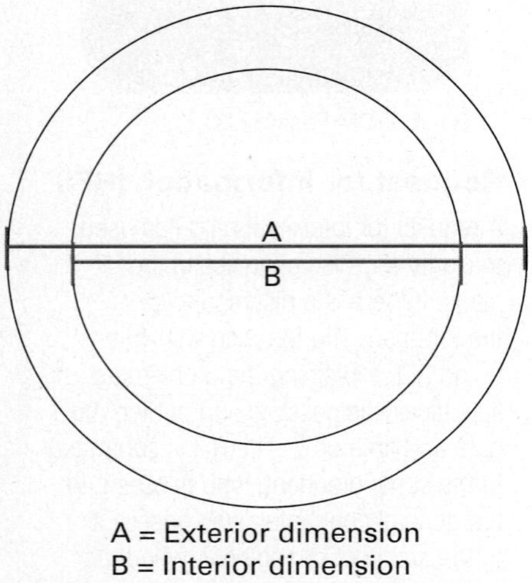

1.3.1 Dimensions

Dimensions are the parts of the drawings that show the size and the placement of the objects that will be built or installed. Dimension lines can have arrowheads or slashes at both ends, with the dimension itself written near the middle of the line. The dimension is a measurement written as a number, and it may be written in inches with fractions (6½"), in feet with inches (1' 2"), or in inches with decimals (3.2"). When the metric system is used, the dimensions are usually expressed in meters, centimeters, or millimeters (9 mm).

To do accurate work, workers need to know how to read dimensions on construction drawings. This means they need to know whether the dimensions measure to the exterior or the interior of an object. To understand the difference, look at *Figure 12*, which shows a piece of pipe. There are two measurements that could be taken to get the pipe's dimensions.

A = Exterior dimension
B = Interior dimension

Figure 12 Exterior and interior dimensions on pipe.

Going Green

Green Construction

Green construction refers to a method of designing and building structures using materials and techniques that help minimize the stress on our natural resources and the environment.

With just under 5 percent of the world's population, the United States manages to consume about 19 percent of the world's energy (buildings account for 40 percent of this consumption) and produces 170 million tons (154,221,406 metric tons) of construction and demolition debris a year.

In response to these numbers, and a general increase in environmental awareness, organizations such as the US Green Building Council (USGBC) created environmental assessment systems, such as the LEED (Leadership in Energy and Environmental Design) Green Building Rating System. Systems like LEED provide green standards for the construction industry to follow, and they officially certify nonresidential structures that meet USGBC's strict criteria.

The LEED program was launched in 1988. It is a voluntary national standard that awards points for incorporating green strategies into areas such as site planning, safeguarding water quality and efficiency, efficiency in energy use and recycling, conservation of resources and materials, and the design, quality, and efficiency of the indoor environment.

Going Green

Recycling Rinker Hall, University of Florida in Gainesville

An example of a LEED Gold rated building is Rinker Hall on the campus of the University of Florida in Gainesville, Florida. During the planning phase of the building, special consideration was given to land use and the use of recycled and recyclable materials. Special attention was also given to planning the incorporation of the most recent green technologies in the areas of water consumption and conservation, heating and cooling energies, and other power needs required to run the various systems of a building.

A few of the green strategies incorporated into the construction of Rinker Hall include:

- Orienting the structure on a pure north-south axis to allow it to use low-angle light for daytime lighting.
- Using rooftop skylights and a central skylight-covered atrium to light specific areas using natural sunlight.
- Using shade walls on the east and west sides of the building.
- Landscaping with native trees and flora that require minimal watering.
- Installing water-free urinals and fixtures that use 20 percent less water than mandated.

With the various green strategies, materials selection, and building techniques employed, Rinker Hall and all of its systems use up to 57 percent less energy than a similar structure designed in minimal compliance with the American Society of Heating, Refrigerating, and Air Conditioning Engineers (ASHRAE).

The first measurement is from the pipe's exterior edge on one side directly across to its exterior edge on the other side. The second measurement is from the pipe's inside (interior) edge on one side directly across to its interior edge on the other side. Even though the difference between these two dimensions may be only a fraction of an inch (the thickness of the pipe), they are still two completely different dimensions. This is important to remember because any dimensioning inaccuracy or miscalculation in one place will affect the accuracy of calculations in other places.

1.3.2 Drawing Scale

The scale of a drawing tells the size of the object drawn compared with the actual size of the object represented. The scale is shown in one of the spaces in the title block, beneath the drawing itself, or in both places. The type of scale used on a drawing depends on the size of the objects being shown, the space available on the paper, and the type of plan.

On a site plan, the scale may read SCALE: 1" = 20' 0". This means that every 1 inch on the drawing represents 20 feet, 0 inches. The scale used to develop site plans is an **engineer's scale**.

On a floor plan, the scale may read SCALE: ¼" = 1' 0". This means that every ¼ inch on the drawing represents 1 foot, 0 inches. Floor plans are developed using an **architect's scale**. This scale is divided into fractions of an inch. Metric floor plans typically use a ratio of 1:50, indicating that each millimeter on the drawing represents 50 meters.

Some drawings are not drawn to scale. A note on such drawings reads **not to scale (NTS)**.

Checking Scales on Drawings

Normally the scale of a drawing is marked directly on the drawing itself. If it is not, however, do not guess what the scale is. Rather, compare different scales by identifying a given length on the drawing (for example, an 8' 6" pipe), and then find the scale that measures the pipe at 8' 6". Check at least two items on the drawing this way before proceeding.

CAUTION

When a plan is marked NTS, workers cannot measure dimensions on the drawing and use those measurements to build the project. Not-to-scale drawings give relative positions and sizes. The sizes are approximate and are not accurate enough for construction.

Engineer's scale: A straightedge measuring device divided uniformly into multiples of 10 divisions per inch so that drawings can be made with decimal values. Used mainly for land measurements on site plans.

Architect's scale: A specialized ruler used in making or measuring reduced scale drawings. The ruler is marked with a range of calibrated ratios for laying out distances, with scales indicating feet, inches, and fractions of inches. Used on drawings other than site plans.

Not to scale (NTS): Describes drawings that show relative positions and sizes only, without scale.

Schematic Drawings

Most plumbing and electrical sketches are single-line drawings or schematic drawings. These drawings illustrate the scale and relationship of the project's components. In a single-line or schematic plumbing drawing, the line represents the centerline of the pipe. In a single-line or schematic electrical drawing, the line represents electrical wiring routing or circuit.

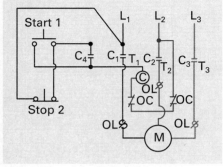

Orthographic Drawings

Orthographic drawings are used for elevation drawings. They show straight-on views of the different sides of an object with dimensions that are proportional to the actual physical dimensions. In orthographic drawings, the designer draws lines that are scaled-down representations of real dimensions. Every 12 inches, for example, may be represented by ¼ inch on the drawing. Similarly, in an example using metric measurements with a ratio of 1:2, every 30 millimeters may be represented by 15 millimeters on the drawing.

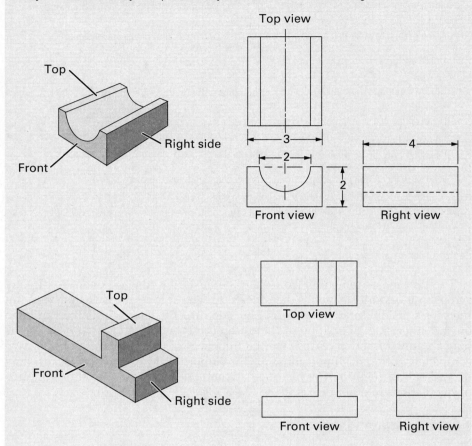

1.4.0 Measuring Scales

Standard and metric rulers and measuring tapes are usually used in a shop or in the field for a variety of measuring tasks. However, there are other types of rulers called scales used by draftsperson to produce drawings. They are also used by many workers to take scaled measurements on a drawing. These include the architect's scale, the **metric scale** (metric architect's scale), and the engineer's scale. Knowing how these scales work makes it easier to understand the information contained in a set of drawings.

1.4.1 Architect's Scale

The architect's scale is often used to create construction drawings. An architect's scale translates the large measurements of real structures (rooms, walls, doors, windows, duct, etc.) into smaller measurements for drawings. Architect's scales are available in several types, but the most common include the triangular and flat scales. A flat scale is shown in *Figure 13*. The triangular architect's scale is most commonly used because it can combine up to 12 different scales on one tool. Each side of the triangular form has two faces, and two scales are combined on each face. Architect's scales are available in 6- and 12-inch lengths.

Metric scale: A straightedge measuring device divided into centimeters, with each centimeter divided into 10 millimeters. Usually used for architectural drawings and sometimes referred to as a metric architect's scale.

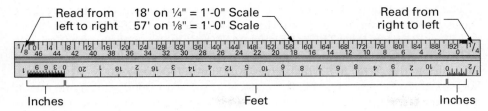

Read from
left to right   18' on ¼" = 1'-0" Scale
57' on ⅛" = 1'-0" Scale   Read from
right to left

Inches   Feet   Inches

Figure 13 The architect's scale.

Each scale on an architect's scale is designated to a different fraction of an inch that equals a foot. These fraction designations appear on the right and left corners of each scale. An architect's scale is read from left to right or right to left, depending on which scale is being read.

Look at the point of measurement in *Figure 13*. It represents 57 feet when read from the left on the ⅛-inch scale and equals 18 feet when read from the right on the ¼-inch scale.

Now refer to *Figure 14*. Using the ⅜-inch scale and reading from the right, it can be determined that the section of duct is 7 feet, 5 inches long. Notice how the 0 point on an architect's scale is not at the extreme end of the measuring line. This is because numbers to the right of the 0 represent fractions of one foot (or inches).

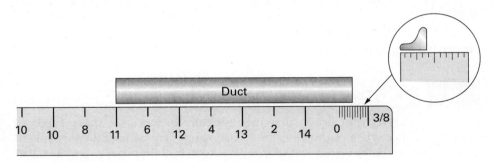

Figure 14 Measuring a section of duct with an architect's scale.

1.4.2   Metric Scale (Metric Architect's Scale)

Similar to the architect's scale is the metric scale, sometimes referred to as a metric architect's scale (*Figure 15*). Common lengths indicated on metric scales are 30 and 60 millimeters. Like the architect's scale, a metric scale can have a number of scales on it and is used to generate drawings. *Figure 15* shows how 10 meters would be shown on a 1:200 scale. The actual length on the plans is 50 mm, equaling 10 meters of real distance on the project.

Metric scales are calibrated in units of 10. Some of the most common metric scales include 1:5, 1:10, 1:20, 1:50, 1:100, 1:200, 1:500, and 1:1,000.

The two common length measurements used with the metric scale on architectural drawings are the meter and millimeter, the millimeter being $\frac{1}{1000}$ of a meter. On drawings drawn to scale between 1:1 and 1:100, the millimeter is typically used. The millimeter symbol (mm) will not be shown, but there should be a note on the drawing indicating that all dimensions are given in millimeters unless otherwise noted.

On drawings with scales between 1:200 (*Figure 15*) and 1:2,000, the meter is generally used. Again, the meter symbol (m) will not be shown, but the drawing will have a note indicating that all dimensions are in meters unless otherwise noted. Land distances shown on site and plot plans, expressed in metric units, are typically given in meters or kilometers (1,000 meters).

Reading a metric scale is easy after identifying the scale to use and the length that the scale increments represent on the drawing. Here's how it works.

50 mm scale measurement = 10 meters

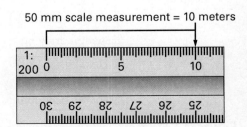

Figure 15 Metric architect's scale.

The unit of length in *Figure 15* is the meter and the object on the drawing starts at 0 and extends out to the 10 on the scale. At a 1:200 ratio this object would be 10 meters long. This is because every millimeter on the scale represents 200 millimeters on the object. For example, if there are 50 millimeters from 0 to 10, multiply the 50 by 200, which gives 10,000 mm (50 mm × 200 = 10,000 mm). Because the unit being used is the meter and there are 1000 mm in one meter, divide the 10,000 mm by 1,000, which results in 10 meters. This same process is also used to determine lengths for other ratios on the scale.

1.4.3   Engineer's Scale

The engineer's scale (*Figure 16*) is used mainly for land measurements on site plans, which means the scale must accommodate very large measurements. Each engineer's scale is set up as multiples of 10 and the measurements are taken in decimals. This is different from the architect's scale in that a unit is represented by a portion of an inch. The most common engineer's scales are 10, 20, 30, 40, 50, and 60, which can all be combined on the triangular engineer's scale.

For each scale, the measurements can represent various units derived from that scale number and a multiple of 10. For example, on a 10 scale, 1 inch can represent 1 foot (10 × 0.10), 10 feet (10 × 1.0), 100 feet (10 × 10.0), and so on. On a 50 scale, an inch can be 5 feet (50 × 0.10), 50 feet (50 × 1.0), or 500 feet (50 × 10.0), and so on. This same process can also be used to determine lengths on the scales.

Figure 16 The engineer's scale.

1.5.0   Six Types of Construction Drawings

A set of building construction drawings typically includes six major types of drawings. Other types of drawings and documents may be used to supplement the drawing set, but their presence depends on each building's usage and installed subsystems. For example, both a fire station and a convenience store drawing package would include the six major drawing types, but the rest of the package would differ significantly. The six major drawing types (*Figure 17*) include the following:

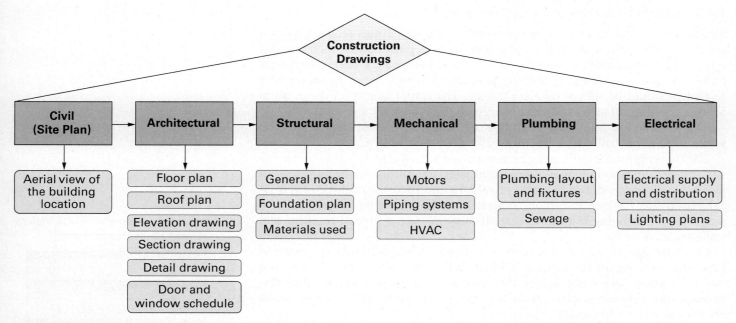

Figure 17 Types of construction drawings.
Source: Croxton Collaborative/Gould Evans Associates

- Civil
- Architectural
- Structural
- Mechanical
- Plumbing
- Electrical

As mentioned previously, other types of drawings and documents may be included to supplement a set of construction drawings depending on the type and use of the building. These could include:

- Fire protection
- Shop drawings
- As-built
- Change order

This section examines the characteristics of each type of construction drawing.

Did You Know?

BIM and Energy Efficient Design

Simulation tools within BIM software allow architects and designers to look at different building designs and test how energy efficient a building will be long before ground is ever broken on the project. These tools allow the user to simulate sunlight and shading throughout different seasons based on the exact location of the project along with existing surrounding landscape features, buildings, and structures. This means that the final design will make the best use of building location and direction, windows, walls, existing site features, etc., to optimize the utilization of natural sunlight and maximize energy efficiency.

1.5.1   Civil Plans

Civil plans are also called *site plans, survey plans,* or *plot plans*. They show the location of the building on the site from an aerial view (*Figure 18*). A civil plan also shows the natural contours of the earth, represented on the plan by **contour lines**, or topographical lines. The vertical data for these contour lines are obtained by a surveyor. Before a topographical survey can take place, a benchmark must be established. A benchmark is a fixed point with a known elevation that is used as the base point for measuring all elevation values (*Figure 19*). Surveyors also use a special coordinate system for identifying location, direction, and distance by utilizing lines of latitude and longitude (*Figure 20*). The civil plans can also include a landscape plan (*Figure 21*) that shows any trees on the property; construction features such as walks, driveways, or utilities; the dimensions of the property; and possibly a legal description of the property.

Civil plans: Drawings that show the location of the building on the site from an aerial view, including contours, trees, construction features, and dimensions.

Contour lines: Solid or dashed lines showing the elevation of the earth on a civil drawing.

1.5.2   Architectural Plans

Architectural plans (also called architectural drawings) show the design of the project. Look at architectural plans first, because all other drawings follow from them. Architectural plans are the most general; they show how all the parts of the project fit together. One part of an architectural plan is a **floor plan**, also known as a plan view (refer to Drawing 1, Residential Floor Plan, in *Appendix D*). Any drawing made looking down on an object is commonly called a plan view. The floor plan is an aerial view of the layout of each room. It provides the most information about the project. It shows exterior and interior walls, doors, stairways, and mechanical equipment. The floor plan shows the floor as someone would see it from above if the upper part of the building were removed.

An architectural plan also includes a **roof plan** (*Figure 22*), which is a view of the roof from above the building. It shows the shape of the roof and the materials that will be used to finish it.

Architectural plans: Drawings that show the design of the project. Also called architectural drawings.

Floor plan: A drawing that provides an aerial view of the layout of each room.

Roof plan: A drawing of the view of the roof from above the building.

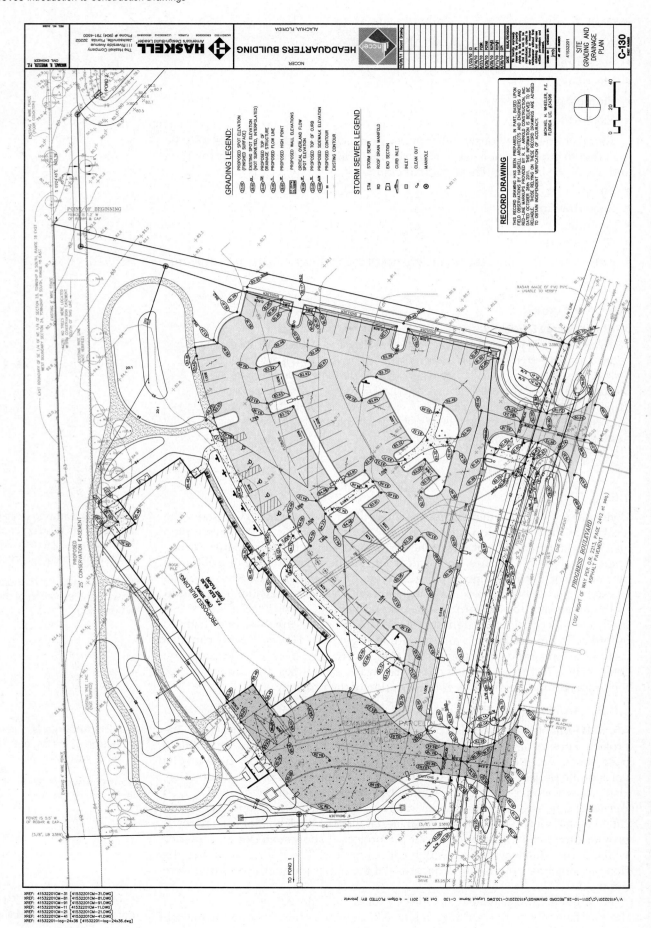

Figure 18 Civil plan, aerial view.

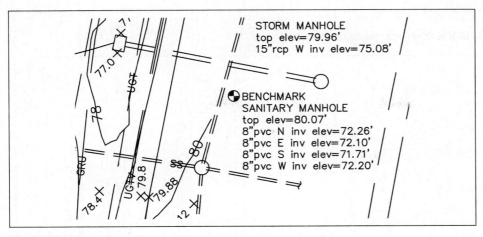

Figure 19 Benchmark.

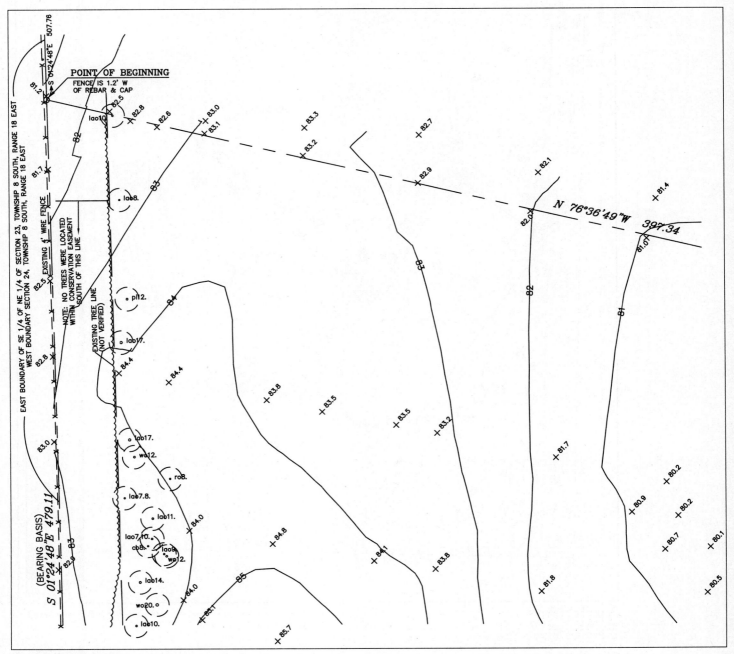

Figure 20 Coordinate system.

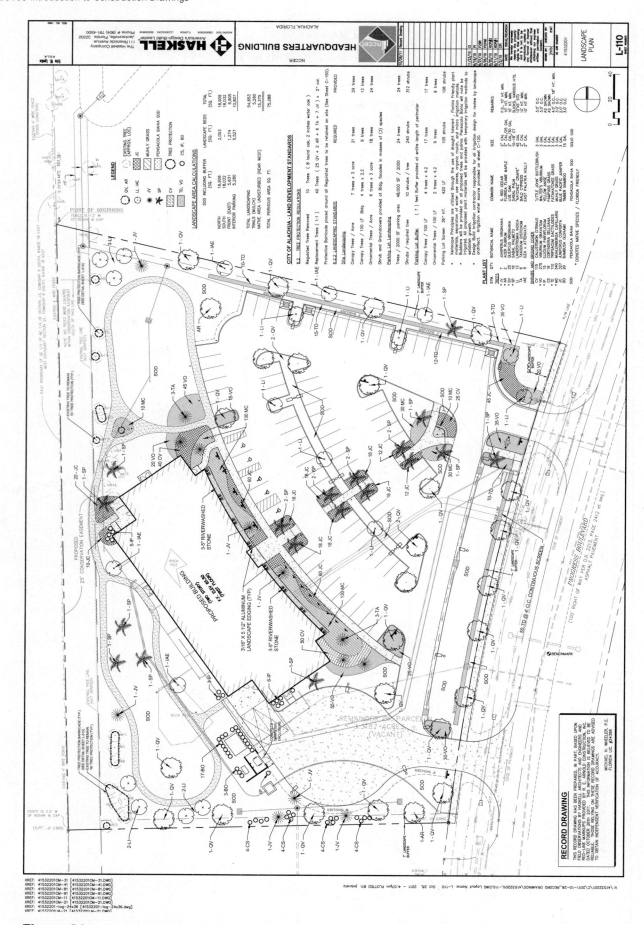

Figure 21 Landscape plan.

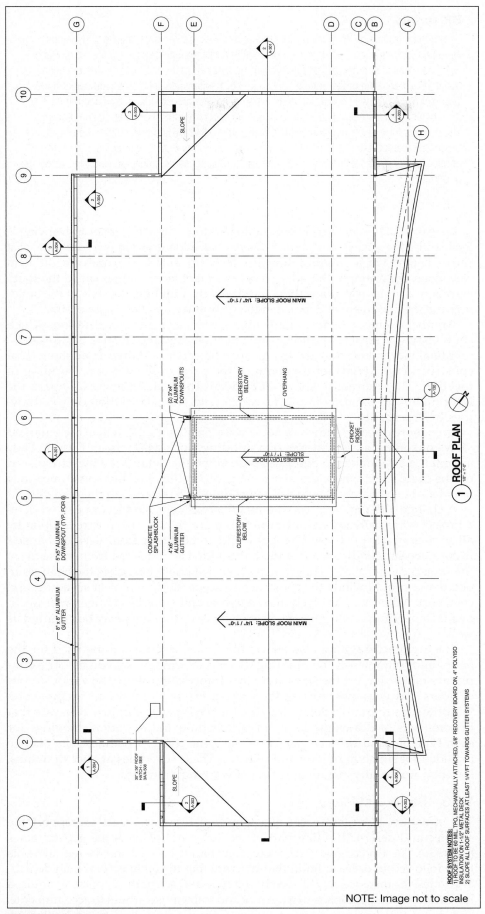

NOTE: Image not to scale

Figure 22 Roof plan.
Source: Croxton Collaborative/Gould Evans Associates

Blueprints

Before the introduction of large-format plotters, printers, and copiers, technical drawings were reproduced using a contact print process that utilized light-sensitive paper to produce what were called blueprints. The name blueprint came from the fact that the reproduction resulted in a negative of the original drawing, shown with white lines on a blue background. This technology allowed construction drawings to be reproduced quickly and accurately. The term blueprints was typically used to describe any set of technical or construction drawings. The process for making blueprints was developed in 1842 by an English astronomer named Sir John F. Herschel and was used widely for over a century until it was replaced by modern printing and copying technology. Today these reproductions are simply referred to as drawings or prints.

Elevation (EL): Height above sea level, or other defined surface, usually expressed in feet or meters.

Elevation drawing: Side view of a building or object, showing height and width.

Section drawing: A cross-sectional view of a specific location, showing the inside of an object or building.

Detail drawings: Enlarged views of part of a drawing used to show an area more clearly.

Structural plans: A set of engineered drawings used to support the architectural design.

Elevation (EL) is another element of architectural drawings (refer to Drawing 2, Residential Elevations, in *Appendix D*). An **elevation drawing** is a side view that shows height. On a building drawing, there are standard names for different elevations. For example, the side of a building that faces south is called the south elevation. Exterior elevations show the size of the building; the style of the building; and the placement of doors, windows, chimneys, and decorative trim.

Another element of the architectural plan is a **section drawing**, which shows how the structure is to be built. Section drawings (*Figure 23*) are cross-sectional views that show the inside of an object or building. They show what construction materials to use and how the parts of the object or building fit together. They normally show more detail than plan views. Compare the information on the drawing in *Figure 23* with the information on the drawings in *Appendix D*.

Even more detail is shown in **detail drawings** (*Figure 24*), which are enlarged views of some special features of a building, such as floors and walls. They are enlarged to make the details clearer. Often the detail drawings are placed on the same sheet where the feature appears in the plan, but sometimes they are placed on separate sheets and referred to by a number on the plan view.

A change order is required when there are changes to the scope of work on a project which already has a contract in place that has been agreed upon by all parties involved. These changes could come about from design changes, unforeseen site conditions, or work that is added to or deleted from the scope of the project by the client. The change order document describes the changes in detail and could include the construction scope, design, project schedule, and even cost. All parties involved must agree to and sign the change order document. Depending on the change, new or revised drawings may be required as well.

The architectural plan also shows the finish schedules to be used for the doors and windows of the building. Door and window schedules, for example, are tables that list the sizes and other information about the various types of doors and windows used in the project. *Figure 25* shows an example of a window schedule and a detail drawing for the Type D1 window. Be aware that in some sets of drawings the window schedule and elevation drawings for each type of window may appear on the same sheet. Schedules may also be included for finish hardware and fixtures. These schedules are not drawings, but they are usually included in a set of working drawings.

1.5.3 Structural Plans

The **structural plans** are a set of engineered drawings used to support the architectural design. The first part of the structural plans is the general notes (*Figure 26*). These notes give details of the materials to be used and the requirements to be followed in order to build the structure that the architectural plan depicts. The notes, for instance, might specify the type and strength of concrete required for the foundation, the loads that the roof and stairs must be built to accommodate, and codes that contractors must follow. General notes may be on a separate general notes sheet or may be part of individual plan sheets.

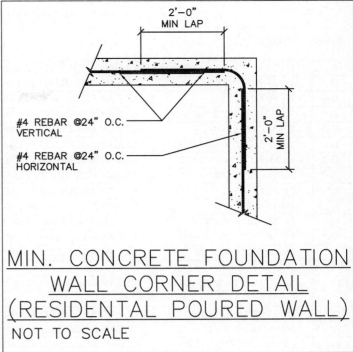

Figure 24 Detail drawing (wall corner detail).

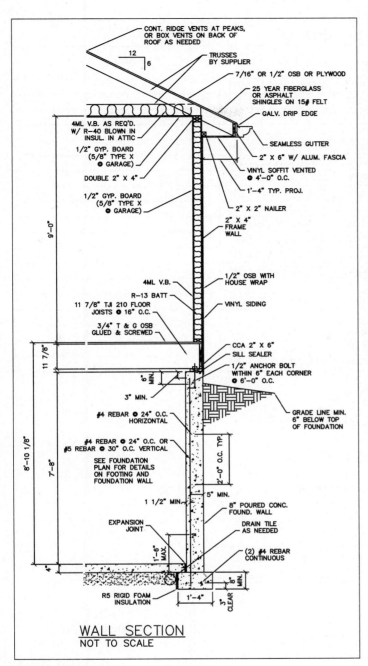

Figure 23 Residential wall section.

The structural plans also include a **foundation plan** (*Figure 27*), which shows the lowest level of the building, including concrete footings, slabs, and foundation walls. They also may show steel girders, columns, or **beams**, as well as detail drawings to show where and how the foundation must be reinforced. Column schedules, spread footing schedules, and foundation notes may be included on the foundation plan. A related element is the structural floor plan, which depicts a wood or metal joist framing and the underlayment of each floor of the structure.

The structural plans show the materials to be used for the walls, whether concrete or masonry, and whether the framing is wood or steel. Structural plans also include a framing plan, showing what kinds of ceiling joists and rafters are to be used and where trusses are to be placed. Notes for the framing plan are usually found on the same sheet as the drawing.

The structural plans include structural section drawings (*Figure 28*), which are similar to the architectural section drawings but show only the structural requirements. Miscellaneous structural details may also be shown in these sections to provide a better understanding of such things as connections and attachments of accessories.

Foundation plan: A drawing that shows the layout and elevation of the building foundation.

Beams: Large, horizontal structural members made of concrete, steel, stone, wood, or other structural material to provide support above a large opening.

WINDOW SCHEDULE

TYPE	FRAME SIZE	FRAME FINISH	GLAZING	REMARKS
A	4' 2" x 4' 2"	CLEAR ANODIZED ALUMINUM	VIRACON VE-7-2M	
B	24' 2-1/2" x 11' 0"	CLEAR ANODIZED ALUMINUM	VIRACON VE-7-2M	
B-1	12' 10" x 11' 0"	CLEAR ANODIZED ALUMINUM	VIRACON VE-7-2M	
C	15'-11-1/2" x 8' 3-1/2"	CLEAR ANODIZED ALUMINUM	VIRACON VE-7-2M	
C-1	15'-11-1/2" x 4' 1-1/2"	CLEAR ANODIZED ALUMINUM	VIRACON VE-7-2M	
D	15 11-1/2'-0" x 11' 1-1/2"	CLEAR ANODIZED ALUMINUM	VIRACON VE-7-2M	
D-1	15 11-1/2'-0" x 11' 1-1/2"	CLEAR ANODIZED ALUMINUM	VIRACON VE-7-2M	
D-2	31' 11-1/2" x 4' 4"	CLEAR ANODIZED ALUMINUM	VIRACON VE-7-2M	
E	14' 4" x 7' 8"	CLEAR ANODIZED ALUMINUM	VIRACON VE-7-2M	
F	15'-11-1/2" x 4' 4 "	CLEAR ANODIZED ALUMINUM	VIRACON VE-7-2M	
H	11' 11-1/2" x 8' 1/4"	CLEAR ANODIZED ALUMINUM	VIRACON VE-7-2M	
H-1	11' 11-1/2" x 10' 1-3/4"	CLEAR ANODIZED ALUMINUM	VIRACON VE-7-2M	
I	15'-11-1/2" x 4' 1-1/2 "	CLEAR ANODIZED ALUMINUM	VIRACON VE-7-2M	
CW - 1	12' 0" x 27' 11-1/4"	CLEAR ANODIZED ALUMINUM	VIRACON VE-7-2M	
CW - 3	15' 11 1/2" x 37' 3 1/4"	CLEAR ANODIZED ALUMINUM	VIRACON VE-7-2M	
CW - 4	36' 10" x 20' 9 1/2"	CLEAR ANODIZED ALUMINUM	VIRACON VE-7-2M	
CW - 5	28' 11-7/8" x 20' 9-1/2"	CLEAR ANODIZED ALUMINUM	VIRACON VE-7-2M	

NOTES

1.   FRAME SIZES HAVE BEEN INDICATED ASSUMING 1/2" JOINT ADJACENT TO METAL PANEL SYSTEM.   ALIGNMENT OF JOINTS W/ METAL PANEL MODULE IS REQUIRED.   INTERMEDIATE MULLIONS TO BE CENTERED ON ADJACENT METAL PANEL JOINTS

2. "SG" INDICATES SAFETY GLAZING (TEMPERED GLASS REQUIRED)

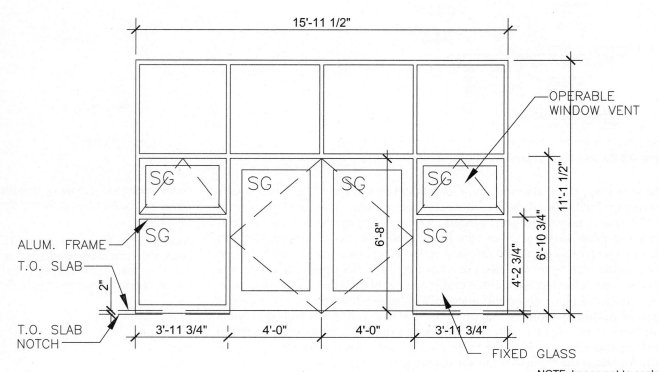

NOTE: Image not to scale

Figure 25 Window schedule and window detail.
Source: Croxton Collaborative/Gould Evans Associates

GENERAL NOTES

I.   DESIGN CRITERIA

  A. GENERAL BUILDING CODE
    The Contract Documents are based on the requirements of the:
    1. Standard Building Code, 1997 edition.

  B. DEAD LOADS

    1. Partitions. An allowance of 20 PSF has been made for partitions as a uniformly distributed dead load.

    2. Hanging Ceiling and Mechanical Loads. An allowance of 10 PSF has been made for hanging ceiling and mechanical equipment loads such as duct work and sprinkler pipes.

  C. LIVE LOADS
    1. Design live loads are based on the more restrictive of the uniform load listed below or the concentrated load listed acting over an area 2.5 feet square.

CATEGORY	UNIFORM LOAD (PSF)	CONCENTRATED LOAD (LB)
1. Roof	20	N/A
2. Elevated Floors	50	0
3. Terraces, Lobbies	100	0
4. Stairways, Exit Facilities	100	0
5. Elevator Machine Rooms	100	Assumed Eqp. Wt.
6. Mechanical Rooms, typical	150	Assumed Eqp. Wt.

    NOTES:
    1. Live Load Reduction. Live loads have been reduced on any member supporting more than 150 square feet, including flat slabs, except for floors in places of public assembly and for live loads greater than 100 pounds per square foot in accordance with the following formula:

$$R = r(A-150)$$

    The reduction, R, shall not exceed 40 percent for members supporting one level only, 60 percent for other members, or R as calculated in the following formula:

$$R = 23.1 \ \frac{(1 + D)}{L}$$

    R = Reduction in percent.
    r = Rate of reduction equal to .08 percent for floors.
    A = Area of floor supported by the member.
    D = Total dead load supported by the member.
    L = Total, unreduced, live load supported by the member.

    2. For storage loads exceeding 100 pounds per square foot, no reduction has been made, except that design live loads on columns have been reduced 20 percent.

  D. ELEVATOR LOADS
    Machine Beam, Car Buffer, Counterweight Buffer, and Guide Rail Loads. Assumed elevator loads to the supporting structure are shown on the drawings, including machine beam reactions, car buffer reactions, counterweight buffer reactions, and horizontal and vertical guide rail loads. The General Contractor shall submit to the Structural Engineer final elevator shop drawings showing all loads to the structure prior to the installation of the elevators for verification of load carrying capacity.

  E. MECHANICAL EQUIPMENT LOADS
    The General Contractor shall submit actual weights of equipment to be used in the project to the Structural Engineer for verification of loads used in the design at least three weeks prior to fabrication and construction of the supporting structure.

  F. WIND LOADS
    1. Wind pressures are based on the American Society of Civil Engineers, Minimum Design Loads for Buildings and Other Structures, ASCE 7–98 with a Wind Speed = 110 MPH (3 sec. gust), Exposure C, Importance Factor 1.15.
    2. Wind pressures used in the design of the cladding are shown on these Drawings.

II.   FOUNDATION

  A. GEOTECHNICAL REPORT
    Foundation design is based on the geotechnical investigation report as follows:
    1. Reports of Geotechnical exploration, M.E. Rinker Sr. Hall (Revised Location) Near the southeast Corner of Newell Drive and Inner Road, Gainesville, Alachua County, Florida. Law Engineering and Environmental Services, Inc. January 2, 2001.

    The geotechnical report is available to the General Contractor upon request to the Owner. The information included therein may be used by the General Contractor for his general information only. The Architect and Engineer will not be responsible for the accuracy or applicability of such data therein.

  B. FOUNDATION TYPE

    1. Spread Footing.

      a. Design Pressures:
        1. All footings have been designed assuming an allowable bearing pressure of 4000 PSF.
        Allowable pressures are increased 33% for combined gravity and wind loads.

  C. SLAB-ON-GRADE
    Radon resistant construction guidlines are being followed on this project. The details and specifications for slab-on-grade construction must be adhered to without deviation.
    Slab-on-Grade shall be immediately underlain by a 8 mil. vapor barrier. Seams shall be lapped 12 inches and sealed with 2" wide pressure sensitive vinyl tape. All penetrations shall be sealed with tape.

  D. CONSTRUCTION DEWATERING
    The Contractor shall determine the extent of construction dewatering required for the excavation. The Contractor shall submit to the Geotechnical Engineer for review the proposed plan for construction dewatering, prior to beginning the excavation.

III.   REINFORCED CONCRETE

  A. CLASSES OF CONCRETE
    All concrete shall conform to the requirements as specified in the table below unless noted otherwise on the drawings.

Usage	28 Day Comp. Strength (PSI)	Conc Type	Max Size Agg.	W/C Ratio
1. Elevated Floors	4000	NWT	3/4"	0.48
2. Spread Footings	3000	NWT	1"	0.55
3. Slab-On-Grade	4000	NWT	1"	0.48
4. Fnd. Walls & Plinths	4000	NWT	1"	0.48

    All concrete shall be proportioned for a maximum allowable unit shrinkage of 0.03% measured at 28 days after curing in lime water as determined by ASTM C 157 (using air storage).

  B. HORIZONTAL CONSTRUCTION JOINTS IN CONCRETE POURS
    There shall be no horizontal construction joints in any concrete pours unless shown on the drawings. The Architect/Engineer shall approve all deviations or additional joints in writing.

  C. REINFORCING STEEL SPECIFICATION

    1. All Reinforcing Steel shall be ASTM A615 Grade 60 unless noted otherwise on the drawings or in these notes.
    2. Welded Reinforcing Steel. Provide reinforcing steel conforming to ASTM A706 for all reinforcing steel required to be welded and where noted on the drawings.
    3. Galvanized Reinforcing Steel. Provide reinforcing steel galvanized according to ASTM A767 Class II (2.0 oz. zinc PSF where noted on the drawings.
    4. Deformed Bar Anchors. ASTM A496 minimum yield strength 70,000 PSI as noted on the drawings. Reinforcing bars shall not be substituted for deformed bar anchors.
    5. Welded Wire Fabric. Welded smooth wire fabric, ASTM A 185, yield strength 65,000 PSI where noted on the drawings. Welded deformed wire fabric, ASTM A 497, yield strength 70,000 PSI where noted on the drawings.

  D. PLACEMENT OF WELDED WIRE FABRIC
    Wherever welded wire fabric is specified as reinforcement, it shall be continuous across the entire concrete surface and not interrupted by beams or girders and properly lapped one cross wire spacing plus 2".

  E. REINFORCEMENT IN TOPPING SLABS
    Provide welded smooth wire fabric minimum 6 x 6 W2.9 x W2.9 in all topping slabs unless specified otherwise on the drawings.

  F. REINFORCEMENT IN HOUSEKEEPING PADS
    Provide welded smooth wire fabric 6 x 6 W2.9 x W2.9 minimum in all housekeeping pads supporting mechanical equipment whether shown on the drawings or not unless heavier reinforcement is called for on the drawings.

  G. REINFORCING STEEL COVERAGE
    Concrete Cover for reinforcement layer nearest to the surface unless specified otherwise on the drawings.
    1. Concrete surfaces cast against and permanently exposed to earth.   3 inches
    2. Concrete surfaces exposed to earth or weather or where noted on the drawings   2 inches.
    3. Concrete surfaces not exposed to weather or in contact with the ground.
      a. #3 to #11 bars 1 inch

  H. SPLICES IN REINFORCING STEEL
    1. All unscheduled splices shall be Class A tension splice.

IV.   STRUCTURAL STEEL

V.   STEEL DECKS

VI.   CURTAIN WALL

VII.   CONCRETE MASONRY

VIII.   MISCELLANEOUS

IX.   SUBMITTALS

X.   DRAWING INTERPRETATION

NOTE: Image not to scale

Figure 26 General notes for structural plans.
Source: Croxton Collaborative/Gould Evans Associates

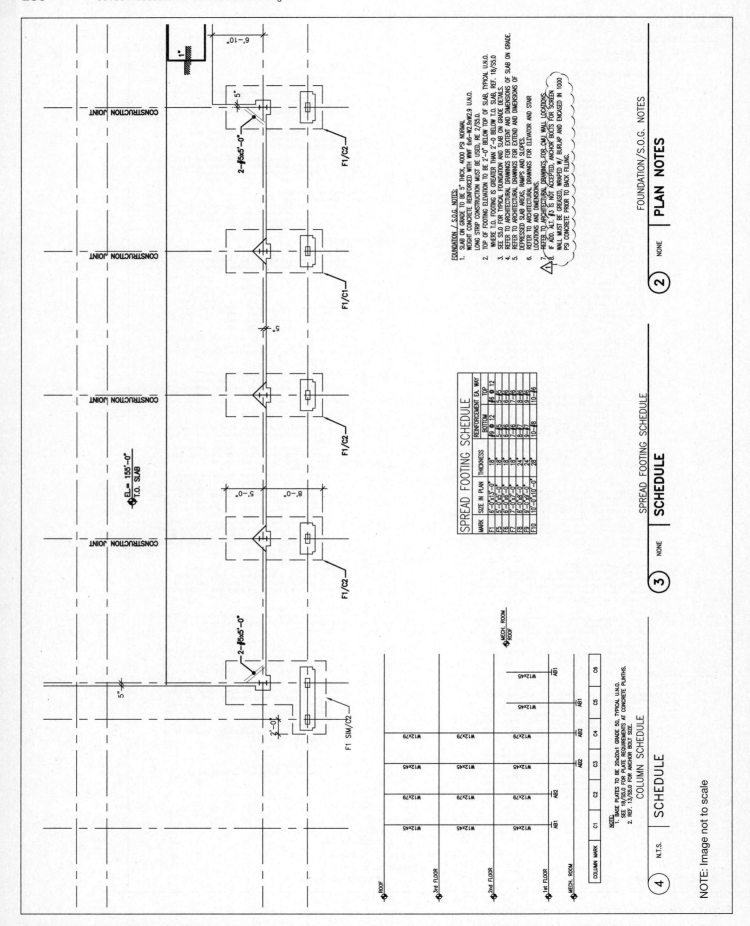

Figure 27 Foundation plan (foundation/slab-on-grade plan). Source: Croxton Collaborative/Gould Evans Associates

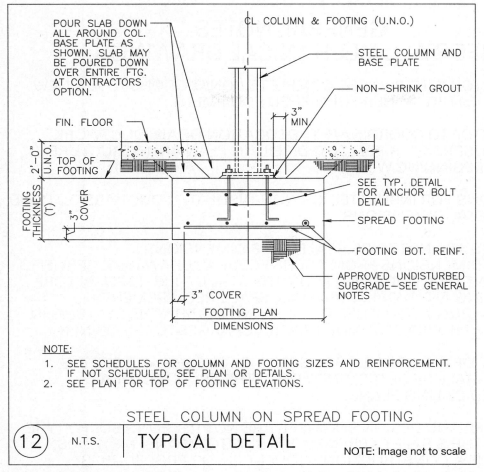

Figure 28 Structural section drawing (foundation details).
Source: Croxton Collaborative/Gould Evans Associates

1.5.4 Mechanical Plans

Mechanical plans are engineered plans for motors, pumps, piping systems, and piping equipment. These plans incorporate general notes (*Figure 29*) containing specifications ranging from what the contractor is to provide to how the contractor determines the location of **grilles** and **registers**. A mechanical **legend** (*Figure 30*) defines the symbols used on the mechanical plans. A list of abbreviations (*Figure 31*) spells out abbreviations found on the plans. Symbols and abbreviations may vary depending on job type and trade. Always refer to the drawings to ensure accuracy when translating symbols and abbreviations.

Piping and instrumentation drawings (P&IDs) (*Figure 32*) are **schematic** diagrams of a complete piping system that show the process flow. They also show all the equipment, pipelines, valves, instruments, and controls needed to operate the system. P&IDs are not drawn to **scale** because they are meant only to give a representation, or a general idea, of the work to be done. Additionally, P&IDs do not indicate north, south, east, and west directions.

For more complex jobs, a separate heating, ventilating, and air conditioning (**HVAC**) plan is added to the set of plans. Piping system plans for gas, oil, or steam heat may be included in the HVAC plan. The mechanical plans include the layout of the HVAC system, showing specific requirements and elements for that system, including a floor, a reflected ceiling, or a roof. HVAC drawings (*Figure 33A* and *Figure 33B*) often include an electrical schematic that shows the electrical circuitry for the HVAC system. HVAC plans can be both mechanical and electrical drawings in one plan. Be aware that a page with a series of mechanical detail drawings may be included in the mechanical plans. These drawings show specific details of certain components within the mechanical system. *Figure 34* is an example of a mechanical detail drawing.

Mechanical plans: Engineered drawings that show the mechanical systems, such as motors and piping.

Grilles: HVAC components that pull air from a room and return it to the heating or cooling system. They are different from a register in that they do not have a damper to control airflow.

Registers: HVAC components that direct air from the heating or cooling system into a room. A damper in the register is used to control the airflow.

Legend: A description of the symbols and abbreviations used in a set of drawings.

Piping and instrumentation drawings (P&IDs): Schematic diagrams of a complete piping system.

Schematic: A one-line drawing showing the flow path for electrical circuitry or the relationship of all parts of a system.

Scale: The ratio between the size of a drawing of an object and the size of the actual object.

HVAC: Heating, ventilating, and air conditioning.

GENERAL NOTES
(FOR ALL MECHANICAL DRAWINGS)

1. CONTRACTOR IS TO PROVIDE COMPLETE CONNECTIONS TO ALL NEW AND RELOCATED OWNER FURNISHED EQUIPMENT.

2. CONTRACTOR TO COORDINATE THE LOCATION OF ALL DUCTWORK AND DIFFUSERS WITH REFLECTED CEILING PLAN AND STRUCTURE PRIOR TO BEGINNING WORK.

3. DIMENSIONS FOR INSULATED OR NON-INSULATED DUCT ARE OUTSIDE SHEET METAL DIMENSIONS.

4. DRAWINGS ARE NOT TO BE SCALED FOR DIMENSIONS. TAKE ALL DIMENSIONS FROM ARCHITECTURAL DRAWINGS, CERTIFIED EQUIPMENT DRAWINGS AND FROM THE STRUCTURE ITSELF BEFORE FABRICATING ANY WORK. VERIFY ALL SPACE REQUIREMENTS COORDINATING WITH OTHER TRADES, AND INSTALL THE SYSTEMS IN THE SPACE PROVIDED WITHOUT EXTRA CHARGES TO THE OWNER.

5. LOCATION OF ALL GRILLES, REGISTERS, DIFFUSERS AND CEILING DEVICES SHALL BE DETERMINED FROM THE ARCHITECTURAL REFLECTED CEILING PLANS.

6. THE OWNER AND DESIGN ENGINEER ARE NOT RESPONSIBLE FOR THE CONTRACTOR'S SAFETY PRECAUTIONS OR TO MEANS, METHODS, TECHNIQUES, CONSTRUCTION SEQUENCES, OR PROCEDURES REQUIRED TO PERFORM HIS WORK.

7. ALL WORK SHALL BE INSTALLED IN ACCORDANCE WITH PLRC'S SAFETY PLAN AND ALL APPLICABLE STATE AND LOCAL CODES.

8. ALL EXTERIOR WALL AND ROOF PENETRATIONS SHALL BE SEALED WEATHERPROOF. REFERENCE SPECIFICATION SECTION 15050.

9. ALL MECHANICAL WORK UNDER THIS CONTRACT IS TO FIVE (5) FEET OUTSIDE THE BUILDING.

Figure 29 Mechanical plan general notes.

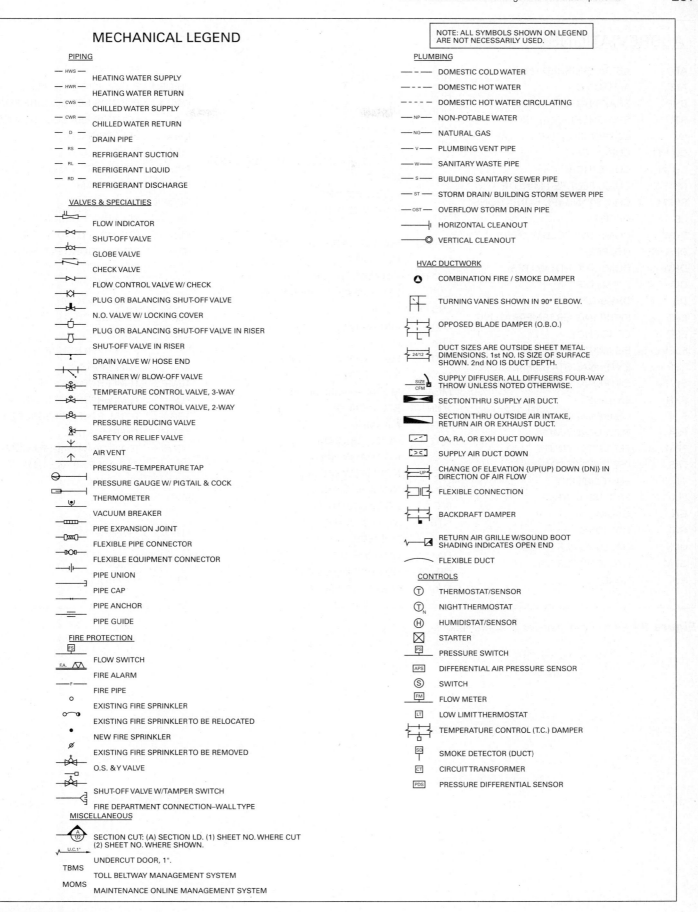

Figure 30 Mechanical plan legend.

ABBREVIATIONS

AFF	ABOVE FINISHED FLOOR	HG	MERCURY	%	PERCENT
ALT	ALTITUDE	HGT	HEIGHT	PH OR f	PHASE (ELECTRICAL)
BHP	BRAKE HORSEPOWER	HORZ	HORIZONTAL	PSF	POUNDS PER SQUARE FOOT
BTU	BRITISH THERMAL UNIT	HP	HORSEPOWER	PSI	POUNDS PER SQUARE INCH
Cv	COEFFICIENT, VALVE FLOW	HR	HOUR(S)	PSIA	PSI ABSOLUTE
CU FT	CUBIC FEET	HWC	HOT WATER CIRCULATING (DOMESTIC)	PSIG	PSI GAUGE
CU IN	CUBIC INCH	HZ	HERTZ	PRESS	PRESSURE
CFM	CUBIC FEET PER MINUTE	ID	INSIDE DIAMETER	RA	RETURN AIR
SCFM	CFM, STANDARD CONDITIONS	IE	INVERT ELEVATION	RECIRC	RECIRCULATE
dB	DECIBEL	IN	INCHES	RH	RELATIVE HUMIDITY
DCW	DOMESTIC COLD WATER	IN W.C.	INCHES WATER COLUMN	RLA	RUNNING LOAD AMPS
DEG OR °	DEGREE	KW	KILOWATT	RPM	REVOLUTIONS PER MINUTE
DHW	DOMESTIC HOT WATER	KWH	KILOWATT HOUR	SL	SEA LEVEL
DIA	DIAMETER	LAT	LEAVING AIR TEMPERATURE	SENS	SENSIBLE
DB	DRY-BULB	LBS OR #	POUNDS	SPEC	SPECIFICATION
EAT	ENTERING AIR TEMPERATURE	LF	LINEAR FEET	SQ	SQUARE
EFF	EFFICIENCY	LRA	LOCKED ROTOR AMPS	STD	STANDARD
ELEV or EL	ELEVATION	LWT	LEAVING WATER TEMPERATURE	SP	STATIC PRESSURE
ESP	EXTERNAL STATIC PRESSURE	MAX	MAXIMUM	SA	SUPPLY AIR
EWT	ENTERING WATER TEMPERATURE	MCA	MINIMUM CIRCUIT AMPS	TEMP	TEMPERATURE
EXH	EXHAUST	MBH	BTU PER HOUR (THOUSAND)	TD	TEMPERATURE DIFFERENCE
F	FAHRENHEIT	MIN	MINIMUM	TSP	TOTAL STATIC PRESSURE
FLA	FULL LOAD AMPS	NC	NOISE CRITERIA	TSTAT	THERMOSTAT
FPM	FEET PER MINUTE	N.O.	NORMALLY OPEN	TONS	TONS OF REFRIGERATION
FPS	FEET PER SECOND	N.C.	NORMALLY CLOSED	VAV	VARIABLE AIR VOLUME
FT	FOOT OR FEET	N/A	NOT APPLICABLE	VEL	VELOCITY
FU	FIXTURE UNITS	NIC	NOT IN CONTRACT	VERT	VERTICAL
GA	GAUGE	NTS	NOT TO SCALE	V	VOLT
GAL	GALLONS	NO	NUMBER	VOL	VOLUME
GPH	GALLONS PER HOUR	OA	OUTSIDE AIR	W	WATT
GPM	GALLONS PER MINUTE	OD	OUTSIDE DIAMETER	WT	WEIGHT
HD	HEAD	PPM	PARTS PER MILLION	WB	WET-BULB

Figure 31 Mechanical plan list of abbreviations.

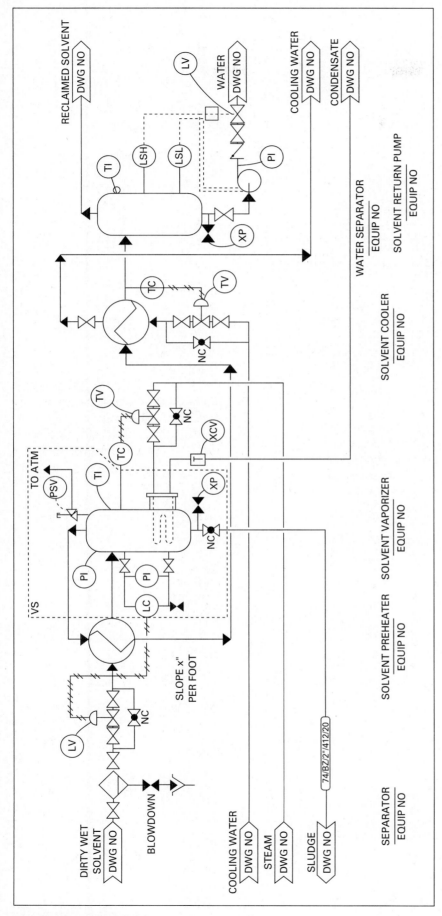

Figure 32 P&ID drawing.

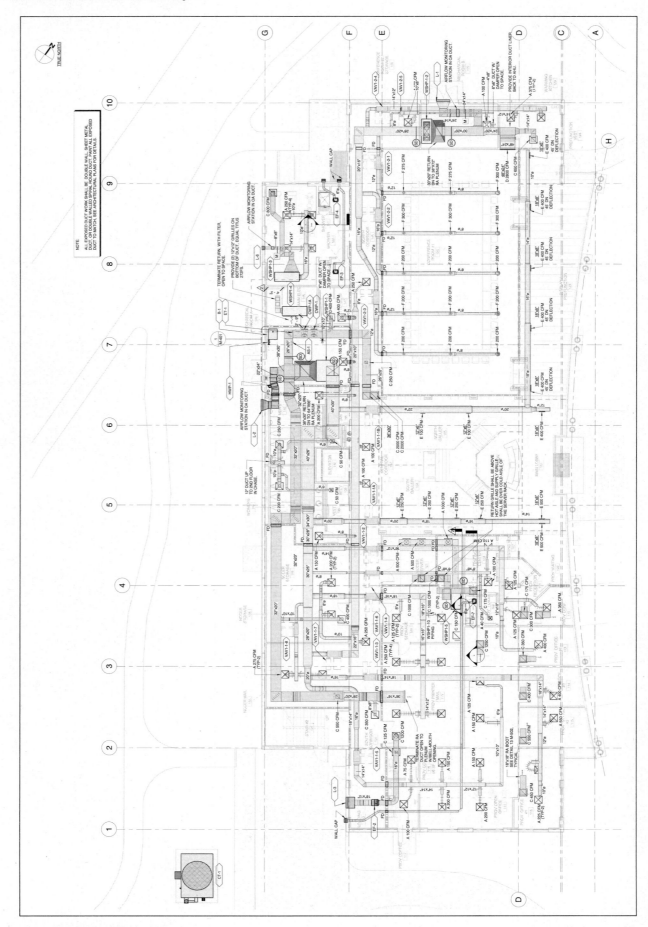

Figure 33A HVAC drawing (1 of 2).

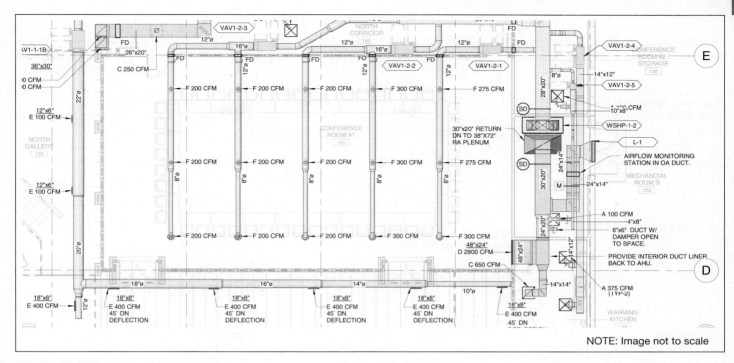

NOTE: Image not to scale

Figure 33B HVAC drawing (2 of 2).

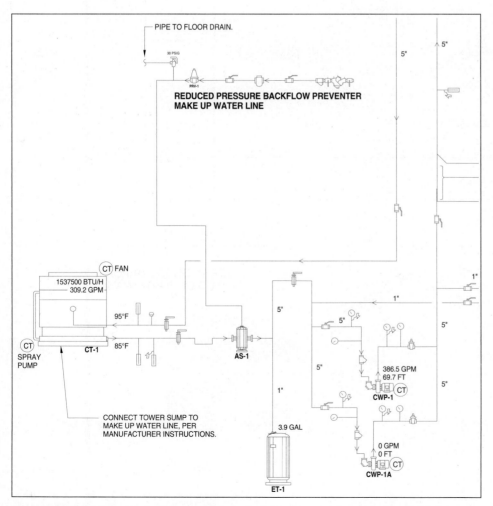

Figure 34 Mechanical detail drawing for hot water riser and drain connections.

Plumbing plans: Engineered drawings that show the layout for the plumbing system.

Plumbing isometric drawing: A type of three-dimensional drawing that depicts a plumbing system.

1.5.5 Plumbing/Piping Plans

Plumbing plans (*Figure 35*) are engineered plans showing the layout for the plumbing system that supplies the hot and cold water, for the sewage disposal system, and for the location of plumbing fixtures. For commercial projects, each system may be on a separate plan.

A **plumbing isometric drawing** is a three-dimensional drawing that depicts the plumbing system and serves as part of the plumbing plan. *Figure 36* shows a plumbing isometric drawing for a sanitary riser system.

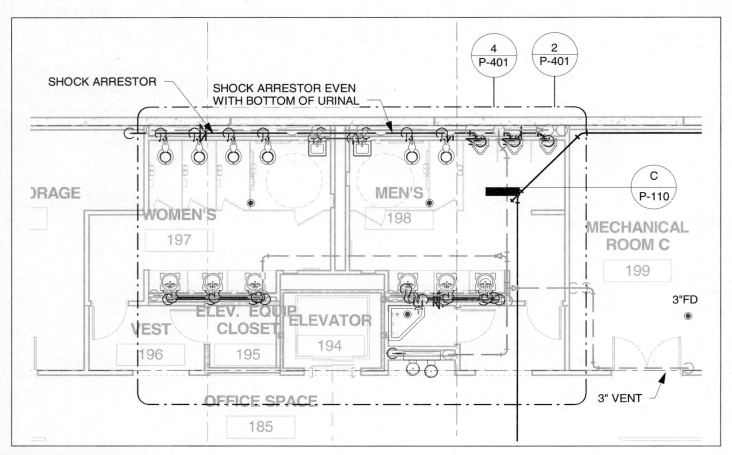

Figure 35 Plumbing plan.

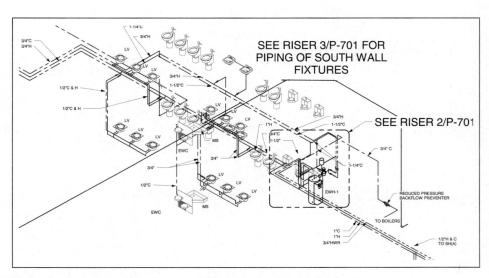

Figure 36 Plumbing isometric drawing (sanitary riser diagram).

Isometric Drawings

An isometric drawing is a type of three-dimensional drawing also known as a pictorial illustration. Isometric construction drawings typically depict objects at a 30-degree angle to provide a three-dimensional perspective rather than a flat, two-dimensional view.

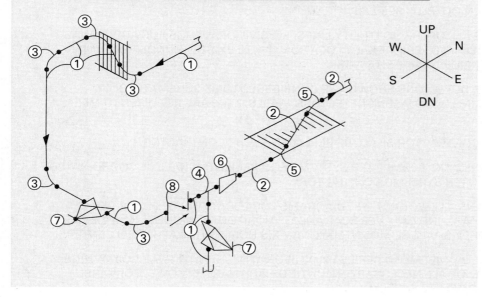

BILL OF MATERIAL

P.M.	REQ'D	SIZE	DESCRIPTION
1		1-1/2"	PIPE SCH/40 ASTM-A-120 GR.B
2		3/4"	PIPE SCH/40 ASTM-A-120 GR.B
3	5	1-1/2"	90° ELL ASTM-A-197 BW
4	1	1-1/2"	TEE ASTM-197 STD
5	2	3/4"	45° ELL ASTM-197 STD
6	1	1-1/2" × 3/4"	BELL RED. CONC.
7	2	1-1/2"	GATE VA. BW ASTM-B62
8	1	1-1/2"	CHECK VA. SWING BW 150#

1.5.6 Electrical Plans

Electrical plans are engineered drawings for electrical supply and distribution. These plans may appear on the floor plan itself for simple construction projects. Electrical plans include locations of the electric meter, distribution panel, switchgear, convenience outlets, and special outlets.

For more complex projects, the information may be on a separate plan added to the set of plans. This separate plan leaves out construction-related details and shows just the electrical layout. More complex electrical plans include locations of switchgear, transformers, main breakers, and motor control centers.

The electrical plans usually start with a set of general notes (*Figure 37*). These notes cover items ranging from main transformers to the coordination of underground penetrations into the building.

The electrical plans can include lighting plans, which show the location of lights and receptacles, power plans (*Figure 38*), and panel schedules (*Figure 39*). Electrical plans have an electrical legend, which defines the symbols (*Figure 40*) used on the plan, and a key to the abbreviations (*Figure 41*) used on the plan. Depending on the size and scope of the drawings, the legend is often on a separate drawing sheet of its own, rather than on each individual drawing.

Electrical plans: Engineered drawings that show all electrical supply and distribution.

Visualize Before Building

It is important for a builder to be able to visualize a finished project before starting it. Detail drawings help builders visualize different parts of the structure long before they measure a board or hammer a nail. By visualizing, builders can plan ahead and anticipate potential problems. This saves time and money on the job.

GENERAL NOTES (FOR ALL ELECTRICAL SHEETS)

1. COORDINATE LOCATION OF LUMINARIES WITH ARCHITECTURAL REFLECTED CEILING PLANS.

2. COORDINATE LOCATION OF ALL OUTLETS WITH ARCHITECTURAL ELEVATIONS, CASEWORK SHOP DRAWINGS AND EQUIPMENT INSTALLATION DRAWINGS.

3. COORDINATE LOCATION OF MECHANICAL EQUIPMENT WITH MECHANICAL PLANS AND MECHANICAL CONTRACTOR PRIOR TO ROUGH-IN.

4. PROVIDE (1) 3/4"C WITH PULL WIRE FROM EACH TELEPHONE, DATA OR COMMUNICATION OUTLET SHOWN, TO ABOVE ACCESSIBLE CEILING, AND CAP.

5. 3-LAMP FIXTURES SHOWN HALF SHADED HAVE INBOARD SINGLE LAMP CONNECTED TO EMERGENCY BATTERY PACK FOR FULL LUMEN OUTPUT. SEE SPECIFICATIONS.

6. SITE PLAN DOES NOT INDICATE ALL OF THE UG UTILITY LINES, RE: CIVIL DRAWINGS FOR ADDITIONAL INFORMATION. CONTRACTOR TO FIELD VERIFY EXACT LOCATION OF ALL EXISTING UNDERGROUND UTILITY LINES OF ALL TRADES PRIOR TO ANY SITE WORK.

7. THE LOCATIONS OF ALL SMOKE DETECTORS SHOWN ARE CONSIDERED TO BE SCHEMATIC ONLY. THE ACTUAL LOCATIONS (SPACING TO ADJACENT DETECTORS, WALLS, ETC.) ARE REQUIRED TO MEET NFPA 72.

8. ANY ITEMS DAMAGED BY THE CONTRACTOR SHALL BE REPLACED BY THE CONTRACTOR.

9. "CLEAN POWER" AND COMMUNICATION/COMPUTER SYSTEM REQUIREMENTS SHALL BE COORDINATED WITH COMMUNICATION/COMPUTER SYSTEMS CONTRACTOR.

10. REFER TO ARCHITECTURAL PLANS, ELEVATIONS AND DIAGRAMS FOR LOCATIONS OF FLOOR DEVICES AND WALL DEVICES. LOCATION WILL INDICATE VERTICAL AND/OR HORIZONTAL MOUNTING. IF DEVICES ARE NOT NOTED OTHERWISE THEY SHALL BE MOUNTED LONG AXIS HORIZONTAL AT +16" TO CENTER.

11. ALL PLUGMOLD SHOWN SHALL BE WIREMOLD SERIES V2000 (IVORY FINISH) WITH SNAPICOIL #V20GB06 (OUTLETS 6" ON CENTER). PROVIDE ALL NECESSARY MOUNTING HARDWARE, ELBOWS, CORNERS, ENDS, ETC. REQUIRED FOR A COMPLETE SYSTEM.

12. ALL EMERGENCY RECEPTACLE DEVICES SHALL BE RED IN COLOR.

13. ALL BRANCH CIRCUITS SHALL BE 3-WIRE (HOT, NEUTRAL, GROUND).

14. COORDINATE EXACT EQUIPMENT LOCATIONS AND POWER REQUIREMENTS WITH OWNER AND ARCHITECT PRIOR TO ROUGH-INS.

15. ADA COMPLIANCE: ALL ADA HORN/STROBE UNITS SHALL BE MOUNTED +90" AFF OR 6" BELOW FINISHED CEILING, WHICH EVER IS LOWER. ELECTRICAL DEVICES PROJECTING FROM WALLS WITH THEIR LEADING EDGES BETWEEN 27" AND 80" AFF SHALL PROTRUDE NO MORE THAN 4" INTO WALKS OR CORRIDORS. ELECTRICAL AND COMMUNICATIONS SYSTEMS RECEPTACLES ON WALLS SHALL BE 15" MINIMUM AFF TO BOTTOM OF COVERPLATE.

16. COORDINATE ALL UNDERGROUND PENETRATIONS INTO THE BUILDING AND TUNNEL WITH STRUCTURAL ENGINEER, DUE TO EXPANSIVE SOILS.

17. ELECTRONIC STRIKES, MOTION DETECTORS AND ALARM SHUNTS ARE PROVIDED BY OTHERS. PROVIDE ALL NECESSARY ROUGH-INS FOR THESE ITEMS. COORDINATE WORK WITH SECURITY SYSTEM PROVIDER.

Figure 37 Electrical plan general notes.

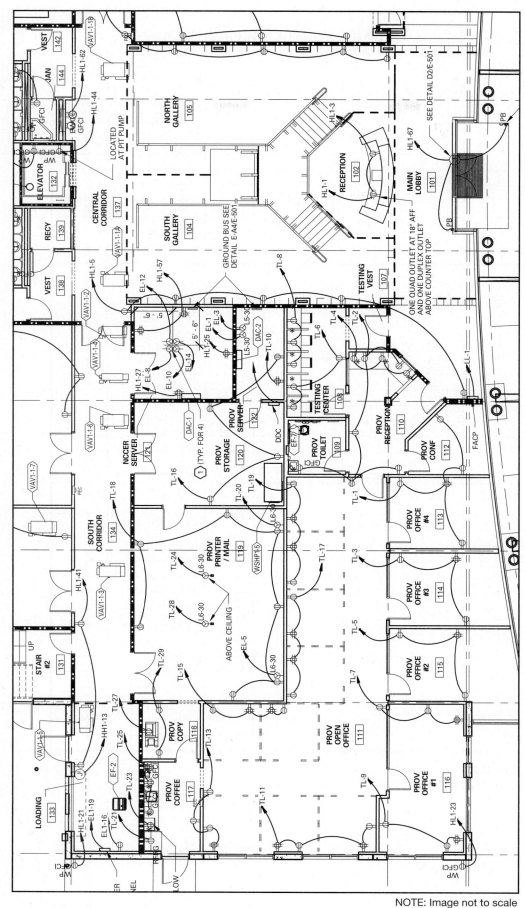

Figure 38 Portion of a power plan.

NOTE: Image not to scale

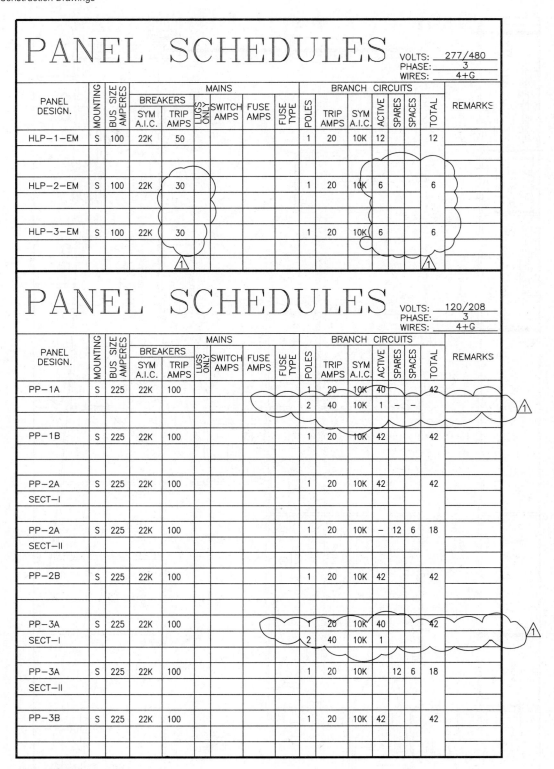

Figure 39 Panel schedules.
Source: Croxton Collaborative/Gould Evans Associates

ELECTRICAL SYMBOLS LIST

Ⓙ/Ⓙ/Ⓙ⊢	JUNCTION BOX. CEILING/FLOOR/WALL MOUNTED
▨	PULL BOX SIZED AS REQUIRED
⊂▢	BELL
‖▢	CHIME
▷▢ ▷◁	HORN (SINGLE/BI- DIRECTIONAL)
▽▢	BUZZER
▪	DOOR PUSHBUTTON
⊤ˢ	SIGNAL TRANSFORMER
DH	MAGNETIC DOOR HOLDER
DHˢ	MAGNETIC TYPE DOOR CLOSER, 'S' DENOTES SMOKE DETECTOR TYPE
WF	SPRINKLER WATER FLOW SWITCH
TS	SPRINKLER SUPERVISED VALVE – TAMPER SWITCH
◔ / S◔ R	INDICATING CLOCK – SINGLE FACE 'S' – DENOTES SKELETON TYPE CLG/WALL MTD 'R' – DENOTES RECESSED TYPE 'D' – DENOTES DOUBLE FACED
Ⓢ / Ⓢ⊢	SPEAKER; CLG/WALL MTD 'S' – DENOTES SURFACE MOUNTED 'D' – DENOTES DOUBLE FACED
Ⓥ	VOLUME CONTROL
Ⓜ / Ⓜ⊢	MICROPHONE OUTLET; CLG/WALL MTD
PA	PUBLIC ADDRESS EQUIPMENT RACK
PPP	ELECT. POWER PATCH PANEL WITH 4 RECEPTS.AND R/S SWITCH
A/V	MUTIMEDIA PATCH PANEL
WS	WINDOW ALARM SWITCH
BS	DOOR PUSHBUTTON
▷CCS	CLOSED CIRCUIT SURVEILLANCE CAMERA 'P' DENOTES PAN – P/T TILT
CCSM	COLSED CIRCUIT SURVEILLANCE MONITOR – NUMBERS DENOTE QUANTITY
TV TV⊢	TELEVISION ANTENNA OUTLET
S 2TV	TELEVISION SPLITTER NUMBER DENOTES QUANTITY OF SPLITS
▷TV⊢	TELEVISION CAMERA OUTLET
TVHE	TELEVISION HEADED EQUIPMENT
▶	TELEVISION OUTLET WITH BUSHED OPENING LETTERS DENOTE: 'H' – HOUSE PHONE 'J' – JACK TYPE, 'P' – PAY (PUBLIC) PHONE, 'W' – WALL TYPE PHONE
TC	TELEPHONE TERMINAL CABINET
◉S	TELEPHONE FLOOR BOX WITH NIPPLE (RISER) EXTENSION. 'S' DENOTES SERVICE FITTING TYPE
▷	DATA OUTLET WITH BUSHED OPENING WALL MOUNTED
▶	COMBINATION TELEPHONE/DATA OUTLET WITH BUSHED OPENING WALL MOUNTED
S	SWITCH
⊂S	BELL
‖S	CHIME
ˢS	DOOR STRIKE
SSP	START/STOP WITH PILOT LIGHT
⋀⋀⋀ ⋀⋀⋀	POWER TRANSFORMER
⊣‖—	GROUND (EARTH)
⊣‖—•—•—	LIGHTNING ARRESTER
⌒	CIRCUIT BREAKER
—G—	GROUND BUS
—N—	NEUTRAL BUS
CR R	CARD READER 'R' – DENOTES RECESSED TYPE
DO	POWER DOOR OPERATOR
I	INTERCOM
Ⓜ	CEILING MOUNTED OCCUPANCY SENSOR
⊢Ⓜ	SURFACE MOUNTED OCCUPANCY SENSOR
Ⓟ	PHOTOSENSOR

Figure 40 Electrical symbols list.
Source: Croxton Collaborative/Gould
Evans Associates

ELECTRICAL ABBREVIATIONS

A	AMPERE(S)	MATV	MASTER ANTENNA TELEVISION SYSTEM
AC	ALTERNATING CURRENT	MC	METAL CLAD CABLE
ACB	AIR CIRCUIT BREAKER	MCC	MOTOR CONTROL CENTER
AFF	ABOVE FINISHED FLOOR	MCM	THOUSAND CIRCULAR MIL(S)
AFG	ABOVE FINISHED GRADE	MCP	MOTOR CONTROL PANEL
AL	ALUMINUM	M.C.	MECHANICAL CONTRACTOR
ALT	ALTERNATE	MH	MANHOLE
ASYM	ASYMMETRICAL	MIC	MICROPHONE
ATS	AUTOMATIC TRANSFER SWITCH	MIN	MINIMUM
AWG	AMERICAN WIRE GAUGE	MS	MAGNETIC STARTER
BC	BOTTOM CONDUIT	MTD	MOUNTED
BD	BUS DUCT	MTG	MOUNTING
BFG	BELOW FINISHED GRADE	MTR	MOTOR
BIL	BASIC IMPULSE LEVEL	MTS	MANUAL TRANSFER SWITCH
BLDG	BUILDING	N	NEUTRAL
BX	ARMORED CABLE	NA	NON-AUTOMATIC
C	CONDUIT	NC	NORMALLY CLOSED
CATV	CABLE ANTENNA TELEVISION SYSTEM	NF	NON-FUSE
CCAB	CONTROL CABINET	N.I.C.	NOT IN CONTRACT
CCTV	CLOSED CIRCUIT TELEVISION	NL	NIGHT LIGHT
CH	CABINET HEATER	NO	NORMALLY OPEN
CKT	CIRCUIT	NP	NETWORK PROTECTOR
CKT BKR/CB	CIRCUIT BREAKER	NTS	NOT TO SCALE
CL	CLOSET	OC	ON CENTER
CLG	CEILING	OL	OVERLOAD ELEMENT
COND	CONDUCTOR	P	POLE
CO	CONDUIT ONLY	PA	PUBLIC ADDRESS
CT	CURRENT TRANSFORMER	PB	PULL BOX
CU	COPPER	P.C.	PLUMBING CONTRACTOR
DB	DUCT BANK	PF	POWER FACTOR
DMB	DIMMER BOARD	Ø	PHASE
DC	DIRECT CURRENT	PL	PILOT (INDICATOR) LIGHT
DH	DUCT HEATER	PNL	PANEL (PANELBOARD)
DIM	DIMMER CONTROL	PP	POWER PANEL
DISC	DISCONNECT	PRI	PRIMARY
DM	DAMPER MOTOR	PT	POTENTIAL TRANSFORMER
DN	DOWN	PVC	POLYVINYL CHLORIDE
DP	DISTRIBUTION POWER PANEL(BOARD)	PWR	POWER
DT	DOUBLE THROW	R	RECESSED
DWG	DRAWING	RC	REMOTE CONTROL
EA	EACH	REC	RECEPTACLE
E.C.	ELECTRICAL CONTRACTOR	S	SURFACE
E.HTR	ELECTRIC HEATER	SC	SEPARATE CIRCUIT
ELEV	ELEVATOR	SDB	SUB-DISTRIBUTION BOARD
EL	ELECTRIC	SEC	SECONDARY
EM	EMERGENCY	SMR	SURFACE METAL RACEWAY
EMT	ELECTRICAL METALLIC TUBING	SP	SINGLE POLE
ENT	ELECTRICAL NON-METALLIC TUBING	SPK	SPEAKER
EWC	ELECTRIC WATER COOLER	ST	SINGLE THROW
EX	EXISTING	SW	SWITCH
F	FUSE	SWBD	SWITCHBOARD
FA	FIRE ALARM	SYM	SYMMETRICAL
FACP	FIRE ALARM CONTROL PANEL	T	THERMOSTAT
FBO	FURNISHED BY OTHERS	TEL	TELEPHONE
FCC	FLAT CONDUCTOR CABLE	TB	TERMINAL BOX
FCU	FAN COIL UNIT	TC	TOP CONDUIT
FDR	FEEDER	TCAB	TELEPHONE CABINET
FL	FLOOR	T.C.C.	TEMPERATURE CONTROL CONTRACTOR
FLUOR	FLUORESCENT	TP	TAMPER PROOF
F.P.C.	FIRE PROTECTION CONTRACTOR	TV	TELEVISION
FS	FUSIBLE SWITCH	TYP	TYPICAL
F.S.C.	FOOD SERVICE CONTRACTOR	UG	UNDERGROUND
FT	FEET OR FOOT	UH	UNIT HEATER
G.C.	GENERAL CONTRACTOR	UNG	UNGROUNDED
GEN	GENERATOR	UON	UNLESS OTHERWISE NOTED
GF	GROUND FAULT	UPS	UNINTERRUPTED POWER SYSTEM
GG	GROUND GRID	V	VOLT(S)
GRD	GROUND	VA	VOLTAMP(S)
HC	HUNG CEILING	VAR	VOLT AMPERES REACTIVE
H.I.D.	HIGH INTENSITY DISCHARGE	VP	VAPORPROOF
HP	HORSEPOWER	W	WATT(S)
H.P.S.	HIGH PRESSURE SODIUM	WP	WEATHERPROOF
HPU	HEAT PUMP UNIT	WT	WATERTIGHT
HT	HEIGHT	XFR	TRANSFORMER
HV	HIGH VOLTAGE	XP	EXPLOSION PROOF
HW	HEAVY WALL RIGID CONDUIT		
HZ	FREQUENCY IN CYCLES PER SECOND		
IC	INTERRUPTING CAPACITY		
IG	ISOLATED GROUND		
IMC	INTERMEDIATE METALLIC CONDUIT		
INC	INCANDESCENT		
JB	JUNCTION BOX		
K	KEY OPERATED		
kVA	KILOVOLT AMPERE(S)		
kVAR	KILOVAR(S)		
kW	KILOWATT(S)		
kWhr	KILOWATT(S) HOUR(S)		
LP	LIGHTING PANEL		
L.P.S.	LOW PRESSURE SODIUM		
LTG	LIGHTING		
LV	LOW VOLTAGE		

Figure 41 Electrical abbreviations.
Source: Croxton Collaborative/Gould Evans Associates

Importance of Specifications

Specifications clarify information that cannot be shown on the drawings. Specifications are very important to the architect and owner to ensure compliance to the standards set. The figure provided shows one page of the specifications for a building's air handling units.

M.E. RINKER SR. HALL
SCHOOL OF BUILDING CONSTRUCTION
BR-191

AIR HANDLING UNITS
SECTION 15760

2.1 AIR HANDLING UNITS

A. Units shall be of the type, size and capacity as set forth in the schedule. The fan outlet velocities and coil and filter face velocities shall be within 5% of the values specified in the schedule. Units shall be <u>double wall</u> McQuay, or approved equal.

B. The units, as assembled, shall be complete with fans, coils, insulated casing, filters, drives and accessories. Each unit, including the fan enclosure, shall have essentially constant cross-sectional dimensions as to width and height. Internal baffles shall be provided as required to prevent bypassing of coils and filters.

C. The casing shall consist of an independent structural steel frame, properly reinforced and braced for maximum rigidity, having individually removable, flush mounted, insulated panels. The casing shall be of sectionalized construction, consisting basically of individual fan section, coil section, access sections, filter section, and drain pan. Sections shall be joined with continuous gasketing to form an air tight closure. Sections shall be so designed that the method of joining can be performed with relative ease and without damage to the insulation and vapor barrier.

The framework shall be constructed of AISC structural rolled shapes having minimum thickness of 1/8" (3 mm) or die formed sheet steel having the minimum gauges set forth in the following schedules:

Maximum Individual Casing Cross-Section	Minimum Framework Gauge
Up to 30 sq.ft. (2.8 sq.m.)	14
30.5 sq.ft. to 47 sq.ft. (2.81 to 4.4 sq.m.)	12
48 sq.ft. (4.45 sq.m.) and up	10

D. Framework shall be designed with recesses suitable to receive enclosure panels, providing neat appearance, airtight enclosure, and ease of panel removal.

E. Enclosure panels 12 sq. ft. in area and larger shall be constructed of not less than 18 gauge die formed sheet steel. Should the sides or top of a casing section exceed 20 sq. ft. in area, the panels shall be fabricated of more than one piece, with the individual panels recessed into intermediate structural members.

Protection for the insulation edges shall be provided around the perimeter of each

GEA 0300-0600

15760 - 2
Revision No. 1
Dec. 19, 2021

Source: Croxton Collaborative/Gould Evans Associates

1.5.7   Other Drawings and Documents

Other types of drawings and documents may be used to supplement the drawing set, depending on each building's type, usage, and installed subsystems.

An important drawing that may be included in a set of drawings for public buildings, multifamily residential buildings, and similar structures is a **fire protection plan**. This drawing shows the piping, valves, heads, and switches that make up a building's fire sprinkler system. A fire sprinkler symbols list is usually included on a separate sheet along with the fire sprinkler specifications, details and assembly drawings, and riser diagrams.

Shop drawings provide additional details about the design of specific building components and prefabricated items, such as sheet metal ductwork, piping, plumbing, and electrical components. They include information and step-by-step instructions on how those components are to be fabricated and installed, and provide greater detail than what is shown on construction (contract) drawings. Shop drawings help to estimate the type and quantity of required materials and identify dimensions and instructions for fabrication and installation, as well as estimate the time required to fully complete the task. For large commercial jobs, a draftsperson creates shop drawings based on a design drafted by an architect or engineer. On smaller jobs, the draftsperson may work from freehand sketches based on field measurements. Shop drawings, like section and detail drawings, are drawn to a larger scale than the engineer's design.

As-built drawings reflect the final as-constructed conditions of a project. Construction projects occasionally encounter unexpected obstacles and conditions that mean changes need to be made on site that conflict with the original design. As-built drawings are usually created by marking on the contract drawings any field-built deviations and then illustrating those on the final drawing set.

Fire protection plan: A drawing that shows the details of the building's sprinkler system.

What Is Reality Capture?

Reality capture is the process of capturing existing site or building conditions with the use of 3D laser scanning and photogrammetry. The scan and photo data can then be brought into the CAD environment, which allows designers to create an accurate 3D model of the existing site and structures to utilize during their design process. This helps to address any site challenges long before the construction process. Benefits from reality capture include:

- Greater design accuracy
- Improved accuracy in the prefabrication process
- Improved scheduling, which can help to decrease overall construction time
- Reduced potential for change orders
- Quick and accurate as-built process

1.0.0 Section Review

1. Revisions to the drawing are entered in the revision block and must include _____.
   a. the project tag and the date the drawing was approved
   b. engineering approvals and the intended date of completion
   c. the date and the initials of the person who made the revision
   d. dates and signatures for customer approval documentation

2. Code books and specifications for a construction project often include the meanings of _____.
   a. drawings
   b. references
   c. glossaries
   d. symbols

3. When a plan is marked NTS, the dimensions as measured on the drawing cannot be used to build the project.
   a. True
   b. False

4. What is the length (in feet and inches) of the section of duct using the architect's scales shown in *Figure SR01*?

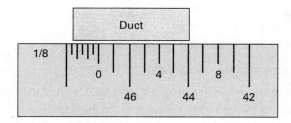

Figure SR01

   a. 6 feet, ¾ inches

   b. 6 feet, 5 inches

   c. 44 feet, ¾ inches

   d. 48 feet, 4 inches

5. A site plan _____.

   a. does not show features such as trees and driveways

   b. shows the location of the building from an aerial view

   c. is drawn after the floor plan is drawn, before the HVAC is designed

   d. includes information about plumbing fixtures and equipment

Module 00105 Review Questions

1. The title block generally contains _____.

   a. revision blocks

   b. special marks used in the drawing

   c. the mechanical plans

   d. the legend

2. The latest revision date on a set of construction drawings can be found _____.

   a. inside the detail

   b. in the schedule

   c. inside the title block

   d. in the legend

3. The Alphabet of Lines consists of _____.

   a. the line types used on a construction drawing

   b. lines that are indicated using letters of the alphabet

   c. lines that match up detail and section drawings

   d. lines that indicate land boundaries on the site plan

4. On a gridline system, a grid divides the area into small parts called _____.

   a. segments

   b. bays

   c. sections

   d. PODS

5. When the metric system is used, dimensions are written in _____.

   a. inches, feet, and yards
   b. degrees, radians, and gradians
   c. CADS, RADS, and KRADS
   d. meters, centimeters, and millimeters

6. If the scale on a site plan reads SCALE: 1" = 20' 0", then every _____.

   a. ¹⁄₂₀ of an inch on the drawing represents 20 feet, 0 inches
   b. 20 inches on the drawing represents 1 foot, 0 inches
   c. inch on the drawing represents 20 feet, 0 inches
   d. 20 inches on the drawing represents 20 feet, 0 inches

7. What is the length in meters at the point indicated by the arrow on the metric scale in *Figure RQ01*? The unit of length used on the scale is the meter.

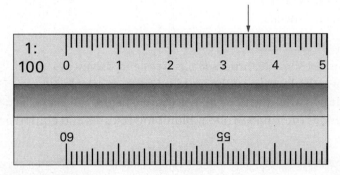

Figure RQ01

   a. 3500
   b. 350
   c. 35
   d. 3.5

8. An engineer's scale is set up in multiples of _____.

   a. 5
   b. 10
   c. 12
   d. 39.37

9. Which type of plan shows the layout of the HVAC system?

   a. Plumbing
   b. Structural
   c. Mechanical
   d. Foundation

10. Electrical plans can include _____.

   a. exterior elevations
   b. section drawings
   c. plumbing isometrics
   d. lighting plans

Answers to Section Review Questions and Module Review Questions are found in *Appendix A*.

Introduction to Basic Rigging

Source: Slingmax, Inc.

Objective

Successful completion of this module prepares you to do the following:

1. Identify and describe various types of rigging slings, hardware, and equipment.
   a. Identify and describe various types of slings.
   b. Describe how to inspect various types of slings.
   c. Identify and describe how to inspect common rigging hardware.
   d. Identify and describe various types of hoists.
   e. Identify and describe basic rigging hitches and the related Emergency Stop hand signal.

Performance Tasks

Under supervision, you should be able to do the following:

1. Demonstrate the proper ASME Emergency Stop hand signal.
2. Demonstrate the ability to report the load capacity of a sling, and if the sling is too damaged to use.

Overview

A common activity at nearly every construction site is the movement of material and equipment from one place to another using various types of lifting gear. The procedures involved in performing this task are known as rigging. Not every worker will participate in rigging operations, but nearly all will be exposed to it at one time or another. This module provides an overview of the various types of rigging equipment, common hitches used during a rigging operation, and the related Emergency Stop hand signal.

Industry Recognized Credentials

If you are training through an NCCER-accredited sponsor, you may be eligible for credentials from NCCER's Registry. The ID number for this module is 00106. Note that this module may have been used in other NCCER curricula and may apply to other level completions. Contact NCCER's Registry at 888.622.3720 or go to **www.nccer.org** for more information.

NOTE

This module is an elective. It is not required for successful completion of *Core.*

1.0.0 Basic Rigging Equipment

Performance Tasks

1. Demonstrate the proper ASME Emergency Stop hand signal.
2. Demonstrate the ability to report the load capacity of a sling, and if the sling is too damaged to use.

Objective

Identify and describe various types of rigging slings, hardware, and equipment.

a. Identify and describe various types of slings.
b. Describe how to inspect various types of slings.
c. Identify and describe how to inspect common rigging hardware.
d. Identify and describe various types of hoists.
e. Identify and describe basic rigging hitches and the related Emergency Stop hand signal.

Slings: Wire rope, alloy steel chain, metal mesh fabric, synthetic rope, synthetic webbing, or jacketed synthetic continuous loop fibers made into forms, with or without end fittings, used to handle loads.

Hoists: A device that applies a mechanical force for lifting or lowering a load.

Qualified person: A person who, through the possession of a recognized degree, certificate, or professional standing, or one who has gained extensive knowledge, training, and experience, has successfully demonstrated his or her ability to solve problems relating to the subject matter, the work, or the project.

Rigging is the planned movement of material and equipment from one location to another, using **slings, hoists,** or other types of equipment. Some rigging operations use a loader to move materials around a job site. Other operations require cranes to lift such loads. Two common types of cranes are overhead cranes (*Figure 1*) and mobile cranes (*Figure 2*). As with other types of equipment, cranes are available in many different configurations and capacities.

Rigging operations can be extremely complicated and dangerous. Do not experiment with rigging operations, and never attempt a lift without the supervision of an officially recognized, **qualified person**. A lift may appear simple while it is in progress. That is because the people performing the lift know exactly what they are doing; there is no room for guesswork. No matter whether rigging operations involve simple or complicated equipment, only qualified persons may perform rigging operations without supervision.

WARNING!

Although this module provides instruction and information about common rigging equipment and hitch configurations, it does not provide any level of certification. Any questions about rigging procedures or equipment should be directed to an instructor or supervisor. Always refer to the manufacturer's instructions for any type of rigging equipment.

Figure 1 Overhead crane.

Figure 2 Mobile cranes.
Source: Mammoet USA South, Inc.

1.1.0 Slings

During a rigging operation, the **load** being lifted or moved must be connected to the apparatus, such as a crane, that will provide the power for movement. The connector—the link between the load and the apparatus—is often a sling made of synthetic, chain, or **wire rope** materials. This section focuses on three types of slings:

- Synthetic slings
- Alloy steel chain slings
- Wire rope slings

1.1.1 Sling Tagging Requirements

All slings are required to have identification tags (*Figure 3*). An identification tag must be securely attached to each sling and clearly marked with the information required for that type of sling. For all three types of slings, that information will include the manufacturer's name or trademark and the **rated capacity** of the type of **hitch** used with that sling. The rated capacity is the maximum load weight that the sling is designed to carry. Rated capacity is technically referred to as working load limit (WLL). The rated capacity, or WLL, of a sling must never be exceeded. Overloading a sling can result in catastrophic failure.

Load: The total amount of what is being lifted, including all slings, hitches, and hardware.

Wire rope: A rope made from steel wires that are formed into strands and then laid around a supporting core to form a complete rope; sometimes called cable.

Rated capacity: The maximum load weight a sling or piece of hardware or equipment can hold or lift. Also referred to as the working load limit (WLL).

Hitch: The rigging configuration by which a sling connects the load to the hoist hook. The three basic types of hitches are vertical, choker, and basket.

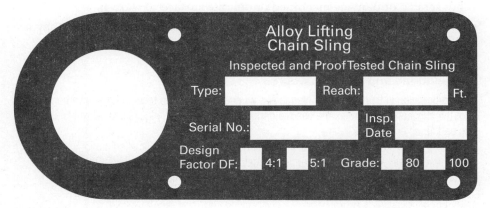

Figure 3 Identification tag.

WARNING!

All slings and hardware have a rated capacity, or working load limit (WLL). Rated capacity, or WLL, is defined as the maximum load weight that a sling or piece of hardware or equipment can hold or lift. Under no circumstances should the rated capacity ever be exceeded. Overloading a piece of rigging equipment can result in catastrophic failure.

The following are the tagging requirements for synthetic slings:

- Manufacturer's name or trademark
- Manufacturer's code or stock number (unique for each sling)
- Rated capacities for the types of hitches used
- Type of synthetic material used in the manufacture of the sling

The following are the tagging requirements for alloy steel chain slings:

- Manufacturer's name or trademark
- Manufactured grade of steel
- Link size (diameter)
- Rated load and the angle on which the rating is based
- **Sling reach**
- Number of **sling legs**

Sling reach: A measure taken from the master link of the sling, where it bears weight, to either the end fitting of the sling or the lowest point on the basket.

Sling legs: The parts of the sling that reach from the attachment device around the object being lifted.

The following are the tagging requirements for wire rope slings:

- Manufacturer's name or trademark
- Rated capacity in a vertical hitch (other hitches optional)
- Diameter of wire rope (optional)
- Manufacturer's code or stock number of sling (optional)

Any sling without an identification tag must be removed from service immediately, since its characteristics and rating can no longer be identified. A qualified person may be able to install a new tag in some cases.

1.1.2 Synthetic Slings

Synthetic slings are widely used to lift loads, especially easily damaged loads. This section covers two types of synthetic slings: synthetic web slings and round slings.

Synthetic web slings provide several advantages over other types of slings:

- They are softer and wider than chain or wire rope slings. Therefore, they do not scratch or damage machined or delicate surfaces (*Figure 4*).
- They do not rust or corrode and therefore will not stain the loads they are lifting.

Regulations and Site Procedures

The information in this module is intended as a general guide. The techniques shown here are not the only methods that can be used to perform a lift. Many techniques can be safely used to rig and lift different loads.

Some of the techniques for certain kinds of rigging and lifting are spelled out in requirements issued by federal government agencies. Some will be provided at the job site, where written site procedures that address any special conditions that affect lifting procedures on that site can be found. Questions about any of these procedures should be directed to the supervisor at the site.

Figure 4 Web slings provide surface protection.
Source: Slingmax, Inc.

Figure 5 Synthetic web sling shaping.
Source: Myphotojumble/Shutterstock

- They are lightweight, making them easier to handle than wire rope or chain slings. Most synthetic slings weigh less than half as much as a wire rope that has the same rated capacity. Some new synthetic fiber slings weigh one-tenth as much as wire rope.
- They are flexible. They mold themselves to the shape of the load (*Figure 5*).
- They are very elastic, and they stretch under a load much more than wire rope. This stretching allows synthetic slings to absorb shocks and to cushion the load.
- Loads suspended in synthetic web slings are less likely to twist than those in wire rope or chain slings.

For all the advantages that synthetic web slings provide, there are some concerns that must be kept in mind when using them. For example, synthetic web slings should not be exposed to temperatures above 180°F (82°C). They are also susceptible to cuts, abrasions, and other wear-and-tear damage. To prevent damage to synthetic web slings, riggers use protective pads (*Figure 6*).

Most synthetic web slings are manufactured with red-core **warning yarns**. These are used to let the rigger know whether the sling has suffered too much wear or damage to be used. When the red yarns are exposed, the synthetic web sling should not be used (*Figure 7*).

CAUTION

If the sling does not come with protective pads, use other kinds of softeners of sufficient strength or thickness to protect the sling where it makes contact with the load. Pieces of old sling, fire hose, canvas, or rubber can be used. Manufactured softeners can also be purchased.

Warning yarns: A component of the sling that shows the rigger whether the sling has suffered too much damage to be used.

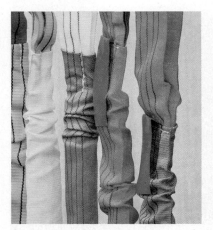

Figure 6 Protective pads.
Source: Chiradech/iStock/Getty Images Plus

Red warning yarns

Figure 7 Synthetic web sling warning yarns.

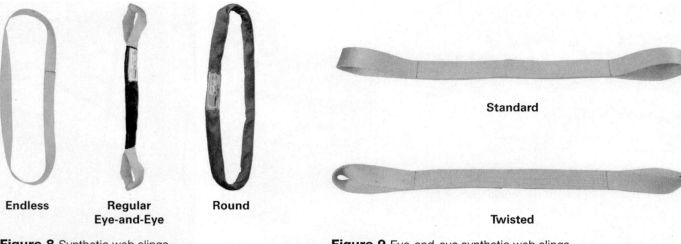

Endless   **Regular Eye-and-Eye**   **Round**

Figure 8 Synthetic web slings.
Source: US Cargo Control

Standard

Twisted

Figure 9 Eye-and-eye synthetic web slings.
Source: US Cargo Control

Synthetic web slings are available in several designs. The most common are the following (*Figure 8*):

- Endless web slings, which are also called grommet slings.
- Synthetic web eye-and-eye slings, which are made by sewing an end of the sling directly to the sling body. Standard eye-and-eye slings (*Figure 9*) have eyes on the same **plane** as the sling material; twisted eye-and-eye slings have eyes at right angles to the main portion of the sling and are primarily used for choker hitches (which are discussed later).
- Round slings, which are endless and made in a continuous circle out of polyester filament yarn. The yarn is then covered by a woven sleeve.

Synthetic web eye-and-eye slings are also available with hardware end fittings instead of fabric eyes. The standard end fittings are made of either aluminum or steel. They come in male and female configurations (*Figure 10*).

Round slings are made by wrapping a synthetic yarn around a set of spindles to form an endless loop. A protective jacket encases the **core** yarn (*Figure 11*). One type of round sling is a Twin-Path® sling. Twin-Path® is a trademarked sling made by Slingmax®, Inc., although the term is sometimes used generically.

Twin-Path® slings are made of a synthetic fiber, such as polyester. The material used to make up the **strands** and the number of wraps in a loop determines

Plane: A surface in which a straight line joining two points lies wholly within that surface.

Core: Center support member of a wire rope around which the strands are laid.

Strands: A group of wires wound, or laid, around a center wire, or core. Strands are laid around a supporting core to form a rope.

Figure 10 Synthetic web sling hardware end fittings.
Source: US Cargo Control

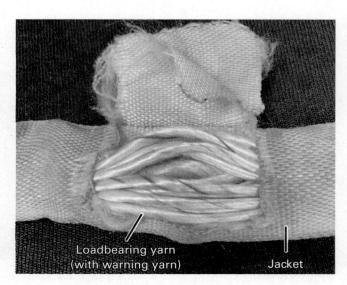

Loadbearing yarn (with warning yarn)   Jacket

Figure 11 Synthetic endless-strand jacketed sling.

Around the World

Heavy Lifters

The world's largest cranes are capable of lifting astonishing amounts of weight. Some of these megacranes are permanently stationary, some can be assembled and disassembled on site, and some are mobile. Taisun, a fixed dual-beam gantry crane in China, can lift over 22,000 tons (over 20,000 metric tons). Mammoet, a manufacturer of heavy-lifting equipment in the Netherlands, has built numerous cranes that are referred to as PTCs (Platform Twinring Containerised). These cranes can be disassembled and packed into shipping containers, unloaded at any dock in the world, and hauled by truck to a job site. This type of crane, as shown here, is assembled onto a stationary ring that enables the crane and counterweights to move in a 360-degree radius. These giant cranes can lift up to 3,200 tons (about 2,900 metric tons).

One of the largest mobile cranes is Liebherr's LR 13000 crawler crane. It has a more conventional design and appearance. Its maximum lifting capacity is over 3,300 tons (about 3,000 metric tons).

Source: Mammoet USA South, Inc.

the rated capacity of the sling, or how much weight it can handle. Aramid round slings, for example, have a greater rated capacity for their size than the web-type slings do.

Twin-Path® slings are also available in a design with two separate wound loops of strand jacketed together side-by-side (*Figure 12*). This design greatly increases the lifting capacity of the sling.

The jackets of these slings are available in several materials for various purposes, including heat-resistant Nomex®, polyester, and bulked nylon (Covermax™) (*Figure 13*). Twin-Path® slings featuring K-Spec® yarn weigh at least 50 percent less than a polyester round sling of the same size and capacity.

Twin-Path® slings are equipped with **tattle-tail** yarns to help riggers determine whether the sling has become overloaded or stretched beyond a safe limit (*Figure 14*). Note that these devices are different from red-core yarn, which is incorporated directly into the fabric of the sling.

These slings are also available with a fiber optic inspection cable running through the strand (*Figure 15*). When a light is directed at one end of the fiber, the other end will light up to show that the strand has not been broken.

The way the sling will be used determines what material is used for the jackets of round slings. Polyester is normally used for light- to medium-duty sling jackets. Covermax™, a high-strength material with much greater resistance to cutting and abrasion, is used to make a sturdier jacket for heavy-duty uses.

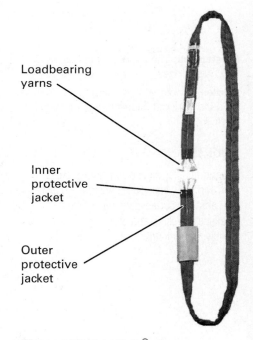

Loadbearing yarns

Inner protective jacket

Outer protective jacket

Figure 12 Twin-Path® sling.
Source: Lift-It Manufacturing Co., Inc.

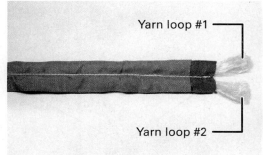

Yarn loop #1

Yarn loop #2

Figure 13 Twin-Path® sling makeup.
Source: Lift-It Manufacturing Co., Inc.

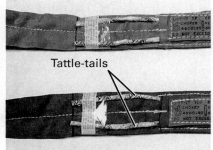

Tattle-tails

Figure 14 Tattle-tails.
Source: Lift-It Manufacturing Co., Inc.

Tattle-tail: Cord attached to the strands of an endless loop sling. It protrudes from the jacket. A tattle-tail is used to determine if an endless sling has been stretched or overloaded.

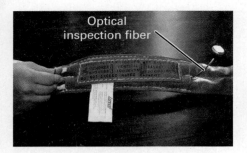

Figure 15 Fiber optic inspection cable.
Source: Lift-It Manufacturing Co., Inc.

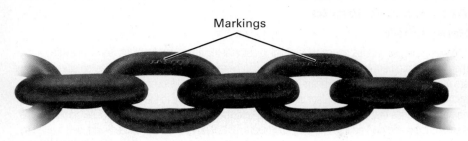

Figure 16 Markings on alloy steel chain slings.
Source: Columbus McKinnon Corporation

1.1.3   Alloy Steel Chain Slings

Alloy steel chain, like wire rope (which is covered later), can be used in many different rigging operations. Chain slings are often used for lifts in high heat or rugged conditions. Chain slings can be adjusted over a center of gravity, making it a versatile sling. They are also the most durable. However, a chain sling weighs much more than a wire rope sling, and it also may be harder to inspect. Workers will encounter both types of slings in the field and must decide which type of sling is best for each situation.

Steel chain slings used for overhead lifting must be made of alloy steel. The higher the alloy grade, the safer and more durable the chain is. Alloy steel chains commonly used for most overhead lifting are marked with the number 8, the number 80, the number 800, or the letter A (see *Figure 16*).

Steel chain slings have two basic designs with many variations:

- Single- and double-basket slings (*Figure 17*) do not require end-fitting hardware. The chain is attached to the **master link** in a permanent basket hitch or hitches.
- Chain **bridle** slings are available with two to four legs (*Figure 18*).

Alloy steel chain slings are used for lifts when temperatures are high or where the slings will be subjected to steady and severe abuse. Most alloy steel chain slings can be used in temperatures up to 500°F (260°C) with little loss in rated capacity.

Even though alloy steel chain slings can withstand extreme temperature ranges and abusive working conditions, these slings can be damaged if loads are dropped on them; if they are wrapped around loads with sharp corners (unless softeners are used); or if they are exposed to intense temperatures.

Master link: The main connection fitting for chain slings.

Bridle: A configuration using two or more slings to connect a load to a single hoist hook.

CAUTION

Never drag alloy steel chain slings across hard surfaces, especially concrete. Friction with abrasive surfaces wears out the chain, weakens the sling, and often damages identifying markings.

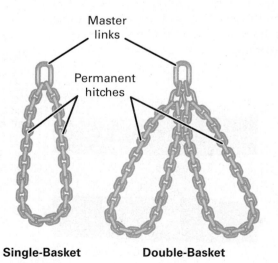

Figure 17 Chain slings.

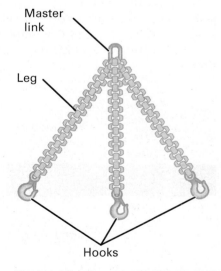

Figure 18 Three-leg chain bridle sling.

Figure 19 Wire rope sling.
Source: US Cargo Control

1.1.4 Wire Rope Slings

Wire rope slings (*Figure 19*) are made of high-strength steel wires that are formed into strands and wrapped around a supporting core. While there are many types of wire rope, they all share this common design. As a general rule, wire rope slings are lighter than chain, can withstand substantial abuse, and are easier to handle than chain slings. They can also withstand relatively high temperatures.

The wire rope that makes up a wire rope sling is designed so that the strand wires and the supporting core interact with one another by sliding and adjusting. This movement compensates for the ever-changing stresses placed upon the rope. It makes it less likely that the wires and the core will be damaged when the rope is bent around **sheaves** or loads, or when it is placed at an angle during rigging. The wires must be able to move the way they were designed to move, so a wire rope's rated capacity depends on it being in good condition.

Figure 20 shows the three basic components of a wire rope: the supporting core, high-grade steel wires, and multiple center wires.

The supporting core enables the strands to keep their original shapes. There are three basic types of supporting cores for wire rope (*Figure 21*):

- *Fiber cores* – Usually made of synthetic fibers, but can also be made from natural vegetable fibers, such as sisal.

- *Independent wire rope cores* – Made of a separate wire rope with its own core and strands; the core rope wires are much smaller and more delicate than the strand wires in the outer rope.

- *Strand cores* – Made by using one strand of the same size and type as the rest of the strands of rope.

Sheaves: A grooved pulley-wheel for changing the direction of a rope's pull; often found on a crane.

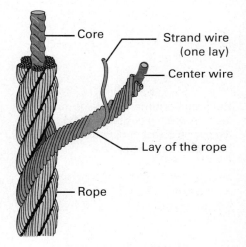

Figure 20 Wire rope components.

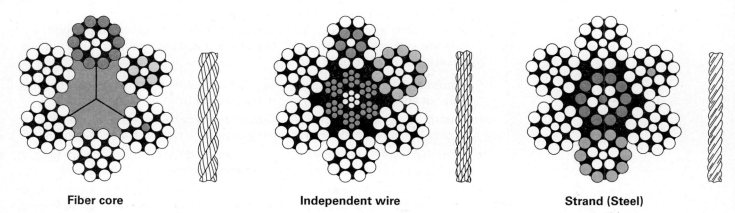

Fiber core   **Independent wire**   **Strand (Steel)**

Figure 21 Wire rope supporting cores.

The materials used for the supporting core have both desirable characteristics and drawbacks, depending on how they are to be used. Fiber core ropes, for example, may be damaged by heat at relatively low temperatures (180°F to 200°F [82°C to 93°C]), as well as by exposure to caustic chemicals.

1.2.0	Sling Inspection

Did You Know?

Sling Cut Protection

Cuts are the primary cause of synthetic sling failure. Often these cuts come from making contact with sharp edges on the load. Besides sling jackets, or wear pads, devices called edge protectors are often placed on the edges of a load to help reduce the likelihood of cutting the sling.

Rigging operations can be dangerous even in the best conditions. If equipment failure occurs, the results can be devastating. Since slings are commonly used to connect loads to lifting equipment, they are susceptible to wear and damage. To ensure that slings are safe to use, it is absolutely critical that they be thoroughly inspected on a regular basis. Inspections must be conducted regardless of any existing markings or tags that are related to previous inspections.

WARNING!

Although this module provides instruction and information about lifting component inspection, it does not provide any level of certification. Any questions about rigging procedures or equipment should be directed to an instructor or supervisor. Always refer to the manufacturer's instructions for any type of rigging equipment.

1.2.1   Synthetic Sling Inspection

Like all slings, synthetic slings must be inspected before each use to determine whether they are in good condition and can be used. They must be inspected along the entire length of the sling, both visually (looking at them) and manually (feeling them). If any **rejection criteria** are met, the sling must be removed from service.

Rejection criteria: Standards, rules, or tests on which a decision can be based to remove an object or device from service because it is no longer safe.

WARNING!

A competent person must inspect synthetic slings before they are used, every time. The inspection must involve a visual and a touch examination. Sometimes defects and damage can be seen; other times they can be felt.

If any synthetic sling meets any of the rejection criteria presented in this section, it must be removed from service immediately. In addition, the rigger has to exercise sound judgment. Along with looking for any single major problem, the rigger must also watch for combinations of relatively minor defects in the sling. Combinations of minor damage may make the sling unsafe to use, even though the specific defects found may not be listed in the rejection criteria.

Workers who help inspect synthetic slings must alert a qualified person if any defects are suspected, especially if any of the following synthetic sling damage rejection criteria are found (*Figure 22*):

- A missing identification tag or a tag that cannot be read. Any synthetic sling without an identification tag must be removed from service immediately.
- Abrasion that has worn through the outer jacket or has exposed the loadbearing yarn of the sling, or abrasion that has exposed the warning yarn of a web sling.
- A cut that has severed the outer jacket or exposed the loadbearing yarn (single-layer jacket) of a round sling, or has exposed the warning yarn of a web sling.
- A tear that has exposed the inner jacket or the loadbearing yarns (single-layer jacket) of a round sling, or has exposed the warning yarn of a web sling.
- A puncture.

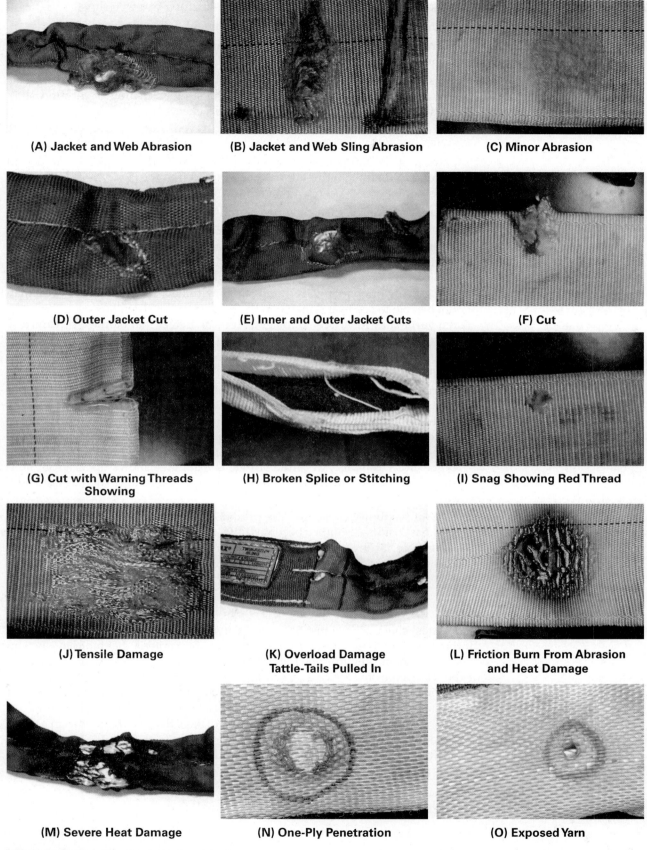

(A) Jacket and Web Abrasion

(B) Jacket and Web Sling Abrasion

(C) Minor Abrasion

(D) Outer Jacket Cut

(E) Inner and Outer Jacket Cuts

(F) Cut

(G) Cut with Warning Threads Showing

(H) Broken Splice or Stitching

(I) Snag Showing Red Thread

(J) Tensile Damage

(K) Overload Damage Tattle-Tails Pulled In

(L) Friction Burn From Abrasion and Heat Damage

(M) Severe Heat Damage

(N) One-Ply Penetration

(O) Exposed Yarn

Figure 22 Sling damage rejection criteria.
Source: Ed Gloninger (22A–22M)

Splice: To join together.

- Broken or worn stitching in the **splice** or stitching of a web sling.
- A knot that cannot be removed by hand in either a web or round sling.
- A snag in the sling that reveals the warning yarn of a web sling or tears through the outer jacket, or exposes the loadbearing yarns of a round sling.
- Crushing of either a web sling or a round sling. Crushing in a web sling feels like a hollow pocket or depression in the sling. Crushing in a round sling feels like a hard, flat spot underneath the jacket.
- Damage from overload (often called tensile damage or overstretching) in a web or round sling. Tensile damage in a web sling is evident when the weave pattern of the fabric begins to pull apart. Twin-Path® slings have tattle-tail features. When the tattle-tail has been pulled into the jacket, it indicates that the sling may have been overloaded.
- Chemical damage, including discoloration, burns, and melting of the fabric or jacket.
- Heat damage, ranging from friction burns to melting of the sling material, the loadbearing strands, or the jacket. Friction burns give the webbing material a crusty or slick texture. Heat damage to the jacket of a round sling looks like glazing or charring. In round slings with heat-resistant jackets, you may not be able to see heat damage to the outside of the sling; however, the internal yarns may have been damaged. You can detect that damage by carefully handling the sling and feeling for brittle or fused fibers inside the sling, or by flexing and folding the sling and listening for the sound of fused yarn fibers cracking or breaking inside.
- Ultraviolet (UV) damage. The evidence of UV damage is a bleaching-out of the sling material, which breaks down the synthetic fibers. Web slings with UV damage will release a powder-like substance when they are flexed and folded. Round slings, especially those made of Cordura® nylon and other specially treated synthetic fibers, have a much greater resistance to UV damage. In round slings, UV damage shows up as a roughening of the fabric texture where no other sign of damage, such as abrasion, can be found.
- Loss of flexibility caused by the presence of dirt or other abrasives. A sling that has lost its flexibility becomes stiff. Abrasive particles embedded in the sling material act like tiny blades that cut apart the internal fibers of the sling every time it is stretched, flexed, or wrapped around a load.

1.2.2 Alloy Steel Chain Sling Inspection

Alloy steel chain slings must be carefully inspected before each use to determine if they are safe to use. There are numerous rejection criteria that indicate when an alloy steel chain sling must be removed from service. A chain sling may also need to be removed from service if something shows up that does not exactly match the rejection criteria. Not all riggers are qualified to make decisions about the condition of the equipment being used. If questions arise about whether a sling is defective or unsafe, a qualified person should be consulted. Note that the chain links in a chain sling cannot be repaired by replacement.

An alloy steel chain sling must be removed from service for any of the following defects (*Figure 23*):

- Missing or illegible identification tag
- Cracks
- Heat damage
- Stretched links; damage is evident when the link grows long and when the barrels—the long sides of the links—start to close up
- Bent links
- Twisted links

Sling Jackets

If only the outer jacket of a sling is damaged, under the rejection criteria, that sling can be sent back to the manufacturer for a new jacket. It must be tested and certified by the manufacturer before it is used again. It is much less costly to do this than to replace the sling. However, slings that are removed because of heat- or tension-related jacket damage cannot be returned for repair and must be disposed of by a qualified person.

Heat Damage and Crack

Overload Damage

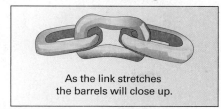

As the link stretches
the barrels will close up.

**Impact Damage
Bent Links**

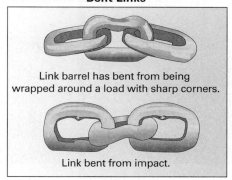

Link barrel has bent from being
wrapped around a load with sharp corners.

Link bent from impact.

Twisted Links

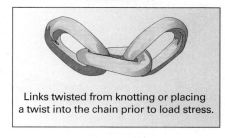

Links twisted from knotting or placing
a twist into the chain prior to load stress.

Excessive Wear

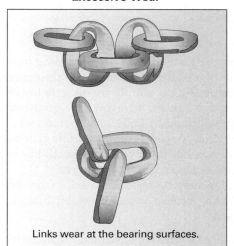

Links wear at the bearing surfaces.

Cuts, Chips, and Gouges

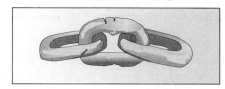

Rust and Corrosion

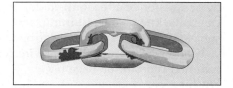

Figure 23 Damage to chains.

- Evidence of replacement links that have been used to repair the chain
- Excessive rust or corrosion, meaning rust or corrosion that cannot be easily removed with a wire brush
- Cuts, chips, or gouges resulting from impact on the chain
- Damaged end fittings, such as hooks, clamps, and other hardware
- Excessive wear at the link-bearing surfaces
- Scraping or abrasion

1.2.3   Wire Rope Sling Inspection

Like other slings, wire rope slings must be inspected before each use by a **competent person**. Remember that a competent person is defined by OSHA as a person who is capable of identifying existing and predictable hazards in the surroundings or working conditions that are unsanitary, hazardous, or dangerous to employees, and who has authorization to take prompt corrective measures to eliminate those conditions.

If the wire rope is damaged, it must be removed from service. Only a competent person can decide whether to use a wire rope in a rigging operation or

Competent person: An individual capable of identifying existing and predictable hazards in the surroundings and working conditions which are unsanitary, hazardous, or dangerous to employees; an individual who is authorized to take prompt corrective measures to eliminate these issues.

(A) Broken Wires   (B) Kinking   (C) Birdcaging

(D) Crushing   (E) Corrosion

Figure 24 Common types of wire rope damage.
Source: Kittipong Prakobsukmak/Shutterstock (24A); SyedMKimi/Shutterstock (24B); Beate Wolter/Shutterstock (24C); hidemozart/Shutterstock (24D); rutpratheep nilpechr/Shutterstock (24E)

discard it if it is damaged. Any worker who suspects that a wire rope sling may be damaged must bring it to the attention of a competent person, stopping an active lift if necessary to do so. If a wire rope sling is simply missing its tag or the tag has become unreadable due to wear, it can be retagged by a qualified person. Remember that a qualified person is defined by OSHA as a person who, by possession of a recognized degree, certificate, or professional standing, or by extensive knowledge, training, and experience, has demonstrated the ability to solve or prevent problems relating to a certain subject, work, or project.

Following an inspection, a wire rope sling may be rejected based on several common types of damage (*Figure 24*), including broken wires, kinks, birdcaging, crushing, corrosion and rust, and heat damage.

One-rope lay: The lengthwise distance it takes for one strand of a wire rope to make one complete turn around the core.

- *Broken wires* – Broken wires in the strands of a wire rope weaken the material strength of the rope and interfere with the interaction among the rope's moving parts. External broken wires usually mean normal fatigue, but internal or severe external breaks should be investigated closely. Internal or severe breaks in a wire rope mean it has been used improperly. Rejection criteria for broken wires consider how many wires are broken in one lay length of rope, or **one-rope lay**. One-rope lay (identifying a single rope lay and not a rope lay made with one rope) is a term that defines the lengthwise distance it takes for one strand of wire to make one complete turn around the core (*Figure 25*). Different wire ropes have different one-rope lays, so it is important for a competent person to inspect each wire rope closely when looking for broken wires.

> **WARNING!**
>
> Broken wires in a sling are extremely sharp and can easily cut or puncture the skin. Never use a bare hand to handle wire rope slings or to inspect them by running a hand along their length.

Figure 25 One-rope lay.

- *Kinks* – Kinking, or distortion of the rope, is a very common type of damage. Kinking can result in serious accidents. Sharp kinks restrict or prevent the movement of wires in the strands at the area of the kink. This means the rope is damaged and must not be used. Ropes with kinks in the form of large, gradual loops in a corkscrew configuration must be removed from service.

- *Birdcaging* – This damage occurs when a load is released too quickly and the strands are pulled or bounced away from the supporting core. The wires in the strands cannot compensate for the change in the stress level by adjusting inside the strands. The built-up stress then finds its own release out through the strands. Birdcaging usually occurs in an area where already-existing damage prevents the wires from moving to compensate for changes in stress, position, and bending of the rope. Any sign of birdcaging is cause to remove the rope from service immediately.

- **Unstranding** – Unstranding describes wire rope strands that have become untwisted. This weakens the rope and makes it easier to break.

- *Signs of core failure* – Failure or breakage within the core is usually spotted by the protrusion of core strands through or into the outer jacket of strands.

- *Crushing* – This results from setting a load down on a sling or from hammering or pounding a sling into place. Crushing of the sling prevents the wires from adjusting to changes in stress, changes in position, and bends. A crushed sling usually results in the crushing or breaking of the core wires directly beneath the damaged strands. If crushing occurs, the sling must be removed from service immediately.

- *Corrosion and rust* – Corrosion and rust of wire rope are the result of improper or insufficient lubrication. Corrosion and rust are considered excessive if there is surface scaling or rust that cannot easily be removed with a wire brush, or if they occur inside the rope. If corrosion and rust are excessive, the rope must be removed from service.

- *Heat damage* – Heat damage makes wires in the strands and core in the affected area become brittle. Heat damage appears as discoloration and sometimes the actual melting of the wire rope. A wire rope that has been damaged by heat must be removed from service.

- *Integrity of end connections* – Any end connections that have been applied to the wire rope must be inspected for any signs of damage or failure, such as marks indicating the connection has slipped or moved.

Unstranding: Describes wire rope strands that have become untwisted. This weakens the rope and makes it easier to break.

1.3.0 Rigging Hardware

Rigging hardware is as crucial as the crane, the slings, or any specially designed lifting frame or hoisting device. If the hardware that connects the slings to either the load or the master link were to fail, the load would drop just as it would if the crane, hoist, or slings were to fail. Hardware failure related to improper attachment, selection, or inspection contributes to a great number of the deaths, serious injuries, and property damage events in rigging accidents. The importance of hardware selection, maintenance, inspection, and proper use cannot be stressed enough. The requirements for rigging hardware are as stringent as those governing cranes and slings.

1.3.1 Shackles

A **shackle** is an item of rigging hardware used to attach an item to a load or to couple slings together. For example, a shackle can be used to couple the end of a wire rope to eye fittings or hooks. It consists of a U-shaped body and a removable pin.

Shackles used for overhead lifting should be made from forged steel, not cast steel. Quenched and tempered steel is the preferred material because of its increased toughness, but at a minimum, shackles must be made of drop- or hammer-forged steel.

Shackle: Coupling device used in an appropriate lifting apparatus to connect the rope to eye fittings, hooks, or other connectors.

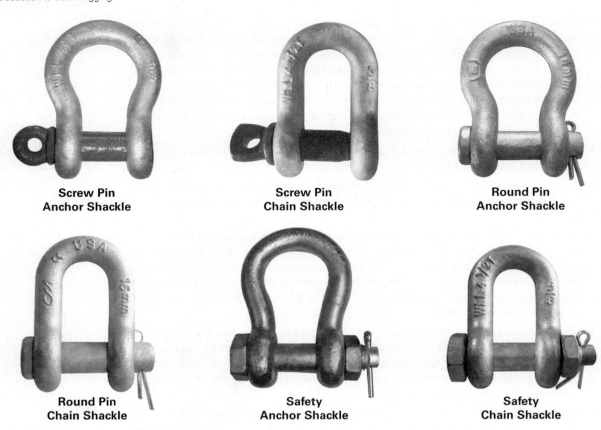

Figure 26 Shackles.
Source: Columbus McKinnon Corporation

Figure 27 Wide-body shackle.
Source: Courtesy of The Crosby Group LLC

Load stress: The strain or tension applied on the rigging by the weight of the suspended load.

All shackles must have a stamp that is clearly visible, showing the manufacturer's trademark, the size of the shackle (determined by the diameter of the shackle's body, not by the diameter of the pin), and the rated capacity of the shackle.

Shackles are available in two basic classes, identified by their shapes: anchor shackles and chain shackles. Both anchor and chain shackles have three basic types of pin designs, each one unique, as shown in *Figure 26*. The screw pin shackle design is the most widely used type in general industry. Shackle pins can be threaded into the shackle (screw pin), be threaded and use a nut to secure them, use a locking pin through the end, or be both threaded and pinned to secure them to the shackle.

Specialty shackles are available for specific applications where a standard shackle would not work well. For example, wide-body shackles (*Figure 27*) are for heavy-lifting applications.

Synthetic web sling shackles (*Figure 28*) are designed with a wide throat opening and a wide bow that is contoured to provide a larger, nonslip surface area to accommodate the wider body of synthetic web slings.

When a screw pin shackle is used, it is important to tighten the pin the proper amount. The pin should be screwed in until it is fully engaged and the shoulder of the pin is in contact with the shackle body (*Figure 29*). If the pin is left loose, the shackle can stretch under load. If the pin is over tightened, the **load stress** from the sling can torque the pin even more. This can stretch and jam the pin's threads.

When any type of shackle is used in a rigging operation, it must be positioned so that the load is centered in the bow of the shackle and not on the shackle pin (*Figure 30*). If the shackle uses a screw pin and the load is positioned on the pin, the shifting of the load could cause the pin to twist and unscrew.

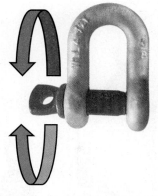

Figure 28 Synthetic web sling shackle.
Source: Courtesy of The Crosby Group LLC

Figure 29 Don't over-tighten the shackle pin.
Source: Columbus McKinnon Corporation

Figure 30 Shackle correctly positioned.
Source: Volvo Construction Equipment

Shackles, like any other type of hardware, must be inspected by the rigger before each use to make sure there are no defects that would make the shackle unsafe. Each lift may cause some degree of damage or may further reveal existing damage.

If any of the following conditions exists, a shackle must be removed from use:

- Bends, cracks, or other damage to the shackle body
- Incorrect shackle pin or improperly substituted pin
- Bent, broken, or loose shackle pin
- Damaged threads on threaded shackle pin
- Missing or illegible capacity and size markings

1.3.2 Eyebolts

An eyebolt is an item of rigging hardware with a **threaded shank**. The eyebolt's shank end is attached directly to the load, and the eyebolt's eye end is used to attach a sling to the load.

Eyebolts for overhead lifting should be made of drop- or hammer-forged steel. Eyebolts are available in three basic designs with several variations (*Figure 31*).

Unshouldered eyebolts are designed for straight vertical pulls only. Shouldered eyebolts have a shoulder that is used to help support the eyebolt

Threaded shank: A connecting end of a fastener, such as a bolt, with a series of spiral grooves cut into it. The grooves are designed to mate with grooves cut into another object in order to join them together.

Unshouldered   **Shouldered**   **Swivel**

Figure 31 Eyebolt variations.
Source: Courtesy of The Crosby Group LLC

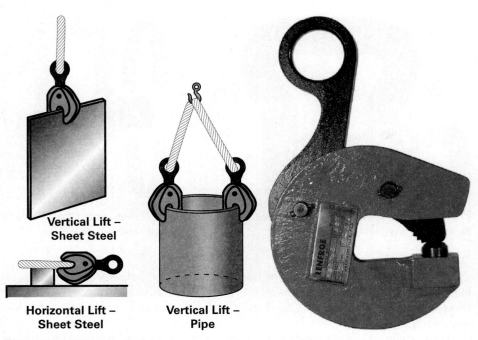

Figure 32 Using lifting clamps.

Figure 33 Basic nonlocking clamp.
Source: J.C. Renfroe & Sons

during pulls that are slightly angular. Swivel eyebolts, also called hoist rings, are designed for angular pulls from 0 to 90 degrees from the horizontal plane of the load.

In a few cases, eyebolts must be torqued to a specific value when they are installed for lifting use.

1.3.3 Lifting Clamps

Lifting clamps are used to move loads such as steel plates, sheet piles, large pipe, or concrete panels without the use of slings. These rigging devices are designed to bite down on a load and use the jaw tension to secure the load (*Figure 32*).

All lifting clamps must be made of forged steel, and they must be stamped with their rated capacity. Some clamps, such as the one in *Figure 33*, use the weight of the load to produce and sustain the clamping pressure; the grip tightens as the load increases. Others use an adjustable cam that is set and tightened to maintain a secure grip on the load.

Lifting clamps are designed to carry one item at a time, regardless of the capacity or jaw dimensions of the clamp or the thickness or weight of the item being lifted. In order for the clamp to hold an item securely, the cam and the jaw must bite or grip both sides of a single item. Placing more than one plate or sheet into the clamp prevents both the cam and the jaw from securing both sides of the plate or sheet. The clamps must be placed to ensure that the load remains balanced. For larger loads, two lifting clamps are typically required.

Lifting clamps are available in a wide variety of designs, so it is important to match the type of clamp with the intended application. *Figure 34* shows some lifting clamps designed for specific uses.

Lifting clamps, like any other type of rigging hardware, must be inspected by the rigger before every use to make sure there are no defects that would make the clamp unsafe. Each lift may cause some degree of damage or may further reveal existing damage.

Lifting clamps: A device used to move loads such as steel plates or concrete panels without the use of slings.

Did You Know?

Shackles and Pins

Pins should never be swapped in shackles. The thread forms of different brands of shackles and different brands of pins may not engage properly with one another. In addition, there is no easy way to tell how much reserve strength is left in a pin or a shackle. Further, the rated capacities of different shackles and pins may not be compatible. The shackle pin should be placed and secured into a single shackle between uses.

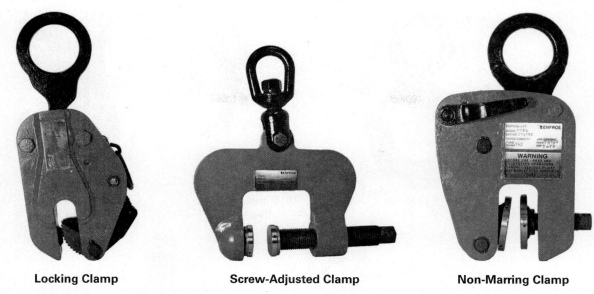

Locking Clamp   **Screw-Adjusted Clamp**   **Non-Marring Clamp**

Figure 34 Lifting clamps.
Source: J.C. Renfroe & Sons

If any of the following conditions exists, a clamp must be removed from use (*Figure 35*):

- Cracks
- Abrasion, wear, or scraping
- Any deformation or other impact damage to the shape that is detectable during a visual examination
- Excessive rust or corrosion, meaning rust or corrosion that cannot be removed easily with a wire brush
- Excessive wear of the teeth
- Heat damage
- Loose or damaged screws or rivets
- Worn springs

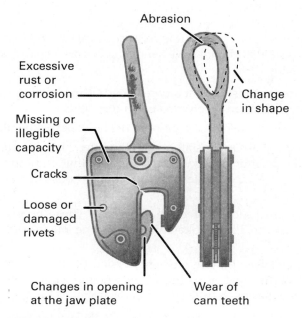

Figure 35 Rejection criteria for lifting clamps.

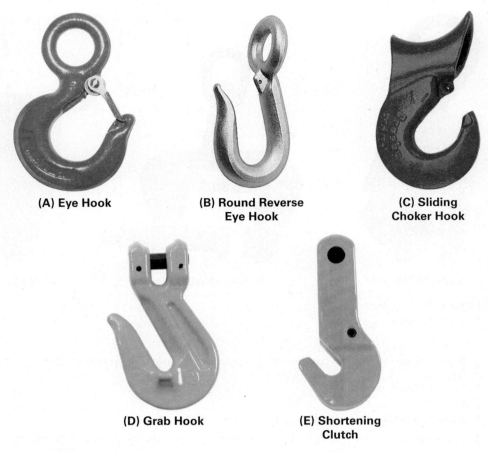

(A) Eye Hook

(B) Round Reverse Eye Hook

(C) Sliding Choker Hook

(D) Grab Hook

(E) Shortening Clutch

Figure 36 Rigging hooks.
Sources: Columbus McKinnon Corporation (36A); Courtesy of The Crosby Group LLC (36B–36C); Gunnebo Johnson Corporation (36D–36E)

1.3.4 Rigging Hooks

Rigging hook: An item of rigging hardware used to attach a sling to a load.

A **rigging hook** is an item of rigging hardware used to attach a sling to a load. There are many classes of rigging hooks used, but most rigging hooks fall into a few basic categories (*Figure 36*). Most rigging hooks have safety latches or gates to prevent slings or other connectors from accidentally coming out of the hook during use. Hooks that do not have safety latches are usually designed so that a latch can be added.

- Eye hooks are the most common type of end fitting hook. A rigging eye hook has a large eye that can accommodate large couplers.

- Reverse eye hooks position the point of the hook perpendicular to the eye.

- Sliding choker hooks are installed onto the sling when it is made. The hooks, which can be positioned anywhere along the sling body, are used to secure the sling eye in a choker hitch. Sliding choker hooks are available for steel chain slings.

- Grab hooks are used on steel chain slings. These hooks fit securely in the chain link, so that choker hitches can be made and chains can be shortened.

- Shortening clutches, a more efficient version of the grab hooks, provide a secure grab of the shortened sling leg with no reduction in the capacity of the chain because the clutch fully supports the links.

Hooks used for rigging must be made of drop- or hammer-forged steel. Although most rigging hooks have safety latches or gates, some hooks used for

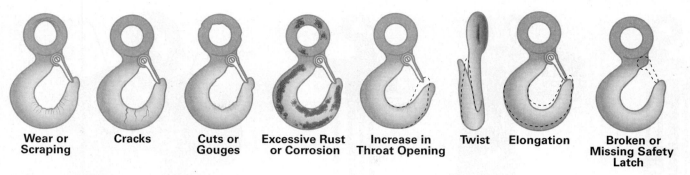

Figure 37 Common rigging hook defects.

special applications may not have them. If a safety latch is installed in a rigging hook, the latch must be in good working condition. Damaged safety latches can be easily replaced. Any damage to a safety latch must be reported to an instructor or supervisor.

When hooks are installed as end fittings, they must be inspected along with the rest of the sling before each use. Slings with hook-type end fittings need to be removed from service for any of the following defects (*Figure 37*):

- Wear, scraping, or abrasion
- Cracks
- Cuts, gouges, nicks, or chips
- Excessive rust or corrosion, meaning rust or corrosion that cannot be easily removed with a wire brush
- An increase in the throat opening of the hook—easy to detect if the hook is equipped with a safety latch, because the latch will no longer bridge the throat opening
- A twist in the hook
- An elongation of the hook
- A broken or missing safety latch

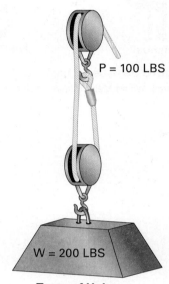

Types of Hoists

Figure 38 Block and tackle hoist system.

1.4.0 Hoists

A hoist is a device that uses pulleys or gears to provide a mechanical advantage for lifting a load, allowing objects to be lifted that cannot be lifted manually. Some hoists are mounted on trolleys and use electricity or compressed air for power.

In this section, a simple hoisting mechanism called a **block and tackle**, and a more complex hoisting mechanism called a chain hoist (*Figure 38*), are discussed.

A block and tackle is a simple rope-and-pulley system used to lift loads. By using fixed pulleys and a wire rope attached to a load, a rigger can raise and lower the load by pulling the rope or by using a winch to lift the load. A block and tackle assembly multiplies the power applied, so that a worker can lift a much heavier load than could be lifted without it.

Chain hoists may be operated manually or mechanically. There are three types of chain hoists (*Figure 39*): manual, electric, and pneumatic. Because electric and pneumatic chain hoists use mechanical power, they are known as powered chain hoists. All chain hoists use a gear system to lift heavy loads (*Figure 40*). The gearing is coupled to a sprocket that has a chain with a hook attached to it. The load is hooked onto a chain and the gearing turns the sprocket, causing the chain to travel over the sprocket and move the load.

Block and tackle: A simple rope-and-pulley system used to lift loads.

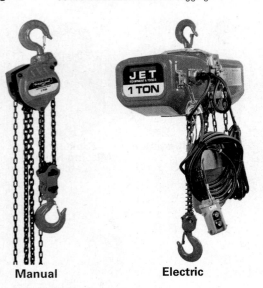

Figure 39 Types of chain hoists.
Source: Walter Meier Manufacturing Americas

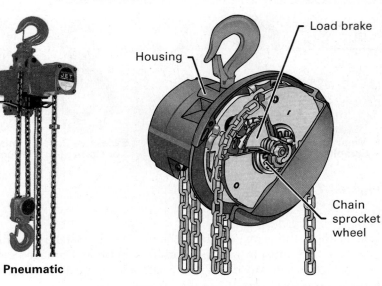

Figure 40 Chain hoist gear system.
Source: Columbus McKinnon Corporation

The hoist can be suspended by a hook connected to an appropriate anchorage point, or it can be suspended from a trolley system (*Figure 41*).

1.4.1 Operation of Chain Hoists

Chain hoists are operated by hand or by electric or pneumatic power. Some of the fundamental operating procedures for hand chain hoists and powered chain hoists are as follows:

- *Hand chain hoist* – To use a hand chain hoist, the rigger suspends the hoist above the load to be lifted, using either the suspension hook or the trolley mount. The rigger then attaches the hook to the load and pulls the hand chain drop to raise the load. The load will either rise or fall, depending on which side of the chain drop is pulled.

- *Powered chain hoist* – To use an electric or a pneumatic powered chain hoist, the rigger positions the chain hoist on the trolley above the load to be lifted, attaches the hook to the load, and uses the control pad to operate the hoist. Only qualified persons may use powered chain hoists.

1.4.2 Hoist Safety and Maintenance

General safety rules are presented in the *Basic Safety* module. In addition to those general safety rules, there are some specific safety rules for working with hoists. Observe the following guidelines:

- Always use the appropriate PPE when working with and around any lifting operations. This includes a hard hat, safety glasses, safety shoes, and a high-visibility vest (at a minimum).

- Make sure that the load is properly balanced and attached correctly to the hoist before a lift is attempted. Unbalanced loads can slide or shift, causing the hoist to fail.

- Keep gears, chains, and ropes clean. Improper maintenance can shorten the working life of chains and ropes.

- Lubricate gears periodically to keep the wheels from freezing up. All such maintenance work must be done by a competent person.

- Never perform a lift of any size without proper supervision.

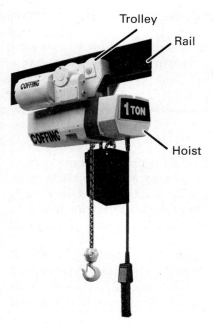

Trolley

Rail

1 TON

Hoist

Ratchet Lever Hoist

Figure 41 Hoist suspended from a trolley system.

Source: Coffing Hoists

Figure 42 Use ratchet lever hoists for vertical lifting.

Source: Walter Meier Manufacturing Americas

Never use a common come-along for vertical overhead lifting. Use a come-along only to move loads horizontally over the ground. Be careful not to confuse a come-along with a ratchet lever hoist. Ratchet lever hoists (*Figure 42*) have both a friction-type holding brake and a ratchet-and-pawl **load control** brake. Cable come-alongs have only a spring-load ratchet that holds the pawl in place to secure the cable. If the ratchet and pawl fail, an overhead load falls. Be certain that any device used to lift is designed and specified for that task. Among manufacturers, the terms *come-along* and *ratchet lever hoist* are often used in the description of a single item, creating confusion in the marketplace and on the job. In most cases, cable come-alongs are not designed for lifting loads.

Load control: The safe and efficient practice of load manipulation, using proper communication and handling techniques.

1.5.0 Hitches

In a rigging operation, the link between the load and the lifting device is often a sling made of synthetic, alloy steel chain, or wire rope material. The way the sling is arranged to hold the load is called the rigging configuration, or hitch. Hitches can be made using just the sling or by using connecting hardware, as well. There are three basic types of hitches:

- Vertical
- Choker
- Basket

One of the most important parts of the rigger's job is making sure that the load is held securely. The type of hitch the rigger uses depends on the type of load to be lifted. Different hitches are used to secure, for example, a load of pipes, a load of concrete slabs, or a load of heavy machinery.

Controlling the movement of the load once the lift is in progress is another important part of a rigger's job. Therefore, the rigger must also consider the intended movement of the load when choosing a hitch. For example, some loads are lifted straight up and then straight down. Other loads are lifted up, turned in midair, and then set down in a completely different place. This section examines how each of the three basic types of hitches is used to both secure a load and control its movement.

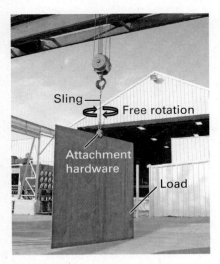

Figure 43 Single vertical hitch.
Source: Courtesy of The Crosby Group LLC

Tag line: Rope that runs from the load to the ground. Riggers hold on to tag lines to keep a load from swinging or spinning during the lift.

Bull ring: A single ring used to attach multiple slings to a hoist hook.

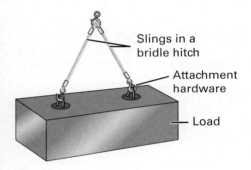

Figure 44 Bridle hitch.

WARNING!

All rigging operations are dangerous, and extreme care must be used at all times. A straight up-and-down vertical lift is every bit as dangerous as a lift that involves rotating a load in midair and moving it to a different place. Only a qualified person may select the hitch to be used in any rigging operation.

1.5.1   Vertical Hitch

The single vertical hitch is used to lift a load straight up. This configuration forms a 90-degree angle between the hitch and the load. With this hitch, some type of attachment hardware, such as a shackle, is needed to connect the sling to the load (*Figure 43*). The single vertical hitch allows the load to rotate freely. To prevent the load from rotating, some method of load control, such as a **tag line**, must be used.

Another classification of hitch is the bridle hitch (*Figure 44*). The bridle hitch consists of two or more vertical hitches attached to the same hook, master link, or **bull ring**. The bridle hitch allows the slings to be connected to the same load without the use of such devices as a spreader beam, which is a stiff bar used when lifting large objects with a crane hook.

The multiple-leg bridle hitch (*Figure 45*) consists of three or four single hitches attached to the same hook, master link, or bull ring. Multiple-leg bridle hitches provide increased stability for the load being lifted. A multiple-leg bridle hitch is always considered to have only two of the legs supporting the majority of the load and the rest of the legs balancing it.

1.5.2   Choker Hitch

The choker hitch is used when a load has no attachment points or when the attachment points are not practical for lifting. The choker hitch is made by wrapping the sling around the load and passing it through one eye to form a constricting loop around the load. In many configurations, the sling is wrapped around the load and passed through a shackle to form the constricting loop. In those situations, it is important that the shackle used in the choker hitch be oriented properly, as shown in *Figure 46*. Remember that the pressure of the load should never be placed on the pin of the shackle.

The choker hitch affects the capacity of the sling, reducing it by a minimum of 25 percent. This capacity reduction must be considered when choosing the

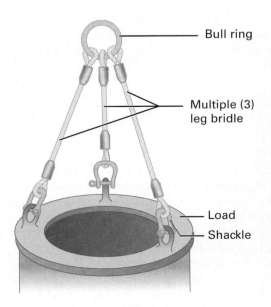

Figure 45 Multiple-leg bridle hitch.
Source: Volvo Construction Equipment (45B photo)

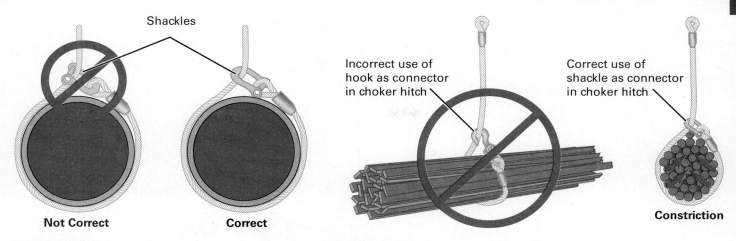

Shackles

Not Correct   **Correct**

Figure 46 Choker hitches.

Incorrect use of
hook as connector
in choker hitch

Correct use of
shackle as connector
in choker hitch

Constriction

Figure 47 Choker hitch constriction.

proper sling. Further, the choker hitch does not grip the load securely. It is not recommended for loose bundles of materials because it tends to push loose items up and out of the choker. Many riggers use the choker hitch for bundles, mistakenly believing that forcing the choke down provides a tight grip. In fact, it serves only to drastically increase the stress on the choke leg (*Figure 47*).

When an item more than 12 feet (≈3.5 m) long is being rigged, the general rule is to use two choker hitches spaced far enough apart to provide the stability needed to transport the load (*Figure 48*).

To lift a bundle of loose items such as pipes and structural steel, or to maintain the load in a certain position during transport, double-wrap choker hitches (*Figure 49*) may be useful. A double-wrap choker hitch is made by wrapping the sling completely around the load, and then wrapping the choke end around again and connecting to the running end like a conventional choker hitch. This enables the load weight to produce a constricting action that binds the load into the middle of the hitch, holding it firmly in place throughout the lift.

Forcing the choke down will drastically increase the stress placed on the sling at the choke point. The double-wrap choker uses the load weight to provide the

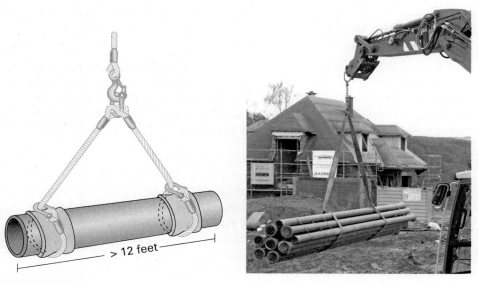

> 12 feet

Figure 48 Double choker hitches.
Source: Volvo Construction Equipment (48B photo)

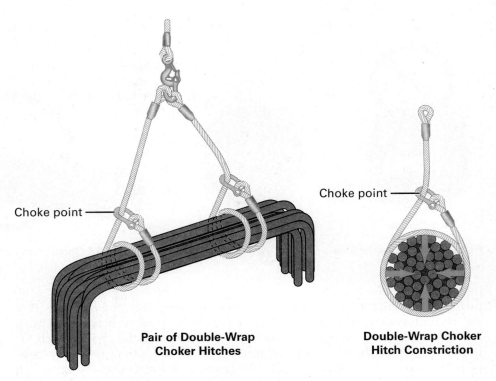

Choke point

Choke point

Pair of Double-Wrap Choker Hitches

Double-Wrap Choker Hitch Constriction

Figure 49 Double-wrap choker hitches.

constricting force, so there is no need to force the sling down into a tighter choke; it serves no purpose. As with a single choker hitch, lifting a load longer than 12 feet (≈3.5 m) requires two double-wrap choker hitches.

1.5.3 Basket Hitch

Basket hitches are very versatile and can be used to lift a variety of loads. A basket hitch is formed by passing the sling around the load (or, in some cases, through the load) and placing both eyes in the hook (*Figure 50*). Placing a sling into a basket hitch has the effect of doubling the capacity of the sling. This is because the basket hitch creates two sling legs from one sling.

A double-wrap basket hitch combines the constricting power of a double-wrap choker hitch with the capacity advantages of a basket hitch. This means

---

Around the World

Crane Hand Signals

Hand signals of all types have different meanings around the world. In most cases, the issue only causes some confusion and embarrassment. However, when it comes to crane operations, incorrect hand signals can lead to disaster. Major lifting and rigging operations take place around the world, and not every member of the lift crew, including the crane operator, has received the same training or uses the same signals. On many projects, team members are not even from the same country.

A new set of hand signals has been published for adoption by the International Organization for Standardization (ISO) to make operations safer and more efficient worldwide. ISO 16715, entitled "Cranes—Hand Signals Used With Cranes," was developed by a subcommittee with international experts from countries such as Russia, Brazil, Australia, and the United States—31 countries in all. However, the objective was not to replace all unique signals worldwide. The real objective was to create a set of standardized signals used on job sites that involve multinational rigging and crane operation teams. Common operations in each country, involving local or regional workers, will likely continue to use the hand signals familiar to that region.

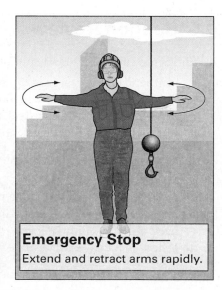

Figure 50 Basket hitch.
Source: SME Brett Richardson, Starcon

Figure 51 Emergency Stop hand signal.

it is able to hold a larger load more tightly. A double-wrap basket hitch requires a considerably longer sling length than a double-wrap choker hitch. If it is necessary to join two or more slings together, the load must be in contact with the sling body only, not with the hardware used to join the slings. The double-wrap basket hitch provides support around the load. As is true of the double-wrap choker hitch, the load weight provides the constricting force for the hitch.

1.5.4   The Emergency Stop Signal

A major aspect of rigging safety involves maintaining clear communication between the crane operator and the designated signal person on the ground. This communication is normally accomplished using common signals—either verbal signals given by radio or hand signals. Hand signals used in rigging operations in the United States have been developed and standardized by the American Society of Mechanical Engineers (ASME) for all cranes. With the exception of the Emergency Stop signal, hand signals can only be given by the designated signal person on the ground.

The Emergency Stop signal used in rigging operations is shown in *Figure 51*. In the event of an emergency, this signal can be given by anyone on the ground within sight of the crane operator. The Emergency Stop signal is made by extending both arms horizontally with palms down, and then quickly moving the arms back and forth by repeatedly extending and retracting them.

> **CAUTION**
>
> A basket hitch should not be used to lift loose materials. Loads placed in a basket hitch should be balanced.

> **WARNING!**
>
> Although this module provides instruction and information about common rigging equipment and hitch configurations, it does not provide any level of certification. Any questions about rigging procedures or equipment should be directed to an instructor or supervisor. Always refer to the manufacturer's instructions for any type of rigging equipment.

1.0.0 Section Review

1. Slings are commonly made out of _____.
   a. synthetic fibers, cast-iron links, and wire rope
   b. fiberglass strands, exotic metals, and synthetic rope
   c. synthetic material, alloy steel, and wire rope
   d. carbon fiber, stainless steel, and twisted pair cable

2. If the identification tag on a synthetic sling is missing or illegible, the sling must be _____.
   a. re-tagged by a qualified rigger on site
   b. removed from service immediately
   c. used only for minimal loads
   d. tested on a load to determine its capacity

3. The size of a shackle is determined by the diameter of its _____.
   a. body
   b. shank
   c. eye
   d. pin

4. A type of rigging equipment that uses a pulley system to provide a mechanical advantage for lifting a load is a _____.
   a. fulcrum
   b. hitch
   c. hoist
   d. come-along

5. When a load has no attachment points or when the attachment points are not practical for lifting, it is best to use a _____.
   a. bridle hitch
   b. vertical hitch
   c. multiple-leg hitch
   d. choker hitch

Module 00106 Review Questions

1. Identification tags for slings must include the _____.
   a. type of protective pads to use
   b. type of damage sustained during use
   c. color of the tattle-tail
   d. manufacturer's name or trademark

2. The type of wire rope core that is susceptible to heat damage at relatively low temperatures is the _____.
   a. fiber core
   b. strand core
   c. independent wire rope core
   d. metallic link supporting core

3. Synthetic slings must be inspected _____.
   a. once every month
   b. visually at the start of each work week
   c. before every use
   d. once wear or damage becomes apparent

4. An alloy steel chain sling must be removed from service if there is evidence that _____.

   a. the sling has been used in different hitch configurations
   b. replacement links have been used to repair the chain
   c. the sling has been used for more than one year
   d. strands in the supporting core have weakened

5. A piece of rigging hardware used to couple the end of a wire rope to eye fittings, hooks, or other connections is a(n) _____.

   a. eyebolt
   b. hitch
   c. shackle
   d. U-bolt

6. A lifting clamp is most likely to be used to move loads such as _____.

   a. steel plates
   b. piping bundles
   c. concrete blocks
   d. plastic tubing

7. Chain hoists are able to lift heavy loads by utilizing a _____.

   a. rope and pulley system
   b. rigger's strength
   c. stationary counterweight
   d. gear system

8. Before attempting to lift a load with a chain hoist, make sure that the _____.

   a. hoist is secured to a come-along
   b. load is properly balanced
   c. tag lines are properly anchored
   d. tackle is connected to its power source

9. A hitch configuration that allows slings to be connected to the same load without using a spreader beam is a _____.

   a. double-wrap hitch
   b. choker hitch
   c. bridle hitch
   d. basket hitch

10. To make the Emergency Stop signal that is used by riggers, extend both arms _____.

   a. horizontally with palms down and quickly move both arms back and forth
   b. directly in front and then move both arms up and down repeatedly
   c. vertically above the head and wave both arms back and forth
   d. horizontally with clenched fists and move both arms up and down

---

Answers to Section Review Questions and Module Review Questions are found in *Appendix A*.

Basic Communication Skills

Source: AzmanL/E+/Getty Images

Objectives

Successful completion of this module prepares you to do the following:

1. Describe the communication, listening, and speaking processes and their relationship to job performance.
   a. Describe the communication process and the importance of listening and speaking skills.
   b. Describe the listening process and identify good listening skills.
   c. Describe the speaking process and identify good speaking skills.
2. Describe good reading and writing skills and their relationship to job performance.
   a. Describe the importance of good reading and writing skills.
   b. Describe job-related reading requirements and identify good reading skills.
   c. Describe job-related writing requirements and identify good writing skills.

Performance Tasks

Under supervision, you should be able to do the following:

1. Perform a given task after listening to oral instructions.
2. Fill out a work-related form provided by your instructor.
3. Read and interpret a set of instructions for properly donning a safety harness and then orally instruct another person on how to don the harness.

Overview

The construction professional communicates constantly. The ability to communicate skillfully will help to make you a better worker and a more effective leader. This module provides guidance in listening to understand, and speaking with clarity. It explains how to use and understand written materials, and it also provides techniques and guidelines that will help you to improve your writing skills.

Industry Recognized Credentials

If you are training through an NCCER-accredited sponsor, you may be eligible for credentials from NCCER's Registry. The ID number for this module is 00107. Note that this module may have been used in other NCCER curricula and may apply to other level completions. Contact NCCER's Registry at 888.622.3720 or go to **www.nccer.org** for more information.

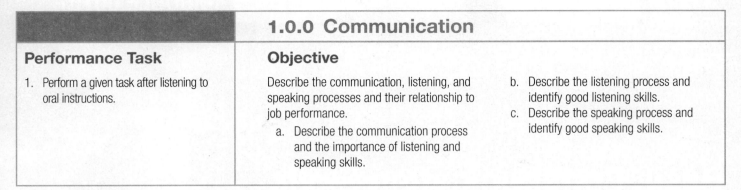

1.0.0 Communication

Performance Task

1. Perform a given task after listening to oral instructions.

Objective

Describe the communication, listening, and speaking processes and their relationship to job performance.

a. Describe the communication process and the importance of listening and speaking skills.

b. Describe the listening process and identify good listening skills.

c. Describe the speaking process and identify good speaking skills.

Every construction professional learns how to use tools. Depending on your trade, the tools you use could include welding machines and cutting torches, press brakes and plasma cutters, or surveyor's levels and pipe threaders. However, some of the most important tools you will use on the job are not tools you can hold in your hand or put in a toolbox. These tools are your abilities to read, write, listen, and speak.

At first, you might think that these are not really construction tools. They are things you already learned how to do in school, so why do you have to learn them all over again? The types of communication that take place in the construction workplace are very specialized and technical, just like the communications between pilots and air traffic control. Good communications result in a job done safely—a pilot hears and understands the message to change course to avoid a storm, and a construction worker hears and understands the message to install a water heater according to the local code requirements. In a way, you are learning another language, a special language that only trained professionals know how to use. Even though you will use a professional language that other people may not understand, the same communication skills apply to all professions, whether doctors, builders, managers, or mechanics.

The following are some specific examples of why these skills are so important in the construction industry:

- *Listening* — Your supervisor tells you where to set up safety barriers, but because you did not listen carefully, you missed a spot. As a result, your co-worker falls and is injured.

- *Speaking* — You must train two co-workers to do a new task, but you mumble, use words they don't understand, and don't answer their questions clearly. Your co-workers do the task incorrectly, and all of you must work overtime to fix the mistakes.

- *Reading* — Your supervisor tells you to read the manufacturer's basic operating and safety instructions for the new drill press before you use it. You don't really understand the instructions, but you don't want to ask him. You go ahead with what you think is correct and damage the drill press.

- *Writing* — Your supervisor asks you to write up a material takeoff (supply list) for a project. You rush through the list and don't check what you've written. The supplier delivers 250 feet of PVC piping cut to your specified sizes instead of 25 feet.

As you can see, good communication on the work site has a direct effect on safety, schedules, and budgets. A good communications toolbox is a badge of honor; it lets everyone know that you have important skills and knowledge. And like a physical toolbox, the ability to communicate well verbally and in writing is something that you can take with you to any job. You will find that good communications skills can help you advance your career. This module introduces you to the techniques you will need to read, write, listen, and speak effectively on the job.

1.1.0 The Communication Process

There are two basic steps to clear communication (*Figure 1*). First, a sender sends a spoken or written message through a communication channel to a receiver (examples of communication channels include meetings, phones, two-way radios, and email). When the receiver gets the message, he or she figures out what it means by listening or reading carefully. If anything is not clear, the receiver gives the sender feedback by asking the sender for more information.

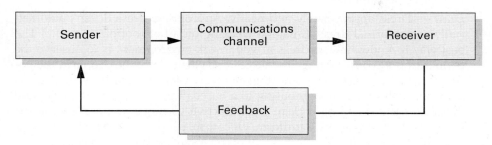

Figure 1 The communication process.

This process is called two-way communication, and it is the most effective way to make sure that everyone understands what's going on. It sounds simple, doesn't it? So why is good communication so hard to achieve? When we try to communicate, a lot of things—called noise—can get in the way. Following are some examples of communication noise:

- The sender uses work-related words, or **jargon**, that the receiver does not understand.
- The sender does not speak clearly.
- The sender's written message is disorganized or contains mistakes.
- The sender is not specific.
- The sender does not get to the point.
- The receiver is tired or distracted or just not paying attention.
- The receiver has poor listening or reading skills.
- Actual noise on the construction site makes it physically hard to hear a message.
- There is a mechanical problem with the equipment used to communicate, such as static on a phone or radio line.

Jargon: Specialized terms used in a specific industry.

1.1.1 Nonverbal Communication

It is obvious that humans communicate with their words, known technically as verbal communication. If a mechanic asks for a ratchet set, he or she expects to get it because the words have a shared meaning between the listener and speaker. However, did you know that people communicate constantly without using any words at all? This kind of communication is called **nonverbal communication**. It is hard to communicate some kinds of information with nonverbal communication. For example, try asking for that ratchet set without any words. For expressing attitudes and emotion, nonverbal communication is a very powerful form of communication.

Imagine that you are talking to a doctor in a clinic. The whole time you are talking, he is yawning, checking his cell phone messages, and cleaning his fingernails. After you finish, he writes out a prescription. Do you trust the diagnosis and the value of the prescription? His nonverbal communication told you that he was bored and inattentive. Perhaps the doctor missed the most important thing you said about your condition, including an allergy to certain medications.

Similarly, you can express feelings and attitudes in a variety of ways without intention. You may show you are nervous by fidgeting in your chair or fiddling with your hands. You may show you are angry by raising your voice, folding your arms, and furrowing your eyebrows. You may show you are happy to

Nonverbal communication: All communication that does not use words. This includes appearance, personal environment, use of time, and body language.

see someone by smiling widely and giving him or her a warm handshake. The ways for a person to communicate nonverbally are limitless.

There are several basic categories of nonverbal communication:

- *Grooming* — Generally, people who maintain an attractive appearance have more successful careers. A groomed appearance communicates self-discipline and awareness. Shaving or keeping a trimmed beard could make the difference between getting promoted at work or being passed up by newcomers. Likewise, messy hair can communicate that someone is incompetent and unable to take on new responsibilities.

- *Dress* — Dress appropriately and neatly. Appropriate dress doesn't automatically mean formal dress. (Imagine working on a hot rooftop in a suit.) The best way to know how to dress in your work environment is to observe the people around you who are most successful, particularly the people in the position that you hope to obtain next. Even on occasions in which casual dress is allowed, such as a company outing to a baseball game, you must make sure that your casual clothing is clean and attractive. Casual does not mean messy. Many workplaces require a uniform. Of course, people working in these environments should always arrive at work with a neat and clean uniform.

- *Condition of one's personal environment* — People make their spaces their own by the way they arrange things even if the space does not belong to them. An office worker can show a sloppy work ethic by having binders, pens, paper clips, and candy wrappers strewn about his or her cubicle. A technician also might show a lack of responsibility by having gloves, coupons, magazines, and bags from fast food restaurants on the dashboard of his or her work vehicle. Whoever sees those environments unconsciously makes judgments about the character of the people who work there.

- *Use of time* — Clearly, people show respect and care by arriving on time or early to their scheduled events. Workers who show up late may be considered lazy or inconsiderate by those around them. Further, they may hinder the productivity of the entire organization by making others wait for them. Not only should you arrive or start on time, but when you hold meetings, you should also try to end on time. Doing so shows a humility on your part that others will appreciate.

- *Facial expressions* — One of the main ways that humans express and read emotions is through facial expressions. A smile can express interest and excitement. A frown could express displeasure or pain. An expressionless face expresses an emotion too, perhaps boredom or the desire to get away from whoever is talking. Of course, eyes are known as the window of the soul. Direct eye contact can express interest, understanding, intelligence, and confidence. On the other hand, a lack of eye contact could show inattention, a lack of confidence, or deceit. A person should not look into the eyes of another person the entire time while speaking, however. Too much eye contact could give the impression of initiating a challenge or trying to dominate. There is no solid rule for what amount of eye contact is appropriate. Different cultures have different rules. For example, in the West, authority figures expect subordinates to look at them when they are reprimanding them. In parts of Asia, however, etiquette requires the one being scolded to look down at the ground as a sign of respect. In the United States, people look at each other's eyes 50 to 60 percent of the time as they communicate.

- *Posture and gestures* — Slouching may feel comfortable, but it may also make people think of you as a sloppy person. Likewise, folding your arms in front of you may make you feel warmer on a cold day, but it will make you seem distant from others and unwilling to talk. Having a confident and powerful posture shows other people that you are confident and powerful. What if you feel timid and weak? Assume a powerful posture anyway. Research shows that people who kept a certain powerful posture for a few minutes—such as leaning slightly over a desk with their hands wide apart on the desk—experienced less stress and actually behaved more boldly when doing a task later. Those who had practiced low-power poses for a few minutes—such as folding their hands in their lap or touching their neck—showed a higher level of stress and behaved more timidly.

- *Physical distance* — Various cultures have different accepted personal distances. People of some countries stand close together while speaking, whereas others stand further apart. Also, different social situations call for a different personal range. People stand closer to their close friends and family members than they do to their business associates. Here is one rule of thumb to follow: let the more powerful person choose the distance. A manager or company owner may politely come and stand close to an employee, but an employee would be rude to do the same to his or her manager.

When it comes to safety and the accuracy of information communicated on the job, there is no doubt that what you say plays a greater role than the way you say it. However, beyond that, your nonverbal communication habits have a major effect on the quality of what is said and how it is received.

1.1.2 Listening and Speaking Skills

Every day on the job can be a learning experience. The more you learn, the more you will be able to help others learn, too (*Figure 2*). An effective method of learning and teaching is through verbal communication—that is, through speaking and listening. As a construction professional, you need to be able to state your ideas clearly. You also need to be able to listen to and understand ideas that other people express. The following are some of the ways that verbal teaching and learning take place on the job:

- *Giving and taking instructions* — One worker may read the steps in a calibration process while a second worker accomplishes the task.
- *Offering and listening to presentations* — Equipment manufacturers may visit the job site or offices to provide operating instruction for a new piece of equipment.
- *Participating in team discussions* — Safety is often discussed among teams; offer your input.
- *Talking with your co-workers and your supervisor* — Listen carefully as they speak, without distracting yourself by thinking of a response too quickly. A slight pause in the conversation to prepare a response after a speaker is finished actually encourages others to listen more carefully to you.
- *Talking with clients* — Again, listening skills are critical.

Before we discuss some of the ways to become a more effective listener and speaker, evaluate your current speaking and listening skills by completing the self-assessment quizzes in *Figure 3* and *Figure 4*.

Figure 2 Teaching and learning are often accomplished by speaking and listening.
Source: Courtesy of Oak Ridge National Laboratory

A Sense of Humor

A special tip: Maintain a sense of humor! A good sense of humor will get you through many situations. As a construction professional, you should always take yourself seriously. This means speaking well and conveying the proper professional attitude. But your listeners will always appreciate you more if you show them that you have a sense of humor and have a light side to you. Humor can diffuse tension and relieve frustration over things that aren't working properly. Remember, though, never to tell off-color or offensive jokes or make jokes at another person's expense. Also, never play practical jokes, as they can lead to accidents and injuries on the job.

Listening in the Classroom

When you are in the classroom, be aware of things that affect your ability to listen well. Take action to correct these problems. Is someone on the other side of the room speaking too softly? Ask other classmates to face the class when they speak and to speak loudly. Is there noise out in the hall? Ask permission to close the door to shut out noise from outside. Did your instructor say something you did not understand? Ask your instructor to explain things you don't understand.

Are You a Good Listener?

Do you have good listening habits? Take the following self-assessment quiz to find out.
Be sure to answer each question honestly.

	Always	Sometimes	Rarely
1. I maintain eye contact when someone is talking to me.	☐	☐	☐
2. I pay attention when someone is talking to me.	☐	☐	☐
3. I ask questions when I don't understand something I hear.	☐	☐	☐
4. I take notes when receiving instructions.	☐	☐	☐
5. I repeat instructions my supervisor has given me to make sure I understand them.	☐	☐	☐
6. I nod my head or say I understand to show others I am listening to them.	☐	☐	☐
7. I let others speak without interrupting.	☐	☐	☐
8. I move to a quieter spot or ask someone to speak up if I am in a noisy location.	☐	☐	☐
9. I put aside what I am doing when someone is speaking to me.	☐	☐	☐
10. I listen with an open mind.	☐	☐	☐

Figure 3 Listening skills self-assessment.

Are You a Good Speaker?

How good are your speaking skills? This self-assessment quiz will help you understand your speaking strengths and weaknesses.

	Always	Sometimes	Rarely
1. When giving instructions to co-workers, I explain words they might not understand.	☐	☐	☐
2. When giving instructions for a task with several steps, I organize my thoughts first, then give the instructions.	☐	☐	☐
3. I give more details when explaining a task to inexperienced co-workers.	☐	☐	☐
4. When giving instructions to others, I try to keep from sounding like a know-it-all.	☐	☐	☐
5. When giving instructions to others, I encourage them to ask questions about anything they don't understand.	☐	☐	☐
6. I am patient and will explain instructions more than once if necessary.	☐	☐	☐
7. I try to speak more carefully when giving instructions to a co-worker for whom English is a second language.	☐	☐	☐
8. When speaking on the phone or over a two-way radio, I repeat instructions and spell out words when necessary.	☐	☐	☐
9. If someone asks me a question I don't know the answer to, I admit it and then try to find the answer.	☐	☐	☐
10. When I give instructions to others, I am confident, upbeat, and encouraging.	☐	☐	☐

Figure 4 Speaking skills self-assessment.

At this stage in your career, you will probably do more listening than speaking. You may be wondering why it is so important to be a good listener. The answer is simple: experience. People learn by listening, not by speaking. You are only beginning to learn how the construction industry works, and there is a lot to learn! Teachers, supervisors, and experienced workers can guide you to make sure you are learning what you need to know (*Figure 5*).

1.2.0 Active Listening on the Job

You might think that listening just happens automatically, that someone says something and someone else hears it. However, real listening, the process not only of hearing, but of understanding what is said, is an active process. You have to be involved and pay attention to really listen. In other words, understanding begins with **active listening**. You must develop good listening skills to be able to listen actively. This section presents some tips and suggestions that you can use to develop good listening skills.

Active listening: A process that involves respecting others, listening to what is being said, and understanding what is being said.

First of all, you should understand the possible consequences of not listening. Poor listening skills can cause mistakes that waste time and money (*Figure 6*), and may even result in injury or loss of life. Recognize that other people may have important things to say. Even if what they are currently saying is boring, they may happen to mention something vitally important, like what to do when an alarm goes off.

Refuse to allow yourself to be distracted. If a work-related conversation or presentation is taking place, be there in the room listening, not darting around from place to place in your mind. If it helps you to stay focused, keep a pad of paper with you. When you are sidetracked by thoughts of something you need to do later, jot down a couple words about it. Then put the matter out of your mind.

One way to stay focused is to show that you are listening with your **body language**, that is your facial expressions, your posture, and your gestures (*Figure 7*). Not only does body language communicate something to other people, it affects you as well. It was mentioned earlier that behaving confidently can actually make you feel more confident. Similarly, acting like you are listening can actually help you listen better. While you listen, maintain eye contact and acknowledge what is being said by nodding or making appropriate sounds to show that you are paying attention.

Body language: A person's facial expression, physical posture, gestures, and use of space, all of which communicate feelings and ideas.

While the other person is speaking, do not interrupt or criticize. Instead, wait until he or she has finished, and then ask plenty of questions, especially if he

Figure 5 Your supervisor can help you learn what you need to know.
Source: M.C. Dean, Inc.

Figure 6 Listen carefully and ask questions to make sure you understand.

Figure 7 Body language shows whether you are paying attention.
Source: Ed Gloninger

or she gave incomplete or unclear instructions. Asking questions can minimize misunderstandings and maximize your memory of the content. Of course, your mind will never remember perfectly, so take notes on what is said, especially if the content is technical in nature—dealing with numbers, times, amounts, temperatures, or other such data.

Paraphrasing: Expressing something heard or read using different words.

Finally, end a conversation by **paraphrasing** what you heard back to the conversation partner. Paraphrasing is the act of repeating what you heard in your own words. By using your own words, you can check the accuracy of your understanding. Therefore, it is important not to simply repeat what the person said exactly. The original speaker must listen carefully to ensure the paraphrased information is accurate.

Imagine how many hours of wasted work you can avoid by just confirming what you think you hear. For example, your supervisor may tell you, "Recalibrate the machines according to the specifications on this paper." You can respond by saying, "So you want me to recalibrate these three machines in this building according to the chart posted here." Then he or she replies, "Oh, no! I mean, recalibrate the machines like these in the building next door, and follow the chart on the back of this paper, since they are slightly different models." Just providing a simple summary in your own words of what was said could save time, money, or even someone's life. Paraphrase instructions whenever you can. It completes the act of listening just as installing a roof completes the construction of a building.

1.2.1 Barriers to Listening

As you have learned, listening well takes some work on your part. However, even when you have mastered effective listening skills, you will still have to overcome some barriers that will keep the message from getting through. Just like the warning signs on a construction site that indicate hazards to be avoided, there are signs that indicate problems in the process of listening and understanding (*Figure 8*). Some of those barriers are listed below, along with tips to overcome them:

Figure 8 Learn to read the warning signs of listening problems.
Source: Courtesy of Oak Ridge National Laboratory

- *Emotion* — When angry or upset, humans tend to stop listening actively. Try counting to 10 or asking the speaker to excuse you for a minute. Go get a drink of water and calm down. When you are ready, come back and focus your mind on the task to be done.

- *Boredom* — Maybe the speaker is dull or overbearing. Maybe you think you know it all already. There is no easy tip for overcoming this barrier. You just have to force yourself to stay focused. Keep in mind that the speaker has important information you need to hear.

- *Distractions* — Anything from too much noise and activity on the site to problems at home can steal your attention. If the problem is noise, ask the speaker to move away from it. If a personal problem is keeping you from listening, concentrate harder on staying focused. In some cases it may help if you explain to your supervisor why you are having trouble concentrating.

- *Your ego* — Do you finish people's sentences for them? Do you interrupt others a lot? Do you think about the things you are going to say next instead of listening? That is your ego putting itself squarely between you and effective listening. Be aware of your ego and try to tone it down a bit so you can get the information you need.

1.3.0 Speaking on the Job

Although you can use the skills presented in this module to give a speech if necessary, the term *speaking skills* does not refer merely to your ability to give a speech or make a presentation to a group of people. It also refers to your ability to communicate effectively, one-on-one, to others on the job every day.

Effective listening depends on effective speaking. After all, you cannot be expected to understand what has not been made clear to you. Look at the following examples of sentences spoken by one worker to another. Which one is the clearest and most effective? Which one would you like to hear if you were the listener?

"Hand me that tool there."

"Hand me the grinder on that bench."

"Hand me the 4-inch angle grinder that's on the bench behind you."

The third example has enough information for you to identify the correct tool and its location. You do not have to ask the speaker, "Which tool? Where is it?" You will not accidentally give the other worker the wrong tool. As a result, time is not wasted trying to clear up confusion. The time it takes for someone to stop what he or she is doing and explain something again because it was not clear the first time is time that the job is not getting done. Time lost this way can add up quickly.

Ten Tips for Dealing with Conflict

Conflict can cause emotions to heat up, it can ruin relationships, and it can decrease productivity. Maybe you don't care to be friends with the person who bothers you at work. However, you still have to get along in order to accomplish your work. Allowing conflicts to get worse can only make your work less pleasurable and less effective. Here are some tips for dealing with conflicts at work:

- *Deal with it quickly.* While it is still a small problem, it will be easier to solve. Further, your frustration will not have increased to the breaking point.

- *Set aside a time to discuss the problem.* Dedicating a time later will help emotions cool and provide a neutral emotional atmosphere. You will also have more time to discuss the issue fully.

- *Discuss the problem with civility.* As the old saying goes, "A soft answer turns away wrath." Getting angry will only make the other person angry as well, escalating the situation.

- *Ask questions first.* If a co-worker had a reason for doing what made you angry, you should know what it is. Maybe just knowing why the person behaved that way will solve the problem.

- *Never say "never."* Do not say "always" either. "You never help me" is probably not true, and it will make the other person feel defensive. "You always criticize me" will also make the other person angry.

- *Always focus on specifics, not generalities.* Notice the difference between, "You are a careless person who hurts the whole team," and, "On the last two jobs we had, you forgot to bring the key for the building. We all had to wait while you went back for it. Is there anything you can do or I can do to make sure this doesn't happen again?"

- *Do not bring up things from the past that do not relate to the current situation.* Bringing up something that the co-worker did last year will only cause his or her temper to flare up.

- *Put yourself in the other person's shoes.* Understand what the other person needs in order to come out of the conflict happy. One of the main things people need is an intact sense of honor. If you publicly insult that person or force him or her with an overwhelming use of authority or even logic, you will implant a grudge that will inevitably hurt you later. Beyond giving the person the honor he or she needs, see if you can also compromise on some of the issues you disagree on.

- *Praise the person's good points.* This will show that you are not attacking his or her character, and it will help focus attention on the issue rather than the interpersonal tension.

- *Take the blame for your part of the problem.* Sincerely accepting some blame for the issue can make a huge difference. However, do not take the blame in a self-righteous or manipulative manner in order to shame the other person into submission.

One of the best ways to learn to speak effectively is to listen to someone who speaks well. Think about what makes that person such an effective speaker. Is it the person's choice of words? Or perhaps their body language? Or their ability to make something complex sound simple? Keep the following things in mind when you speak, and they will make a difference for you:

- Think about what you are going to say before you say it, but not at the expense of listening actively.

- Follow the old advice about giving speeches, "Tell them what you are going to tell them. Tell them. Then tell them what you told them." This means you should introduce your topic with a brief summary, then share all of the detailed information, and finally end with a paraphrased summary of what you said.

- As with writing, take time to organize your ideas logically.

- Choose an appropriate place and time. For example, if you need to give detailed assembly instructions to your team, pick a quiet place, and do not hold the meeting just before lunch or the end of the shift.

- Encourage your listeners to take notes if necessary.

- Do not over-explain if people are already familiar with the topic.

- Always speak clearly, and maintain eye contact with the person or people you are speaking to.

- Do not talk on the phone, send text messages, or listen to music while communicating with the work crew.

- When using jargon be sure that everyone knows what the term means.

- Give your listeners enough time to ask questions, and take the time to answer questions thoroughly.

- When you are finished, make sure that everyone understands what you were saying.

Keep these things in mind when you speak, and they will make a difference for you.

1.3.1 Placing Telephone Calls

You may remember when telephones were anchored to walls and desks. To make or receive a phone call, you had to stop what you were doing and go to the telephone. Today, cell phones allow you to make and receive calls from just about anywhere. A cell phone can be a useful tool on the job site, but keep in mind the following guidelines:

- Cell phones can distract you from your job, so never make or receive personal calls while working.

- Wait until a designated break time to make or receive calls.

- Do not operate cell phones where they would pose a safety hazard, such as while operating a piece of machinery, a power tool, or driving a vehicle.

Going Green

Recycling Used Cell Phones

Recycling your mobile devices (cell phones, pagers, PDAs) is a great way to save energy and protect the environment. These mobile devices contain toxic materials. Recycling keeps them out of landfills, cutting down on air and water pollution and greenhouse gas emissions. These devices contain precious metals, copper, and plastics, which require energy to manufacture and mine.

Over 100 million cell phones are retired each year. The EPA estimates that the energy saved from recycling cell phones alone could power more than 194,000 households for one year.

Your community benefits from cell phone donations. Cell phones can be reconditioned for reuse. They are often donated to charities and law enforcement, and for emergency use by individuals. Recycling is free and some organizations will even pay for your old cell phone.

- Be aware of the regulations regarding cell phone use at your workplace. While some companies allow cell phone use, many others do not even allow a phone to be in your possession. Companies make these rules to avoid accidents and wasted time.

- Recognize that cell phone cameras can be a potential threat to intellectual property. Workers may photograph secret equipment or data and sell the photographs, causing the company to lose its competitive edge. This is especially worth noting on projects for the government or military. Obviously, the military fiercely protects its secrets for good reason, and may ban employees and contractors from taking cell phones into certain areas.

When you speak to people face-to-face, you can see them and judge how they react to what you say. When you are on the telephone, you don't have these clues. Effective speaking is all the more important in such cases.

When making a call, keep the following points in mind:

- Start by identifying yourself and ask who you are speaking to.
- Speak clearly and explain the purpose of your call.
- Take notes to help you remember the conversation later (*Figure 9*).

If you leave a message for someone, remember the following:

- Keep it brief.
- Prepare your message ahead of time so you will know what to say.
- Be sure to leave a number where you can be reached and the best time to reach you.

Figure 9 Take notes to help you remember important details.

1.3.2   Receiving Telephone Calls

How you answer a phone call is just as important as how you place a phone call. Remember to be professional and courteous when answering your phone because you don't know who is going to be on the other end of the line. When you receive a phone call, remember the following guidelines:

- Don't just say "Hello." Identify yourself immediately by giving your name and the company name.
- Don't keep people on hold for long. People generally resent it. Instead, ask the caller if you can call back at a later time.
- Transfer calls courteously and introduce the caller to the recipient.
- Keep your calls brief.
- If the call is of a personal nature, continue working and do not answer. There are other opportunities throughout the day when you can return the call without interfering with the job or creating a safety hazard.
- Finally, never talk on the phone in front of co-workers, supervisors, or customers. This is usually considered rude and unprofessional.

Did You Know?

Cultural Interpretation

Communication is culturally diverse. Our experiences and surroundings, along with context, individual personality, and mood, help to form the ways we learn to speak and give nonverbal messages. For instance, in Europe the correct form for waving hello or goodbye is to keep your hand and arm stationary, palm out, with fingers moving up and down. In America, the common wave is the whole hand in motion moving back and forth. In Europe, the common American hand gesture for a wave means no, and in Greece it is considered an insult. It is important to remember that not all people communicate in the same way, so be aware of your surroundings when choosing your words and actions.

1.0.0 Section Review

1. Making meetings last longer than they were scheduled is a good way to show that you are a diligent person that wants to encourage others to work hard, too.

   a. True
   b. False

2. If the content of a speech or conversation is technical in nature, including lots of data such as numbers, be sure to _____.

   a. take notes
   b. ask others to summarize
   c. check to see if the information is true
   d. have the speaker repeat the key information

3. When is the best place and time to give assembly instructions to the team?

   a. In a quiet place when they are fresh and not too tired
   b. In the lunchroom while they are eating
   c. In the work area right at the end of their shift before they go home
   d. In the break room while people are getting snacks and watching TV

4. What of the following is banned in some workplaces because it can cause wasted time, accidents, and the theft of intellectual property?

   a. Pagers
   b. Watches
   c. Hearing aids
   d. Cell phones

5. When making a telephone call, start by _____.

   a. making small talk
   b. explaining the purpose of your call
   c. identifying yourself
   d. asking who you are talking to

2.0.0 Reading and Writing

Performance Tasks

2. Fill out a work-related form provided by your instructor.
3. Read and interpret a set of instructions for properly donning a safety harness and then orally instruct another person on how to don the harness.

Objective

Describe good reading and writing skills and their relationship to job performance.

a. Describe the importance of good reading and writing skills.
b. Describe job-related reading requirements and identify good reading skills.
c. Describe job-related writing requirements and identify good writing skills.

Imagine for a moment that you wake up one morning and head to work just like any other day, except for one difference: writing has not yet been invented. You grab a diagram (which of course has no notes on it) for installing a heating system, but you do not understand one part of the drawing. You head back to the office to ask the engineer what he meant. He starts to explain, but then he cannot clearly remember what he had discussed with the client. The discussion with the client was so long, and of course, there is no formal written contract or even an informal list of what the client wanted. Therefore, you and the engineer go to the client's office together to resolve the issue.

Although the Inca existed during the Bronze Age without a system of writing, in our complex modern society, the written word is indispensable. Reading and writing is not just for academics. Almost all Americans read, if not write, dozens of times per day regardless of whether they are plumbers, professors, statisticians, or stay-at-home parents.

2.1.0   The Importance of Reading and Writing Skills

The construction industry, just like any other, depends on written materials of all kinds to carry on business, from routine office paperwork to construction drawings and building codes (*Figure 10*). Written documents allow workers to follow instructions accurately, help project managers ensure that work is on schedule, and enable the company to meet its legal obligations. As a construction professional, you need to be able to read and understand the written documents that apply to your work, whether they are printed in a book or letter, or on a computer. Further, as you gain professional experience, you will eventually take on responsibilities for writing documents.

Other forms of written construction documents that you are very familiar with are textbooks and codebooks (*Figure 11*). A textbook is an instructional document that contains information that a reader needs to know to carry out a task or a series of tasks. A codebook is a guide that provides the codes and standards for certain areas of construction, such as electrical and building codes. Instruction manuals for tools, and installation manuals from manufacturers, are other common examples of documents used on the job. It is crucial for workers to be able to read and understand these important documents. When a person's reading skills are poor, it typically causes them to avoid reading technical data whenever possible. This can lead to errors and accidents. This section reviews some techniques that you may find helpful in reading and writing construction documents.

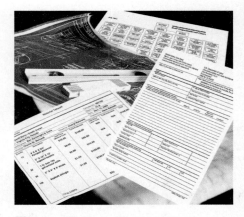

Figure 10 The written word is extremely important in the construction trade.
Source: Ed Gloninger

Figure 11 Codebooks are critical resources for your job.

2.2.0   Reading on the Job

When your company issues new safety guidelines, or the latest version of the local building code contains new standards that affect how you do your job, you need to be able to understand these documents to perform your job safely and effectively. You should not always rely on someone else to tell you how to do your job. Reading is an essential skill for construction professionals.

You may think that you do not have to read on the job, but, in fact, the written word is at the center of the construction trade. The following are some typical examples of things construction workers read on the job:

- Safety instructions
- Construction drawings

- Manufacturer's installation instructions
- Materials lists
- Signs and labels
- Work orders and schedules
- **Permits**
- Specifications
- **Change orders**
- Industry magazines and company newsletters
- Emails

This section reviews some simple tips and techniques that will help you read faster and more efficiently on the job. Further, some of these tips and techniques will help you become a better writer as well. Because you have been reading for some time now, it is probably something you do without even thinking. With practice, these guidelines will become second nature, too.

You should always have a purpose in mind when you read. This will help you find the information faster. For example, say you are looking for a specific installation procedure in a manual. Do not waste time reading other parts of the manual; go straight to the section that deals with that procedure. Read slowly enough to be able to concentrate on what you are reading. Often, technical publications are packed with detailed information; you do not want to misunderstand it because you rushed.

Most books have special features that can help you locate information, including the following:

- **Table of contents**
- **Index**
- **Glossary**
- **Appendixes**
- **Tables** and **graphs**

Tables of contents (*Figure 12*) are lists of chapters or sections in a book. They are usually at the front of a book. This is the first place experienced readers look when they want to know about the book in general or when they want to find some specific information in the book.

Permits: Legal documents that allow a task to be undertaken.

Change orders: A written order by the owner of a project for the contractor to make a change in time, amount, or specifications.

Table of contents: A list of book chapters or sections, usually located at the front of the book.

Index: An alphabetical list of topics, along with the page numbers where each topic appears.

Glossary: An alphabetical list of terms and definitions.

Appendixes: Source of detailed or specific information placed at the end of a section, a chapter, or a book.

Tables: A way to present important text and numbers so they can be read and understood at a glance.

Graphs: Information shown as a picture or chart. Graphs may be represented in various forms, including line graphs and bar charts.

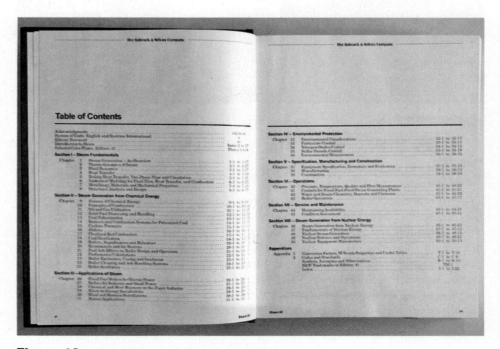

Figure 12 A table of contents lists the chapters or sections in a book.

Indexes are alphabetical listings of topics with page numbers to show where those topics appear. Indexes are also used to list information in documents other than books—for example, in construction drawings (*Figure 13*). Say you want to look up information about transistors. Whereas the table of contents will tell you where to find the chapter that focuses on transistors, the index will tell you every page that mentions transistors even if the page is about another topic and mentions transistors only briefly.

Glossaries are alphabetical lists of terms used in a book, along with definitions for each term. A glossary is usually located at the back of a publication.

Appendixes are sources of additional information placed at the end of a section, chapter, or book. Appendixes are separate because the information in them is more detailed or specific than the information in the rest of the book. If the information were placed in the main part of the book, it could distract readers or slow them down too much. Some of the NCCER *Core* trainee guide modules contain appendixes located in the end of the book.

Tables and graphs summarize important facts and figures in a way that lets the reader understand them at a glance. Tables usually contain text or numbers, and graphs use images or symbols to convey their meaning.

Given the importance of written language to your job, make sure to pay close attention when reading. This section details some techniques you can use to maximize your comprehension of written material.

The noise and activity on a construction site or in a shop can easily distract you. While you are reading, try to avoid distractions. Radios, cell phones, televisions, conversations, and nearby machinery and power tools can all affect your concentration.

Take notes or use a highlighter to help you remember and find important text. Your notes on the blank pages at the beginning of a book can be an invaluable resource. If you find the information useful now, you may want to look it up again later, so make a note of the topic and record the page number. You can

Did You Know?

Text Presentation

Text can be modified using different styles and fonts. These elements can be changed according to the purpose of the writer. Styles can be used to make words stand out more clearly. Here are the basic styles commonly used.

- ALL CAPITAL LETTERS. Use this sparingly. Writing in capital letters is the written equivalent of shouting and can seem rude.
- **Bold.** Use to emphasize important words. Notice that in this book the trade terms are bold and in color to show their importance.
- *Italics.* Some common uses for italics are for the names of books, magazines, ships, and other vehicles.
- Underline. The uses for underlining are almost the same as for italics, so the writer should choose one or the other and be consistent in use.

In typography, font refers to the shape of the letters. The two main categories of fonts are serif and sans serif. A serif is a small line attached to the end of a stroke in a letter or symbol. The letters of sans serif fonts have no small lines. The strokes are simply straight. (The word *sans* means "without.")

Serif fonts, being easily readable, are best for text in books and printed documents. One of the most popular serif fonts is Times New Roman, but there are many other serif fonts available to use in your documents.

Sans serif fonts can also be used as body text, but they are most commonly used as headlines or titles. One of the most common sans serif fonts in America is Arial.

There are thousands of creative specialty fonts available as free internet downloads, but the construction professional should use them sparingly. Unless carefully selected by a graphic designer, they can make your document look unprofessional.

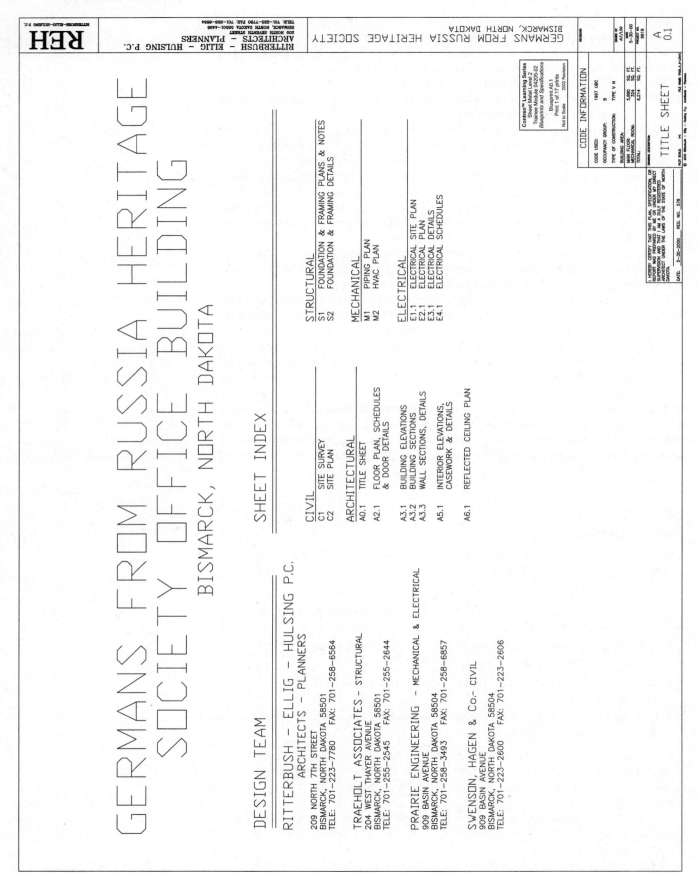

Figure 13 Indexes are often used in construction drawings to identify key information.
Source: Ritterbush-Ellig-Hulsing PC

find it in an instant later. This old-fashioned technique is faster than pulling out your smartphone and doing a web search. Also, highlighting a main sentence of an especially useful section will help you find the section quickly when you flip through the book.

When reading an unfamiliar book, skim or scan the chapter titles and section headings before you start to read. This will help you organize the material in your mind. Look for visual clues that indicate important material. For example, a bold **font** or **italics** indicate important words or information. Bold fonts are letters and numbers that are heavier and darker than the surrounding text. Italics are letters and numbers that lean to the right (like *this*), rather than stand straight up.

When reading instructions or a series of steps for performing a task, such as turning on a welding machine, imagine yourself performing the task; you may find the steps easier to remember that way. Be sure to read all the directions through before you begin to follow them. You will then be able to understand how all of the steps work together.

Always reread what you have just read to make sure you understand it. This is especially important when the reading material is complex or very long. And finally, take a break every now and then. If you read slowly or you find that you are getting tired or frustrated, put the material down and do something else to give your mind a break. Reading to understand is hard work and does require effort.

Font: The type style used for letters and numbers.

Italics: Letters and numbers that lean to the right rather than stand straight up.

2.3.0   Writing on the Job

At this stage in your career, you will probably do more reading on the job than writing. But no matter what job you have in the construction industry, you'll eventually have to write something. Writing skills are essential if you want to succeed. Construction workers write work orders, health and safety reports, **punch lists** (written lists that identify deficiencies requiring correction at completion), memoranda (informal office correspondence, or **memos** for short), emails, work orders or change orders, and a whole range of other documents as part of their job (see *Figure 14A* and *Figure 14B*).

Writing skills will allow you to perform these tasks, too, as you gain experience on the job. They will help you move into positions of greater responsibility. It is never too early to start brushing up on your writing skills, so that when the opportunity comes, you will be ready. This section reviews the writing process and then explains the characteristics of good writing.

First, before you begin to write, do some prewriting. Organize in your mind what you want or need to write. One prewriting technique is brainstorming, which is simply writing down a list of ideas. Do not pressure yourself to make a good list of ideas; just get ideas down on paper. Next, cross out rejected ideas and arrange the better ideas in a logical order to create an outline. After you have built a foundation of basic concepts and have a general sense of what you want to write about, do the necessary research. It may be necessary to ensure that the statements are appropriate for the situation and not something you may regret writing later.

After you have researched, write a rough draft. A rough draft is called this because it is rough, not polished. Do not attempt to write a masterpiece at this time; doing so will merely make you feel discouraged and overwhelmed. Just follow your outline, incorporate your ideas, and include your research. You can improve it at a later point.

The third step is to simply rest or do something else for a while. When you come back to the project later, your mind will be fresh and critical. Maybe you thought something was understandable when you were writing it, but when you read it a day later, you may see that it is unclear. On the other hand, you could also discover that some important information is missing.

Next, revise the paper. Come back to your writing and make major changes. You may need to rearrange some sections, add information or try to simplify them, or delete unnecessary sections. Look for wordy phrases.

Finally, proofread. Now that you have fixed the major problems in the paper, look for the smaller errors, like grammar or spelling mistakes. Watch for the

Punch lists: Written lists that identify deficiencies requiring correction at completion.

Memos: Informal written correspondence. Another term for memorandum (plural: *memoranda*).

Burning - Welding - Hot Work Permit

Valid from _____ to _____, _____   Master Card No._____
    (am/pm) (am/pm)   DATE

1. Work Description
- Equipment Location or Area _____
- Work to be done:

2. Gas Test

	Test Results	Other Tests	Test Results
☐ None Required			
☐ Instrument Check			
☐ Oxygen 20.8% Min			
☐ Combustible % LFL			

Gas Tester Signature _____   Date   Time

3. Special Instructions   ☐ None   ☐ Check with issuer before beginning work

4. Hazardous Materials   ☐ None   What did the line / equipment last contain?

5. Personal Protection   ☐ Standard Equipment: welder's hood with long sleeves; cutting goggles

☐ Goggles or Face Shield   ☐ Respirator   ☐ Forced Air Ventilation

☐ Standby Man   ☐ Other, specify: _____

6. Fire Protection   ☐ None Required   ☐ Portable Fire Extinguisher

☐ Fire Watch   ☐ Fire Blanket   ☐ Other, specify: _____

7. Condition of Area and Equipment

Required

Yes	No	THESE KEY POINTS MUST BE CHECKED
		a. Lines disconnected & blanked or if disconnecting is not possible, blinds installed?
		b. Lines steamed, purged, or otherwise properly cleared of combustibles?
		c. Area and equipment satisfactorily clean of oil or combustibles?
		d. Trenches, catch basins & sewer connections properly covered or sealed?
		e. Immediate area and/or area under the work barricaded or roped off?
		f. Adjoining equipment & operations checked to have any effect on the job?
		g. Area fire suppression (fire water and sprinkler system) in service?

Comments

Figure 14A A hot work permit is a typical written product in a construction project (1 of 2).

Burning - Welding - Hot Work Permit

8. Approval

	Permit Authorization			Permit Acceptance		
	Area Supv.	Date	Time	Maint. Supv./Engineer Contractor Supv.	Date	Time
Issued by						
Endorsed by						
Endorsed by						

9. Individual Review

I have been instructed in the proper Hot Work Procedures

	Signed	Signed
Persons Authorized to Perform Hot Work	_____	_____
	_____	_____
	_____	_____
	_____	_____
	_____	_____
Fire Watch	_____	_____

10. Job Completion

☐ Yes   ☐ No   Is the work on the equipment completed?

☐ Yes   ☐ No   Has the work site been cleaned and made safe?

Worker answering above questions _____

Issuer's Acceptance _____

Forward to Production Superintendent within 7 days of job completion

Figure 14B A hot work permit is a typical written product in a construction project (2 of 2).

errors that spell-check cannot find. If you write a different word than the one you meant to (like *four* instead of *for*) spell-check will not notice.

After you have finished writing, it may be helpful to ask a friend or a co-worker to read it. Another reader can often spot mistakes or problems that you have missed. For example, you might ask a friend to review your job application or resume. You can ask a co-worker to check over a work permit or material takeoff to make sure you haven't left out any information or made any math mistakes. Don't be embarrassed to ask for this type of help, and don't be upset if your friend or co-worker spots a mistake. Professionals are always glad to catch mistakes before those mistakes create bigger problems. Note that how you view the material may affect the accuracy of proofreading, too. Writers often find errors when they read a piece printed on paper rather than on a computer display.

Now that you understand the process of writing, we can examine the characteristics of good writing. Being a good writer is not as hard as you might think. You do not need to be a master of English composition and grammar, or have a college degree, to write well. You just need to remember these five characteristics of good writing: clear, concise, correct, complete, and considerate of the reader. Let's examine them one by one.

Bullets: Large, vertically aligned dots that highlight items in a list.

- *Be clear* — Make your main point easy to find, and include all the necessary details so the reader can understand what you are saying. One way to highlight important information is to list it. Lists often use **bullets**, like this one does, or numbers. Several different bullet formats can be used. When you finish writing, reread what you have written to make sure it is clear and accurate. Check for mistakes and take out words you don't need.

- *Be concise* — More words do not mean better writing. Instead of writing "In order to prevent unwanted electrical shocks, workers should make sure they disconnect the main power supply before work commences," write "To avoid being shocked, disconnect the main power supply before starting." Instead of writing "at the present time," simply write "now." It is also important to think about the reading level of the intended audience.

- *Be correct* — Accuracy is vital in the construction industry. For example, if you are filling out a form to order supplies from a manufacturer's catalog, make sure you use the correct terms and quantities. When writing material takeoffs or other documents that will be used to purchase equipment and assign workers, you need to get it right. Get in the habit of looking over your writing, whether it's a handwritten note, a supply list, a permit, or an email. Errors cost money and time.

- *Be complete* — When writing, ask yourself the following questions:
  - Have I identified myself to the reader?
  - Have I said why I'm writing this?
  - Will the reader know what to do and, if necessary, how to do it?
  - Will the reader know when to do it?
  - Will the reader know where to do it?
  - Will the reader know whom to call with questions?

  You should always ask these questions when you are writing something for others to read, even if you are writing something short and simple.

- *Be considerate of the reader* — Not everyone who reads what you write will be a professional in the same trade that you are in, and not everyone will understand the same words and concepts that you do. For example, you may need to write a report to a client to describe a technical problem. In that report you must use words that intelligent, yet uninformed, people would understand. Similarly, not every professional in your field will know all of the inside information about your company or the history of the problem you are writing about.

The Journalist's Questions

Make sure that you cover all of the necessary information in your writing by asking yourself the following journalist's questions:

- **Who?** Who told you that? Who wants it done? Who are you?
- **What?** What is the problem? What do you need? What do you want to do?
- **When?** When did it happen? When do you want it finished?
- **Where?** Where is it taking place? Where are you going? Where can you be found?
- **Why?** Why do you want to do that? Why is this problem happening? Why did you decide to write about it now?
- **How?** How will you fix it? How can you be reached? How can they help?

2.3.1   Emails

Email has become the standard way to send written information, files, and pictures as long as the files are not too large. Like printed letters and memos, email is used to ask and answer questions, and to provide information quickly. However, remember that a direct telephone call may be in order if email is not providing a timely result.

There are advantages and disadvantages to email. Email delivery is much faster than paper-based delivery, and emails can be stored permanently on a computer. They can be replicated quickly and distributed to many people at the same time. However, because of their ability to be copied and sent so easily, emails are not as private as paper-based documents. Business emails should not include private, sensitive, or confidential information. They can easily be sent by accident to people who are not authorized to read them. Information such as this should be sent in a more secure fashion. Remember that emails may be stored permanently, and can represent a binding agreement or contract. Negative, aggressive emails can lead to dismissal, just like a verbal attack on someone.

Another advantage to transmitting information electronically is that it reduces the amount of paper, printing materials, and shipping or processing fees associated with written documents. In some cases, however, an email is not considered legally binding. Written documents with handwritten signatures remain the standard for contracts and other formal agreements. However, this situation is changing with the advent of **electronic signatures**.

Because email is now used daily in business to communicate with associates, clients, supervisors, and co-workers, it is important to know the proper way to write an email (see *Figure 15* and *Figure 16*). When writing an email, the nonverbal part of communication (facial expressions, body language, tone of voice) is lost, so people may misinterpret the message you are trying to convey if you are not careful with

Electronic signatures: Signatures that are used to sign electronic documents by capturing handwritten signatures through computer technology and attaching them to the document or file.

From:	JQSmith@smithcontracting.com	Sent: Mon 3/15/2021 1:47 PM
To:	WJones@paintersplus.com	
Cc:		

Subject: Quick Note
Attachments:

View As Web Page

Here are some paint colors and faucets available for your bathroom: Color #1415, Soft Jade; Color #1416, Garden Moss; and Color #1417, Forest Glen. All are available in semi-gloss or eggshell finish. There are also three faucet sets: the Meridian (single handle) $109.88; the Mermaid (dual handles) $83.50; and the Monitor (dual handles) $95.75. All are available in polished brass or polished chrome. I've included paint samples and photos of the faucets. Please tell me your choices by Friday. If you have any questions, call me at 703-555-1212.

Figure 15 Functional but unattractive email example.

From: JQSmith@smithcontracting.com	Sent: Mon 3/15/2021 1:47 PM

To: WJones@paintersplus.com

Cc:

Subject: Bathroom Paint and Faucet Options
Attachments:

View As Web Page

Dear Mr. Jones,

The paint colors and faucets available for your bathroom are listed below (photos of faucets and paint colors are attached to this email). Please let me know what you decide by 5:00 pm on Friday, March 19th. If you have any questions, please do not hesitate to contact me at 703-555-1212.

Paint Colors (Available in semi-gloss or eggshell finish)
- #1415 – Soft Jade
- #1416 – Garden Moss
- #1417 – Forest Glen

Faucet Sets (Available in polished brass or polished chrome)

Model	Price	Handle Style
• Meridian	$109.88	Single
• Mermaid	$ 83.50	Dual
• Monitor	$ 95.75	Dual

Regards,
John Q. Smith
Smith Contracting

Figure 16 Better example for an email when a sale is at stake.

words. Many of the rules that apply to writing paper-based documents still apply to emails, but there is a certain etiquette involved. Business emailing standards exist so that emails are composed in a professional, efficient, and responsible manner.

The following are some general rules for writing a proper business email. Keep in mind that some rules will vary according to the nature of the business and company culture.

- Write business email the same way you would write a formal business letter or memo.
- Make sure you are sending the email to the correct individual(s) to maintain confidentiality.
- Always start with a clear subject line that indicates the purpose of the message.
- Begin the email by addressing the recipient.
- Try to keep the body of the email brief and to the point, typically no longer than one screen length, so that the recipient does not have to scroll when reading.
- Use a concise format, such as numbered or bulleted points, to clarify what the email is about and what response or action is required of the recipient.
- Write in a positive tone and avoid using negative or blaming statements.
- Do not type in all capital letters, as it gives the impression of shouting.
- Use italicized, bold, or underlined text if you need to stress a point.
- Avoid sarcasm, as it can be easily misunderstood.
- Do not forward junk mail.
- Do not address private issues and concerns.
- When sending an attachment, state the name of the file, and its format.
- Double-check the email for spelling and grammatical mistakes before sending. Remember, once an email is sent it cannot be retrieved.
- Include additional contact information other than your email address, such as a phone number or a mailing address.

Going Green

Reducing Your Carbon Footprint

Many companies are taking part in the paperless movement. They reduce their environmental impact by reducing the amount of paper they use. Using email helps to reduce the amount of paper used, and there are even postscripts on emails that ask you to reconsider printing the email unless necessary.

- Email is not a substitute for face-to-face interaction. Know when to pick up the phone or schedule a meeting if you cannot express your thoughts or concerns through an email. Never send bad news through email, and don't use email to avoid your responsibilities.

2.3.2　Texting

Like email, texting has become nearly universal. According to a 2019 Pew Research Center study, 96 percent of American adults own a cell phone. Tablet computers are also common and used for both texting and email. Texting is also an immediate form of communication. Although an email may take several hours or more to be seen, a text message is more likely to be seen right away. This depends on whether or not an individual's cell phone is set up to receive emails as well as text messages. Texting has become an important part of business and industry today. Therefore, the construction professional must be able to text effectively as well. However, how you write a text to your friend is likely quite different than a text you may write to your supervisor.

Remember that both texting and email can create a safety hazard. Safety must always be given a higher priority than a text message. Be familiar with and follow company policy on the use of phones and tablets on the job. If an immediate response is not vital, focus on the job at hand first.

You must consistently ask yourself if sending a text message is the best way to communicate. Text messages should not be used when the message is long or complex. Entering text into a cell phone is much slower than discussing the same information over the phone or even typing it into an email. The average speaking speed is 130 words per minute. Therefore, before you start a long text-message conversation, consider calling. Text messages should not be used to convey highly emotional messages, apologies, or criticism. Emotion cannot be effectively communicated this way.

Under what circumstances should text messages be used? Extremely brief messages that should be viewed immediately are best communicated with a text message. That's why text messages are also called short messages. Any snippet of information that does not require much interaction can be texted: for example, "I'm leaving now. I'll be at the factory in 10 minutes." Texts are also great for confirming information after a telephone or face-to-face conversation. For example, "Looking forward to meeting you at 3:30 Friday afternoon."

When sending a text message, be sure that it is accurate. Texting poses a special problem in the form of auto-corrections. Most smartphone systems correct typos automatically; for example, "sprll" instantly becomes "spell" after hitting the space key. This feature saves time, but it also frequently causes embarrassment. Websites commonly post autocorrect incidents that completely changed the meanings of the messages or added offensive words, but these mistakes are avoidable. Always take a moment and proofread before sending a text.

He Said *What*?

Being careful about the words you use when writing is especially important when dealing with clients. Always make sure that the person you are communicating with understands the slang that you are using. To appreciate the humor in the following story, you have to know a couple of terms used in the electrical trade. Two types of electrical wires enter a building: drops (overhead wires) and laterals (underground wires).

A power company customer called to report a power outage. The service technician looked at the problem, wrote up the work order, gave the customer a copy, and left. The customer took one look at the work order, got angry, and called the company to complain that its service technician was rude. What made the customer angry? Here is what the service technician wrote on the work order:

INVESTIGATED POWER OUTAGE – DROP DEAD

The technician described the problem accurately, but the customer took it as an insult!

Source: Adapted from *Tools for Success: Soft Skills for the Construction Industry*, Steven A. Rigolosi. 2009. Upper Saddle River, NJ: Pearson Learning.

Not only should your text message be accurate, but it must be clear. Many people use shorthand in their texts in order to save time. The use of abbreviations such as LOL (laugh out loud), TTYL (talk to you later), and IDK (I don't know) is controversial. There are a few factors to consider when deciding whether to use shorthand. If the message is informal, shorthand is usually acceptable. This is especially true if your work group has its own abbreviations for certain phrases. If the person you are sending the message to is a friend or close colleague of the same level in the company, you can use shorthand. If the message is to a stranger or someone in upper management, it may be better to send messages with correct spelling and grammar in order to leave a good impression. If you are sure that the person you are texting knows the shorthand, it is fine to use it. Otherwise, using abbreviations that your friend or co-worker does not understand could cause frustration.

The warnings in the previous section about making phone calls while operating machinery apply to sending texts as well. According to research by the Federal Motor Carrier Safety Administration, typing or sending a text takes a driver's eyes off the road for an average of 4.6 seconds. At 55 miles per hour, the car will travel the length of a football field in that time. Never send texts while driving or operating machinery.

2.0.0 Section Review

1. When a person's reading skills are poor, it typically causes him or her to _____.

   a. rewrite the material themselves
   b. avoid reading technical information
   c. have another worker read to them
   d. file for disability

2. Where can a reader look to discover if a book has a chapter on a particular topic?

   a. Table of contents
   b. Index
   c. Glossary
   d. Appendix

3. Bold fonts can be used to indicate that words _____.

   a. have recently been changed
   b. need to be checked for accuracy
   c. are important
   d. are illustrated below

4. After you finish writing a rough draft, you should _____.

   a. ask a colleague to proofread it for you
   b. do some research
   c. check for spelling or grammar errors
   d. have a rest and then review and revise

5. Which of the following is the best advice for sending an email?

   a. Treat a business email the same way you would treat a formal business letter.
   b. Type in all capital letters in order to emphasize the importance of your email and set it apart from others.
   c. Send bad news or emotional information via email so that the recipient can read the message in privacy.
   d. Spelling and grammar do not matter in email communication since it is informal.

1. Good communication on the job site _____.

   a. affects safety, schedules, and budgets
   b. will make you popular
   c. takes too much time
   d. cannot be learned

2. Nonverbal communication is best for communicating _____.

   a. feelings and attitudes
   b. complex ideas
   c. what a person desires
   d. instructions for installing new equipment

3. Which of the following are examples of positive nonverbal communication?

   a. Sitting up straight, keeping a clean workspace, and looking at people who are talking to you.
   b. Keeping an expressionless face while looking into the eyes of the speaker the entire time.
   c. Speaking with an even tone and looking away most of the time during a conversation.
   d. Telling people all of the information they need in order to do a task properly and answering questions.

4. Which of the following is an example of positive verbal communication on the job?

   a. Talking over or interrupting another person.
   b. Giving and taking instructions.
   c. Tuning out when someone is speaking.
   d. Ducking out of team discussions.

5. Real listening is _____.

   a. tedious
   b. an active process
   c. unnecessary
   d. an art

6. A co-worker asks you to hand her a tool, but does not specify which tool she needs. The most appropriate response would be: _____

   a. "Get it yourself."
   b. "How am I supposed to figure out which tool you need?"
   c. "I'm sorry; which tool do you need?"
   d. "I am too busy; ask someone else."

7. A supervisor wants an apprentice to insert a plug into a water supply line for a pressure test. The clearest way to instruct the apprentice would be for the supervisor to say, _____.

   a. "Put the plug in"
   b. "Insert that one plug into the pipe"
   c. "Get the water supply line ready for the pressure test"
   d. "Insert the plug into the end of the water supply line so I can do a pressure test"

8. If you are in the middle of a task and you receive a personal phone call, _____.
   a. put the caller on hold until you complete the task
   b. continue working and do not answer
   c. tell the caller you don't have time to talk
   d. hand the phone to a co-worker

9. In the construction industry, a codebook provides _____.
   a. the keypad entry codes for locked work areas
   b. codes and standards, such as building codes or electrical codes
   c. passwords for company web services, such as databases and training software
   d. keys for understanding blueprint symbols

10. A reader that wants to find any mention of a certain topic in a book can look in the _____.
    a. table of contents
    b. glossary
    c. appendixes
    d. index

11. Which is a typical example of an item a construction worker might read for work on a regular basis?
    a. Newspaper
    b. Roadmap
    c. Materials list
    d. Summons

12. When you are looking for a specific installation procedure in a manual, the best way to find the information is to _____.
    a. identify the correct section and go straight there
    b. start reading the book from the beginning
    c. study the terms in the glossary
    d. skim the book and take notes on all important information

13. Which of the following is one disadvantage to using email versus paper-based delivery?
    a. Email is an expensive way to communicate.
    b. The spell-checker is unreliable.
    c. Emails cannot easily be replicated.
    d. Emails are not as private as paper-based documents.

14. Many of the rules that apply to writing paper-based documents also apply to _____.
    a. writing emails
    b. leaving voice messages
    c. having a face-to-face conversation
    d. taking orders

15. Text messages are best used for _____.
    a. long conversations since a record of what was said will be preserved
    b. telling people how one feels
    c. detailed or complex information
    d. brief messages that should be viewed immediately

Answers to Section Review Questions and Module Review Questions are found in *Appendix A*.

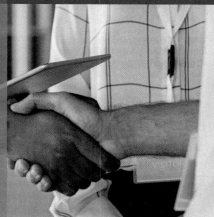

Basic Employability Skills

Objectives

Successful completion of this module prepares you to do the following:

1. Describe the opportunities in the construction businesses and how to enter the construction workforce.
    a. Describe the construction business and the opportunities offered by the trades.
    b. Explain how workers can enter the construction workforce.
2. Explain the importance of critical thinking and how to solve problems.
    a. Describe critical thinking and barriers to solving problems.
    b. Describe how to solve problems using critical thinking.
    c. Describe problems related to planning and scheduling.
3. Explain the importance of social skills and identify ways good social skills are applied in the construction trade.
    a. Identify good personal and social skills.
    b. Explain how to resolve conflicts with co-workers and supervisors.
    c. Explain how to give and receive constructive criticism.
    d. Identify and describe various social issues of concern in the workplace.
    e. Describe how to work in a team environment and how to be an effective leader.

Performance Tasks

This is a knowledge-based module. There are no Performance Tasks.

Overview

Becoming gainfully employed in the construction industry takes more preparation than simply filling out a job application. It is essential to understand how the construction industry and potential employers operate. Your trade skills are extremely important, but all employers are also looking for those who are eager to advance and demonstrate positive personal characteristics. This module discusses the skills needed to pursue employment successfully.

Industry Recognized Credentials

If you are training through an NCCER-accredited sponsor, you may be eligible for credentials from NCCER's Registry. The ID number for this module is 00108. Note that this module may have been used in other NCCER curricula and may apply to other level completions. Contact NCCER's Registry at 888.622.3720 or go to **www.nccer.org** for more information.

<table>
<tr><td colspan="2"></td><td>

1.0.0 Opportunities in the Construction Industry

</td></tr>
</table>

Performance Tasks

There are no performance tasks in this section.

Objective

Describe the opportunities in the construction businesses and how to enter the construction workforce.

a. Describe the construction business and the opportunities offered by the trades.

b. Explain how workers can enter the construction workforce.

Whether a laborer (blue collar) or professional (white collar), the construction business offers an abundance of opportunities. It is up to each individual to take the necessary steps to advance their own career.

To establish a career in your trade of choice, it is important to know where to look for the opportunities and be prepared to act when opportunities arise.

1.1.0 The Construction Business

The construction industry is made up of a wide variety of specialized skills (*Figure 1*). This diverse industry is made up of both laborers and professionals working together. Industry opportunities have a global presence; skilled tradespeople are needed in areas such as war zones and major construction sites around the world. The trades are not limited solely to building construction. There are also global opportunities for many craftworkers such as wind turbine technicians, welders, powerline workers, crane operators, and instrumentation technicians that are not directly related to new construction.

The construction industry consists of independent companies of all sizes that specialize in one or more types of work. For example, a smaller company might install HVAC systems in residences, whereas a larger mechanical contractor might be responsible for the HVAC and piping systems associated with a hospital.

The following are some examples of the workforce needed to build a large-scale project:

- Architects to create and design the shape, color, and spaces of the structure.
- Civil engineers to analyze the architects' drawings to determine how to make the construction design possible and suggest modifications if it is not. They are also responsible for determining suitable building materials.
- Project managers to plan, coordinate, and control a project's progress or a portion thereof.
- Heavy equipment operators to excavate the site.
- Concrete pourers to pour the foundation.
- Crane operators to lift the steel beams.
- Ironworkers to cut and fit the steel beams.
- Welders to secure the framework of buildings and other metal structures.
- Electricians to install an array of electrical devices and wiring.
- Plumbers to properly install fixtures and other plumbing equipment.
- Heating, ventilation, and air conditioning (HVAC) technicians to install building comfort systems.
- Pipefitters to install chilled and hot water systems, as well as process piping.
- Carpenters to frame the walls.
- Instrumentation technicians to install measuring and control instruments.
- Landscape designers to decide on the aesthetics of the project.

(A) Wind Turbine Technician

(B) Ironworker

(C) Welder

(D) HVAC Technician

(E) Surveyor

(F) Crew Leader/Supervisor

Figure 1 The construction industry offers many career options.

Sources: Mike Casey, National Science Foundation (1A); Lincoln Electric Company, Cleveland, OH, USA (1C); Courtesy of Mark S. Knoke (1D); Trimble Navigation Limited (1E); Roanoke Land Surveying (1F)

Mission statement: A statement of how a company does business.

Many companies have a **mission statement**, which explains how a company does business. The company may describe its philosophy in an employee handbook or other materials during a new-hire orientation. However a company describes its role in the construction industry, it is very important to become familiar with your company's policies and procedures to know what is expected of you. In fact, the time to learn about a potential employer is before an application is ever submitted. The more you know about a company and what is important to them, the better you can plan your responses for an interview.

1.2.0 Entering the Construction Workforce

Upon completion of your training, it will be time to find a job where your newly aquired skills can be put to use. The first step of a job search is writing a resume that will get an employer's attention.

A good resume can make the task of finding a job much easier. The following are some guidelines for resume writing:

- Use a readable font, such as Times New Roman or Arial, for the text of the resume.
- Write your objective using key words found in the job description to create a relationship between the two.
- Format your resume chronologically, putting your current or most recent job first.
- Use short sentences and bulleted lists to describe your skills, again using key words found in the job description.
- Include certifications and training credentials in specific trade areas. Credentials could include such items as NCCER training certificates or wallet cards, or an automated external defibrillator (AED) certification.
- Make sure that all contact information is current, accurate, and easy to find on the resume.
- Personal information, such as hobbies or volunteer work and commitments, can be listed. Employers often appreciate employees who have such interests, as they indicate a well-rounded individual. However, such information should be kept brief.
- Check your resume for inaccurate or missing dates. The timeline of your career should be easy to follow. Large gaps of undocumented time on a resume may require a reasonable explanation.
- Have someone proofread your resume to make sure it is free of spelling errors, typos, and poor grammar.
- Provide clear personal references or indicate that such a list can be provided upon request.

Look for a job that will match your skill level. Typically, a job posting will include a summary of the position, the required qualifications, and an overview of benefits. Applying for positions well above your skill level usually leads to disappointment.

Most employers advertise openings on the Internet and require online applications. To perform an Internet search, use a website dedicated to job searching or go directly to a company's website. If a computer is not available, libraries and unemployment offices generally have computer stations for public use. Job openings can also be found in local newspapers and trade magazines.

Today, one of the most important and productive methods to identify job opportunities is through networking. Studies indicate that as many as 70 percent of available positions are not publicly advertised. At the same time, job seekers are spending roughly 70 percent of their time looking through public job postings. Obviously, there are many more opportunities available than job seekers are able to identify through postings. There are

Did You Know?

The High Cost of Benefits

While the pay rates for craftworkers vary widely, the actual cost of an employee to a company averages 35 percent more than the employee's wage. For example, if a worker's hourly wage is $25.00, the company will spend another $8.75 per hour to pay for benefits such as healthcare and vacation time. Even higher percentages are not uncommon.

many ways, including social media, to share your interest in finding a specific type of job. Identify companies that you would like to work with, regardless of job postings, and then look for contacts within the company that could help you identify opportunities. Express your interest in finding a job to family and friends. Networking is quite often the most productive method of finding and landing a job.

Ask your current supervisor, your co-workers, and friends that know you well if they would act as **references** for you. Current and accurate contact information for references must be provided. Both personal and professional references are recommended.

Ensure that your resume is up-to-date, well organized, and easy to read (*Figure 2*). Apply the effective writing techniques that you learned in *Basic Communication Skills* when writing a resume.

By carefully selecting jobs that you are qualified for and by submitting an accurate, well-written resume, you improve your chances of being called for interviews (*Figure 3*). The effective communication skills that you learned in *Basic Communication Skills* will help you present yourself well.

References: People who can confirm to a potential employer that you have the skills, experience, and work habits that are listed in your resume.

John Q. Smith

Address: 123 Main Street, Anytown, MD 67890
Home phone: (555) 111-2233 **Cell phone:** (555) 333-4455
Email address: johnqsmith@email.com

EMPLOYMENT OBJECTIVE	• To obtain a position within a construction company where I can use my carpentry and custom woodworking skills.
SKILLS AND QUALIFICATIONS	• Able to work independently and with a crew. • Practical knowledge of carpentry hand and power tools, measuring tools, and woodworking tools. • Trained and certified to use powder-actuated tools. • Qualified to read residential and commercial construction drawings. • Experience with energy efficient and conservation building techniques. • Knowledge of LEED standards.

WORK EXPERIENCE	**Journeyman Carpenter**
	May 2021 – present
	LJL Construction, 123 Hammer Heights Road, Fairfax, VA
	800-222-2222
	  • Built custom cabinets and closet systems
	  • Framed single family homes
	  • Trained apprentice carpenters
	
	Carpenter
	June 2015 – May 2021
	Hammer Company, 456 Lathe Lane, Philadelphia, PA
	866-555-1234
	  • Built closets and storage systems for commercial storage units
	
	Carpenter
	February 2012 – June 2015
	Blue Ridge Carpenters, 789 Mountain Road, Harrisburg, PA
	888-123-4567
	  • Framed condo units
	  • Built fences
	  • Raise wood walkways

CONSTRUCTION EDUCATION	• Carpentry apprenticeship. MK Builders Inc., Gettysburg, PA. Certificate, 2012. • Professional carpentry certification program. ABC Training Corp., Philadelphia, PA. Certificate, 2010. • Powder-actuated tools certification program. All States Community College, Marietta, PA. Certificate, 2010.

Figure 2 Resume.

Figure 3 Job interviews are an important part of the hiring process.
Source: Ed Gloninger

Before accepting an offer, ask yourself the following questions to ensure the position will be the right fit:

- Is the salary enough to meet my needs?
- Does the company offer a benefits package that covers what I need?
- Will the work be interesting and challenging, but not more than I can handle?
- Does the company have a good reputation in the industry?
- Do the people appear to be nice to work with?
- Is travel required and am I able to agree to that?
- Does the company offer training?
- Are advancement opportunities within the organization evident?
- What is the company's safety record?

Before an offer is made, some trades require credentials or qualification testing to assess the applicant's skill level. To ensure workplace safety, many companies also require drug testing as part of the pre-employment process. Appropriate identification, such as a driver license, US Social Security card, national identification card, and/or a valid passport must be given to the employer. If a position outside of your country of citizenship is applied for, check the hiring and documentation requirements for that specific country and what may be required of non-citizens.

After accepting a position, and even before the interview process begins, it is wise to obtain an organizational chart (*Figure 4*). Organizational charts are used to help all employees understand the structure of the company and what position and/or individual has responsibility for various areas. Understanding the organizational structure and being able to recall some important details about a prospective employer during an interview can make a strong impression.

TWIC® NexGen

In the United States, the Transportation Security Administration (TSA) and the US Coast Guard have worked together to create the guidelines for the Transportation Worker Identification Credential (TWIC). The tamper-resistant, biometric-based card is required for all workers who require unescorted access into secure areas of port facilities, outer continental shelf facilities, and all vessels regulated under the Maritime Transportation Security Act of 2002, as well as for all US Coast Guard-credentialed merchant mariners. In 2018, the new TWIC® NexGen card was released enhancing many security features including an integrated circuit chip that stores the holder's information, photograph, and biometric data for positive identification. This credential will likely expand to other forms of transportation in the future.

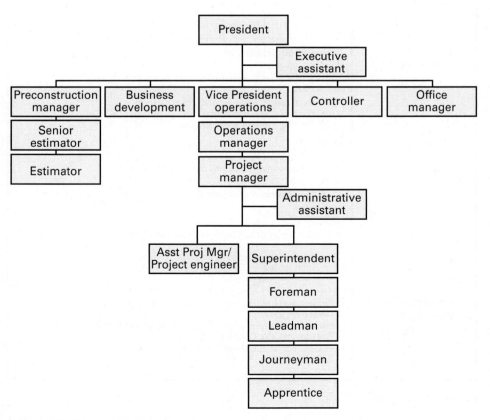

Figure 4 An organizational chart.

Going Green

Green-Collar Jobs

In recent years, there has been a big shift to environmentally friendly construction. This has created many new sectors in the US economy, including everything from recycling to energy retrofits; green building; solar installation/maintenance; water retrofits; and whole house performance (HVAC). These jobs are in green businesses and all involve work that directly affects environmental quality. Any job that involves the design, manufacture, installation, operation, and/or maintenance of renewable energy and energy-efficient technologies are considered green-collar jobs. Many of these jobs have opened new doors or removed barriers to employment. In the construction industry, opportunities exist in the following areas:

- Installation, construction, maintenance, and repair of energy retrofits, HVAC, solar panels, and whole home performance
- Installation, construction, maintenance, and repair of water conservation and adaptive grey water reuse
- Construction, carpentry, and demolition in green building and recycling

Applications: Then and Now

Before computers and the Internet became part of daily life, job applicants approached the job market in a very different way. Instead of searching websites and sending their resumes off with the click of a mouse, more footwork was involved. Reading through the classifieds of the local newspaper or going door-to-door to fill out applications were arduous, time-consuming tasks.

1.0.0 Section Review

1. Who plans, coordinates, and controls a project's progress, or a portion thereof?
   a. Civil engineer
   b. Architect
   c. Project manager
   d. The funding bank

2. Which of the following is a valid statement about a resume?
   a. Always use a highly creative and unusual font to set your resume apart from all others.
   b. Format your resume chronologically, putting your current or last job first.
   c. Do not allow others to proofread your resume, as they will likely be critical of it.
   d. Resumes should not include certifications; present them only if called for an interview.

2.0.0 Critical Thinking and Problem Solving

Performance Tasks

There are no performance tasks in this section.

Objective

Explain the importance of critical thinking and how to solve problems.

a. Describe critical thinking and barriers to solving problems.

b. Describe how to solve problems using critical thinking.

c. Describe problems related to planning and scheduling.

Having the ability to solve problems using critical thinking skills is a valuable skill in the workplace. There will be situations in which timely job completion is at risk because a problem suddenly arises. Using the tools of critical thinking skills can get the project moving forward and will also demonstrate your problem-solving capabilities.

2.1.0 Critical Thinking and Barriers

Throughout your construction career, you will encounter a variety of problems that must be solved. Critical thinking skills allow you to solve such problems effectively. Critical thinking means evaluating information and then using it to reach a conclusion or to make a decision. Critical thinking allows you to draw sound conclusions and make good decisions when you use the following approach:

- Do not let personal feelings get in the way of fairly evaluating information; put them aside and remain objective.

- Determine the cause (why it happened) and effect (what happened as a result).

- Think about the expertise or experience of people who are sources of information. Ask experts and people you trust for their advice. There is no reason to think through a problem in isolation.

- Compare new information with what you already know. If it does not fit, question it and try to separate fact from fiction.

- Weigh the merits of each option and alternative, and consider if one or more can be justified.

2.1.1   Barriers to Problem Solving

When searching for the solution to a problem, it is easy to fall into a trap that prevents you from making the best possible decision. The following are the most common barriers to effective problem solving:

- Closed-mindedness

- Personality conflicts

- General fear of change

To be closed-minded is to distrust any new ideas. Closed-minded people may also resist any suggested changes. Effective problem solving, however, requires you to be open to new ideas. Sometimes the best solution is one that you would have never considered on your own. Remember that other people have good ideas, too. You must be willing to listen and evaluate their input fairly.

Sometimes you may fail to appreciate the value of information or advice simply because you do not get along with the person offering it. One of the most important skills one can master is the ability to separate the message from the messenger. Weigh the value of the information separately from your feelings about the individual. This ability will show people that you are a true professional.

People often fear change when they believe it threatens them somehow, while it is often the lack of change that is the problem. In addition, it is often difficult to clearly see the path to implementing change or to envision the final result. When the path toward, or results of, a proposed change is difficult to see, the fear of change is combined with a fear of the unknown. If you embrace the concept of change, you will never stop finding new ways to solve problems. An ancient Greek philosopher named Heraclitus is credited for first speaking the words "change is the only constant in life."

2.2.0 Solving Problems Using Critical Thinking

Problems arise when there is a difference between the way something is and the way it should (or needs to) be. You might feel frustrated or intimidated by a problem, or you might feel that you do not have enough time to solve it. Your reaction might be to simply ignore the problem. Instead, try to look at it as an opportunity to demonstrate your skills. By actively seeking solutions to problems, you will demonstrate to your colleagues and supervisors that you are responsible and capable. Encountering and conquering a challenging problem can have a significant effect on your reputation, in addition to building your confidence and level of experience.

To solve problems that arise on the job, use the following five-step process (*Figure 5*). Reviewing each of these steps in detail will help you understand how to apply them.

Step 1 Define the problem. Before you can solve a problem, you need to know exactly what it is. This step might seem obvious, but people often find themselves trying to solve the wrong problem or one that doesn't even exist.

Step 2 Analyze and explore alternatives. Once you have defined the problem, consider different ways to solve it. Collect information from a wide range of sources, and ask people for their opinions. Perhaps you have a solution in mind, or maybe just one step towards the solution. It can be very productive to seek opinions on solving small portions of a problem rather than seeking opinions on the big picture.

Teamwork is an important part of this step. The more suggestions you get from experts and co-workers, the better your chances are of finding the right answer. Identify and compare the alternatives. Look for solutions that will be the most cost-effective, take the least time, and ensure the highest-quality result. Try to identify the short- and long-term consequences, both good and bad, of each possible solution. Eventually, you will have created several possible options and given each one some thought.

Step 3 Choose a solution and plan its implementation. Move forward by choosing the solution that you think will work best. Next, develop a plan to put the solution into action. The plan should include all necessary tools and materials. It should also specify all the tasks involved and who is responsible for them, and estimate how long each task will take.

Step 4 Put the solution into effect and monitor the results. Once the plan is in progress, follow it closely to ensure that it is progressing toward the desired results.

Step 5 Evaluate the final result. If your solution does not work, go back to Step 3 and choose another solution, as shown in *Figure 5*. Then develop and implement a plan suited to the new solution.

Remember that not every solution will be successful. In fact, life and the workplace have a way of turning what appears to be a logical solution into a new disaster. This happens to everyone; do not allow such a failure to destroy your confidence. Go back, think through the options once again, and move forward with a new solution.

When the problem is finally solved, look back over the steps you took and see what lessons can be learned from the process. Perhaps there was something you could have done better. Maybe you could have saved time or money by

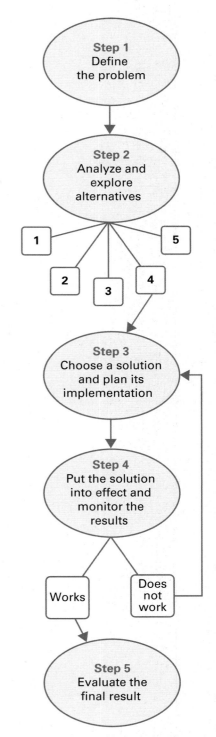

Figure 5 The five-step problem-solving process.

Information on the Internet

Everybody knows that the Internet provides access to information about almost everything. The trick is to know whether the answers you find there are the right ones for you. How do you evaluate resources that you find on the Internet?

One of the easiest and most effective ways is to look at the two- or three-letter extension at the end of the website address. Addresses ending in .gov or .us are official local, state, or federal government web pages. Use those pages to find accurate information about codes and regulations that apply to your work. Addresses ending in .org are generally nonprofit organizations. If the organization is affiliated with your industry or sets the standards for it, you can also trust its information.

Manufacturers put a wide variety of information, such as product specifications and catalogs, on their websites. When looking for this type of information, go straight to the manufacturer's website. The websites of companies that sell products from many different manufacturers may not be as complete or as up-to-date as the manufacturer's own site.

changing a small part of the plan. When you are satisfied with the solution, remember what worked and what did not. When you face a similar problem in the future, your experience will help you find a solution.

Review the following hypothetical case. Imagine that you have personal responsibility for finding a way to prevent the situation below from repeating itself.

While on a job site, your co-workers ask you to retrieve more welding electrodes. This particular welding task must be finished by the end of the day, and that is going to cut it close. You walk to the nearest tool room, but the attendant only has 60 electrodes in stock and you need several hundred. The attendant checks with the other tool room and they are fully stocked with the electrode you need. It will take some time to get to the other tool room, though. What can be done to solve the immediate problem? What can be done to ensure that the shortage doesn't happen again?

The following sections describe an example approach you may take in this situation.

2.2.1   Defining the Problem

The proper type of welding electrodes must be readily available in order to keep the job running smoothly. It is critical that the task is complete by the end of the shift because the next shift needs to start the next phase. If the welding is not complete, it will impact the schedule and budget. There is a short-term shortage of the proper welding electrode.

2.2.2   Analyzing the Alternatives

You could take the 60 electrodes in stock, walk back to the task site, and then take a bicycle to the other tool room for the remaining electrodes. Or you could take a bicycle now to the other tool room, but it would take at least 30 minutes to complete the trip.

You could also contact the other tool room attendant that has the electrodes and explain the predicament. Then you could ask the attendant to bring the electrodes to you. This will save you (and the welders) the time involved in making the trip. While you are waiting, you can take the 60 electrodes to your co-workers so the job does not come to a halt.

What are some of the benefits and drawbacks of each alternative?

2.2.3   Choose a Solution and Plan

After weighing each option, you decide to contact the other tool room attendant, ask for assistance, and then deliver the sixty electrodes to your co-workers. This will keep the job going and will save some of your time away from the task by having the electrodes delivered. However, the delivery requires the attendant to close the stock room while making the trip, which could affect others needing important materials.

To keep the tool rooms properly stocked with all consumable materials from now on, you suggest to your supervisor that the local stock attendants should take an inventory at the end of each shift and restock as appropriate.

2.2.4 Implement and Monitor

At the end of each shift, the attendants now inventory consumable materials to ensure a full stock of materials for the next day. They previously took inventory every three days. After a few days, someone in authority returns to the local stock rooms to see how the change is working and how the attendants feel about the new process. Perhaps they might have fresh input and suggest additional improvements to the process.

2.2.5 Evaluate the Final Result

Since implementation, no tool room has run low on consumables. No one has had to scramble and search for materials, and an attendant does not have to close the tool room to deliver materials.

Remember, the word *critical* means important or indispensable. Critical thinking skills are important and indispensable tools in your personal tool box (*Figure 6*). As with any other tool, you must learn how to use these skills correctly and safely. When you do, you will find that one of the most rewarding experiences you can have as a trade professional is to face a problem and solve it yourself.

Figure 6 Critical thinking is just as important as any tool you carry.
Source: Ed Gloninger

2.3.0 Planning and Scheduling Problems

Every task has a logical position in the overall project schedule. Each task must occur before or after other tasks, or at the same time as other tasks. Before beginning each task, ensure that it is understood what you are expected to do and what is required to accomplish the task. You must know where your work begins and ends and which materials to install. Divide the task into individual steps and determine the sequence for performing the work.

No matter how carefully you plan ahead, unexpected problems can suddenly appear. Usually, extra time is built into project schedules so that simple problems will not cause a delay. However, more complex problems can throw the project off track.

Project managers and/or site planners (when the position is assigned) are generally responsible for ensuring that a project stays on schedule; therefore, it is common for them to use project management software (*Figure 7*). Nevertheless, as a team member who has been working on a particular job, you may be called upon to evaluate and help solve problems. Schedules must change to accommodate all types of problems that arise. It is important to be familiar with the types of problems that can disrupt a job, including regulatory issues. Typically, problems will fall under one or more of the following categories:

- Materials
- Equipment
- Tools
- Labor

2.3.1 Materials

The materials required for a job are identified during the estimating stages of a project. Once the contract has been completed, they are then ordered from suppliers and delivered according to a prearranged schedule. Many job sites do not have sufficient room to store all the needed materials for a structure at one time. Problems with materials can include errors of quantity or type, delays in delivery, and unavailability due to backorders or even to abrupt business closure. A shortage of materials due to waste, theft, or vandalism is another source of project delays.

In many cases, material problems lead to changes in the material to be used. If the new material was not listed as an option in the contract specifications, a significant amount of time and effort may be required for the change request to make its way through the approval process.

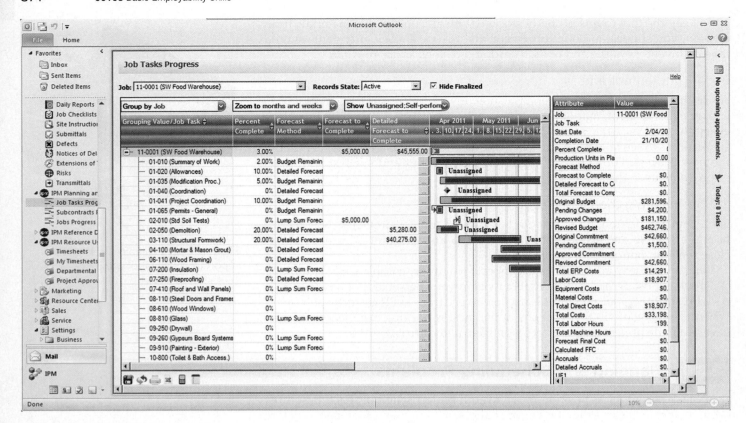

Figure 7 Project management software.
Source: IPM Global

2.3.2 Equipment

Major construction equipment is usually selected and scheduled well before the need arises. If a site planner is a member of the construction team, he or she may be responsible for ensuring that equipment is on site at the right time. Otherwise, project managers and other supervisors may be assigned this responsibility. This generally depends on the scope of the project.

Often, multiple pieces of equipment must be scheduled at the same time: backhoes to excavate a foundation and dump trucks to haul away the excavated dirt, for example. Problems with equipment can include unavailability on the scheduled day(s), lack of qualified operators, mechanical breakdown, and extended maintenance.

The lack of major equipment can create significant delays. In many cases, multiple crafts and subcontractors are relying on the equipment to move forward with their responsibilities. For example, if a crane is required to lift steel but it is not available for any reason, ironworkers, welders, and possibly other crafts find themselves at a standstill.

2.3.3 Tools

Workers and supervisors alike are responsible for ensuring that the appropriate tools are available. Workers are often expected to have common but specific tools when they are employed. Special tools that are expensive or unique are usually the responsibility of the employer. Both parties must ensure that they do their part to provide and maintain proper and functional tools. Imagine how a house project would stall if all the carpenters arrived for a day's work without hammers. On the other hand, if pneumatic nail guns are being used extensively, but the supervisor for the contractor does not ensure an air compressor is on site, the project comes just as quickly to a standstill. Other common problems related to tools include breakage or damage, loss or theft, or a lack of skilled users.

2.3.4 Labor

Labor—the men and women who perform the work on the job site—is the most important component of a project. Labor is often one of the most expensive project

Going Green

Banner Bank Building in Boise

What makes the Banner Bank Building one of the greenest buildings in the US? It was designed to do more with less. It uses many modular components that can be reused over and over again as changes are made. Modular wall systems, along with modular power and data components, are easy to relocate without waste. The pressurized raised floor system delivers air at a higher temperature and a lower velocity than common overhead systems. It also eliminates a great deal of ductwork, and allows the location of supply air grilles to be changed at will. The structural components were designed to use less steel and concrete than most buildings in its class. Materials with a high level of recycled components were used where available and when quality would not be compromised.

   The building was intentionally constructed close to public transportation nodes to encourage their use by employees. It also has inside storage for bicycles and showers to accommodate those who want to take green living a step further. Water is reclaimed and a geothermal heating system was installed. The building is even equipped with backup generators that run on biodiesel derived from used vegetable oil.

Source: The Banner Bank Building

components as well, depending upon the country, project, and type of workforce needed. Project planners estimate the number of people needed for each day and identify the range of skills required of those workers. Foremen ensure that the right people are available and working on the right tasks. Common problems related to labor include **tardiness** and **absenteeism**, lack of experience or qualifications to perform a given task, and not enough available workers. Idle time due to a lack of guidance or laziness can also contribute to delays.

Tardiness: Arriving late for work.

Absenteeism: A consistent failure to show up for work.

2.3.5   Handling Delays

Always be aware of possible delays as you carry out your assigned tasks. Look as far forward into the portion of a project that you are responsible for, and consider the types of delays that could happen. If you see a potential cause of delay, notify your supervisor immediately. Having a plan in place for delays, especially delays that occur regularly or happen often, enables a solution to be applied promptly. Always know clearly what your responsibilities are in such situations. Be prepared to solve the problem yourself if that is part of your assignment. Even then, always keep your supervisors updated on delicate situations involving delays.

2.0.0 Section Review

1. In critical thinking, if new information you receive about an issue does not fit with what you already know, _____.
   a. question it and try to separate fact from fiction
   b. change what you previous held as factual in your mind
   c. berate the person and let them know you don't appreciate that kind of input
   d. thank them and forget about the information they offered

2. Why is it important to follow a plan through once it is implemented?
   a. To be sure there are no other options
   b. To be sure no one asks questions
   c. To look for other opinions
   d. To be sure the results are what you are looking for

3. If material that was not specified in contract documents must be substituted for another type, the change will likely need to _____.
   a. pass through an approval process
   b. pass through a testing procedure
   c. be agreeable to all other trades
   d. be considered after the project is done

3.0.0 Relationship and Social Skills

Performance Tasks

There are no performance tasks in this section.

Objective

Explain the importance of social skills and identify ways good social skills are applied in the construction trade.

a. Identify good personal and social skills.
b. Explain how to resolve conflicts with co-workers and supervisors.
c. Explain how to give and receive constructive criticism.
d. Identify and describe various social issues of concern in the workplace.
e. Describe how to work in a team environment and how to be an effective leader.

A relationship results from the process of interacting with another person, or a group of people. Relationships are affected both by real actions and the perceptions of individuals. Every day, you interact with co-workers, supervisors, and members of the public who see you working. No one wants to work with, hire, or spend time with an unprofessional person.

Craftworkers need to be aware of the appropriate professional conduct for work situations, and follow that conduct at all times. Your actions reflect on your own professional status, that of your colleagues, the company you work for, and the image of your profession as seen by the public.

3.1.0 Personal and Social Skills

Self-presentation: The way a person dresses, speaks, acts, and interacts with others.

Work ethic: Work habits that are the foundation of a person's ability to do his or her job.

Proper **self-presentation**—the way you dress, speak, act, and interact—is a vital part of any successful work relationship. Self-presentation involves developing good personal and work habits. Personal habits apply to your appearance and general behavior. Work habits, also known as your **work ethic**, apply to how you do your job. Your personal habits and work ethic make powerful impressions on your colleagues, supervisors, and potential employers. Something as simple as filling out a job application with no errors and good penmanship will give an employer a good first impression.

When you introduce yourself, look the other person in the eyes and stand with confidence. In the United States and a number of other countries, a firm and sincere handshake has long been considered an indicator of a person's character and personality, especially among men. Recent studies indicate that there actually may be a connection. A firm grip (but not so firm as to imply a competition is at hand) and making eye contact at the appropriate moment make a good impression, regardless of gender. Once formed, first impressions are hard to change, so make sure that the first impression you make is a good one.

3.1.1 Personal Habits

Co-workers and supervisors like people who are dependable. Co-workers know that they can trust dependable people to pull their own weight, take their responsibilities seriously, and look after one another's safety. Supervisors know that dependable people will do their best to finish a job correctly and on schedule. When dependable workers say they will do a task, they follow through on that commitment. It also means showing up for work on time every day and not stretching out lunch hours and breaks.

Organizational skills are important as well. Keep your tools and your work areas clean and organized. Know which tools you are responsible for and keep them in good working order. This applies to your personal tools as well as those belonging to your employer. Always plan what you need to do each day before you begin. Although your best plan is likely to change as the day unfolds, a failure to plan is a plan for failure. Approach your work in an organized fashion and follow the schedule that you have been assigned.

Ensure that you are technically qualified to perform your duties and specific tasks. This means that you know how to use tools, equipment, and machines safely. Take advantage of opportunities to expand your technical knowledge

The War Against Absenteeism

During World War II, factories across America were humming along, many producing products needed for the war effort. To combat absenteeism in the workplace, employers like Remington Arms related the issue to patriotism and the crucial role that America's workers played in the war effort.

Source: National Archives and Records Administration

through classes, books and trade periodicals, and mentoring by experienced colleagues. Being technically qualified also means that you will not attempt a task that you are unfamiliar with when any level of safety hazard exists.

Offer to pitch in and help whenever you can. The best workers are willing to take on new tasks and learn new ideas. Supervisors notice employees who are willing, but be careful not to take on more tasks than you can handle; you will not impress anyone if you cannot fulfill what you promised.

Honesty is one of the most important personal habits you can have. Do not abuse your company's system by calling in sick just to take a day off or by leaving early and asking someone else to punch you off the clock. Never steal from the company or from your co-workers; this includes everything from tools and equipment to simple office supplies. If you are struggling with a problem, don't hide it; speak with your supervisor about it truthfully. Be willing to look actively for a solution to problems you are facing.

Professionalism means that you approach your work with integrity and a professional manner. As you learned in *Basic Safety*, there is no place for horseplay or irresponsible behavior on the construction site. Employers want workers who respect the rules, and who understand that rules exist to keep people safe and projects on schedule. A professional employee always respects company **confidentiality**. In some cases, workers are required to sign documents indicating that they understand the importance of confidentiality and that they agree not to discuss certain information that is shared with them.

Professionalism: Integrity and work-appropriate manners.

Confidentiality: Privacy of information.

Most companies have a dress code, and often they have additional special requirements for specific jobs. Always follow these requirements. Pushing the limits of such policies never impresses anyone of importance. Also, do not forget to attend to the basics of good grooming. People who pay attention to their appearance and develop positive personal habits are more likely to be considered for a job over people who do not have good grooming habits. On the job site, you are a reflection of your trade and employer, as well as yourself.

3.1.2 Work Ethic

Along with good personal habits, a strong work ethic is essential for getting hired and being promoted. Employers look for people who they believe will give them a fair day's work for a fair day's pay. Having a strong work ethic means that you enjoy working and that you always try to do your best on each task. When work is important to you, you believe that you can make a positive contribution to any project you are working on.

Construction work requires people who can work without constant supervision. When there is a problem that you can solve, solve it without waiting for someone to tell you to do it. If you finish your task ahead of time, look for another task that can be finished in the remaining time or ask co-workers if they need help. This type of positive action is called taking **initiative**. Colleagues and supervisors will respect you more when you demonstrate initiative on the job.

Initiative: The ability to work without constant supervision and solve problems independently.

An important part of taking initiative, however, is to know when and how to implement it. Suppose a supervisor tells you to perform a task. He demonstrates five steps required to complete the task. Later, as you perform the work, you realize that one of the steps is unnecessary. Leaving out the step could save time—but should you? No; that would not be the right initiative. Instead, tell your supervisor about your discovery, and ask what you should do. Bringing the options to your supervisor is showing initiative. Your supervisor may realize that you are right and allow you to leave out the extra step. After that, every time anyone performs that task, it will take less time and save the company money. Or, your supervisor may explain why the step is important. Then you will have learned something more about the work you are doing. The result of taking the initiative in either case is a positive one.

3.1.3 Tardiness and Absenteeism

The two most common problems supervisors face on the job are tardiness and absenteeism. Tardiness is when a worker habitually shows up late to work. Absenteeism is when a worker consistently fails to show up for work at all, with or without excuses. People with a strong work ethic are rarely late or absent.

To make a profit and stay in business, construction companies operate under tight schedules. These schedules are primarily built around workers. If you are late or do not show up regularly, expensive adjustments become necessary. Your employer may decide that the best way to save money is to stop wasting it on you.

Consider the following suggestions for improving or maintaining your record of punctuality and attendance:

- Think about what would happen if everyone on the job were late or absent frequently.
- Think about how being late or absent affects your co-workers.
- Know and follow your company's policy for reporting legitimate absences or lateness.
- Keep your supervisor informed if you need to be out for more than a day.
- Allow yourself enough time to get to work.
- Explain a late arrival to your supervisor as soon as you get to work.
- Do not abuse lunch and break-time privileges.

3.2.0 Conflict Resolution

Conflict resolution is an important relationship skill, because conflict can happen anywhere, anytime. Conflicts between you and your co-workers can arise because of disagreements over work habits; different attitudes about the job or the company; differences in personality, appearance, culture, or age; or distractions caused by problems at home. Conflicts between you and your supervisor can happen because of a disagreement over workload; lateness or absenteeism; or criticism of mistakes and inefficiencies. Significant disagreements with a supervisor should be reported to the Human Resources (HR) office of your employer.

Most of the time, people are not trying to turn disagreements into conflicts. People are often unaware of the effects of their behavior. Before reacting negatively to a co-worker's behavior, remember to be **tactful**. This means considering how the other person will feel about what you say or do. Do not accuse, embarrass, or threaten the person. This behavior rarely leads to an acceptable solution to anything.

Tactful: Being aware of the effects of your statements and actions on others.

Never let a disagreement or conflict affect job performance, team morale, or—most importantly—site safety. The goal is to keep events from turning into conflicts in the first place. If that is not possible, then the next best thing is to address the conflict quickly and resolve it professionally. If a disagreement with a co-worker is getting out of hand, try one of the following techniques to cool the situation down:

- Think before you react.
- Walk away, explaining that you need some time to think through the situation.
- Try not to take it personally.
- Avoid being drawn into others' disagreements.

Do not let a conflict simmer and then boil over before you take action. Not only is it unprofessional, tempers may flare unexpectedly and make the situation worse. Being proactive when an issue becomes uncomfortable is usually better than being reactive. This is true unless tempers are already out of bounds. In that case, it may be best to allow some time to pass. If, despite your best efforts, the conflict escalates, walk away and notify your supervisor immediately.

Keep in mind, however, that there are important differences between the way you resolve conflicts with your co-workers and the way you resolve conflicts with your supervisor. The following sections discuss how to handle these types of conflicts.

3.2.1 Resolving Conflicts with Co-Workers

Remember to have respect for the people you disagree with. After all, they believe they are right, too. Be clear, rational, respectful, and open-minded at all times. Begin by admitting to each other that there is a conflict. Then analyze and discuss the problem. Allow everyone to describe his or her own perception of the conflict. You may realize that the whole problem was simply miscommunication. Begin by asking the following questions:

- How did the conflict start?
- What is keeping the conflict going?
- Is the conflict based on personality issues or a specific event?
- Has this problem been building up for a while, or did it start suddenly?
- Did the conflict start because of a difference in expectations?
- Could the problem have been prevented?
- Do both sides have the same perception of what is happening?

Once you have analyzed the situation, discuss the possible solutions. You will probably have to **compromise** to find a solution that everyone agrees on. When you agree on the solution, act on it, and see if it works. If it does not, then consult your supervisor for help in resolving the conflict. Notice that this process is similar to the problem-solving techniques discussed earlier. If you find that the conflict happened because you were wrong, apologize. An apology is not a sign of weakness; it is a sign of respect.

Once an agreement or compromise has been reached, stand by it in every possible way, and do so with a positive attitude.

Compromise: When people involved in a disagreement make concessions to reach a solution that everyone agrees on.

3.2.2 Resolving Conflicts with Supervisors

You can usually approach co-workers as equals, because you are working on the same job. However, on the job, your supervisor is in charge of you and your work. This means that you must use a different approach to resolve conflicts with your supervisor.

Before going to your supervisor, take some time to think about the cause of the conflict. Consider writing down your thoughts; organizing the information this way often puts things into perspective. When you approach your supervisor, do so with respect. Wait until your supervisor has a free moment, and then ask if you could arrange a time to talk about something important; or leave a message or note for your supervisor asking to meet. Be willing to meet at a time that is convenient for your supervisor. Remember, supervisors have many responsibilities. They do not have much free time during the regular workday, and you may have to meet before or after work.

When meeting with your supervisor, speak calmly and clearly. Do not be emotional, sarcastic, or accusatory. Do not confront your supervisor in a threatening or angry way. State only the facts as you see them; never say anything that you cannot prove. Do not mention the names of your co-workers unless they are directly involved. If you want to suggest changes or solutions, explain them clearly and discuss how they will benefit the people involved.

Once you have made your case, allow your supervisor to make a decision. You should accept and respect your supervisor's final decision. It may not be the one you wanted, but it will be the one that must be followed.

3.3.0 Giving and Receiving Criticism

As a construction professional, you will always be learning something new about your job. New technologies, materials, and methods appear all the time. For example, talking to an experienced construction worker can help you learn a new way to perform a task. As your skills improve with practice, you will be able to use tools and methods that you were not able to use before. This is a common way of learning on the job (*Figure 8*).

Constructive criticism: A positive offer of advice intended to help someone correct mistakes or improve actions.

Figure 8 A good employee never stops learning on the job.

Another way to learn on the job is through **constructive criticism**. Constructive criticism is advice designed to help you correct a mistake or improve an action. Constructive criticism can improve your job performance and relations with co-workers. You have probably heard the word *criticism* used in a negative way to indicate fault or blame; that is not the type of criticism discussed here. Constructive criticism does not mean that colleagues and supervisors think little of you; in fact, it means exactly the opposite. Colleagues and supervisors who offer constructive criticism do so because they believe it is worth their time to help you improve your skills.

As you gain experience on the job, you will be able to give constructive criticism as well as receive it. To make sure that someone does not mistake your constructive criticism for blame, you need to know how to offer it in a positive manner. The following sections offer some general advice on offering and receiving constructive criticism.

3.3.1   Offering Constructive Criticism

When you are training a less-experienced person or working with co-workers, you might find yourself offering some constructive criticism. How should you offer it? Before you say anything, think about the rules of effective speaking that you learned in *Basic Communication Skills*. Use positive, supportive words and offer facts, not opinions.

Constructive criticism works best when offered occasionally. Do not constantly comment on another person's work or methods; they may block out your criticism or even become angry. Never criticize people in front of their co-workers or supervisors; they may feel embarrassed and will likely resent you. Constructive criticism should include suggested alternatives. Do not criticize how a co-worker does something unless you can suggest another way. Above all, limit your comments to the person's work or methods, without targeting them as a person. There is a great deal of difference in how people react to a statement such as "your pipe-fitting work is unacceptable" compared to "this threaded joint is unacceptable."

Point out improper behavior or incorrect work techniques the first time they occur. If not addressed, the behavior or technique could become a bad habit. Bad habits, in turn, can lead to accidents. Gently and firmly offer constructive criticism for improper behavior as soon as it is noticed.

Remember to compliment the person you are criticizing. Compliments have a genuinely positive effect on the person receiving them, and they are easy to offer. You do not have to wait to compliment someone until you are prepared to offer constructive criticism. However, if a compliment is to be offered along with constructive criticism, offer the compliment first.

Try to offer compliments on a regular basis. Your appreciative and respectful approach will help keep team spirits high and your supervisor will appreciate your professional attitude.

3.3.2   Receiving Constructive Criticism

Not only can constructive criticism mean the difference between success and failure, but in cases where workplace safety is involved, it can also be the difference between life and death. Think of constructive criticism as a chance to learn and to improve your skills. Take constructive criticism from your supervisor seriously. Treat constructive criticism from experienced co-workers with the same respect; even though they may not be your supervisor, their experience gives them a level of expertise. When someone presents you with constructive criticism in the proper manner, learn from their approach as well.

Always take responsibility for your actions. Never be overly defensive or dispute criticism. If you respond negatively, you might offend a person who is simply trying to help you. As a result, that person will either lose respect for you or will no longer be willing to help. If that happens, the quality of your work, and that of everybody who works with you, will suffer.

When someone offers you constructive criticism, demonstrate your positive personal habits and work ethic. If the criticism is vague, ask the person to suggest specific changes that you should try to make. If you do not understand the criticism, ask for more details. The deepest respect you can show is to improve your performance based on their advice.

> **It's All In How You Say It**
>
> Constructive criticism should be given in a positive, non-confrontational way. People can be sensitive, and when approached about how they do their jobs, it becomes very personal to them.
>
> When delivering constructive criticism, it is not a good idea to start the sentence off with "You" because it will immediately put the person on the defense. Instead, try starting out with "In the future." This approach will sound less like an attack on their abilities and they will react and listen better. Talk about the work, not the person.

You have a right to disagree with criticism that is unacceptable or incorrect. Someone might honestly misunderstand your situation and offer incorrect advice. Or a co-worker might criticize you simply because he thinks he performs a task better than you do. In such cases, clearly and respectfully give your reasons for disagreeing, and explain why you believe that you are correct.

3.3.3 Destructive Criticism

Constructive criticism has positive effects. However, you may also encounter destructive criticism. Destructive criticism, as its name suggests, is designed to hurt, not help, the person receiving it. The destructive criticism could actually be about something that needs improvement, but because it is offered negatively, there is no desire to listen.

If you receive destructive criticism, the professional thing to do is to stay calm. Do not get into a fight by replying in a negative tone. Find a way to let the person know how the criticism made you feel. If there is a legitimate criticism beneath a destructive tone, ask the person to offer positive suggestions for ways you can improve. By taking a positive approach to negative criticism, you can set a good example for others.

3.4.0 | Social Issues in the Workplace

The modern construction workplace is a cross section of our society (*Figure 9*). A typical construction project involves men and women from all walks of life, of many ethnic and racial backgrounds, and often from many different countries. Many workers speak more than one language. Workers have grown up in, and currently live in, many different income brackets.

Construction workers face a wide range of mental and physical demands every day on the job. These demands can sometimes feel overwhelming, and

Figure 9 Today's construction workplace is a cross section of our society.
Source: Ed Gloninger

people may seek to escape pressure through illegal drugs or alcohol abuse. As a construction professional, you need to be aware of these issues. You also need to know what to do if someone is behaving inappropriately.

You may encounter the following issues in the workplace:

- **Harassment** or **bullying**
- Stress
- Drug and alcohol abuse

3.4.1 Harassment

Harassment: A type of discrimination that can be based on race, age, disabilities, sex, religion, cultural issues, health, or language barriers.

Harassment is a broad term describing negative social behavior that can be based on race, age, disabilities, gender, religion, cultural issues, health, or language barriers. Harassment often takes the form of ethnic slurs, racial jokes, offensive or derogatory comments, and other verbal or physical conduct. It can create an intimidating, hostile, or offensive working environment. It can also interfere with an individual's work performance.

One type of harassment commonly reported and talked about is **sexual harassment**. While other forms of harassment may not have been talked about as much or as openly in the past, this issue has made headlines for many years.

Bullying: Unwanted, aggressive behavior that involves a real or perceived power imbalance. This form of harassment may include offensive, persistent, insulting, or physically threatening behavior directed at an individual.

When someone makes unwelcome sexual advances, or requests or exhibits verbal or physical behavior with sexual overtones, he or she is guilty of sexual harassment. Contrary to popular belief, sexual harassment can happen between members of the opposite sex or even those of the same gender. The harasser may threaten to fire the victim to keep him or her quiet. Sexual harassment is illegal, and if you experience it, you should report it to your supervisor immediately. Usually, sexual harassment is a pattern of behavior repeated over a period of time, so if you do not report it, you run the risk of experiencing it over and over again.

Sexual harassment: A type of discrimination that results from unwelcome sexual advances, requests for sexual favors, or other verbal or physical behavior with sexual overtones.

Bullying may be more common than either sexual harassment or racial discrimination on the job, but it is not reported as often. The term *bullying* is perceived by most people to be related to school-age children. The United States government defines bullying as the unwanted, aggressive behavior among school-aged children that involves a real or perceived power imbalance. That definition covers a lot of ground. When bullying is discussed among adults, it is typically referred to as hazing, stalking, or simply harassment. Most adults are uncomfortable using the term *bullying* when reporting such behavior, as they also perceive it to be related to school-age children. Using the term makes adult victims feel childish to report it as a result. Bullies may target those that appear to be a threat to them and feel the need to control the threat. In some cases, it occurs simply because the individual feels stronger and wishes to exert control to build on their feeling of superiority.

Workers should understand that dealing with an adult bully can be very much like trying to work things out with a school-age bully. Adult bullies are often lifetime bullies, having started the practice at an earlier age. They are not typically interested in compromise or improvement. Bullying makes them feel important and in control. Adult bullies are hard to change through simple interaction and positive discussions.

Workplace bullying can come from an individual or even a group. Bullying, just as any other form of harassment, can cause serious emotional and physical illnesses such as anxiety, depression, and hypertension among victims.

With the use of today's electronic technology, bullies can even cause harm through social media and other public formats; this is referred to as cyberbullying. Once emailed rumors, malicious text messages, embarrassing pictures, videos, or fake profiles are posted on social media, the damage is difficult to undo. While the most common target for cyberbullying is school children, adults are not immune to it.

Once a bullying situation has been reported to your employer, the situation may not change immediately. Your supervisor and/or employer must work within the confines of the law. Careful documentation of events is very important, as it is in all forms of harassment. Clear documentation allows the situation to be handled properly by those in authority. Since there are laws prohibiting such behavior and your co-workers are not school children, legal action will often occur based on careful documentation.

3.4.2 Drug and Alcohol Abuse

Drinking is a common way to avoid or forget about stress. Alcohol is an accepted, legal part of modern social life. The moderate use of alcohol is socially acceptable and is generally believed to cause little or no harm to most people. Alcohol abuse, or habitually drinking to excess, not only is socially unacceptable, but also poses a serious health and safety risk to everyone involved. Although some countries and US states have legalized or decriminalized the recreational and/or medical use of marijuana—and others are contemplating it—it too can be abused and have the same effects as alcohol abuse. Regardless of whether alcohol and marijuana are considered socially acceptable, there is no room in the workplace for either substance. Among employers, this is referred to as **zero tolerance** (*Figure 10*).

Illegal drug use is never acceptable, nor is the misuse of legal prescription drugs. Many companies require a clean drug test before offering a job or allowing an employee to begin work. In fact, many employers in the construction industry are often required to share records related to drug testing as part of their obligations to their clients. Drug testing can also be ordered randomly, to test the current workforce and ensure compliance. Further, an accident in the workplace often leads to immediate drug testing to determine if illegal substances were a factor. There are a number of situations that can lead to a drug test beyond a pre-employment situation. This matter is taken very seriously by the industry as a whole.

WARNING!

Working under the influence of drugs or alcohol can have serious consequences. It can lead to fatal accidents and serious injuries on the job site affecting you and your co-workers, and can cause serious errors affecting the integrity of the work. In addition, causing harm to people or property in the workplace while under the influence of substances can also lead to personal liability for the results of a poor choice.

Types of drugs that are sometimes used inappropriately by workers include the following:

- **Amphetamines**
- **Methamphetamines**
- **Barbiturates**
- **Hallucinogens**
- **Opiates**
- **Synthetic drugs**

Amphetamines and methamphetamines are addictive stimulants that affect the central nervous system. Also called uppers, they are used to prolong wakefulness and endurance, and they produce feelings of euphoria, or excitement. They can disturb vision; cause dizziness and an irregular heartbeat; cause the loss of coordination; and can even cause physical collapse or unconsciousness. Cocaine, crack, and crystal meth are examples of illegal amphetamine drugs.

Barbiturates are a sedative, which basically means they cause you to relax. They are also called downers. They can create slurred speech; slow reactions; cause mood swings; and cause a loss of inhibition, which means that they make a person feel less shy or self-conscious. Many abused barbiturates are controlled-substance prescription drugs.

Hallucinogens distort the perception of reality to the point where people see things that are not there (hallucinations). The effects of hallucinogens can range from ecstasy to terror. When someone is hallucinating, they put themselves and others at great risk because they cannot react to situations the right way. Hallucinogens can also cause chills, nausea, trembling, and weakness. Mescaline and LSD are two examples of illegal hallucinogens.

Zero tolerance: The policy of applying laws or penalties to even minor infringements of a code in order to reinforce its overall importance, typically related to drug and alcohol abuse when applied to the workplace.

Figure 10 Signs like this are common in the construction workplace.

Amphetamines: A class of drugs that causes mental stimulation and feelings of euphoria.

Methamphetamines: Highly addictive crystalline drugs, derived from amphetamines, that affect the central nervous system.

Barbiturates: A class of drugs that induces relaxation, slowing the body's ability to react.

Hallucinogens: A class of drugs that distorts the perception of reality and cause hallucinations.

Opiates: Narcotic painkillers derived from the opium poppy plant or synthetically manufactured. Heroin is the most commonly used opiate.

Synthetic drugs: Drugs with properties and effects similar to known substances but having slightly altered chemical structures. Such drugs are often not illegal since they are somewhat different than well-defined restricted or illegal substances. The two typical categories are cannabinoids (lab-produced THC or marijuana substitutes) and cathinones, which are designed to mimic the effects of cocaine or methamphetamines.

Derived from opium, opiates are used to kill pain. Morphine is the primary psychoactive chemical found in opium, and therefore the basis of most opiates. While opiate use creates a feeling of euphoria and relieves pain, it can cause muscle spasms, cramps, anxiety, nausea, fever, and diarrhea over time. Some prescription-drug forms of opiates are Oxycontin and Vicodin.

Synthetic drugs are relatively new. They are defined as drugs with properties and effects similar to well-defined and familiar restricted or illegal substances, but having a slightly altered chemical structure. Such drugs are often not illegal since they are somewhat different from the illegal substances. However, due to the health risks and other problems associated with them, legislation has begun in some areas to control them. The two typical categories are **cannabinoids** (lab-produced THC or marijuana substitutes) and cathinones, which are designed to mimic the effects of cocaine or methamphetamines. The cathinones are more commonly known as bath salts (but they are not used for this purpose). Synthetic drugs have quickly become readily available for retail sale. They are often deceptively packaged and do not generally list all the ingredients. Synthetic drugs can be extremely hazardous and even deadly, as users have no idea what their true composition is. Lab testing has shown the potency of some formulas can be as much as 500 times as potent as marijuana.

If you abuse drugs or alcohol, you will probably not be able to get or keep the job you want. If you are discovered using illegal drugs, you can be fired on the spot because of the safety risks related to drug use. Addiction affects your well-being and state of mind, and you cannot do your best on the job or at home. It can also jeopardize your relationships with friends, family, and co-workers. If people you know are addicted to alcohol or drugs, seek help for them or encourage them to get help for themselves. If you are having trouble with alcohol or drugs, talk to your supervisor. Many companies provide substance abuse counseling and will support you with periodic drug testing to help you remain sober and drug-free.

Cannabinoids: A diverse category of chemical substances that repress neurotransmitter releases in the brain. Cannabinoids have a variety of sources; some are created naturally by the human body, while others come from cannabis (marijuana). Still others are synthetic.

3.5.0 Teamwork and Leadership

Every day, you will interact with members of your work crew and your supervisor. Although the ability to get along with people is an important skill, it is not quite enough. The success of any team depends on all of its members filling their roles competently. Everyone must contribute to the effort to ensure that the team achieves its goals. This cooperation is referred to as teamwork.

From the moment your career begins, you will be assigned to different teams for different reasons. Teams can be as small as two people and as large as your entire company. Teams are often made up of people from different trades who work together to complete a task. For example, a team that is assigned the task of burying water and electrical lines might include a backhoe operator to dig the trench, pipefitters to lay and join the pipe, and electricians to run the electrical cable.

Always show respect for the other members of your team. Although you have a specific job to do, always be willing to help a co-worker. Support your co-workers when they need it and they will support you in return.

Good team members are goal-oriented, and those goals are shared among team members. Being goal-oriented means making sure that all activities benefit the team's final objective. Whatever the end result is—fabricating fittings, laying pipe, or framing a house—everyone on the team knows the desired result in advance and concentrates on achieving it.

As a team member, use your skills and strengths to the team's advantage. At the same time, accept your personal limitations and the limitations of others. To be a good team member, strive to do the following:

- Follow your team leader's and/or supervisor's directions.
- Accept that others might be better at some tasks than you.
- Keep a positive attitude when you work with other people.
- Recognize that the work you do is for the benefit of the entire company, not for you personally.
- Learn to work with people who work at different speeds.

- Accept goals that are set by someone else, not by you.
- Trust other members of the team to perform their tasks, just as you perform yours.
- Appreciate the work of others as much as you appreciate your own.

Keep in mind that you are not the only person working hard. Everyone on your team should be focusing on the goal, too. Offer praise and encouragement to your co-workers, and they will do the same for you. This mutual respect will help you feel confident that your team will reach its goal. Share the credit for good work, and be willing to take responsibility for your mistakes and errors. These actions will help you earn the respect of your co-workers.

If goal-oriented teamwork is practiced, the team should be able to meet deadlines and keep projects on schedule. This translates into time and money saved by your company that can be used for pay increases. There will be times when your co-workers or other trades cannot start their tasks until you have completed yours. Out of respect for them and your company's reputation, always finish your work in a timely manner. The success of a team reflects on you, just as your personal behavior and characteristics reflect on the team.

An important part of teamwork is training. As an apprentice, you will receive guidance and advice from more experienced colleagues. As you gain experience, you will be able to help less-experienced colleagues. Training also offers an excellent opportunity to build good relationships.

Being asked to teach someone is an honor. It means that someone believes that you do your job well enough to teach it to others. Such confidence should inspire you to be the best teacher you can be. Be patient with the person you are teaching, and teach by example. Offer encouragement and give constructive criticism, as you have learned in this module. Teach people in the same manner that you would like to be taught. Keep in mind that your teachers in the workplace are not likely to be trained educators. As a result, their approach may be less than perfect. Discount their limitations and focus on the skill or knowledge they are sharing with you.

3.5.1 Leadership Skills

As you gain experience and earn credentials, you will assume positions of greater responsibility. These positions are earned through hard work and dedication. Starting as an apprentice, you can work your way up to team leader, foreman, supervisor, and project manager. Someday, with enough hard work, dedication, and ambition, you may even choose to run your own company.

Leadership: The ability to set an example for others to follow by exercising authority and responsibility.

To progress steadily through your career, you will need to develop **leadership** skills and learn how to use them. Leaders set an example for others to follow. Because of their skills, leaders are trusted not only with the authority to make decisions, but also with the responsibility to carry them out.

You can become a leader at any stage in your career. As an apprentice, you have the authority to perform the task given to you by your supervisor, and your supervisor expects you to be a responsible worker. By carrying out your task quickly, correctly, and independently, you are setting an example for others to follow. In doing so, you are demonstrating leadership skills.

People with the ability to become leaders often exhibit the following characteristics:

- They lead by example.
- They have a high level of drive, determination, and persistence.
- They are effective communicators.
- They can motivate their team to do its best work.
- They are organized planners.
- They have self-confidence.

The functions of a leader vary with the environment, the group of workers being led, and the tasks to be performed. However, certain functions are common to all situations. Some of these functions include the following:

- Organizing, planning, staffing, directing, and controlling.
- Empowering team members with authority and responsibility.
- Resolving disagreements before they become problems.
- Enforcing company policies and procedures.
- Accepting responsibility for failures as well as for successes.
- Representing the team to different trades, clients, and others.

Leadership styles vary. If leadership styles are classified according to the way a leader makes decisions, the result is three broad categories of leadership: autocratic, democratic, and hands-off. An autocratic leader makes all decisions independently, without seeking recommendations or suggestions from the team. A democratic leader involves the team in the decision-making process; such a leader takes team members' recommendations and suggestions into account before making a decision. A hands-off leader leaves all decision making to the team members themselves.

Select a leadership style that is appropriate to the situation. Leading with a single style at all times generally results in some real problems. You will need to consider your authority, experience, expertise, and personality each time a style is chosen. Leaders need to have the respect of people on the team; otherwise, they will be unable to set an example others are willing to follow.

Leaders have to make sure that the decisions they make are ethical. Success in the construction industry demands the highest standards of ethical conduct. The three types of ethics that you will encounter on the job are business or legal ethics, professional ethics, and situational ethics. Business or legal ethics involve adhering to all relevant laws and regulations. Professional ethics involve being fair to everybody, and situational ethics involve appropriate responses to a particular event or situation.

Effective leaders motivate, or inspire, people to do their best. People are motivated by different things at different times. The following are some common ways to motivate people on the job:

- Recognize and praise a job well done.
- Allow people to feel a sense of accomplishment.
- Provide opportunities for advancement.
- Encourage people to feel that their job is important.
- Provide opportunities for change to prevent boredom.
- Reward people for their efforts.

How Do Your Co-Workers See You?

Do you exhibit any of the following unpopular behaviors on the job? If so, it is likely they are keeping you from being an effective team member. Develop an action plan to deal with them. Your action plan should be a two- or three-step process that will allow you to correct the behavior. Be honest with yourself!

Being a loner	Bad-mouthing the company or your boss
Taking yourself too seriously	Being unable to take a joke
Being uptight	Bragging
Always needing to be the best at everything	Taking credit for others' work
Holding a grudge	Being sarcastic
Arriving late to work	Refusing to listen to other people's ideas
Being inconsiderate	Looking down on other people
Taking breaks that are too long	Being unwilling to pitch in and help
Gossiping about others	Being stingy with a compliment
Sticking your nose into other people's business	Having a chip on your shoulder
Acting as a spy and reporting on the behavior of others to your supervisor	Horsing around when others are trying to work
Saying or doing things to create tension or unhappiness	Thinking you work harder than everyone else
Complaining constantly	Manipulating people

Action Plan for Improvement

Example:

Problem: Gossiping about others

Action Plan: 1. Walk away when people start gossiping or bad-mouthing others.
2. Don't repeat what I hear.
3. Focus on my own work, not on that of others.

Problem: _____

Action Plan: _____

_____

Problem: _____

Action Plan: _____

_____

Problem: _____

Action Plan: _____

_____

Going Green
Green Employers

Some employers are doing their part to reduce their environmental footprint. In 2016, the United Parcel Service (UPS) began the selective use of electric hybrid trucks in an effort to gain significant fuel economy. UPS already had a history of using alternative energies including propane, liquid natural gas, and compressed natural gas. Earlier, UPS revised delivery routes of its drivers to minimize left-hand turns. Doing so had made making the deliveries safer (they no longer have to cross traffic) and quicker (due to the ability to make a right on red and green turn arrows). As a result, the company shaved millions of miles off its deliveries and saved the cost of millions of gallons of gas. It also reduced UPS truck emissions. UPS recognized the upfront investment versus bottom line costs but noted their commitment to reduce the environmental impact of its fleet.

Leaders who can motivate people are more likely to have a team with high morale and a positive work attitude. Morale and attitude are key components of a successful company with satisfied workers, and such a company is more likely to be successful.

Building a Team

In this exercise, you will practice building a team. Consider the following hypothetical situation, and then select an appropriate team. Discuss your selections with your instructor and with the other trainees.

Situation

You are the team leader on a job site where a new single-family home is going to be built. Shortly before construction begins, the architect changes the plans to add a sink to one of the rooms. When you review the plans, you see that an electrical conduit is located where the drain for the new sink should be. In addition, a cabinet needs to be built for the new sink. You may choose up to five of the following construction professionals to be on your team. Whom would you choose and why?

Mason	Painter
Carpenter	Roofer
Plumber	Landscape designer
Welder	Secretary
Electrician	Cabinetmaker

3.0.0 Section Review

1. If you finish a task ahead of time, looking for another task that can be finished in the remaining time is referred to as _____.

   a. stealing someone else's job
   b. being tardy
   c. showing off
   d. taking initiative

2. To find a solution everyone can agree on, you may have to _____.

   a. compromise
   b. discuss it with your supervisor
   c. let someone else solve the problem
   d. come up with more than one solution

3. When does constructive criticism work best?

   a. When offered often
   b. When offered occasionally
   c. When offered by a member of management
   d. When offered loudly in front of co-workers

4. To put an end to personal bullying, one of the most critical things an employee can do is _____.

   a. confront the person one-on-one
   b. stage an intervention
   c. carefully document the events when they happen
   d. avoid the person in spite of work assignments

5. Before displaying their skills as a leader, all craftworkers must wait until they have a great deal of experience in their chosen craft.

   a. True
   b. False

1. Someone who can vouch for your skills, experience, and work habits is called a(n) _____.

   a. mission statement
   b. entrepreneur
   c. interviewer
   d. reference

2. One of the most important and productive methods to identify job opportunities is by _____.

   a. networking with family and friends
   b. posting a "Seeking Employment" ad on Facebook
   c. trying to call Human Resource offices directly
   d. checking the Help Wanted section of the online local newspaper

3. A common barrier to effective problem solving includes _____.

   a. overwork
   b. fear of change
   c. inadequate supervision
   d. procrastination

4. After putting a solution to a problem into effect, it is important to _____.

   a. stop considering other solutions
   b. insist it become a company procedure
   c. monitor the results
   d. develop a business plan for it

5. Depending on the type of project and region of work, one of the most expensive components of a project is often _____.

   a. equipment rental
   b. power tools
   c. hand tools
   d. labor

6. Positive or negative interactions affect _____.

   a. managerial skills
   b. tool use
   c. relationships
   d. work schedules

7. A person who works without constant supervision is showing _____.

   a. initiative
   b. fortitude
   c. respect
   d. self-presentation

8. The process of solving disagreements between co-workers is called _____.

   a. tactfulness
   b. conflict resolution
   c. relationship crafting
   d. addressing conflict

9. Constructive criticism should not be given unless _____.

   a. someone asks your opinion
   b. the supervisor tells you to do so
   c. you feel it is your business
   d. you can also be complimentary

10. A broad term identifying negative social behavior that can be based on race, age, disabilities, sex, religion, cultural issues, health, or language barriers is _____.

   a. absenteeism
   b. racism
   c. cyberbullying
   d. harassment

11. Zero tolerance refers to an employer's policy regarding _____.

   a. being sick
   b. training
   c. alcohol and drug abuse
   d. overtime

12. Barbiturates typically cause a person to _____.

   a. hallucinate
   b. accelerate their reaction times
   c. slow their reaction times
   d. be deprived of oxygen

13. When everyone in a group is focused on the final objective, it is called _____.

   a. being goal-oriented
   b. being team players
   c. taking initiative
   d. being leaders

14. To progress steadily through your career, you will need to develop and learn how to use _____.

   a. leadership skills
   b. classroom skills
   c. studying skills
   d. diversity skills

15. The three categories of leadership are hands-off, democratic, and _____.

   a. automatic
   b. liberal
   c. autocratic
   d. conservative

Answers to Section Review Questions and Module Review Questions are found in *Appendix A*.

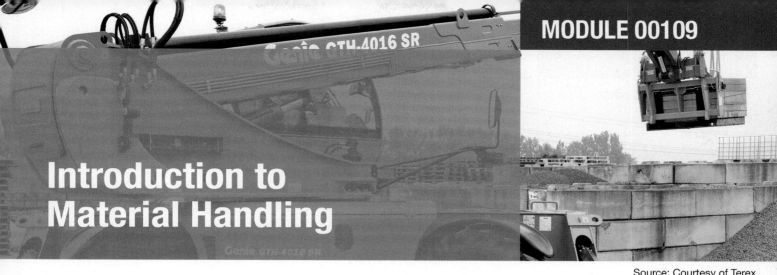

Introduction to Material Handling

Objectives

Successful completion of this module prepares you to do the following:

1. Identify the basic concepts of material handling and common safety precautions.
    a. Describe the basic concepts of material handling and manual lifting.
    b. Identify common material handling safety precautions.
    c. Identify and describe how to tie knots commonly used in material handling.
2. Identify various types of material handling equipment and describe how they are used.
    a. Identify non-motorized material handling equipment and describe how they are used.
    b. Identify motorized material handling equipment and describe how they are used.

Performance Tasks

Under supervision, you should be able to do the following:

1. Demonstrate safe manual lifting techniques.
2. Demonstrate how to tie two of the following common knots:

    - Square
    - Bowline
    - Half hitch
    - Clove hitch

Overview

Lifting, stacking, transporting, and unloading materials such as brick, pipe, and various supplies are routine tasks on a job site. Whether performing these tasks manually or with the aid of specialized equipment, workers must follow basic safety guidelines to keep themselves and their co-workers safe. This module provides guidelines for using the appropriate PPE for the material being handled and using proper procedures and techniques to carry out the job.

1.0.0 Material Handling

Performance Tasks

1. Demonstrate safe manual lifting techniques.
2. Demonstrate how to tie two of the following common knots:
   - Square
   - Bowline
   - Half hitch
   - Clove hitch

Objective

Identify the basic concepts of material handling and common safety precautions.

a. Describe the basic concepts of material handling and manual lifting.
b. Identify common material handling safety precautions.
c. Identify and describe how to tie knots commonly used in material handling.

Manual material handling is a common task on most construction sites, and it is one of the leading causes of non-fatal injuries in the construction industry. According to the Bureau of Labor Statistics, each year more than one million workers suffer from back injuries, and one of every five workplace injuries or illnesses is associated with back injuries.

Most tasks performed in construction involve the handling of some type of material or load, such as wood, brick, lumber, pipe, or other supplies, on a daily basis. A load is simply the quantity of material that can be carried, transported, or relocated at one time by a machine, vehicle, piece of equipment, or person. Because the risk of injury is present whenever materials are being handled, workers should always be conscious of what they are doing and follow basic material handling safety rules.

1.1.0 Material Handling Basics

To reduce the risk of injury when manually handling material, plan the task before doing it, wear the appropriate personal protective equipment (PPE), and follow proper lifting procedures. Also be aware of hazards when working from heights or working near suspended loads.

1.1.1 Pre-Task Planning

When it comes to material handling, it is just as important to be mentally fit as it is to be physically fit. Most material handling accidents occur when workers are new and relatively inexperienced at a job, or when very experienced workers think that accidents cannot happen to them. Before handling any material, consciously think about what you are doing. Before attempting to lift any material, always assess the situation by doing the following:

- Check to make sure the load is not too big, too heavy, or too hard to grasp.
- Make sure the load does not have protruding nails, wires, or sharp edges.
- Make sure the material is something that you can lift by yourself. If not, ask a co-worker for assistance.
- Inspect the path of travel. Look for slip, trip, or fall hazards. If there are hazards in the path of travel, move them or go around them.
- Always read the warning labels or instructions on materials before they are moved, and be aware of the potential dangers associated with mishandling a particular product.

1.1.2 Personal Protective Equipment

Wear the proper clothing and personal protective equipment (PPE) when moving or handling materials. Remember the following safety guidelines when dressing to perform material handling operations:

- Do not wear loose clothing that can get caught in moving or protruding parts.
- Be sure to button shirt sleeves and tuck in shirt tails.
- Remove all rings and jewelry.
- Tie back and secure long hair underneath your hard hat.
- If wearing a wristwatch, wear one that will easily break away if it gets caught in machinery.
- Wear gloves whenever cuts, splinters, blisters, or other hand injuries are possible. Select gloves that fit properly. Gloves that are too tight may increase hand fatigue. However, the gloves must fit snug and offer reasonable dexterity. Loose gloves reduce grip strength and may get caught on moving objects or machinery.
- Remove gloves when working with rotating machinery and equipment with exposed moving parts.
- Follow the policies of your employer and the job site at large.

1.1.3 Proper Lifting and Lowering Procedures

To reduce the risk of back injuries, use proper lifting techniques. Determine the weight of the load prior to lifting and plan the lift. Know where it is to be unloaded and remove any slipping or tripping hazards from the path of travel.

When lifting objects, avoid unnecessary physical stress and strain. Know your limits and what you are able to handle physically when lifting a load. If the load is too heavy for you to lift on your own, ask a co-worker for help.

As you lift an object, make sure you have firm footing. Bend your knees and get a good grip. Lift with your legs, keep your back straight, and keep your head up. Keep the load close to your body (*Figure 1*). Do not reach out with the load. Never turn or twist until you are standing straight, then pivot your feet and body.

When unloading materials, use your leg muscles to set the load down. Space your feet far enough apart to maintain good balance and control of the load. When placing the load down, move your fingers out of the way to avoid pinching or crushing them. When possible, attach handles to loads to reduce the chance of injuring your fingers.

Ask a co-worker for assistance in the following situations:

- The load is too heavy (weighing 50 pounds or more) for one person to carry alone.
- The objects to be handled are longer than 10 feet, such as lumber, conduit, pipe, and scaffolding poles.
- The objects to be handled, such as plywood and tarps, can be affected by wind gusts.

Sometimes it may not be the weight of a load that is troublesome, but its configuration. Secure long or cumbersome objects. Floppy rods, pipe, or lumber should be tied together in several places before you try to carry them. Never carry more pieces than you can control.

When handling long objects, such as scaffold poles or planks, be aware of objects and persons that may be struck by them as you turn your body. Never swing around quickly or without checking the surroundings. Never pass any material to another worker unless the person is looking directly at you and expecting to receive it.

Most workers have had training about the dangers of trying to lift heavy loads. Likewise, it is important to talk about lowering heavy loads from overhead. Reaching and lowering heavy loads from overhead can be very dangerous. When you are lifting something, you can always stop and put it down if you find that it is too heavy to handle alone. However, if you are lowering something from

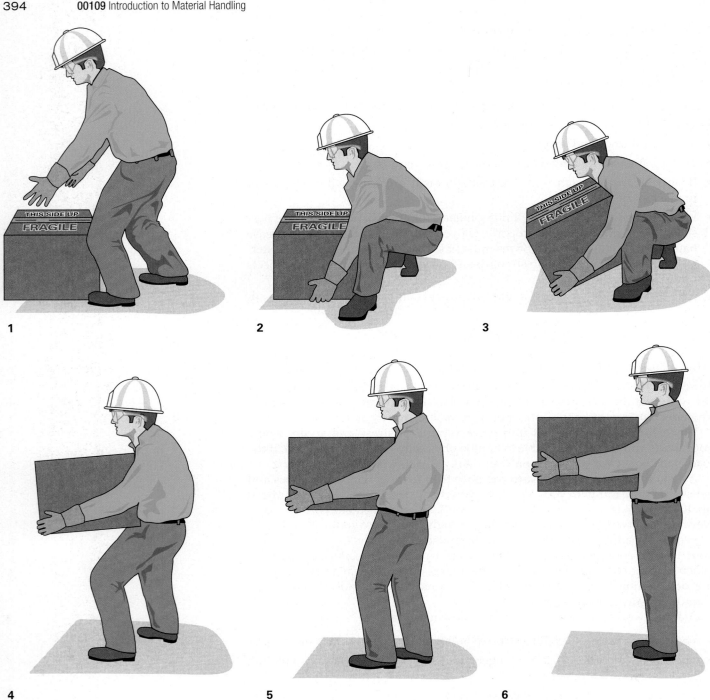

Figure 1 Proper lifting technique.

overhead, it is usually too late by the time you realize it is too heavy. Also, reaching above shoulder height can cause stress to the shoulders and back, and often results in awkward hand and wrist postures.

Before attempting to lower overhead loads, consider the following:

- Size up the load. If it looks too heavy to have been lifted by one person to its current location, then it is probably too heavy for one person to take down.

- Ask yourself, "How did it get up there?" Was it put there by a lift truck or by more than one person? The way it got up is probably the best way to get it back down.

- Lower objects down the same way you would lift them up. Keep your knees bent and your back straight. If you have to place a load to one side or another, move your feet instead of twisting your body.

- Treat the area that the material is being lowered into as a fall zone. Do not allow people to enter the area while the load is being lowered.

Back injuries can easily occur when lifting and lowering materials. Keep in mind that exercise, good posture, and a healthy weight and diet can help prevent such injuries. You can also ask your doctor about muscle-strengthening techniques for your back.

1.2.0 Material Handling Safety

When working on a job site, make sure that all workers, the materials, and the equipment are safe from unexpected movement such as falling, slipping, tipping, rolling, blowing over, or any other uncontrolled motion.

1.2.1 Stacking and Storing Materials

To work efficiently and effectively, workers should properly stack and secure materials. Taking care to stack and secure materials correctly can save space and time, and helps avoid accidents and injuries. Stacking materials keeps them from interfering with other activities and makes them more organized and readily accessible for use. Stacking materials also helps to create safe storage without danger to others and allows more room in the work area or on the job site.

Always properly store materials and equipment when they are not in current use. Learn your company- and site-specific rules, as well as the guidelines of the US Occupational Safety and Health Administration (OSHA) when working in the United States. Failing to store materials properly can lead to accidents, injuries, wasted materials, increased job costs, and project delays. For example, failing to remove nails from reusable lumber prior to storing it can cause puncture wounds when workers take the lumber out again. Stacking materials on the ground when they belong on a pallet can result in having to move them again to transport them safely and can also damage the materials. Leaving tools out when they should be stored can create an untidy work site, which can lead to an accident, tool damage, and misplaced items. Take the time to learn where and how to store materials properly. If you are unsure about something, ask your supervisor.

Observe the following guidelines when stacking and storing materials:

- Keep aisles and passageways clear in the storage area.
- Keep materials that are stored in cartons away from rain and moisture.
- Never stack cartons higher than the height listed on the carton (*Figure 2*).
- Stack lumber on level, solidly supported surfaces, and remember to remove all nails.
- Stack pipes neatly and chock them so that they cannot roll (*Figure 3*).

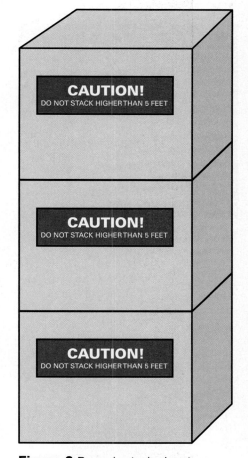

Figure 2 Properly stacked cartons.

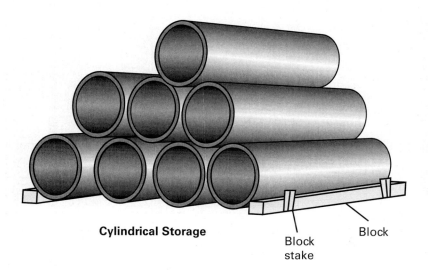

Cylindrical Storage

Block stake

Block

Figure 3 Properly stacked pipes.

- Stack pipes and fittings according to size so that workers do not have to dig to find the piece required for the job. Make sure pipes are properly chocked.
- Chock all material and equipment such as pipes, drums, tanks, reels, trailers, and wagons, as necessary to prevent shifting or rolling. Make sure the chock size and shape are appropriate for the material.
- Tie down or band all light, large surface area materials that might be moved by the wind.
- Do not stack bagged material the same width all the way to the top of the pile. Stack bagged material by stepping back the layers and cross-keying the bags at least every 10 bags high (*Figure 4*).
- Stack bricks no higher than seven feet. Taper brick stacks back two inches for every foot above four feet (*Figure 5*).
- Taper masonry blocks back one-half block per tier above six feet (*Figure 6*).
- Store flammables, such as gasoline, in a well-ventilated area away from danger of ignition or in an approved flammable storage cabinet.

Figure 4 Properly stacked bags.

1.2.2 Working from Heights

Moving material to a higher location requires good planning and safety awareness. Make sure you use a safety harness with a fall-arrest lanyard connected to an appropriate anchor point when working from heights over six feet; a positioning belt with a lanyard can be added if required. Never throw tools or materials up to a worker on a higher level, or down to a worker on a lower level. Instead, use a rope to raise or lower tools and materials to and from an elevated work area. Once material is lifted to an elevated level, tie off the rope and secure the material to

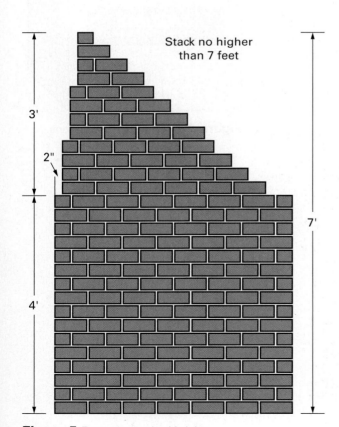

Figure 5 Properly stacked bricks.

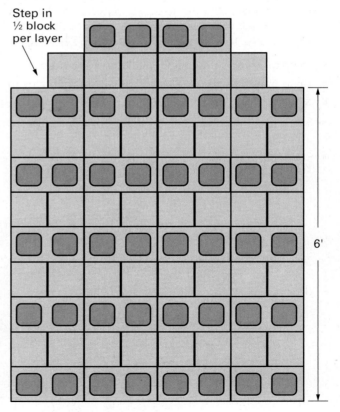

Figure 6 Properly stacked masonry blocks.

keep it from falling. Knots that are commonly used in material handling will be covered later in this module.

When climbing up or down a ladder, check your pockets and tool belt to ensure that no objects will fall out of them or get caught on something along the way. Always keep your hands free when climbing a ladder. Never carry tools or materials by hand up a ladder, or throw material up to or down from a work platform. Place objects in a bucket to convey them to and from the platform.

Never stack or store materials on scaffolds or runways. Stacking materials on scaffolds or storing materials in high-traffic areas can increase the chances of falling object hazards.

1.3.0 Knots for Material Handling

Knots and hitches are common ways of fastening ropes or lines either to each other or to another object, such as a beam or container. Most knots are meant to be permanent; they have to be unwoven to be unfastened. However, some knots, which are generally referred to as hitches, are not permanent. They can be undone by simply pulling the ropes in the reverse direction from the direction in which the knots are meant to hold.

Knots and hitches are used for many different material handling tasks, such as securing loads, stabilizing scaffolding, and lifting and lowering tools and supplies. Using the correct type of knot for the task at hand is critical. This section will cover four knots that are commonly used in material handling: the **square knot**, the **bowline**, the **half hitch**, and the **clove hitch**.

First, some knot-tying terminology needs to be reviewed. The **standing end** is the long end of the rope; the end that is not being knotted. The **working end** is the short end; the end of the rope that is being used to tie the knot. The **standing part** is the portion of the rope that lies between the standing end and the working end. Finally, to **capsize** means to change the form and rearrange the parts of a knot, usually by pulling on specific ends of the knot. This may be done intentionally to undo a hitch. On the other hand, if certain knots are used inappropriately they can easily—even spontaneously—capsize, resulting in serious safety hazards.

There are two other basic points to keep in mind when learning how to tie knots. First, many knots can be tied in more than one way. As long as the chosen technique is followed correctly, the final structure of the knot will be the same, no matter which technique was used to tie it. Second, knots can be tied right-handed or left-handed, depending on the dominant hand of the person who is doing the tying. A knot that is tied left-handed will be the mirror image of the same knot tied right-handed. Note, however, that a few knots may have a right-handed element and a left-handed element. Failing to maintain these elements while tying can produce the wrong knot.

1.3.1 The Square Knot

A square knot (*Figure 7*), also called a reef knot, can be used to tie the ends of a rope around an object, such as a parcel or the neck of a sack. It is commonly used to join two lengths of rope together in low-strain applications. The two ropes should be of roughly the same diameter; otherwise, the knot is more likely to slip.

WARNING!

Square knots can capsize easily and should never be used for critical loads. Additional half knots are often added to square knots for increased security.

Square knot: A knot made of two reverse half-knots and typically used to join the ends of two ropes of similar diameters; also called a reef knot.

Bowline: A knot used to form a loop that neither slips nor jams; sometimes referred to as a rescue knot or the king of knots.

Half hitch: A knot tied by passing the working end of a rope around an object, across the standing part of the rope, and then through the resulting loop; often used as an element in forming other knots or added to make other knots more secure.

Clove hitch: A knot that consists of two half hitches made in opposite directions; used to temporarily secure a rope to an object.

Standing end: The end of a rope that is not being knotted.

Working end: The end of a rope that is being used to tie a knot.

Standing part: The portion of a rope that is between the standing end and the working end.

Capsize: To change the form and rearrange the parts of a knot, usually by pulling on specific ends of the knot.

Figure 7 Square knot.

Follow these steps (see *Figure 8*) to tie a square knot:

Step 1 Bring the two working ends together with one end oriented to the left (red rope) and the other to the right (blue rope).

Step 2 Cross them left over right. (The red rope goes left over the blue.)

Step 3 Pass the left-oriented (red) rope under the right. (The red rope goes under the blue.)

Step 4 This forms a half knot.

Step 5 Repeat the over-and-under process but go right over left this time. (The red rope goes right over the blue rope.)

Step 6 This forms a second half knot. Note that this is a key step in tying a square knot. The working end that starts out on the top when you tie the first half knot must continue to be on top when you switch directions to tie the second half knot.

Step 7 Pull the ends tight to make a neat, symmetrical square knot.

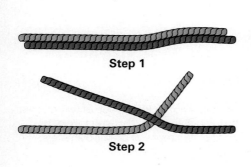

Step 1

Step 2

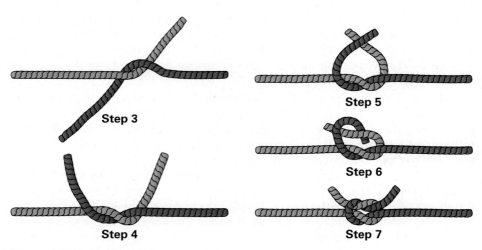

Step 3

Step 4

Step 5

Step 6

Step 7

Figure 8 Tying a square knot.

Figure 9 Granny knot.

A very common error when attempting to tie a square knot is to tie the second half knot in the wrong direction. This produces a highly insecure knot called the granny knot (*Figure 9*). Compare the granny knot to the properly tied square knot in *Figure 7*.

1.3.2   The Bowline

The bowline (*Figure 10*) is used to form a secure loop in the end of a rope. It is sometimes called a rescue knot, or the king of knots, because it is reliable enough to be used for rescue work (usually backed up with a stopper knot for extra security, or tied twice into a double bowline for extra strength). The bowline is useful for many smaller jobs such as securing a sack of material or a bucket of tools to be hauled up to workers on a platform. Two bowlines can also be linked together to join two ropes when the strain is more than a square knot is capable of bearing.

Figure 10 Bowline.

> **The Bowline**
>
> The bowline was originally used to secure a ship's square sail forward, closer to the wind. This knot was mentioned in *The Seaman's Dictionary* of 1644.
>
> The bowline reduces the strength of the rope that it is tied in by as much as 40 percent. Bowlines require a long tail for safety. A rarely practiced rule of thumb is that the loose end (tail) of a completed bowline should be as long as 12 times the circumference. This would translate to a tail of more than 18 inches for a rope with a one-half inch diameter.

A bowline does not slip or bind when under load, but it is easy to untie when there is no load. Because a bowline cannot be tied, or untied, when there is a load on the standing end, avoid using this knot if the rope may have to be released while it is under load.

Follow these steps (see *Figure 11*) to tie a bowline:

Step 1   Form a small loop in the rope, leaving a long enough tail to make the desired size of main loop.

Step 2   Pass the tail of the rope through the small loop.

Step 3   Pass the tail under the standing end.

Step 4   Pass the tail back down through the small loop.

Step 5   Adjust the main loop to the required size and tighten the knot, making sure to maintain a sufficiently long tail.

<table>
<tr><td>

Bowline Trivia

Some people use this saying to help them remember how to tie a bowline: "The rabbit comes out of his hole, around a tree, and back into the hole."

</td></tr>
</table>

Figure 12 Half hitch.

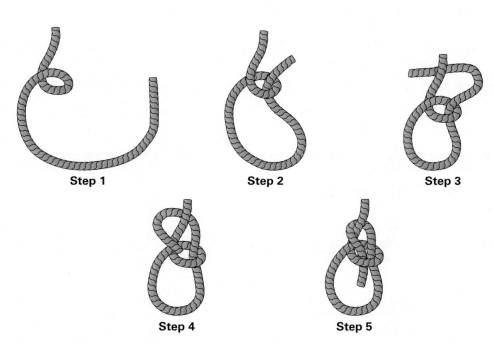

Figure 11 Tying a bowline.

1.3.3   The Half Hitch

A half hitch (*Figure 12*) ties a rope around an object such as a rail, bar, post, or ring. It is commonly used for tasks such as suspending items from overhead beams and carrying light loads that have to be removed easily. The half hitch is also widely used in making other knots, such as the clove hitch. Because the half hitch is not, by itself, a very stable knot, it is not suitable for heavy loads or tasks in which safety is paramount. It is often used to back up and secure another knot that has already been tied.

Follow these steps (*Figure 13*) to tie a half hitch:

Step 1   Form a loop around the object.

Step 2   Pass the tail of the rope around the standing part and through the loop.

Step 3   Tighten the hitch by pulling on the working end and the standing part of the rope simultaneously.

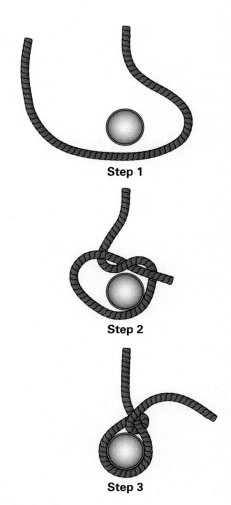

Figure 13 Tying a half hitch.

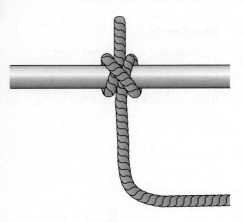

Figure 14 Clove hitch.

1.3.4 The Clove Hitch

A clove hitch (*Figure 14*), sometimes called a builder's hitch, is one of the most widely used general hitches. It is typically used to make a quick and secure tension knot on a fixed object that serves as an anchor, such as a post, pole, or beam. A clove hitch can also be used as the first knot when lashing items together.

Because it is a tension knot, a clove hitch comes loose as soon as tension is removed from the rope. On the other hand, it can also bind. For these reasons, a clove hitch should not be used by itself. Additional half hitches can be added to make a clove hitch more secure.

There are alternate techniques for tying a clove hitch. Regardless of the technique, however, the structure of the resulting knot consists of two half hitches made in opposite directions.

A common technique for tying a clove hitch that will be used to attach a rope to a ring or post is to thread the end, as described in the following steps (see *Figure 15*):

Step 1 Pass the working end of the rope over the object.

Step 2 Pass the working end back over the standing part and then over the object. This forms a half hitch.

Step 3 Thread the working end back under itself and up through the loop to form a second half hitch.

Step 4 Pull both the working and standing ends evenly to tighten the hitch.

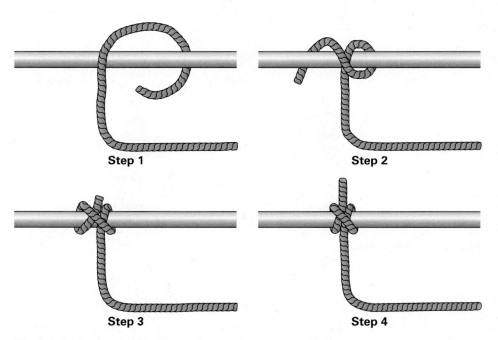

Figure 15 Tying a clove hitch—threading the end technique.

The stacked loops technique for tying a clove hitch allows the rope to be dropped quickly over a stake or post, instead of tying the knot around the object. It also allows the hitch to be tied at any point in a rope, not only at an end.

These are the basic steps of the stacked loops technique (see *Figure 16*):

Step 1 Form two identical loops in a rope, one in the right hand and one in the left.

Step 2 Cross the loops one above the other to form a knot.

Step 3 Place the knot over the stake or post.

Step 4 Tighten the knot by pulling simultaneously on both ends of the rope. The completed clove hitch consists of two half hitches stacked on each other.

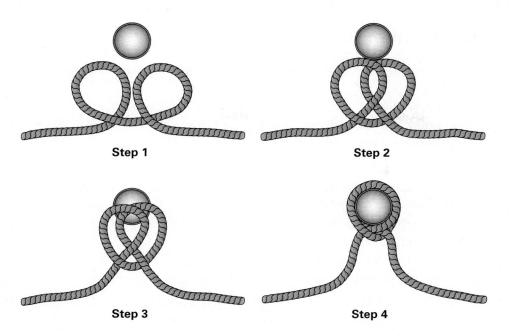

Figure 16 Tying a clove hitch—stacked loops technique.

Another common technique for tying a clove hitch around a post or timber is to use two half hitches, as described in the following steps (see *Figure 17*):

Step 1   Form a loop in the working end of the rope.

Step 2   Place the loop over the post (this makes the first half hitch).

Step 3   Form a second loop identical to the first.

Step 4   Place this loop over the post as well (making a second half hitch) and tighten to complete the clove hitch.

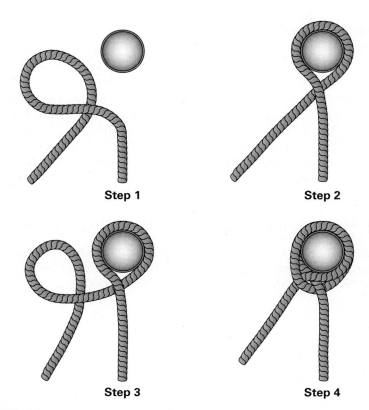

Figure 17 Tying a clove hitch—two half hitches technique.

1.0.0 Section Review

1. For material handling tasks, it is just as important to be mentally fit as it is to be _____.
   a. physically fit
   b. physically aggressive
   c. closely supervised
   d. over 200 pounds

2. When stacking or storing light materials that have a large surface area and could easily be moved by the wind, _____.
   a. keep the stack height less than four feet
   b. weight the top of the stack with clamps
   c. tie down or band the materials
   d. secure the materials in a flameproof cabinet

3. Which of the following is a type of knot that is often used to join the ends of two ropes in non-critical, low-strain applications?
   a. Bowline
   b. Clove hitch
   c. Half hitch
   d. Square knot

2.0.0 Material Handling Equipment

Performance Tasks

There are no performance tasks in this section.

Objective

Identify various types of material handling equipment and describe how they are used.

a. Identify non-motorized material handling equipment and describe how they are used.

b. Identify motorized material handling equipment and describe how they are used.

Instead of physically lifting heavy loads at work, use material handling devices when they are available to reduce your risk of back injury. Material handling devices help you to work smarter instead of harder, easing stress on your muscles and joints. These work-saving devices also increase productivity and reduce the chance of dropping and damaging equipment or supplies. There are two types of material handling equipment: non-motorized and motorized.

2.1.0 Non-Motorized Material Handling Equipment

When using non-motorized, or manual, material handling equipment, you use the force and strength of your body to move the equipment. Examples of manual material handling equipment include the following:

- **Material carts**
- **Hand trucks**
- **Cylinder carts**
- **Drum dollies**
- **Drum carts**
- **Roller skids**
- **Wheelbarrows**

Material carts: Four-wheeled devices used to transport materials around a job site.

Hand trucks: Two-wheeled carts that are used to transport large, heavy loads; also known as a dolly.

Cylinder carts: Two-wheeled carts that are used to transport cylinders, or bottles, of compressed gases.

- Pipe mules
- Pipe transports
- Pallet jacks

2.1.1 Material Carts

Material carts, also known as platform trucks, are platforms or boards that are laid horizontally on four caster wheels. These types of carts are typically used to transport materials around a job site (*Figure 18*). When using a material cart, follow these safety guidelines:

- Before using carts with caster wheels, inspect them to ensure that the casters will roll and swivel freely during transport.

- When using a cart with pneumatic tires, check the tires' air pressure before use. Tires that will not hold pressure or those that have flat spots must be replaced.

- Check to make sure that the surface has adequate traction to avoid slips.

- When moving a cart, keep your hands away from the edges of the handles to avoid pinching, crushing, or cutting your hands.

- Make sure your load is centered and secured so it does not roll or fall off.

- If materials are protruding out in front of the cart, have a **spotter** walk ahead of you to help you avoid contact with objects or other workers.

- Use caution when moving a cart on an inclined or declined surface, and never load a cart past its labeled weight capacity. If the load is too heavy on a decline, it may drag you down with it. If the load is too heavy on an incline, it may back over you. The weight capacity must be marked in plain view on the side of the cart.

Drum dollies: Wheeled circular platforms or caddies with a handle that are used to transport lighter-weight 55-gallon drums/barrels.

Drum carts: Wheeled carts that are used to transport heavier-weight 55-gallon drums/barrels.

Roller skids: Devices that include a surface table and two, three, or four roller skids. Materials that are to be moved are placed on the table surface and then pushed on the skids.

Wheelbarrows: One- or two-wheeled vehicles with handles at the rear that are used to carry small loads.

Pipe mules: Two-wheeled devices used to transport medium-length pieces of pipe, tubing, or scaffolding; sometimes referred to as a tunnel buggy.

Pipe transports: Wheeled devices similar to a pipe mule, but used to move larger pieces of pipe.

Pallet jacks: Devices used to lift and move heavy or stacked pallets; also known as a pallet truck.

Spotter: A person who walks in front of another worker who is carrying or transporting a long load to ensure there is a clear, unobstructed path.

Figure 18 Material cart.

2.1.2 Hand Trucks

Hand trucks, also known as dollies, are two-wheeled carts that are used to transport large, heavy loads, such as appliances or stacks of heavy boxes. Hand trucks are vertical material handling devices that have a metal blade at the bottom that is inserted beneath the load. Many hand trucks are equipped with ratchet straps to secure loads. The entire assembly tilts backward so that it may be easily pulled or pushed (*Figure 19*).

Figure 19 Hand truck.
Source: Northern Tool and Equipment

Before using a hand truck, inspect the framework for stress fractures, and check the tires. When loading a hand truck, tilt the object to be moved forward and slide the blade of the hand truck under the object. Hold the object against the hand truck's framework as you tilt back on the wheels to load the object onto the hand truck. Use a ratchet strap to secure the load; rubber or bungee-style cords may not hold the load if the weight shifts. Simply buckle the strap around the load and tighten the belt snugly. Then tilt the hand truck back and move the load.

2.1.3 Cylinder Carts

High-pressure gas cylinders, or bottles, should be transported on cylinder carts, which are designed specifically to transport compressed gases (*Figure 20*). The pressurized gas contained in cylinders can pose explosion risks, even if the gas itself is inert (meaning it is not chemically reactive).

Follow these guidelines when transporting compressed gas cylinders:

- Check the label on the cylinder to identify the gas that it contains and verify that the cylinder is approved for transporting the gas.
- Never transport a gas cylinder without the safety cap screwed all the way down to the body of the cylinder.
- Never lift a gas cylinder by the safety cap.
- Never transport a gas cylinder with the regulator attached.
- Use chains or straps to secure the cylinder to the cart and keep the cylinder upright.
- Many sites require a partition on the cart to separate the two bottles, as shown in *Figure 20*.
- Wear the appropriate PPE whenever handling gas cylinders.

Figure 20 Cylinder cart.
Source: Vestil Manufacturing

WARNING!

If the tank of a compressed gas cylinder ruptures or cracks during transport, the tank can become a deadly projectile or it could explode into fragments of shrapnel.

2.1.4 Drum Dollies and Carts

Chemicals, lubricants, and other materials are commonly stored in 55-gallon drums or barrels. Drums that are filled with materials can be very heavy and unwieldy to move. Two methods for safely transporting loaded drums around the workplace are drum dollies and drum carts.

A drum dolly (*Figure 21*) is used to move relatively light drums. The simplest type of drum dolly is a flat wheeled platform that the drum sits on. The dolly may be equipped with a strap to secure it and prevent movement, or a handle to pull and control the load.

A drum cart (*Figure 22*) is typically used to move heavier loads, such as sealed oil drums. One common type is a rotating drum cart. Its main features include prongs, handles, wheels, and a set of heavy duty casters.

To use a rotating drum cart follow these steps:

- With the cart held vertically, slide the two prongs of the cart under the drum.
- Allow the cart to roll back into the horizontal position and sit on the wheels and casters.
- With the drum resting horizontally in the well of the cart, use the cart handles to maneuver the load to its destination.
- Once at the destination, tip the cart into the vertical position to allow the drum contents to be unloaded.

Figure 21 Drum dolly.
Source: Vestil Manufacturing

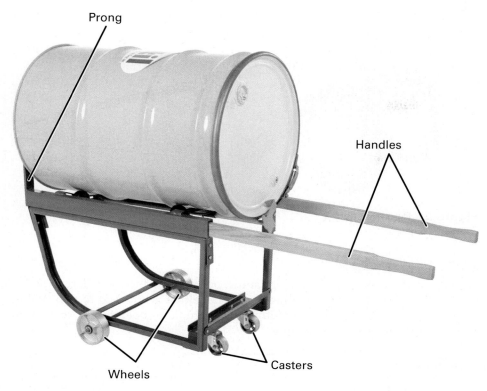

Figure 22 Drum cart.
Source: Vestil Manufacturing

2.1.5   Roller Skids

Roller skids move materials by pushing them on a table surface that is placed on top of two, three, or four roller skids. These devices are available in different capacities, depending on their intended usage. Roller skids used for moving heavy equipment use steel rollers or steel chain rollers, while those used for moving lighter equipment may use polyurethane or nylon rollers. The table surface of roller skids may differ from manufacturer to manufacturer. Some come equipped with a rotating table surface, some have spikes on the table surface for a better grip, and some have a plain surface like any other table. Many roller skids have a pull-push handle attachment for operators to move them easily (*Figure 23*). As with other material handling equipment, inspect the roller skids before use and use roller skids only for their intended purpose and weight capacity.

Roller skids must be used on a relatively smooth surface. The rollers can easily be stopped by small rocks or hardware on the floor surface. Ensure that the route ahead is swept, if necessary, prior to the move.

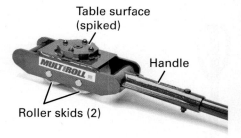

Figure 23 Roller skids.
Source: Cherry's Industrial Equipment Corp.

2.1.6   Wheelbarrows

A wheelbarrow is a one- or two-wheeled vehicle with handles at the rear, used to carry small loads (*Figure 24*). It is recommended that only two-wheeled wheelbarrows be used on job sites, since they are more stable with heavy loads. Remember to check the air pressure in both tires, and check the body of the wheelbarrow for cracks, breaks, or punctures before using it. Use proper lifting techniques when moving a wheelbarrow. Keep your head up, your back straight, bend your knees, and lift with your legs.

2.1.7   Pipe Mule

A pipe mule is a two-wheeled device used to transport medium-length pieces of pipe, tubing, or scaffolding (*Figure 25*). It is sometimes referred to as a tunnel buggy. It may also be called a V-cart or a grasshopper, which is a brand name. When loading or unloading a pipe mule, use proper lifting techniques. The proper loading method is to have one worker lift one end of the load while another worker rolls the pipe mule under the load. Place the wheels of the pipe mule underneath the center of the load. As with any long load, use a spotter when moving a load with a pipe mule.

Figure 24 Wheelbarrow.
Source: GAR-BRO Manufacturing Co., Heber Springs, Arkansas

Figure 25 Pipe mule.
Source: BE & K

Figure 26 Pipe transport.
Source: Sumner

2.1.8 Pipe Transport

A pipe transport is similar to a pipe mule, but it is used to move larger pieces of pipe. The pipe is slung underneath the frame that is attached to two rubber wheels (*Figure 26*). This device requires two people to operate it. Use caution when moving the pipe into position.

WARNING!

When using a pipe transport, beware of pinch points when securing chains around the pipe. Do not overload the equipment. Follow the manufacturer's recommendations. Keep both hands on the handle at all times to maintain safe control.

2.1.9 Pallet Jack

A pallet jack, also known as a pallet truck, is a device that typically uses hydraulics to lift and move heavy or stacked pallets (*Figure 27*). Both motorized and non-motorized versions of this device are available. Motorized pallet jacks are sometimes called walkies.

Remember the following safety guidelines when using a pallet jack:

Figure 27 Pallet jack.
Source: Northern Tool and Equipment

- Plan and review the route before using a pallet jack; be very careful about using it on uneven surfaces or ramps.

- Inspect pallet jacks before use, looking carefully for malfunctions or missing parts.

- Check for oil leaks around fittings or hoses of hydraulic pallet jacks, and repair any leaks before the jack is put into service.

- Never use a pallet jack to lift or transport a load that exceeds the jack's rated capacity. The load rating should be clearly posted on the body of the jack.

- Inspect the wood pallet to ensure it is in good condition to adequately handle the load, and make sure the load is centered between the pallet jack forks.

2.2.0 Motorized Material Handling Equipment

Motorized material handling equipment is powered by gasoline or electric motors. Operators must be trained, certified, and authorized to operate motorized material handling equipment. Training courses must consist of a combination of formal instruction, hands-on training, and evaluation of the operator's performance in the workplace for each type of material handling equipment being used. All operator training and evaluation must be conducted by a person who has knowledge, training, and experience.

When lifting materials mechanically, carefully secure the materials that are being loaded. Also be aware of any obstructions, such as people or equipment, that are in the path and have the potential to cause accidents.

When handling materials with a machine such as a pallet jack or forklift, always follow these guidelines:

- Know the weight of the object to be handled or moved.

- Know the capacity of the handling device that will be used and never exceed the capacity.

- Ensure that the handling equipment is in good working order and free of damage.

Motorized material handling equipment discussed in this section includes the following:

- **Powered wheelbarrow**
- **Concrete mule**
- **Freight elevator**
- **Industrial forklift**
- **Rough terrain forklift**

2.2.1 Powered Wheelbarrow

A powered wheelbarrow, also known as a power buggy, is similar to a manual wheelbarrow, but is powered by an electric or gas motor (*Figure 28*). Before using a powered wheelbarrow, conduct a thorough inspection of the machine. Check the brakes and brake linkages for proper operation and adjustment. Inspect tires and wheels for damage and proper tire pressure. Test the directional controls, dumping controls, and speed controls to verify they are all working properly. Be sure to read safety labels and decals for proper safety instructions. Safety labels should be clearly visible and readable on the side of the machine.

Keep the following safety guidelines in mind when operating a powered wheelbarrow:

- Avoid pinch points on dumping mechanisms.

- Be sure the dump bucket is securely down at all times when not dumping.

- Do not move the buggy with the bucket in the dumping position.

- Do not mount or dismount the buggy while it is moving.

- Shut off the engine and lock the parking brake when you are finished using the machine.

- Do not allow riders.

- Avoid any condition of slope and/or grade which could cause the buggy to tip.

- Do not exceed load limits in weight or height.

- Follow all procedures outlined on the safety labels and decals.

NOTE

Nameplates must be posted on each material handling device. The nameplate must indicate the capacity of the device, the approximate weight, and any instructional information.

Powered wheelbarrow: A vehicle similar to a manual wheelbarrow, but powered by an electric or gas motor; also known as a power buggy.

Concrete mule: A wheeled device used when a concrete pour is in a location that a concrete delivery truck or pump cannot reach; sometimes referred to as a Georgia buggy.

Freight elevator: An elevator used to transport materials from floor to floor.

Industrial forklift: A vehicle with a power-operated pronged platform that can be raised and lowered for insertion under a load to be lifted and moved.

Rough terrain forklift: Similar to an industrial forklift, but designed to be used on rough surfaces. Rough terrain forklifts are characterized by large pneumatic tires, usually with deep treads that allow the vehicle to grab onto the roughest of roads or ground cover without sliding or slipping.

Figure 28 Powered wheelbarrow.
Source: Power Barrows

Tugs

Some companies are reducing their use of forklifts and pallet jacks and are increasing the use of wheeled carts, called tugs, to transport loads throughout a facility. Material handling tugs are battery-powered devices that can push, as well as pull, loads. One reason the use of tugs has grown is that they can increase productivity by reducing the manpower required to move heavy loads. A single worker operating a tug can safely move loads that weigh hundreds, or even thousands, of pounds. In addition, unlike forklift operators, workers who operate tugs are not required to have special training and certification before they are authorized to operate the device.

Tugs are designed to distribute weight ideally and maximize torque, which minimizes the effort required for operators to pull or push loads. The body of the tug acts as a wedge, slightly lifting the load at an angle and transferring all the weight to the tug's drive wheels. Tugs are equipped with safety features to prevent unintended propulsion, including an emergency reverse switch that automatically stops the tug to prevent it from pinning or crushing the operator when a load is being pulled in reverse.

Source: Power Pusher Company, A division of Nu-Star Inc.

Figure 29 Concrete mule.
Source: Multiquip Inc.

2.2.2   Concrete Mule

A concrete mule, sometimes referred to as a Georgia buggy, is a wheeled device used when a concrete pour is in a location that a concrete delivery truck or pump cannot reach. Concrete mules (*Figure 29*) are designed to carry concrete, sand, and gravel, or materials of that nature, and should not be used to transport undesignated materials.

Before using a concrete mule, you must conduct a thorough inspection of the following:

- Tires
- Fluids
- Mechanical hinges and joints
- Throttle cables and steering mechanisms

When refueling a concrete mule, or any other gas-powered material handling machine, always be sure that the engine is turned off and has cooled down before refueling. Use only fuel and oil types that are specified by the manufacturer.

When discharging concrete mix from a concrete mule near an open pit, use a tire stop board or chocks to prevent the mule from rolling forward into the pit during the emptying process.

When operating a concrete mule, make sure there are no obstacles or other workers in the turning radius of the rear platform. Always look behind the mule before operating in reverse, and never allow anyone other than the operator to ride on the machine.

2.2.3   Freight Elevator

On multi-level sites, workers may use a freight elevator to transport materials from floor to floor. Typically, when freight elevator doors close they travel from the top and bottom simultaneously (*Figure 30*). These doors do not have the safety bumpers that are found on a commercial passenger elevator. Rather, as a person pulls the top door down, by way of a strap or handle, the bottom door rises to meet the top door. When these doors close, there is normally metal-against-metal contact, and this can cause injury. Make sure that all body parts such as arms, hands, and fingers are clear of the door before it closes. Check to make sure that tool belts, body harnesses, and all materials are also clear of the door. Always check the weight capacity shown on the elevator wall before placing material on an elevator.

Figure 30 Freight elevator.
Source: Cianbro Corporation

Figure 31 Industrial forklift.
Source: Taylor Machine Works

2.2.4   Industrial Forklift

An industrial forklift is a vehicle with a power-operated pronged platform that can be raised and lowered for insertion under a load to be lifted and moved (*Figure 31*). Forklifts are typically used to lift, lower, and transport large or heavy loads in areas with a smooth terrain, such as warehouses or shops.

When working in the vicinity of a forklift, remember the following safety guidelines:

- All workers must keep a safe distance from the machine. This perimeter is known as the **work zone**; it is the area in which the machine may come in contact with objects or people either with the rear end or the front forks.

- Workers must also stay clear of the fall zone. This area includes the area directly beneath a load, as well as any portion of the area that suspended materials could possibly fall into.

- Stay in designated walkways and make eye contact with the forklift operator to ensure that he or she sees you. Always avoid the driver's blind spot. Be aware that forklifts pivot on the front wheels and turn with the back wheels, which could cause a potential pinch point.

- Plan a route of escape. Never get caught in a position where you are trapped in a corner or on the side of a truck.

Trained and certified forklift operators must observe certain traffic safety rules and regulations while driving.

Before operating a forklift, put on the proper PPE and fasten the seat belt. If you are not wearing a seat belt and the forklift tips, you can easily be pitched under the forklift, or have some of the load land on you. Falling objects are common when stacking and lifting with a forklift, so wear a hard hat. As with most machinery, when working with a forklift do not wear baggy clothes that can catch on controls, cargo, or stacked objects.

Never play on a forklift or drive recklessly. Accidents often happen because workers forget to respect equipment. A forklift is not a toy and should not be treated like one.

Work zone: The area in which a forklift may come in contact with objects or people, either with the rear end or the front forks.

Going Green

Fuel Cell Technology

Increasingly, hydrogen fuel cell power units are replacing lead-acid batteries in forklifts, automated guided vehicles (AGVs), and other material handling vehicles. More than 4,000 vehicles powered by fuel cells were in operation in the United States in 2012, at companies such as Walmart, Coca Cola, BMW, FedEx, Lowe's, Mercedes-Benz, and Stihl Inc. In 2012, approximately 30,000 fuel cell systems were shipped worldwide.

A fuel cell is an electrochemical device that combines hydrogen and oxygen to produce electricity. Since the conversion of the fuel to energy takes place using an electrochemical process, and not combustion, the process is clean, quiet, and highly efficient. Fuel cell technology is estimated to be two to three times more efficient than fuel burning. Furthermore, the cost of fuel cells has dropped more than 50 percent since 2006.

One major way that material handling equipment powered by fuel cells increases productivity is by reducing downtime. The equipment only needs refueling once or twice a day, depending on use, and can be refueled in a few minutes—in contrast to traditional battery-powered equipment that typically has to be taken out of operation about every four to six hours to have the battery replaced or recharged, which usually takes several hours. In addition, unlike lead-acid batteries, hydrogen-powered fuel cells are not adversely affected by low temperatures in the work environment.

For many years, fuel cells were installed in forklifts only as a replacement for lead-acid batteries. Now, however, some lift truck manufacturers have begun producing forklifts that are originally designed to be powered by hydrogen fuel cells, rather than simply being retrofitted with this alternative power source.

Source: Power Plug, Inc.

Before operating the forklift each day, inspect the following:

- Tires
- Fluid levels
- Battery/fuel level
- Fire extinguisher
- Brakes
- Deadman control
- Warning lights
- Horn
- Backup alarm

Use a checklist, if one is available. If you find any defects during the inspection, report them to your supervisor.

Forklift operators with the fewest accidents drive with slow deliberation, take their time, and do not make mistakes. Do not attempt to cut corners to speed up your work. Drive slowly and keep your arms, hands, and legs inside the cage of the forklift. Forklifts are top-heavy and, when driven recklessly, can tip over on a curve, drive off the edge of a dock, or fail to stop in time. Drive at a speed that is appropriate and safe for the location and surface condition and use extra caution on hills, corners, and ramps. Keep the forklift to the right on roadways and in wide aisles, and stop at all designated stop signs. Always start and stop smoothly. This helps to preserve the equipment by putting less wear and tear on the forklift. It also keeps loads from shifting, objects from falling, and helps prevent damage to cargo, equipment, and operators.

Remember that pedestrians have the right-of-way. Forklifts, especially electric models, are often so quiet that they are difficult to hear. Pedestrians may not be aware that you are there, so be careful when you are driving near people, especially if they have their back to you. Make sure that the backup alarm is functional. Driving in reverse presents special dangers to pedestrians, especially if you are watching the load in front as you back up.

Workers sometimes attempt to make modifications to forklifts so that they can transport heavier or wider loads than the equipment is designed to lift. This introduces the hazards of broken masts or forks, dropped loads, and tipovers. Never modify a forklift without obtaining the consent from the manufacturer or from a mechanical engineer familiar with the equipment.

2.2.5 Rough Terrain Forklift

Rough terrain forklifts are designed to be used on irregular surfaces. They are typically used for jobs where there are no paved surfaces. Rough terrain forklifts are characterized by large pneumatic tires, usually with deep treads that allow the vehicle to grab onto the roughest of roads or ground cover without sliding or slipping (*Figure 32*). Be sure to apply the same safety guidelines to rough terrain forklifts as you would to industrial forklifts.

Figure 32 Rough terrain forklift.
Source: Taylor Machine Works

Forklift Automation

Automated guided vehicles (AGVs) have long been used successfully in manufacturing. Now automated forklifts, which are retrofitted with AGV technology, are poised to generate significant improvements in material handling productivity. Tasks that may be appropriate for forklift automation are similar to the tasks that AGVs have long handled in manufacturing—tasks involving repetitive, horizontal travel along the same route for relatively long distances.

The evolving AGV technology holds particular promise for material handling in the warehouse environment. One application that has emerged as a good fit for forklift automation is high-volume picking where the operator is able to remotely advance the forklift, eliminating the need for the operator to climb on and off the vehicle multiple times while working within an aisle.

One of the most common accidents involving rough terrain forklifts is tipovers. This is because these types of forklifts are often operated on uneven ground. To prevent tipovers with rough terrain forklifts, observe the following safety guidelines:

- Study load charts carefully and only operate the forklift within its stability limits.
- Before placing a load, apply the brakes, shift to neutral, level the frame, and engage the stabilizers, if so equipped.
- Carry the load low. Avoid driving with an elevated load except for short distances.
- Move carefully on slopes and only operate within specific grade limits.
- Retract the boom and lower the forks before moving a forklift.

Figure 33 Common forklift hand signals.

2.2.6 Hand Signals

The noise generated by motorized lifting equipment, such as forklifts, and the distance between workers when they are working with the equipment can make verbal communication difficult. Standard hand signals (*Figure 33*) allow both the workers and the forklift operator to go through the lifting, loading, moving, and unloading procedures at a more controlled setting, and help to avoid potential misunderstandings and accidents.

Be aware, however, that there is no globally accepted standard for hand signals. The signals shown here are simply some of the most commonly used ones. The differences between other versions are likely to be very minor. Your employer may have a specific set of hand signals that may or may not fully align with the signals presented here. In any case, prior to performing a task the equipment operator and the other workers must agree on the signals that will be used and the exact meaning for each signal.

For instance, it is important for the operator to know when to raise or lower the tines (the L-shaped parts that support the material load) of a forklift to avoid obstructions that may be in their path. Another very important signal for the operator to know is Dog (or pause) Everything. This signal means to pause, and can be used when potentially risky situations arise, such as when it starts raining, when the load does not fit the space for which it was planned, or when a bystander gets too close.

When providing hand signals, make sure to keep eye contact with the operator at all times. There may be special operations that require adaptations of the basic hand signals, so be sure to review the signals for your work site from time to time. If there is a misunderstanding regarding hand signals, go over the signals with the operator before proceeding.

2.0.0 Section Review

1. Which of the following are often used to secure loads on hand trucks?
   a. Dolly spotters
   b. Bungee cords
   c. Wheel chocks
   d. Ratchet straps

2. When refueling a concrete mule, always be sure the _____.
   a. work zone is at least twice the load height
   b. dump bucket is in the dumping position
   c. engine is turned off and has cooled down
   d. boom is in the fully retracted position

Module 00109 Review Questions

1. Loose gloves may get caught on moving objects, and they reduce _____.
   a. positive friction
   b. worker mobility
   c. manual torque
   d. grip strength

2. When lifting a load, keep the load _____.
   a. at shoulder height
   b. at arm's length
   c. close to your body
   d. balanced at your hips

3. The area into which a load is being lowered should be treated as a(n)
_____.

   a. hot work location
   b. fall zone
   c. primary access route
   d. aerial drop spot

4. How many inches should stacked bricks be tapered for every foot above
four feet in the stack?

   a. Two
   b. Four
   c. Six
   d. Twelve

5. The proper way to get tools to a worker on a higher level is to _____.

   a. toss them up carefully
   b. carry them by hand up a ladder
   c. climb up with them in your pocket
   d. hoist them up with a rope and bucket

6. A knot that is *NOT* meant to be permanent is generally referred to as
a _____.

   a. hitch
   b. slip
   c. runner
   d. bypass

7. The knot shown in Review Question *Figure RQ01* is a _____.

Figure RQ01

   a. square knot
   b. bowline
   c. half hitch
   d. clove hitch

8. To help make sure the path is clear when your view is obstructed while
handling materials, use a _____.

   a. mirror
   b. spotter
   c. tow rope
   d. step ladder

9. Before using a hand truck, inspect the framework for _____.

   a. oil residue
   b. popped nails
   c. stress fractures
   d. seal leaks

10. Roller skids for moving heavy equipment use rollers made of _____.

    a. aluminum
    b. brass
    c. polyurethane
    d. steel

11. Which of these devices requires two people to operate it?

    a. Roller skids
    b. Rotating drum cart
    c. Powered wheelbarrow
    d. Pipe transport

12. Before a worker uses a motorized material handling device to move a load, the worker must consider the _____.

    a. moisture content of the load
    b. fall zone for the load
    c. maximum capacity of the device
    d. universal transport hand signals

13. Which of the following is a motorized material handling device?

    a. Concrete mule
    b. Roller skid
    c. Porta-Power®
    d. Pipe mule

14. An industrial forklift pivots on the front wheels and turns with the back wheels, which could cause a _____.

    a. gravity shift
    b. pinch point
    c. rollover
    d. fall zone

15. When the sound of motorized lifting equipment makes verbal communication difficult, workers should use _____.

    a. written notes
    b. shouted commands
    c. hand signals
    d. a relay person

Answers to Section Review Questions and Module Review Questions are found in *Appendix A*.

APPENDIX A Answer Keys

MODULE 00100

Answer	Section Head	Objective

Section 1.0.0
Answer	Section Head	Objective
1. d	1.1.2	1a
2. d	1.1.3	1a
3. b	1.2.0	1b

Section 2.0.0
Answer	Section Head	Objective
1. a	2.1.1	2a
2. b	2.2.2	2b
3. c	2.2.3	2b

Section 3.0.0
Answer	Section Head	Objective
1. b	3.1.0	3a
2. d	3.1.0	3a
3. a	3.2.0	3b
4. c	3.2.6	3b

Section 4.0.0
Answer	Section Head	Objective
1. a	4.1.0	4a
2. b	4.2.0	4b
3. b	4.3.2	4c

Module Review
Answer	Section Head
1. c	1.1.2
2. a	1.1.3
3. a	1.1.3
4. a	1.1.3
5. d	1.2.1
6. b	2.1.1
7. d	2.2.1
8. c	2.2.3
9. d	3.1.0
10. b	3.1.0
11. a	3.1.0
12. c	3.2.0
13. d	3.2.3
14. b	3.2.4
15. b	4.2.0

MODULE 00101

Section 1.0.0
Answer	Section Head	Objective
1. c	1.0.0	1a
2. d	1.2.1	1b
3. a	1.3.5	1c

Section 2.0.0
Answer	Section Head	Objective
1. c	2.1.2	2a
2. a	2.2.1	2b
3. b	2.3.1	2c
4. c	2.4.2	2d

Section 3.0.0
Answer	Section Head	Objective
1. a	3.1.2	3a
2. d	3.2.1	3b

Section 4.0.0
Answer	Section Head	Objective
1. b	4.1.1	4a
2. b	4.2.0	4b

Section 5.0.0
Answer	Section Head	Objective
1. a	5.1.2	5a
2. d	5.2.1	5b

Section 6.0.0
Answer	Section Head	Objective
1. b	6.1.0	6a
2. c	6.2.5	6b
3. a	6.3.1	6c
4. b	6.4.1	6d
5. b	6.5.0	6e

Module Review
Answer	Section Head
1. b	1.1.1
2. b	1.2.1
3. a	1.2.9
4. c	1.2.9
5. a	1.3.0
6. a	1.3.0
7. c	1.3.4
8. c	2.3.1
9. b	2.3.1
10. a	2.4.1
11. d	3.1.1
12. b	3.1.1
13. b	3.1.3
14. b	3.2.1
15. d	3.2.1
16. b	3.2.1
17. d	3.2.1
18. b	3.2.2
19. a	4.1.4
20. c	4.1.4
21. a	4.2.0
22. c	5.1.2
23. c	5.2.3
24. b	6.1.0
25. b	6.2.7
26. a	6.3.2
27. c	6.3.2
28. b	6.4.1
29. a	6.4.4; Figure 79
30. a	6.5.0

MODULE 00102

The math calculations for these answers are provided on pages 419–420.

Section 1.0.0
Answer	Section Head	Objective
1. d	1.1.0	1a
2. c*	1.2.0	1b
3. b*	1.3.0	1c

Section 2.0.0
Answer	Section Head	Objective
1. b*	2.1.0	2a
2. b*	2.2.0	2b
3. c*	2.3.0	2c
4. d*	2.3.0	2c
5. a*	2.4.0	2d
6. c*	2.4.0	2d

Section 3.0.0
Answer	Section Head	Objective
1. c	3.1.0	3a
2. a	3.1.0	3a
3. b*	3.2.1	3b
4. a*	3.2.1	3b
5. d*	3.2.2	3b
6. a*	3.2.3	3b
7. c	3.3.1	3c
8. c*	3.3.2	3c
9. a*	3.3.2	3c

Section 4.0.0
Answer	Section Head	Objective
1. a	4.1.2	4a
2. b	4.2.3	4b

Section 5.0.0
Answer	Section Head	Objective
1. b	5.1.2	5a
2. d	5.2.1	5b
3. c	5.3.2	5c
4. b	5.4.0	5d

Section 6.0.0
Answer	Section Head	Objective
1. d	6.1.0	6a
2. b	6.2.2; 6.2.4	6b
3. c*	6.3.0	6c
4. c*	6.4.1	6d

Module Review
Answer	Section Head
1. b	1.1.0
2. a*	1.2.0

Answer	Section Head	Objective
3. c*	1.2.0	
4. d*	1.3.0	
5. c*	1.3.0	
6. c*	2.1.1	
7. a	2.1.3	
8. d*	2.3.0	
9. a*	2.3.0	
10. b	3.1.2	
11. a*	3.2.1	
12. b*	3.2.2	
13. b*	3.2.2	
14. a*	3.2.3	
15. c	3.3.1	
16. c*	3.3.3	
17. a	4.1.1	
18. d	4.1.2	
19. c	4.2.1	
20. a	5.0.0	
21. a	5.1.2	
22. c*	5.1.3	
23. b	5.2.3	
24. d*	5.2.3	
25. a*	5.4.0	
26. d	6.1.0	
27. b	6.2.1	
28. a	6.3.0	
29. d*	6.4.3	
30. a*	6.4.4	

MODULE 00103

Section 1.0.0

Answer	Section Head	Objective
1. c	1.1.1	1a
2. a	1.2.1	1b
3. b	1.3.0	1c
4. c	1.4.2	1d
5. d	1.5.5	1e

Section 2.0.0

Answer	Section Head	Objective
1. d	2.1.3	2a
2. a	2.2.5	2b

Section 3.0.0

Answer	Section Head	Objective
1. d	3.1.0	3a
2. d	3.2.0	3b
3. c	3.3.1	3c
4. a	3.4.0	3d

Module Review

Answer	Section Head
1. d	1.1.1
2. b	1.1.1

Answer	Section Head	Objective
3. a	1.1.5	
4. c	1.2.1	
5. b	1.2.1	
6. a	1.2.2	
7. d	1.3.0	
8. c	1.4.1	
9. b	1.4.2	
10. c	1.4.3	
11. a	1.5.0	
12. b	2.1.4	
13. a	2.2.0; 2.2.1	
14. c	2.2.1	
15. b	2.2.5	
16. d	3.1.0	
17. c	3.2.0	
18. b	3.3.1	
19. b	3.3.2	
20. d	3.4.0	

MODULE 00104

Section 1.0.0

Answer	Section Head	Objective
1. a	1.1.0	1a
2. b	1.2.5	1b
3. d	1.3.0	1c
4. c	1.4.2	1d

Section 2.0.0

Answer	Section Head	Objective
1. a	2.1.2	2a
2. d	2.2.1	2b
3. c	2.3.0	2c
4. b	2.4.1	2d
5. c	2.5.2	2e

Section 3.0.0

Answer	Section Head	Objective
1. c	3.1.0	3a
2. a	3.2.0	3b
3. a	3.3.0	3c

Section 4.0.0

Answer	Section Head	Objective
1. d	4.1.0	4a
2. d	4.2.1	4b

Module Review

Answer	Section Head
1. a	1.0.0
2. a	1.1.0
3. c	1.1.0
4. c	1.2.0
5. d	1.2.1
6. b	1.2.4
7. c	1.2.6
8. b	1.3.0

Answer	Section Head	Objective
9. a	1.3.3	
10. a	1.4.3	
11. b	2.1.2	
12. d	2.2.2	
13. a	2.2.3	
14. b	2.2.4	
15. a	2.3.1	
16. c	2.4.1	
17. a	2.4.2	
18. c	3.1.0	
19. a	3.1.0	
20. b	4.1.0	
21. c	4.2.0	
22. c	4.2.0	

MODULE 00105

Section 1.0.0

Answer	Section Head	Objective
1. c	1.1.4	1a
2. d	1.2.2	1b
3. a	1.3.2	1c
4. b	1.4.1	1d
5. b	1.5.1	1e

Module Review

Answer	Section Head
1. a	1.1.1
2. c	1.1.4
3. a	1.2.1
4. b	1.2.3
5. d	1.3.1
6. c	1.3.2
7. d	1.4.2
8. b	1.4.3
9. c	1.5.4
10. d	1.5.6

MODULE 00106

Section 1.0.0

Answer	Section Head	Objective
1. c	1.1.0	1a
2. b	1.2.1	1b
3. a	1.3.1	1c
4. c	1.4.0	1d
5. d	1.5.2	1e

Module Review

Answer	Section Head
1. d	1.1.1
2. a	1.1.4
3. c	1.2.1
4. b	1.2.2

Answer	Section Head	Objective
5. c	1.3.1	
6. a	1.3.3	
7. d	1.4.0	
8. b	1.4.2	
9. c	1.5.1	
10. a	1.5.4	

MODULE 00107

Section 1.0.0

Answer	Section Head	Objective
1. b	1.1.1	1a
2. a	1.2.0	1b
3. a	1.3.0	1c
4. d	1.3.1	1c
5. c	1.3.1	1c

Section 2.0.0

Answer	Section Head	Objective
1. b	2.1.0	2a
2. a	2.2.0	2b
3. c	2.2.0	2b
4. d	2.3.0	2c
5. a	2.3.1	2c

Module Review

Answer	Section Head
1. a	1.0.0
2. a	1.1.1
3. a	1.1.1
4. b	1.1.2
5. b	1.2.0
6. c	1.2.0
7. d	1.3.0
8. b	1.3.2
9. b	2.1.0
10. d	2.2.0

Answer	Section Head	Objective
11. c	2.2.0	
12. a	2.2.0	
13. d	2.3.1	
14. a	2.3.1	
15. d	2.3.2	

MODULE 00108

Section 1.0.0

Answer	Section Head	Objective
1. c	1.1.0	1a
2. b	1.2.0	1b

Section 2.0.0

Answer	Section Head	Objective
1. a	2.1.0	2a
2. d	2.2.0	2b
3. a	2.3.1	2c

Section 3.0.0

Answer	Section Head	Objective
1. d	3.1.2	3a
2. a	3.2.1	3b
3. b	3.3.1	3c
4. c	3.4.1	3d
5. b	3.5.1	3e

Module Review

Answer	Section Head
1. d	1.2.0
2. a	1.2.0
3. b	2.1.1
4. c	2.2.0
5. d	2.3.4
6. c	3.0.0
7. a	3.1.2
8. b	3.2.0
9. d	3.3.1
10. d	3.4.1

Answer	Section Head	Objective
11. c	3.4.2	
12. c	3.4.2	
13. a	3.5.0	
14. a	3.5.1	
15. c	3.5.1	

MODULE 00109

Section 1.0.0

Answer	Section Head	Objective
1. a	1.1.1	1a
2. c	1.2.1	1b
3. d	1.3.1	1c

Section 2.0.0

Answer	Section Head	Objective
1. d	2.1.2	2a
2. c	2.2.2	2b

Module Review

Answer	Section Head
1. d	1.1.2
2. c	1.1.3
3. b	1.1.3
4. a	1.2.1
5. d	1.2.2
6. a	1.3.0
7. b	1.3.2
8. b	2.1.1
9. c	2.1.2
10. d	2.1.5
11. d	2.1.8
12. c	2.2.0
13. a	2.2.2
14. b	2.2.3
15. c	2.2.6

MODULE 00102 MATH CALCULATIONS

Section 1.0.0 Section Review Calculations

2. $46 + 8 + 8 + 30 + 8 + 10 + 20 + 10 + 8 + 10 + 8 + 10 + 10 + 38 = 224$ feet

3. 224 feet $\times$ 12 inches per foot = 2688 inches $\div$ 16 inches per block = 168 blocks per course
A 24-inch-high foundation requires 3 courses of 8-inch block; therefore, 3×168 blocks = 504 blocks.

Section 2.0.0 Section Review Calculations

1. $1/2 \times {}^{32}/_{32} = {}^{32}/_{64}$

2. ${}^{76}/_{64} = 76 \div 64 = 1$ with a remainder of ${}^{12}/_{64}$

   reduce ${}^{12}/_{64}$ by dividing both numerator and denominator by 2, giving ${}^{6}/_{32}$

   since $1{}^{6}/_{32}$ does not equal $1{}^{9}/_{32}$, the answer is False.

3. Convert ${}^{3}/_{8}$ to a common denominator with ${}^{11}/_{16}$: ${}^{3}/_{8} \times {}^{2}/_{2} = {}^{6}/_{16}$
   ${}^{11}/_{16} + {}^{6}/_{16} = {}^{17}/_{16}$, or $1{}^{1}/_{16}$

4. Convert $1{}^{1}/_{4}$ to a common denominator with ${}^{7}/_{16}$: $1{}^{1}/_{4} = {}^{5}/_{4}$; ${}^{5}/_{4} \times {}^{4}/_{4} = {}^{20}/_{16}$
   ${}^{20}/_{16} - {}^{7}/_{16} = {}^{13}/_{16}$

5. ${}^{1}/_{8} \times {}^{1}/_{10} = (1 \times 1)/(8 \times 10) = {}^{1}/_{80}$

MODULE 00102 MATH CALCULATIONS (continued)

6. Invert the divisor and multiply:

   $\frac{7}{16} \div \frac{7}{8} = \frac{7}{16} \times \frac{8}{7} = (7 \times 8)/(16 \times 7) = \frac{56}{112}$

   Using a factor of 56, reduce to lowest terms:
   $(56 \div 56)/(112 \div 56) = \frac{1}{2}$

Section 3.0.0 Section Review Calculations

3. $3.625 + 4.9 = 8.525$

4. $42.58 - 7.577 = 35.003$

5. $9.64 \times 12 = 115.68$

6. $123.82 \div 6.5 = 19.04923$

8. $\frac{7}{8} = 7 \div 8 = 0.875$

9. $\frac{2}{3} = 2 \div 3 = 0.666.$

   Convert 0.666 to a percentage by moving the decimal 2 places to the right: 66.6%
   66.6% is greater than 65%

Section 6.0.0 Section Review Calculations

3. Area = length × width

   Area = 27.3 m × 9.3 m

   Area = 253.89 m^2

4. Volume = length × width × depth

   Volume = 26 ft × 13.66 ft × 3 ft

   Volume = 1,065.48 ft^3

Module Review Question Calculations

2. Total bricks laid = 649 bricks + 632 bricks + 478 bricks

   Total bricks laid = 1,759

3. 1,478 feet − 489 feet installed = 989 feet of cable remaining

4. 15 scaffolds × 26 sites = 390 scaffolds

5. 400 rolls insulation ÷ 5 sites = 80 rolls per site

6. $\frac{3}{8} \times \frac{8}{8} = \frac{24}{64}$

8. $\frac{3}{8} \times \frac{2}{2} = \frac{6}{16}$

   $\frac{6}{16} + \frac{9}{16} = \frac{15}{16}$

9. $\frac{2}{8} \times \frac{4}{4} \times \frac{8}{32}$

   $\frac{11}{32} - \frac{8}{32} = \frac{3}{32}$

11. 51.5 nanometers + 89.7 nanometers = 141.2 nanometers

12. $3.53 \times 9.75 = 34.4175$, rounded up to 34.42

13. 864.5 square feet × $2.37 per square foot = $2,048.865, rounded up to $2,048.87

14. $89.435 \div 0.05 = 1,788.7$

16. $14.75 = \frac{1475}{100} = 1475 \div \frac{25}{100} \div 25 = \frac{59}{4}$

22. 1 inch = 2.54 centimeters

    67 inches × 2.54 centimeters per inch = 170.18 centimeters

24. 1 ounce = 28.35 grams

    49 ounces × 28.35 grams = 1389.15 grams

25. $°C = \frac{5}{9}(15°F - 32°) = \frac{5}{9} \times -17° = -9.4°C$

29. Volume of cylinder = pi × r^2 × h

    = 3.14 × (6.25 feet)^2 × 23 feet

    = 3.14 × 39.06 × 23

    = 2,821 cu ft

30. Volume of triangular prism = 0.5 × base × height × depth

    = 0.5 × 7 centimeters × 4 centimeters × 3 centimeters

    = 42 cm^3

Answers to Study Problems

The math calculations for these answers are provided on pages 421–428.

Section 1.1.1 Place Values of Whole Numbers

1. a	4. b
2. b	5. c
3. b	

Section 1.2.1 Adding and Subtracting Whole Numbers

*1. $1,480	*4. $216
*2. R$2,278	*5. Rub891,135
*3. 156 hours	

Section 1.3.2 Multiplying and Dividing Whole Numbers

*1. 12 bags

*2. a. (520) 2 × 4s
   b. (360) 2 × 8s
   c. (200) 2 × 10s

*3. 13 lengths of pipe needed; 3 meters left over

*4. a. $2,400
   b. $8,400
   c. $18,100

Section 2.1.4 Finding Equivalent Fractions

1. b*	6. $\frac{1}{8}$*	11. c*
2. c*	7. $\frac{1}{4}$*	12. b*
3. d*	8. $\frac{3}{8}$*	13. d*
4. a*	9. $\frac{1}{2}$*	14. b*
5. c*	10. $\frac{1}{16}$*	15. d*

Section 2.2.1 Changing Improper Fractions to Mixed Numbers

*1. $4\frac{3}{8}$	*4. $2\frac{4}{5}$
*2. $6\frac{1}{16}$	*5. 3
*3. $3\frac{1}{4}$	

Section 2.3.1 Adding and Subtracting Fractions

1. $\frac{3}{8}$*	6. $\frac{1}{16}$*	11. $7\frac{1}{4}$*
2. $\frac{7}{8}$*	7. $\frac{1}{16}$*	12. $11\frac{3}{8}$*
3. $1\frac{1}{4}$*	8. $\frac{5}{12}$*	13. d*
4. 1*	9. $\frac{5}{12}$*	14. b*
5. $1\frac{1}{4}$*	10. $\frac{3}{16}$*	15. a*

MODULE 00102 MATH CALCULATIONS (continued)

Section 2.4.1 Multiplying and Dividing Fractions

1. $\frac{5}{32}$*
2. $\frac{21}{32}$*
3. $3\frac{3}{4}$*
4. 21*
5. $\frac{1}{4}$*
6. $\frac{1}{8}$*
7. $1\frac{1}{4}$*
8. 2*
9. a*
10. d*

Section 3.1.3 Working with Decimals

1. b
2. c
3. a
4. c
5. d
6. c
7. d
8. a
9. b
10. b

Section 3.2.5 Decimals

1. 11.7*
2. 10.48*
3. 92.83*
4. b*
5. b*
6. c*
7. a*
8. b*
9. 2.52*
10. 0.25*
11. 0.03*
12. 0.41*
13. 5.2*
14. 20*
15. 216.88*
16. 220*
17. 40.8*
18. 520*
19. 4.91*
20. 10.51*
21. 22*
22. 40.8*
23. 52*
24. 53.74
25. 0.91
26. 123.13
27. 1.98
28. 38.4
29. b*
30. d*
31. d*
32. c*
33. b*

Section 3.3.5 Converting Different Values

1. 62%*
2. 47.5%*
3. 70%*
4. 0.72*
5. 0.125*
6. 0.25*
7. 0.75*
8. 0.125*
9. 0.3125*
10. 0.3125*
11. $\frac{1}{2}$*
12. $\frac{3}{25}$*
13. $\frac{1}{8}$*
14. $\frac{4}{5}$*
15. $\frac{9}{20}$*
16. 0.75*
17. 0.83*
18. 0.17*
19. 0.33*
20. 1.42*

Section 3.3.6 Practical Applications

1. a. 343 feet
   b. $406.98
   c. $48.84
   d. $358.14
   e. $21.49
   f. $379.63

2. a. 50.2%
   b. 43.0%
   c. 6.8%

Section 4.1.3 Reading Rulers

1. $\frac{4}{8}$ in
2. $1\frac{7}{8}$ in
3. $2\frac{6}{8}$ in
4. $3\frac{6}{8}$ in
5. $4\frac{5}{8}$ in
6. $\frac{10}{16}$ in
7. $1\frac{9}{16}$ in
8. $2\frac{5}{16}$ in
9. $3\frac{4}{16}$ in
10. $4\frac{7}{16}$ in
11. 0.1 cm
12. 1.1 cm
13. 1.9 cm
14. 29 mm
15. 38 mm

Section 4.2.4 Reading Measuring Tapes

1. $8\frac{13}{16}$ in
2. $9\frac{13}{16}$ in
3. $10\frac{14}{16}$ in (or $10\frac{7}{8}$")
4. $3\frac{4}{16}$ in (or $3\frac{1}{4}$")
5. $4\frac{11}{16}$ in
6. 4.4 cm
7. 5.9 cm
8. 9.9 cm
9. 11 cm
10. 11.8 cm

Section 5.1.4 Converting Measurements

1. 45 cm*
2. 108 in*
3. 12 yards*
4. 2.5 yards*
5. 0.01 m*
6. 167.64 cm*
7. 14.33 m*
8. 1.79 feet*
9. 17.37 m*
10. 185.04 in*

Section 5.2.4 Converting Weight Units

*1. 22.68 kilograms
*2. 110.25 pounds
*3. 450.77 grams
*4. 3.32 ounces

Section 5.3.4 Converting Volume Units

*1. 6.7 cubic feet
*2. 1,900,000.0 cubic centimeters
*3. 669.7 cubic yards
*4. 198,218.0 cubic centimeters
*5. 7.1 cubic feet

Section 5.4.1 Converting Temperatures

*1. 82.2°C
*2. 18.9°C
*3. −14.8°F
*4. 159.8°F

Section 6.3.1 Calculating Area

1. c*
2. a*
3. c*
4. a*
5. c*

Section 6.4.5 Calculating Volume

1. d*
2. d*
3. b*
4. a*
5. b*

Section 6.4.6 Practical Applications

*1. 653 in^3 (rounded)
*2. 109 ft^3

Study Problem Calculations

Section 1.2.1 Adding and Subtracting Whole Numbers

1. $$\begin{array}{r} 850 \\ 460 \\ +\ 170 \\ \hline 1,480 \end{array}$$

2. $10,236 − (2,477 + 2,263 + 3,218)$

    $10,236 − 7,958 = 2,278$

MODULE 00102 MATH CALCULATIONS (continued)

3. Crew 1: $10 + 9 + 11 + 12 + 9 = 51$

   Crew 2: $9 + 12 + 12 + 9 + 9 = 51$

   Crew 3: $12 + 12 + 10 + 9 + 11 = 54$

   Total: $51 + 51 + 54 = 156$

4. Step 1.
$$\begin{array}{r} 874 \\ 67 \\ +\ 93 \\ \hline 1{,}034 \end{array}$$

   Step 2.
$$\begin{array}{r} 1{,}250 \\ -\ 1{,}034 \\ \hline 216 \end{array}$$

5. Step 1. $850{,}360 + 119{,}980 = 970{,}340$

   Step 2. $970{,}340 - 79{,}205 = 891{,}135$

Section 1.3.2 Multiplying and Dividing Whole Numbers

1.
$$\begin{array}{r} 12 \\ 15\overline{)180} \\ -15 \\ \hline 30 \\ -30 \\ \hline 0 \end{array}$$

2. Step 1.
$$\begin{array}{r} 4 \\ \times 2 \\ \hline 8 \text{ locations} \end{array}$$

   Step 2.
$$\begin{array}{r} 65 \\ \times 8 \\ \hline 520 \end{array} \ 2 \times 4\text{s}$$

   Step 3.
$$\begin{array}{r} 45 \\ \times 8 \\ \hline 360 \end{array} \ 2 \times 8\text{s}$$

   Step 4.
$$\begin{array}{r} 25 \\ \times 8 \\ \hline 200 \end{array} \ 2 \times 10\text{s}$$

3. Step 1. $45 + 30 = 75$
   Step 2. $75 \div 6 = 12.5$; round up to 13
   Step 3. $13 \times 6 = 78$
   Step 4. $78 - 75 = 3$

4. a. $3 \times 800 = 2{,}400$
   b. Step 1. $2 \times 3{,}400 = 6{,}800$
      Step 2. $2 \times 800 = 1{,}600$
      Step 3. $6{,}800 + 1{,}600 = 8{,}400$
   c. Step 1. $1 \times 10{,}500 = 10{,}500$
      Step 2. $2 \times 3{,}400 = 6{,}800$
      Step 3. $1 \times 800 = 800$
      Step 4. $10{,}500 + 6{,}800 + 800 = 18{,}100$

Section 2.1.4 Finding Equivalent Fractions

1. $\dfrac{1 \times 4}{4 \times 4} = \dfrac{4}{16}$

2. $\dfrac{2 \times 2}{16 \times 2} = \dfrac{4}{32}$

3. $\dfrac{3 \times 2}{4 \times 2} = \dfrac{6}{8}$

4. $\dfrac{3 \times 16}{4 \times 16} = \dfrac{48}{64}$

5. $\dfrac{3 \times 2}{16 \times 2} = \dfrac{6}{32}$

6. $\dfrac{2 \div 2}{16 \div 2} = \dfrac{1}{8}$

7. $\dfrac{2 \div 2}{8 \div 2} = \dfrac{1}{4}$

8. $\dfrac{12 \div 4}{32 \div 4} = \dfrac{3}{8}$

9. $\dfrac{4 \div 4}{8 \div 4} = \dfrac{1}{2}$

10. $\dfrac{4 \div 4}{64 \div 4} = \dfrac{1}{16}$

11. $\dfrac{2}{6}, \dfrac{3}{4}$

    $\dfrac{2 \times 2}{6 \times 2} = \dfrac{4}{12}$

    $\dfrac{3 \times 3}{4 \times 3} = \dfrac{9}{12}$

    $\dfrac{4}{12}, \dfrac{9}{12}$

    12

12. $\dfrac{1}{4}, \dfrac{3}{8}$

    $\dfrac{1 \times 2}{4 \times 2} = \dfrac{2}{8}$

    $\dfrac{2}{8}, \dfrac{3}{8}$

    8

13. $\dfrac{1}{8}, \dfrac{1}{2}$

    $\dfrac{1 \times 4}{2 \times 4} = \dfrac{4}{8}$

    $\dfrac{1}{8}, \dfrac{4}{8}$

    8

14. $\dfrac{1}{4}, \dfrac{3}{16}$

    $\dfrac{1 \times 4}{4 \times 4} = \dfrac{4}{16}$

    $\dfrac{4}{16}, \dfrac{3}{16}$

    16

MODULE 00102 MATH CALCULATIONS (continued)

15. $\dfrac{4}{32}, \dfrac{5}{8}$

$\dfrac{4 \div 4}{32 \div 4} = \dfrac{1}{8}$

$\dfrac{1}{8}, \dfrac{5}{8}$

8

Section 2.2.1 Changing Improper Fractions to Mixed Numbers

1. $^{35}\!/_8 = 35 \div 8 = 4$, r3 or $4\frac{3}{8}$

2. $^{97}\!/_{16} = 97 \div 16 = 6$, r1 or $6\frac{1}{16}$

3. $^{13}\!/_4 = 13 \div 4 = 3$, r1 or $3\frac{1}{4}$

4. $^{14}\!/_5 = 14 \div 5 = 2$, r4 or $2\frac{4}{5}$

5. $^{48}\!/_{16} = 48 \div 16 = 3$

Section 2.3.1 Adding and Subtracting Fractions

1. $\dfrac{1}{8} + \dfrac{4}{16}$

$\dfrac{4 \div 2}{16 \div 2} = \dfrac{2}{8}$

$\dfrac{1}{8} + \dfrac{2}{8} = \dfrac{3}{8}$

2. $\dfrac{4}{8} + \dfrac{6}{16}$

$\dfrac{6 \div 2}{16 \div 2} = \dfrac{3}{8}$

$\dfrac{4}{8} + \dfrac{3}{8} = \dfrac{7}{8}$

3. $\dfrac{2}{4} + \dfrac{3}{4} = \dfrac{5}{4}$

$5 \div 4 = 1$, r1 or $1\frac{1}{4}$

4. $\dfrac{3}{4} + \dfrac{2}{8}$

$\dfrac{2 \div 2}{8 \div 2} = \dfrac{1}{4}$

$\dfrac{3}{4} + \dfrac{1}{4} = \dfrac{4}{4}$

$\dfrac{4}{4} = 1$

5. $\dfrac{14}{16} + \dfrac{3}{8}$

$\dfrac{14 \div 2}{16 \div 2} = \dfrac{7}{8}$

$\dfrac{7}{8} + \dfrac{3}{8} = \dfrac{10}{8}$

$\dfrac{2 \div 2}{8 \div 2} = \dfrac{1}{4}$

$1\frac{1}{4}$

6. $\dfrac{3}{8} - \dfrac{5}{16}$

$\dfrac{3 \times 2}{8 \times 2} = \dfrac{6}{16}$

$\dfrac{6}{16} - \dfrac{5}{16} = \dfrac{1}{16}$

7. $\dfrac{11}{16} - \dfrac{5}{8}$

$\dfrac{5 \times 2}{8 \times 2} = \dfrac{10}{16}$

$\dfrac{11}{16} - \dfrac{10}{16} = \dfrac{1}{16}$

8. $\dfrac{3}{4} - \dfrac{2}{6}$

$\dfrac{3 \times 3}{4 \times 3} = \dfrac{9}{12}$

$\dfrac{2 \times 2}{6 \times 2} = \dfrac{4}{12}$

$\dfrac{9}{12} - \dfrac{4}{12} = \dfrac{5}{12}$

9. $\dfrac{11}{12} - \dfrac{4}{8}$

$\dfrac{11 \times 2}{12 \times 2} = \dfrac{22}{24}$

$\dfrac{4 \times 3}{8 \times 3} = \dfrac{12}{24}$

$\dfrac{22}{24} - \dfrac{12}{24} = \dfrac{10}{24}$

$\dfrac{10 \div 2}{24 \div 2} = \dfrac{5}{12}$

10. $\dfrac{11}{16} - \dfrac{1}{2}$

$\dfrac{1 \times 8}{2 \times 8} = \dfrac{8}{16}$

$\dfrac{11}{16} - \dfrac{8}{16} = \dfrac{3}{16}$

11. $8 - \frac{3}{4}$

$7 + 1 - \frac{3}{4}$

$7 + \frac{4}{4} - \frac{3}{4}$

$7 + \frac{1}{4} = 7\frac{1}{4}$

12. $12 - \frac{5}{8}$

$11 + 1 - \frac{5}{8}$

$11 + \frac{8}{8} - \frac{5}{8}$

$11 + \frac{3}{8} = 11\frac{3}{8}$

MODULE 00102 MATH CALCULATIONS (continued)

13. Two punches $= 4\frac{1}{64} + 4\frac{3}{32}$

$$\frac{3}{32} \times \frac{2}{2} = \frac{6}{64}$$

$4\frac{1}{64} + 4\frac{6}{64} = 8\frac{7}{64}$

$9\frac{7}{16} - 8\frac{7}{64}$

$$\frac{7}{16} \times \frac{4}{4} = \frac{28}{64}$$

$9\frac{28}{64} - 8\frac{7}{64} = 1\frac{21}{64}$

14. $20\frac{3}{4} - 12\frac{1}{16}$

$$\frac{3}{4} \times \frac{4}{4} = \frac{12}{16}$$

$20\frac{12}{16} - 12\frac{1}{16} = 8\frac{11}{16}$

15. $36\frac{3}{8} - 35\frac{15}{16}$

$$\frac{3}{8} \times \frac{2}{2} = \frac{6}{16}$$

$36\frac{6}{16} - 35\frac{15}{16}$

$35 + 1 + \frac{6}{16}$

$35 + \frac{16}{16} + \frac{6}{16}$

$35 + \frac{22}{16}$

$35\frac{22}{16} - 35\frac{15}{16} = \frac{7}{16}$

Section 2.4.1 Multiplying and Dividing Fractions

1. $\frac{4}{16} \times \frac{5}{8} = \frac{20}{128} \div \frac{4}{4} = \frac{5}{32}$

2. $\frac{3}{4} \times \frac{7}{8} = \frac{21}{32}$

3. $\frac{2}{8} \times \frac{15}{1} = \frac{30}{8} \div \frac{2}{2} = 3\frac{3}{4}$

4. $\frac{3}{7} \times \frac{49}{1} = \frac{147}{7} \div \frac{7}{7} = \frac{21}{1} = 21$

5.

$$\frac{8}{16} \times \frac{32}{64}$$

$\frac{8 \div 8}{16 \div 8} = \frac{1}{2}$ $\frac{32 \div 32}{64 \div 32} = \frac{1}{2}$

$$\frac{1}{2} \times \frac{1}{2} = \frac{1}{4}$$

or $\frac{8 \times 32}{16 \times 64} = \frac{256}{1,024} \div \frac{256}{256} = \frac{1}{4}$

6. $\frac{3}{8} \div 3 = \frac{3}{8} \times \frac{1}{3} = \frac{3}{24} \div \frac{3}{3} = \frac{1}{8}$

7. $\frac{5}{8} \div \frac{1}{2} = \frac{5}{8} \times \frac{2}{1} = \frac{10}{8} \div \frac{2}{2} = \frac{5}{4}$ or $1\frac{1}{4}$

8. $\frac{3}{4} \div \frac{3}{8} = \frac{3}{4} \times \frac{8}{3} = \frac{24}{12} \div \frac{12}{12} = \frac{2}{1} = 2$

9. $8\frac{1}{2} \div \frac{1}{4} = \frac{17}{2} \div \frac{1}{4} = \frac{17}{2} \times \frac{4}{1} = \frac{68}{2} \div \frac{2}{2} = \frac{34}{1} = 34$

10. $7 \div \frac{7}{8} = \frac{7}{1} \times \frac{8}{7} = \frac{56}{7} \div \frac{7}{7} = \frac{8}{1} = 8$

Section 3.2.5 Decimals

1.
```
   2.50
   4.20
 + 5.00
  11.70
```

2.
```
    ¹
   1.82
   3.41
 + 5.25
  10.48
```

3.
```
     ¹
    6.43
 + 86.40
   92.83
```

4.
```
      ¹
   0.078
 + 0.250
   0.328
```

5.
```
   6.7
 − 2.3
   4.4
```

6.
```
     3
 × 0.3
   0.9
```

7.
```
   6.18
 − 0.90
   5.28
```

8.
```
      128
   × 4.75
      640
     8960
 + 51200
   608.00
```

9.
```
        2.52
   18) 45.36
      − 36
        093
      − 090
        0036
      − 0036
        0000
```

10.
```
        0.252,  rounded to 0.25
   18) 4.536
      − 36
        093
      − 090
        0036
      − 0036
        0000
```

11.
```
        0.0252, rounded to 0.03
   18) 0.4536
      − 00
        045
      − 036
        0093
      − 0090
        00036
      − 00036
        00000
```

MODULE 00102 MATH CALCULATIONS (continued)

12.
```
      0.408, rounded to 0.41
25) 10.200
   − 100
    0020
   − 0000
    00200
   − 00200
    00000
```

13.
```
     5.2
6) 31.2
  − 30
   012
  − 012
   000
```

14.
```
                    20
14.1) 282  = 141) 2820
                 − 282
                   00
                 − 00
                    0
```

15.
```
                216.875, rounded to 216.88
3.2) 694  = 32) 6940.000
               − 64
                054
               − 032
                0220
               − 0192
                00280
               − 00256
                000240
               − 000224
                0000160
               − 0000160
                0000000
```

16.
```
                 220
0.45) 99  = 45) 9900
               − 90
                90
               − 90
                00
               − 00
                 0
```

17.
```
                 40.8
2.5) 102  = 25) 1020.0
               − 100
                0020
               − 0000
                00200
               − 00200
                00000
```

18.
```
                 520
0.6) 312  = 6) 3120
              − 30
               12
              − 12
               00
              − 00
                0
```

19.
```
                  4.910
4.24) 20.82  = 424) 2082.000
                  − 1696
                   03860
                  − 03816
                   000440
                  − 000424
                   0000160
                  − 0000000
                    160r
```

20.
```
                 10.513, rounded to 10.51
3.7) 38.9  = 37) 389.000
                − 37
                 019
                − 000
                 0190
                − 0185
                 00050
                − 00037
                 000130
                − 000111
                  19r
```

21.
```
                 22
0.45) 9.9 = 45) 990
               − 90
                90
               − 90
                 0
```

22.
```
                  40.8
0.25) 10.20  = 25) 1020.0
                 − 100
                  0020
                 − 0000
                  00200
                 − 00200
                  00000
```

23.
```
                52
0.6) 31.2  = 6) 312
              − 30
               12
              − 12
               0
```

29.
```
                  24.13, rounded to 24.1
3.75) 90.5  = 375) 9050.00
                 − 750
                  1550
                 − 1500
                  00500
                 − 00375
                  001250
                 − 001125
                  000125r
```

30.
```
                    6.59, rounded to 6.6
151.8) 1001  = 1518) 10010.00
                  − 09108
                   009020
                  − 007590
                   0014300
                  − 0013662
                    160r
```

MODULE 00102 MATH CALCULATIONS (continued)

31.
$$
\begin{array}{r}
28.127, \text{ rounded to } 28.13 \\
4.30)\overline{120.95} = 430)\overline{12095.000}
\end{array}
$$
$$
\begin{array}{r}
- 0860 \\
\hline
03495 \\
- 03440 \\
\hline
000550 \\
- 000430 \\
\hline
0001200 \\
- 0000860 \\
\hline
00003400 \\
- 00003010 \\
\hline
390r
\end{array}
$$

32.
$$
\begin{array}{r}
311.83, \text{ rounded to } 311.8 \\
0.37)\overline{115.38} = 37)\overline{11538.00}
\end{array}
$$
$$
\begin{array}{r}
- 111 \\
\hline
0043 \\
- 0037 \\
\hline
00068 \\
- 00037 \\
\hline
000310 \\
- 000296 \\
\hline
0000140 \\
- 0000111 \\
\hline
29r
\end{array}
$$

33.
$$
\begin{array}{r}
240.375, \text{ rounded to } 240.4 \\
0.48)\overline{115.38} = 48)\overline{11538.000}
\end{array}
$$
$$
\begin{array}{r}
- 096 \\
\hline
0193 \\
- 0192 \\
\hline
00018 \\
- 00000 \\
\hline
000180 \\
- 000144 \\
\hline
0000360 \\
- 0000336 \\
\hline
00000240 \\
- 00000240 \\
\hline
00000000
\end{array}
$$

Section 3.3.5 Converting Different Values

1. $0.62 = 0.62 \times 100 = 62\%$

2. $0.475 = 0.475 \times 100 = 47.5\%$

3. $0.7 = 0.7 \times 100 = 70\%$

4. $72\% = 72 \div 100 = 0.72$

5. $12.5\% = 12.5 \div 100 = 0.125$

6.
$$
\begin{array}{r}
0.25 \\
4)\overline{1.00} \\
- 08 \\
\hline
020 \\
- 020 \\
\hline
000
\end{array}
$$

7.
$$
\begin{array}{r}
0.75 \\
4)\overline{3.00} \\
- 28 \\
\hline
020 \\
- 020 \\
\hline
000
\end{array}
$$

8.
$$
\begin{array}{r}
0.125 \\
8)\overline{1.000} \\
- 08 \\
\hline
020 \\
- 016 \\
\hline
0040 \\
- 0040 \\
\hline
0000
\end{array}
$$

9.
$$
\begin{array}{r}
0.3125 \\
16)\overline{5.0000} \\
- 48 \\
\hline
020 \\
- 016 \\
\hline
0040 \\
- 0032 \\
\hline
00080 \\
- 00080 \\
\hline
00000
\end{array}
$$

10.
$$
\begin{array}{r}
0.3125 \\
64)\overline{20.0000} \\
- 192 \\
\hline
0080 \\
- 0064 \\
\hline
00160 \\
- 00128 \\
\hline
000320 \\
- 000320 \\
\hline
000000
\end{array}
$$

11. $0.5 = \dfrac{5 \div 5 = 1}{10 \div 5 = 2}$

12. $0.12 = \dfrac{12 \div 4 = 3}{100 \div 4 = 25}$

13. $0.125 = \dfrac{125 \div 125 = 1}{1000 \div 125 = 8}$

14. $0.8 = \dfrac{8 \div 2 = 4}{10 \div 2 = 5}$

15. $0.45 = \dfrac{45 \div 5 = 9}{100 \div 5 = 20}$

16. $\dfrac{9 \div 3 = 3}{12 \div 3 = 4}$
$$
\begin{array}{r}
0.75 \\
4)\overline{3.00} \\
- 28 \\
\hline
020 \\
- 020 \\
\hline
000
\end{array}
$$

17. $\dfrac{10 \div 2 = 5}{12 \div 2 = 6}$
$$
\begin{array}{r}
0.833 \\
6)\overline{5.000} \\
- 48 \\
\hline
020 \\
- 018 \\
\hline
0020 \\
- 0018 \\
\hline
2r
\end{array}
$$
Rounded to the nearest hundredth $= 0.83$

MODULE 00102 MATH CALCULATIONS (continued)

18. $\dfrac{2 \div 2}{12 \div 2} = \dfrac{1}{6}$

```
        0.166
    6) 1.000
     − 06
       040
      − 036
       0040
      − 0036
           4r
```
Rounded to the nearest hundredth = 0.17

19. $\dfrac{4 \div 4}{12 \div 4} = \dfrac{1}{3}$

```
        0.333
    3) 1.000
     − 09
       010
      − 009
          1r
```
Rounded to the nearest hundredth = 0.33

20.
```
        1.416
   12) 17.000
     − 12
       050
      − 048
       0020
      − 0012
       00080
      − 00072
           8r
```
Rounded to the nearest hundredth = 1.42

Section 5.1.4 Converting Measurements

1. $.45 \times 100 = 45$

2. 3×36 in/yd $= 108$

3. $36 \div 3$ ft/yd $= 12$

4. $90 \div 36$ in/yd $= 2.5$

5. $1 \div 100 = 0.01$

6. 66×2.54 conversion factor $= 167.64$

7. 47×0.3048 conversion factor $= 14.33$

8. 54.5×0.03281 conversion factor $= 1.79$

9. 19×0.9144 conversion factor $= 17.37$

10. 4.7×39.37 conversion factor $= 185.04$

Section 5.2.4 Converting Weight Units

1. 50×0.4536 conversion factor $= 22.68$

2. 50×2.205 conversion factor $= 110.25$

3. 15.9×28.35 conversion factor $= 450.77$

4. 94×0.03527 conversion factor $= 3.32$

Section 5.3.4 Converting Volume Units

1. 11600×0.0005787 conversion factor $= 6.7$

2. $1.9 \times 1,000,000$ conversion factor $= 1,900,000$

3. 512×1.308 conversion factor $= 669.7$

4. $7 \times 28,316.85$ conversion factor $= 198,218$

5. 0.2×35.315 conversion factor $= 7.1$

Section 5.4.1 Converting Temperatures

1. $°C = \frac{5}{9}(180 − 32)$

   $°C = \frac{5}{9}(148)$

   $°C = \frac{5}{9} \times \frac{148}{1} = \frac{740}{9}$

   $°C = 82.2$

2. $°C = \frac{5}{9}(66 − 32)$

   $°C = \frac{5}{9}(34)$

   $°C = \frac{5}{9} \times \frac{34}{1} = \frac{740}{9}$

   $°C = 18.9$

3. $°F = (\frac{9}{5} \times −26) + 32$

   $°F = (\frac{9}{5} \times \frac{−26}{1}) + 32$

   $°F = \frac{−234}{5} + 32$

   $°F = −46.8 + 32$

   $°F = −14.8$

4. $°F = (\frac{9}{5} \times 71) + 32$

   $°F = (\frac{9}{5} \times \frac{71}{1}) + 32$

   $°F = \frac{639}{5} + 32$

   $°F = 127.8 + 32$

   $°F = 159.8$

Section 6.3.1 Calculating Area

1. Area $=$ length $\times$ width $= 8 \times 4 = 32$ sq ft

2. Area $=$ length $\times$ width $= 16 \times 16 = 256$ sq cm

3. radius $= \frac{1}{2}$ diameter

   radius $= \frac{1}{2} \times 14 = 7$ ft

   Area $=$ pi $\times$ radius$^2 = 3.14 \times 7 \times 7 = 3.14 \times 49$

```
      1
      1 3
      3.14
    ×  49
     2826
   + 12560
    153.86 sq ft
```

4. Area $= 0.5 \times$ base $\times$ height $= 0.5 \times 4 \times 6 = 0.5 \times 24 = 12$ sq cm

```
     2
     24
   × 0.5
   12.0
```

5. Area $=$ base $\times$ height $= 14 \times 5 = 70$ sq m

Section 6.4.5 Calculating Volume

1. Volume $=$ length $\times$ width $\times$ depth $= 5 \times 6 \times 13 = 30 \times 13 = 390$ cu ft

```
      30
    × 13
      90
   + 300
     390
```

MODULE 00102 MATH CALCULATIONS (continued)

2. Volume = length × width × depth = 3 × 3 × 3 = 27 cu cm

3. Volume = 0.5 × base × height × depth = 0.5 × 6 × 2 × 4 = 24 cu in

4. Radius = ½ diameter = ½ × 6 = ⁶⁄₂ ÷ ²⁄₂ = ³⁄₁ = 3 m

   Height = 60 cm = 0.6 m

   Volume = pi × radius^2 × h = 3.14 × 3 × 3 × 0.6 = 3.14 × 9 × 0.6

   $$\begin{array}{r} 9 \\ \times\ 0.6 \\ \hline 5.4 \end{array}$$

   Volume = 3.14 × 5.4

   $$\begin{array}{r} {\scriptstyle 1 \atop \scriptstyle 2} \\ 3.14 \\ \times\ 5.4 \\ \hline 1256 \\ +\ 15700 \\ \hline 16.956 \end{array}$$

   Volume = 16.956 cu m

5. Depth = 6 in

   ⁶⁄₁₂ = ½ = 0.5 ft

   Volume = length × width × depth = 17 × 17 × 0.5 =

   $$\begin{array}{r} {\scriptstyle 4} \\ 17 \\ \times\ 17 \\ \hline 119 \\ +\ 170 \\ \hline 289 \end{array}$$

   Volume = 289 × 0.5

   $$\begin{array}{r} {\scriptstyle 4\ 4} \\ 289 \\ \times\ 0.5 \\ \hline 144.5 \end{array}$$

   Volume = 144.5 cu ft

   1 yd = 3 ft

   1 cu yd = 3 × 3 × 3 = 27 cu ft

   144.5 cu ft × $\dfrac{1 \text{ cu yd}}{27 \text{ cu ft}}$ = 144.5 ÷ 27

   $$\begin{array}{r} 5.35 \\ 27{\overline{\smash{\big)}\,144.50}} \\ \underline{-\ 135} \\ 95 \\ \underline{-\ 81} \\ 140 \\ \underline{-\ 135} \\ 5r \end{array}$$

   5.35 cu yds

Section 6.4.6 Practical Applications

1. Volume π × 4^2 × 13 in = 653.12 cu in

2. Area of upper rectangle = 39.666 feet × 6 feet = 237.996

   Area of remaining rectangle = 6 feet × 13 feet = 78

   Area of remaining triangle = ½ × 4 × 6 = 12

   Total is 327.996 sq ft × 0.333 feet (thickness of the concrete) = 109 cu ft

APPENDIX B 00101 Basic Safety

OSHA Mission and Standards

The mission of the Occupational Safety and Health Administration (OSHA) is to save lives, prevent injuries, and protect the health of America's workers. To accomplish this, federal and state governments work in partnership with millions of working men and women who are covered by the Occupational Safety and Health Act (OSH Act) of 1970.

Nearly every worker in the US comes under OSHA's jurisdiction. There are some exceptions, such as miners, transportation workers, many public employees, and the self-employed. These specific groups are not covered by OSHA standards, but most are covered by standards developed by the specific industry.

The Code of Federal Regulations

The *Code of Federal Regulations (CFR)* Part 1910 covers the OSHA standards for general industry. *CFR* Part 1926 covers the OSHA standards for the construction industry. Either or both may apply to you, depending on where you are working and what you are doing. If a job-site condition is covered in the *CFR* book, then that standard must be used. However, if a more stringent requirement is listed in *CFR* 1910, it should also be met. Check with your supervisor to find out which standards apply to your job.

29 *CFR* 1926 is divided into subparts A through Z. As you progress in task-specific training, you will learn about all the subparts applicable to your work. *Subpart C of* 29 *CFR* 1926 applies to all construction and maintenance work. It outlines the general safety and health provisions for the construction industry. It covers the following topics:

- Safety training and education
- Injury reporting and recording
- First aid and medical attention
- Housekeeping
- Illumination
- Sanitation
- PPE
- Standards incorporated by reference
- Definitions
- Access to employee exposure and medical records
- Means of egress
- Employee emergency action plans

For assistance in identifying parts, sections, paragraphs, and subparagraphs of an OSHA standard, refer to *Appendix B Figure 1.*

All of OSHA's safety requirements in the Code of Federal Regulations apply to residential as well as commercial construction. In the past, OSHA enforced safety only at commercial sites. The increasing rate of incidents at residential sites led OSHA to enforce safety guidelines for the building of houses and townhomes. Today, however, OSHA still focuses its enforcement efforts on commercial construction.

The General Duty Clause

If a standard does not specifically address a hazard, the general duty clause must be invoked. Failing to adhere to the general duty clause can result in heavy fines for your employer. The general duty clause reads as follows:

An OSHA Standard reference may look like this:

29 CFR 1926.501 (a)(1)(i)(A)

and breaks down like this:

29	=	Title (Labor)
CFR	=	Code of Federal Regulations
1926	=	Part (Construction)
.501	=	Section
(a)	=	Paragraph
(1)	=	Subparagraph
(i)	=	Subparagraph
(A)	=	Subparagraph

Figure 1 Reading OSHA standards.

In practice, OSHA, court precedent, and the review commission have established that if the following elements are present, a general duty clause citation may be issued:

- The employers failed to keep the workplace free of a hazard to which employees of that employer were exposed.
- The hazard was recognized. (Examples might include: through your safety personnel, employees, organization, trade organization, or industry customs.)
- The hazard was causing or was likely to cause death or serious physical harm.
- There was a feasible and useful method to correct the hazard.

Employee Rights and Responsibilities

While it is the employer's responsibility to keep workers safe by complying with the General Duty Clause and all other OSHA regulations, workers have certain rights and responsibilities on the job site as well. First and foremost, workers must follow their employers' safety rules. While workers cannot be cited or fined by OSHA, they can be disciplined for violating their employer's safety rules. Workers must also wear the provided personal protective equipment. Workers should also inform their foreman about health and safety concerns on the job.

Section 11(c) of the OSH Act prohibits employers from disciplining or discriminating against any worker for practicing their rights under OSHA, including filing a complaint. You have the right to file a complaint if you do not think that your employer is protecting your health and safety at work. You may submit a written request to OSHA asking for an inspection of your work site. Workers who file a complaint have the right to have their names withheld from their employers, and OSHA will not reveal this information.

Workers who would like an on-site inspection must submit a written request. You have the following rights when job site inspection is conducted:

- You must be informed of imminent dangers. An OSHA inspector must tell you if you are exposed to an imminent danger. An imminent danger is one that could cause death or serious injury now or in the near future. The inspector will also ask your employer to stop any dangerous activity.
- You have the right to accompany the OSHA inspector in the walk-around inspection. Walk-around activities include all opening and closing conferences related to the conduct of the inspection.
- You have the right to be told about citations issued at your workplace. Notices of OSHA citations must be posted in the workplace near the site where the violation occurred and must remain posted for three days or until the hazard is corrected, whichever is longer.

After an inspection has been performed, OSHA will give the employer a date by which any hazards cited must be fixed. Employers can appeal these dates, and appeals must be filed within 15 days of the citation. Workers have the right to meet privately with the OSHA inspector to discuss the results of the inspection.

If you have been discriminated against for asserting your OSHA rights, you have the right to file a complaint with the OSHA area office within 30 days of the incident. Make sure you file your complaint as soon as possible, as the time limit is strictly enforced.

You also have the right to see and copy any medical records about you that the employer has obtained. Your employer is required by OSHA 29 *CFR* 1926.33 and OSHA 29 *CFR* 1910.1020 to maintain your medical records for 30 years after you leave employment. If you are employed for less than one year, the employer can maintain your records or give them to you when you leave the job.

Inspections

OSHA conducts six types of inspections to determine if employers are complying with standards:

- *Imminent danger inspections* — OSHA's top priority for inspection, conducted when workers face an immediate risk of death or serious physical harm.

- *Catastrophe inspections* — Performed after an incident that requires hospitalization of three or more workers. Employers are required to report fatalities and catastrophes to OSHA within eight hours.

- *Worker complaint and referral inspections* — Conducted due to complaints by workers or a worker representative, or a referral from a recognized professional.

- *Programmed inspection* — Aimed at high-risk areas based on OSHA's targeting and priority methods.

- *Follow-up inspection* — Completed after citations to assure employer has corrected violations.

- *Monitoring inspection* — Used for long-term abatement follow-up or to assure compliance with variances.

Before beginning an inspection, OSHA staff must be able to determine from the complaint that there are reasonable grounds to believe that a violation of an OSHA standard or a safety or health hazard exists. If OSHA has information indicating the employer is aware of the hazard and is correcting it, the agency may not conduct an inspection after obtaining the necessary documentation from the employer.

Complaint inspections are typically limited to the hazards listed in the complaint, although other violations in plain sight may be cited as well. The inspector may decide to expand the inspection based on professional judgment or conversations with workers.

Complaints are not necessarily inspected in first-come, first-served order. OSHA ranks complaints based on the severity of the alleged hazard and the number of workers exposed. That is why lower-priority complaints can often be handled more quickly using the phone/fax method than through on-site inspections.

Inspections are typically performed by conducting a walk-around. During a walk-around inspection, the inspector typically does the following:

- Observes conditions of the job site.
- Talks to workers.
- Inspects records.
- Examines posted hazard warnings and signs.
- Points out hazards and suggests ways to reduce or eliminate them.

After the walk-around inspection, there is typically a closing conference held between the inspector and the site contractor or company managers. During this conference, inspectors discuss their findings, citing specific violations and suggested abatement methods. Inspectors may also conduct interviews with the employers, workers, and representatives at this point.

Violations

Employers who violate OSHA regulations can be fined. The fines are not always high, but they can harm a company's reputation for safety. Fines for serious safety violations can cost up to $7,000. Fines for each violation that was done willfully can cost up to $70,000. In 2012, roughly $260,000,000 in fines were issued against employers. In the decade prior to 2012, annual fines averaged approximately $150,000,000. Significant increases and decreases in annual fines tend to be dependent on the current federal administration and OSHAs budget provided by federal lawmakers.

Compliance

Just as employers are responsible to OSHA for compliance, employees must comply with their company's safety policies and rules. Employers are required to identify hazards and potential hazards within the workplace and eliminate them, control them, or provide protection from them. This can only be done through the combined efforts of the employer and employees. Employers must provide written programs and training on hazards, and employees must follow the procedures. You, as the employee, must read and understand the OSHA poster at your job site explaining your rights and responsibilities. If you are unsure where the OSHA poster is, ask your supervisor.

To help employers provide a safe workplace, OSHA requires companies to provide a competent person to ensure the safety of the employees. In OSHA 29 *CFR* 1926, OSHA defines a competent person as follows:

> Competent person means one who is capable of identifying existing and predictable hazards in the surroundings or working conditions which are unsanitary, hazardous, or dangerous to employees, and who has authorization to take prompt corrective measures to eliminate them.

In comparison, OSHA 29 *CFR* 1926 defines a qualified person as follows:

> Someone who, by possession of a recognized degree, certificate, or professional standing, or who by extensive knowledge, training, and experience, has successfully demonstrated his ability to solve or resolve problems relating to the subject matter, work, or the project.

In other words, a competent person is experienced and knowledgeable about the specific operation and has the authority from the employer to correct the problem or shut down the operation until it is safe. A qualified person has the knowledge and experience to handle problems. A competent person is not necessarily a qualified person.

These terms will be an important part of your career. It is important for you to know who the competent person is on your job site. OSHA requires a competent person for many of the tasks you may be assigned to perform, such as confined space entry, ladder use, and trenching. Different individuals may be assigned as a competent person for different tasks, according to their expertise. To ensure safety for you and your co-workers, work closely with your competent person and supervisor.

APPENDIX C 00102 Construction Math

Multiplication Table

Trace across and down from the numbers being multiplied and find the answer at the intersection. In the example, $7 \times 7 = 49$.

	2	3	4	5	6	7	8	9	10	11	12
2	4	6	8	10	12	14	16	18	20	22	24
3	6	9	12	15	18	21	24	27	30	33	36
4	8	12	16	20	24	28	32	36	40	44	48
5	10	15	20	25	30	35	40	45	50	55	60
6	12	18	24	30	36	42	48	54	60	66	72
7	14	21	28	35	42	49	56	63	70	77	84
8	16	24	32	40	48	56	64	72	80	88	96
9	18	27	36	45	54	63	72	81	90	99	108
10	20	30	40	50	60	70	80	90	100	110	120
11	22	33	44	55	66	77	88	99	110	121	132
12	24	36	48	60	72	84	96	108	120	132	144

Conversion Factors and Common Formulas

Common Measures

WEIGHT UNITS
1 ton = 2,000 pounds
1 pound = 16 dry ounces

LENGTH UNITS
1 yard = 3 feet
1 foot = 12 inches

VOLUMES
1 cubic yard = 27 cubic feet
1 cubic foot = 1,728 cubic inches
1 gallon = 4 quarts
1 quart = 2 pints
1 pint = 2 cups
1 cup = 8 fluid ounces

AREA UNIT
1 square yard = 9 square feet
1 square foot = 144 square inches

Prefix	Symbol	Number	Multiplication Factor
Giga	G	Billion	$1{,}000{,}000{,}000 = 10^9$
Mega	M	Million	$1{,}000{,}000 = 10^6$
Kilo	k	Thousand	$1{,}000 = 10^3$
Hector	h	Hundred	$100 = 10^2$
Deka	da	Ten	$10 = 10^1$
			Base Units $1 = 10^0$
Deci	d	Tenth	$0.1 = 10^{-1}$
Centi	c	Hundredth	$0.01 = 10^{-2}$
Milli	m	Thousandth	$0.001 = 10^{-3}$
Micro	μ	Millionth	$0.000001 = 10^{-6}$
Nano	N	Billionth	$0.000000001 = 10^{-9}$

WEIGHT UNITS
1 kilogram = 1,000 grams
1 hectogram = 100 grams
1 dekagram = 10 grams
1 gram = 1 gram
1 decigram = 0.1 gram
1 centigram = 0.01 gram
1 milligram = 0.001 gram

LENGTH UNITS
1 kiloliter = 1,000 meters
1 hectometer = 100 meters
1 dekameter = 10 meters
1 meter = 1 meter
1 decimeter = 0.1 meter
1 centimeter = 0.01 meter
1 millimeter = 0.001 meter

VOLUME UNITS
1 kiloliter = 1,000 liters
1 hectoliter = 100 liters
1 dekaliter = 10 liters
1 liter = 1 liter
1 deciliter = 0.1 liter
1 centiliter = 0.01 liter
1 milliliter = 0.001 liter

Inch-Pound to Metric Conversions

WEIGHTS
1 ounce = 28.35 grams
1 pound = 453.6 grams or 0.4536 kilograms
1 (short) ton = 907.2 kilograms

LENGTHS
1 inch = 2.540 centimeters
1 foot = 30.48 centimeters
1 yard = 91.44 centimeters or 0.9144 meters
1 mile = 1.609 kilometers

AREAS
1 square inch = 6.452 square centimeters
1 square foot = 929.0 square centimeters or 0.0929 square meters
1 square yard = 0.8361 square meters

VOLUMES
1 cubic inch = 16.39 cubic centimeters
1 cubic foot = 0.02832 cubic meter
1 cubic yard = 0.7646 cubic meter

LIQUID MEASUREMENTS
1 (fluid) ounce = 0.0296 liter or 28.35 grams
1 pint = 473.2 cubic centimeters
1 quart = 0.9464 liter
1 gallon = 3,785 cubic centimeters or 3.785 liters

TEMPERATURE MEASUREMENTS
To convert degrees Fahrenheit to degrees Celsius, use the following: $°C = \frac{5}{9} \times (°F - 32°)$

Metric to Inch-Pound Conversions

WEIGHTS
1 gram = 0.03527 ounces
1 kilogram = 2.2.05 pounds
1 metric tonne = 2,205 pounds

LENGTHS
1 millimeter = 0.00155 square inches
1 centimeter = 0.3937 inches
1 meter = 3.281 feet or 1.0936 yards
1 kilometer = 0.6214

AREAS
1 square millimeter = 0.00155 square inches
1 square centimeter = 0.155 square inches
1 square meter = 10.76 square feet or 1.196 square yards

VOLUMES
1 cubic centimeter = 0.06102 cubic inches
1 cubic meter = 35.31 cubic feet or 1.308 cubic yards

LIQUID MEASUREMENTS
1 cubic centimeter = 0.06102 cubic inches
1 liter (1,000 cm^3) = 1.057 quarts, 2.113 pints, or 61.02 cubic inches

TEMPERATURE MEASUREMENTS
To convert degrees Celsius to degrees Fahrenheit, use the following: $°F = (9/_5 \times °C) + 32°$

Common Formulas

PRESSURE
Absolute pressure = gauge pressure − atmospheric pressure. The atmospheric pressure at sea level is typically accepted as 14.7 psi.

Gauge pressure to static pressure: P = (h × d)/144, where P = the pressure in psi, h = the height of the column in feet, and d = the density of the liquid in pounds per cubic foot.

CIRCLE
Area = $\pi \times r^2$, where r is the radius
Circumference = $\pi \times d$, where d is the diameter

SQUARES/RECTANGLES
Area = length × width
Volume = length × width × height

TEMPERATURE CONVERSIONS
To convert degrees Celsius to Fahrenheit, use the following: $°F = (9/_5 \times °C) + 32°$

To convert degrees Fahrenheit to Celsius, use the following: $°C = 5/_9 \times (°F − 32°)$

SEQUENCE OF OPERATIONS
PEMDAS = 1. Parentheses 2. Exponents 3. Multiplication 4. Division 5. Addition 6. Subtraction

TRIANGLES
Area = ½ × base × height
Pythagorean theorem for right triangles:

$a^2 + b^2 = c^2$, or $c = \sqrt{a^2 + b^2}$, where c is the hypotenuse of the right triangle. The hypotenuse is the side opposite the right angle. Sides a and b are adjacent to the right angle.

PYRAMID
Volume = (length × width × height)/3

Common Formulas

CYLINDER
Volume $= \pi \times r^2 \times h$, where r is the radius of the base and h is the height

CONE
Volume $= (\pi \times r^2 \times h)/3$, where r is the radius of the base and h is the height

SPHERE
Volume $= (4 \times \pi \times r^3)/3$, where r is the radius

Inches Converted to Decimals of a Foot

Inches	Decimals of a Foot	Inches	Decimals of a Foot	Inches	Decimals of a Foot
$1/16$	0.005	$2\ 1/16$	0.172	$4\ 1/16$	0.339
$1/8$	0.010	$2\ 1/8$	0.177	$4\ 1/8$	0.344
$3/16$	0.016	$2\ 3/16$	0.182	$4\ 3/16$	0.349
$1/4$	0.021	$2\ 1/4$	0.188	$4\ 1/4$	0.354
$5/16$	0.026	$2\ 5/16$	0.193	$4\ 5/16$	0.359
$3/8$	0.031	$2\ 3/8$	0.198	$4\ 3/8$	0.365
$7/16$	0.036	$2\ 7/16$	0.203	$4\ 7/16$	0.370
$1/2$	0.042	$2\ 1/2$	0.208	$4\ 1/2$	0.374
$9/16$	0.047	$2\ 9/16$	0.214	$4\ 9/16$	0.380
$5/8$	0.052	$2\ 5/8$	0.219	$4\ 5/8$	0.385
$11/16$	0.057	$2\ 11/16$	0.224	$4\ 11/16$	0.391
$3/4$	0.063	$2\ 3/4$	0.229	$4\ 3/4$	0.396
$13/16$	0.068	$2\ 13/16$	0.234	$4\ 13/16$	0.401
$7/8$	0.073	$2\ 7/8$	0.240	$4\ 7/8$	0.406
$15/16$	0.078	$2\ 15/16$	0.245	$4\ 15/16$	0.411
1	0.083	3	0.250	5	0.417
$1\ 1/16$	0.089	$3\ 1/16$	0.255	$5\ 1/16$	0.422
$1\ 1/8$	0.094	$3\ 1/8$	0.260	$5\ 1/8$	0.427
$1\ 3/16$	0.099	$3\ 3/16$	0.266	$5\ 3/16$	0.432
$1\ 1/4$	0.104	$3\ 1/4$	0.271	$5\ 1/4$	0.438
$1\ 5/16$	0.109	$3\ 5/16$	0.276	$5\ 5/16$	0.443
$1\ 3/8$	0.115	$3\ 3/8$	0.281	$5\ 3/8$	0.448
$1\ 7/16$	0.120	$3\ 7/16$	0.286	$5\ 7/16$	0.453
$1\ 1/2$	0.125	$3\ 1/2$	0.292	$5\ 1/2$	0.458
$1\ 9/16$	0.130	$3\ 9/16$	0.297	$5\ 9/16$	0.464
$1\ 5/8$	0.135	$3\ 5/8$	0.302	$5\ 5/8$	0.469
$1\ 11/16$	0.141	$3\ 11/16$	0.307	$5\ 11/16$	0.474
$1\ 3/4$	0.146	$3\ 3/4$	0.313	$5\ 3/4$	0.479
$1\ 13/16$	0.151	$3\ 13/16$	0.318	$5\ 13/16$	0.484
$1\ 7/8$	0.156	$3\ 7/8$	0.323	$5\ 7/8$	0.490
$1\ 15/16$	0.161	$3\ 15/16$	0.328	$5\ 15/16$	0.495
2	0.167	4	0.333	6	0.500

Inches	Decimals of a Foot	Inches	Decimals of a Foot	Inches	Decimals of a Foot
6 $\frac{1}{16}$	0.505	8 $\frac{1}{16}$	0.672	10 $\frac{1}{16}$	0.839
6 $\frac{1}{8}$	0.510	8 $\frac{1}{8}$	0.677	10 $\frac{1}{8}$	0.844
6 $\frac{3}{16}$	0.516	8 $\frac{3}{16}$	0.682	10 $\frac{3}{16}$	0.849
6 $\frac{1}{4}$	0.521	8 $\frac{1}{4}$	0.688	10 $\frac{1}{4}$	0.854
6 $\frac{5}{16}$	0.526	8 $\frac{5}{16}$	0.693	10 $\frac{5}{16}$	0.859
6 $\frac{3}{8}$	0.531	8 $\frac{3}{8}$	0.698	10 $\frac{3}{8}$	0.865
6 $\frac{7}{16}$	0.536	8 $\frac{7}{16}$	0.703	10 $\frac{7}{16}$	0.870
6 $\frac{1}{2}$	0.542	8 $\frac{1}{2}$	0.708	10 $\frac{1}{2}$	0.875
6 $\frac{9}{16}$	0.547	8 $\frac{9}{16}$	0.714	10 $\frac{9}{16}$	0.880
6 $\frac{5}{8}$	0.552	8 $\frac{5}{8}$	0.719	10 $\frac{5}{8}$	0.885
6 $\frac{11}{16}$	0.557	8 $\frac{11}{16}$	0.724	10 $\frac{11}{16}$	0.891
6 $\frac{3}{4}$	0.563	8 $\frac{3}{4}$	0.729	10 $\frac{3}{4}$	0.896
6 $\frac{13}{16}$	0.568	8 $\frac{13}{16}$	0.734	10 $\frac{13}{16}$	0.901
6 $\frac{7}{8}$	0.573	8 $\frac{7}{8}$	0.740	10 $\frac{7}{8}$	0.906
6 $\frac{15}{16}$	0.578	8 $\frac{15}{16}$	0.745	10 $\frac{15}{16}$	0.911
7	0.583	9	0.750	11	0.917
7 $\frac{1}{16}$	0.589	9 $\frac{1}{16}$	0.755	11 $\frac{1}{16}$	0.922
7 $\frac{1}{8}$	0.594	9 $\frac{1}{8}$	0.760	11 $\frac{1}{8}$	0.927
7 $\frac{3}{16}$	0.599	9 $\frac{3}{16}$	0.766	11 $\frac{3}{16}$	0.932
7 $\frac{1}{4}$	0.604	9 $\frac{1}{4}$	0.771	11 $\frac{1}{4}$	0.938
7 $\frac{5}{16}$	0.609	9 $\frac{5}{16}$	0.776	11 $\frac{5}{16}$	0.943
7 $\frac{3}{8}$	0.615	9 $\frac{3}{8}$	0.781	11 $\frac{3}{8}$	0.948
7 $\frac{7}{16}$	0.620	9 $\frac{7}{16}$	0.786	11 $\frac{7}{16}$	0.953
7 $\frac{1}{2}$	0.625	9 $\frac{1}{2}$	0.792	11 $\frac{1}{2}$	0.958
7 $\frac{9}{16}$	0.630	9 $\frac{9}{16}$	0.797	11 $\frac{9}{16}$	0.964
7 $\frac{5}{8}$	0.635	9 $\frac{5}{8}$	0.802	11 $\frac{5}{8}$	0.969
7 $\frac{11}{16}$	0.641	9 $\frac{11}{16}$	0.807	11 $\frac{11}{16}$	0.974
7 $\frac{3}{4}$	0.646	9 $\frac{3}{4}$	0.813	11 $\frac{3}{4}$	0.979
7 $\frac{13}{16}$	0.651	9 $\frac{13}{16}$	0.818	11 $\frac{13}{16}$	0.984
7 $\frac{7}{8}$	0.656	9 $\frac{7}{8}$	0.823	11 $\frac{7}{8}$	0.990
7 $\frac{15}{16}$	0.661	9 $\frac{15}{16}$	0.828	11 $\frac{15}{16}$	0.995
8	0.667	10	0.833	12	1.000

APPENDIX D 00105 Construction Drawings

Residential Drawings

Drawing 1 Residential Floor Plan   **440**

Drawing 2 Residential Elevations   **441**

A scale drawing shows accurate sizes reduced by a certain amount. Construction drawings are nearly always presented to scale. Each of the drawings in this appendix are to scale.

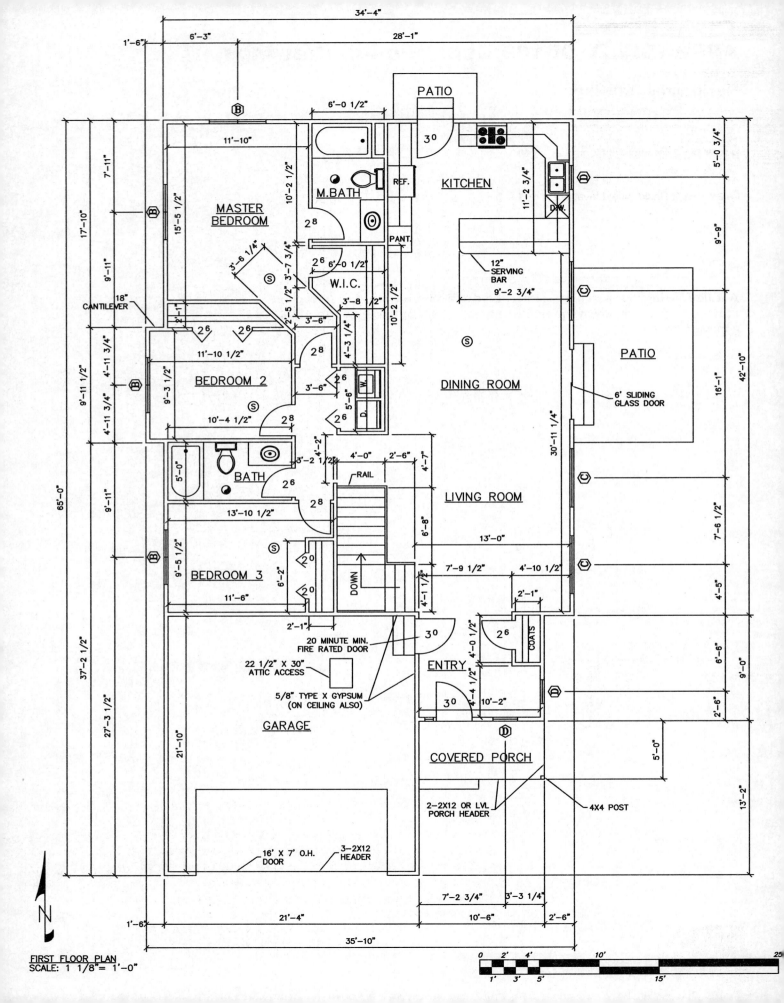

FIRST FLOOR PLAN
SCALE: 1 1/8"= 1'-0"

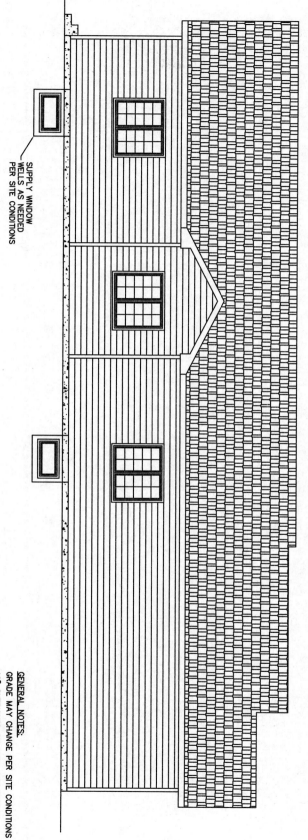

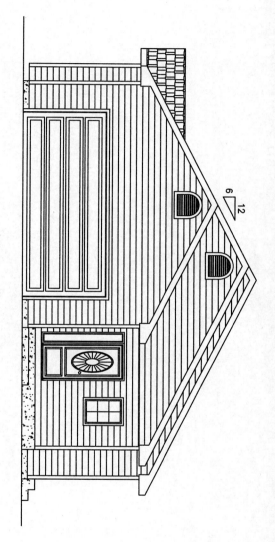

SOUTH ELEVATION
SCALE: 1 1/8" = 1'-0"

WEST ELEVATION
SCALE: 1 1/8" = 1'-0"

6 12

SUPPLY WINDOW
WELLS AS NEEDED
PER SITE CONDITIONS

GENERAL NOTES:
GRADE MAY CHANGE PER SITE CONDITIONS
16" OVERHANG ON EVES AND GABLES
GUTTERS (NOT SHOWN) DO NOT DRAIN ON THE ROOF

0
1'
2'
3'
4'
5'
10'
15'
25'

GLOSSARY

Abrasive: A substance, such as sandpaper, that is used to wear away material.

Absenteeism: A consistent failure to show up for work.

Accident: According to the US Occupational Safety and Health Administration (OSHA), an unplanned event that results in personal injury and/or property damage.

Active listening: A process that involves respecting others, listening to what is being said, and understanding what is being said.

Acute angle: Any angle greater than 0° and less than 90°.

Adjacent angles: Angles that have the same vertex and one side in common.

Alternating current (AC): The common power supplied to most all wired devices, where the current reverses its direction many times per second. AC power is the type of power generated and distributed throughout settled areas.

Amphetamine: A class of drugs that causes mental stimulation and feelings of euphoria.

Amp hour (Ah): A rating that describes the maximum amps a battery can provide continuously for up to 60 minutes. Drawing more amps than the battery's rating is possible, but it reduces the amount of time before the battery is drained.

Angle: The shape made by two straight lines coming together at a point. The space between those two lines is measured in degrees.

Appendix: A source of detailed or specific information placed at the end of a section, a chapter, or a book.

Apprenticeship/craft training: A program in which you gain valuable skills on the job under the instruction of more experienced professionals, often making money as you obtain training.

Arbor: The end of a circular saw shaft where the blade is mounted.

Architect: A qualified, licensed person who creates and designs drawings for a construction project.

Architect's scale: A specialized ruler used in making or measuring reduced scale drawings. The ruler is marked with a range of calibrated ratios for laying out distances, with scales indicating feet, inches, and fractions of inches. Used on drawings other than site plans.

Architectural plans: Drawings that show the design of the project. Also called architectural drawings.

Arc welding: The joining of metal parts by fusion, in which the necessary heat is produced by means of an electric arc.

Area: The amount of space contained in a two-dimensional object such as a rectangle, circle, or square.

Auger bit: A drill bit with a spiral cutting edge for boring holes in wood and other materials.

Barbiturate: A class of drugs that induces relaxation, slowing the body's ability to react.

Base: As it relates to triangles, the base is the line forming the bottom of the triangle.

Beams: Large, horizontal structural members made of concrete, steel, stone, wood, or other structural material to provide support above a large opening.

Bevel: The inclination of a surface that intersects another surface at any angle less than 90 degrees; also used as a verb, meaning to cut or trim an edge to an angled profile.

Bisect: To divide into two equal parts. For example, when an angle is bisected, the two resulting angles are equal.

Block and tackle: A simple rope-and-pulley system used to lift loads.

Blueprints: A traditional method of reproducing technical drawings using a contact print process and light-sensitive paper that resulted in white lines on a blue background. Also, the traditional name used to describe construction drawings.

Body language: A person's facial expression, physical posture, gestures, and use of space, which communicate feelings and ideas.

Bowline: A knot used to form a loop that neither slips nor jams; sometimes referred to as a rescue knot or the king of knots.

Brazing: A process using heat in excess of 800°F (427°C) to melt a filler metal that is drawn into a connection. Brazing is commonly used to join copper pipe.

Bridle: A configuration using two or more slings to connect a load to a single hoist hook.

Building information modeling (BIM): A process in which software reads computer files containing building data to create a virtual picture of a facility before it is built. BIM objects represent many dimensions of construction, and allow builders to adjust the plans prior to beginning actual construction.

Bullets: Large, vertically aligned dots that highlight items in a list.

Bull ring: A single ring used to attach multiple slings to a hoist hook.

Bullying: Unwanted, aggressive behavior that involves a real or perceived power imbalance. This form of harassment may include offensive, persistent, insulting, or physically threatening behavior directed at an individual.

Cannabinoids: A diverse category of chemical substances that repress neurotransmitter releases in the brain. Cannabinoids have a variety of sources; some are created naturally by the human body, while others come from cannabis (marijuana). Still others are synthetic.

Capsize: To change the form and rearrange the parts of a knot, usually by pulling on specific ends of the knot.

Carbide: A very hard material made of tungsten carbide and heavy metals such as cobalt. Commonly baked on to the tips of masonry bits and to the edges of saw blades.

Career and technical education (CTE): Programs offered in high school and technical education centers that allow students to explore career options available in construction.

Change order: A written order by the owner of a project for the contractor to make a change in time, amount, or specifications.

Charge cycle: Describes one complete charge and discharge cycle for a rechargeable battery.

Cheater bar: A length of pipe or rod used to increase the leverage on a wrench or similar tool. The increased leverage provided by a cheater bar often exceeds the tool's capacity and strength, resulting in a damaged tool as well as damaged workpieces.

Chronic obstructive pulmonary disease (COPD): Lung disease(s) that results in the obstruction of lung airflow and interferes with normal breathing. Chronic bronchitis and emphysema generally fall under the diagnosis of COPD.

Chuck: A clamping device that holds an attachment; for example, the chuck of the drill holds the drill bit.

Chuck key: A small, T-shaped steel piece used to open and close the chuck on power drills. Cordless drills typically do not have a chuck key.

Circles: A closed curved line drawn at a constant distance around a central point. A circle measures 360°.

Circumference: The distance around the curved line that forms the circle.

Civil plans: Drawings that show the location of the building on the site from an aerial view, including contours, trees, construction features, and dimensions.

Clean rooms: A room or area that is used for scientific research or specialized processes, such as the manufacture of pharmaceuticals and semiconductors. The cleanliness level of a clean room is generally based on the number of particles measured per cubic meter of area.

Clove hitch: A knot that consists of two half hitches made in opposite directions; used to temporarily secure a rope to an object.

Combustible: Capable of easily igniting and rapidly burning; used to describe a fuel with a flash point at or above 100°F (38°C).

Competent person: A person who is capable of identifying existing and predictable hazards in the surroundings or working conditions that are unsanitary, hazardous, or dangerous to employees, and who has authorization to take prompt corrective measures to eliminate them. Also, an individual capable of identifying existing and predictable hazards in the surroundings and working conditions which are unsanitary, hazardous, or dangerous to employees; an individual who is authorized to take prompt corrective measures to eliminate these issues.

Compromise: When people involved in a disagreement make concessions to reach a solution that everyone agrees on.

Computer-aided design (CAD): The use of computers to digitally sketch construction designs, allowing other people to see how aspects of a structure would look and work together when built.

Computer-aided drafting (CAD): The making of a set of construction drawings with the aid of a computer.

Concave: Having a shape that curves inward, such as the interior of a circle. *Convex* is an antonym, meaning a shape that curves outward.

Concrete mule: A wheeled device used when a concrete pour is in a location that a concrete delivery truck or pump cannot reach; sometimes referred to as a Georgia buggy.

Confidentiality: Privacy of information.

Confined space: A work area large enough for a person to work in, but with limited means of entry and exit and not designed for continuous occupancy. Tanks, vessels, silos, pits, vaults, and hoppers are examples of confined spaces. Also see *permit-required confined space*.

Constructive criticism: A positive offer of advice intended to help someone correct mistakes or improve actions.

Contour lines: Solid or dashed lines showing the elevation of the earth on a civil drawing.

Contractor: Oversees many parts of a construction project, including resource management, budgeting, code adherence, quality assurance, and construction materials.

Core: Center support member of a wire rope around which the strands are laid.

Countersink: A bit or drill used to set the head of a screw at or below the surface of the material.

Craft/trade: A general term referring to one of the many construction specialties.

Craft professional: A trained and skilled individual who designs and builds things. Craft professionals conform to the technical and ethical standards of their trade.

Crew leader: Also called *foreman*, this supervisory role oversees a crew of craft professionals. It is the crew leaders' job to make sure that work is completed correctly and on time. Crew leaders are responsible for the safety and work of those under them.

Cross-bracing: Braces (metal or wood) placed diagonally from the bottom of one rail to the top of another rail to add support to a structure.

Cube: A three-dimensional object whose sides are all the same length.

Cylinder cart: A two-wheeled cart that is used to transport cylinders, or bottles, of compressed gases.

Decimal: The part of a number represented by digits to the right of a point, called a decimal point. For example, in the number 1.25, .25 is the decimal portion of the number. In this case, it represents the fraction $^{25}/_{100}$.

Degree: A unit of measurement for angles. For example, a right angle measures 90°.

Denominator: The part of a fraction below the dividing line. For example, the 2 in ½ is the denominator. It is equivalent to the divisor in a long division problem.

Detail drawings: Enlarged views of part of a drawing used to show an area more clearly.

Diagonal: Line drawn from one corner of a rectangle or square to the farthest opposite corner.

Diameter: The length of a straight line that crosses from one side of a circle, through the center point, to a point on the opposite side. The diameter is the longest straight line that can be drawn inside a circle.

Difference: The result of subtracting one number from another. For example, in the problem 8 − 3 = 5, 5 is the difference between the two numbers.

Digit: Any of the numerical symbols 0 to 9.

Dimension line: A line on a drawing with a measurement indicating length.

Direct current (DC): An electric power supply where the current flows in one direction only. DC power is supplied by batteries and by transformer-rectifiers that change AC power to DC.

Dividend: In a division problem, the number being divided is the dividend. For example, in the problem 16 ÷ 8 = 2, 16 is the dividend.

Divisor: In a division problem, the number that is divided into another number is called the divisor. For example, in the problem 16 ÷ 8 = 2, 8 is the divisor.

Dowel: A pin or rod, usually round, that fits into a corresponding hole to fasten or align two workpieces.

Drawing revision control: A system for recording, documenting, and identifying drawing revisions by using a numeric or alphabetic sequence.

Drum cart: A wheeled cart that is used to transport heavier-weight 55-gallon drums/barrels.

Drum dolly: A wheeled circular platform or a caddy with a handle that is used to transport lighter-weight 55-gallon drums/barrels.

Drywall: A large, flat board made of layers of fiberboard and gypsum used primarily for wall construction and finishing. Drywall, also known as *sheet rock* and *wallboard*, is typically manufactured in 4' × 8' panels and readily accepts paint.

Earn as You Learn: An approach to education that pays trainees wages for work completed as part of an on-the-job learning (OJL) program.

Electrical plans: Engineered drawings that show all electrical supply and distribution.

Electronic signature: A signature that is used to sign electronic documents by capturing handwritten signatures through computer technology and attaching them to the document or file.

Elevation (EL): Height above sea level, or other defined surface, usually expressed in feet or meters.

Elevation drawing: Side view of a building or object, showing height and width.

Emery cloth: A material made with a thin cloth backing, coated with an abrasive that is tightly bonded to the backing; typically sold in sheets or narrow rolls.

Engineers: People who apply scientific principles in design and construction.

Engineer's scale: A straightedge measuring device divided uniformly into multiples of 10 divisions per inch so that drawings can be made with decimal values. Used mainly for land measurements on site plans.

Equation: A mathematical statement indicating that the value of two mathematical expressions, such as 2×2 and 1×4, are equal. An equation is written using the equal sign in this manner: $2 \times 2 = 1 \times 4$.

Equilateral triangle: A triangle that has three equal sides and three equal angles.

Equivalent fractions: Fractions having different numerators and denominators but still having equal values, such as the fractions $\frac{1}{2}$ and $\frac{2}{4}$.

Excavation: Any man-made cut, cavity, trench, or depression in an earth surface, formed by removing earth. It can be made for anything from basements to highways. Also see *trench*.

Fire protection plan: A drawing that shows the details of the building's sprinkler system.

Flammable: Capable of easily igniting and rapidly burning; used to describe a fuel with a flash point below $100°F$ ($38°C$).

Flash burn: The damage that can be done to eyes after even brief exposure to ultraviolet light from arc welding. A flash burn requires medical attention.

Flash point: The temperature at which fuel gives off enough gases (vapors) to burn.

Flats: The flat, straight sides of a wrench opening; also, the sides on a nut or bolt head.

Floor plan: A drawing that provides an aerial view of the layout of each room.

Flush: As an adjective, to be even or at the same level with another object. For example, when a nail is fully seated, the top of the head is flush (even) with, or slightly beyond, the surface of the wood.

Font: The type style used for letters and numbers.

Force: A push or pull on a surface. In this module, a force is the weight of an object or a fluid.

Formula: A mathematical process used to solve a problem. For example, the formula for finding the area of a rectangle is Side A times Side B = Area, or $A \times B =$ Area.

Forstner bit: A bit designed for use in wood or similar soft material. The design allows it to drill a flat-bottom blind hole in material.

Foundation plan: A drawing that shows the layout and elevation of the building foundation.

Fraction: A portion of a whole number represented by two numbers. The upper number of a fraction is called the numerator, while the bottom number is called the denominator.

Freight elevator: An elevator used to transport materials from floor to floor.

Glossary: An alphabetical list of terms and definitions.

Graph: Information shown as a picture or chart. Graphs may be represented in various forms, including line graphs and bar charts.

Green construction: Construction practices that involve the creation of structures that adhere to sustainability principles.

Grilles: HVAC components that pull air from a room and return it to the heating or cooling system. They are different from a register in that they do not have a damper to control airflow.

Grit: A granular, sand-like material used to make sandpaper and similar materials abrasive. Grit is graded according to its texture. The grit number indicates the number of abrasive granules in a standard size (per in or per cm). The higher the grit number, the more particles in a given area, indicating a finer abrasive material.

Ground: The conducting connection between electrical equipment or an electrical circuit and the earth.

Ground fault: An unintentional, electrically conducting connection between an ungrounded conductor of an electrical circuit and the normally noncurrent-carrying conductors, metal objects, or the earth. Also, incidental grounding of a conducting electrical wire.

Ground fault circuit interrupter (GFCI): A device that interrupts and de-energizes an electrical circuit to protect a person from electrocution. Also, a circuit breaker designed to protect people from electric shock and to protect equipment from damage by interrupting the flow of electricity if a circuit fault occurs.

Ground fault protection: Protection against short circuits; a safety device cuts power off as soon as it senses any imbalance between incoming and outgoing current.

Guarded: Enclosed, fenced, covered, or otherwise protected by barriers, rails, covers, or platforms to prevent dangerous contact.

Half hitch: A knot tied by passing the working end of a rope around an object, across the standing part of the rope, and then through the resulting loop; often used as an element in forming other knots or added to make other knots more secure.

Hallucinogen: A class of drugs that distort the perception of reality and cause hallucinations.

Hand line: A line attached to a tool or object so a worker can pull it up after climbing a ladder or scaffold.

Hand truck: Two-wheeled cart that is used to transport large, heavy loads; also known as a dolly.

Harassment: A type of discrimination that can be based on race, age, disabilities, sex, religion, cultural issues, health, or language barriers.

Hazard Communication Standard (HAZCOM): The Occupational Safety and Health standard that requires contractors to educate employees about hazardous chemicals on the job site and how to work with them safely.

Hidden line: A dashed line showing an object obstructed from view by another object.

High-efficiency particulate arrestance (HEPA) filters: Special filters designed to capture very small particles. HEPA filters have a minimum efficiency of 99.97 percent, capturing particles that are 0.3 microns (0.0000117 inch) or larger in size.

Hitch: The rigging configuration by which a sling connects the load to the hoist hook. The three basic types of hitches are vertical, choker, and basket.

Hoist: A device that applies a mechanical force for lifting or lowering a load.

HVAC: Heating, ventilating, and air conditioning.

Hydraulic: Powered by fluid under pressure.

Improper fraction: A fraction whose numerator is larger than its denominator. For example, $\frac{8}{4}$ and $\frac{6}{3}$ are improper fractions.

Incident: According to the US Occupational Safety and Health Administration (OSHA), an unplanned event that does not result in personal injury but may result in property damage or is worthy of recording.

Index: An alphabetical list of topics, along with the page numbers where each topic appears.

Industrial forklift: A vehicle with a power-operated pronged platform that can be raised and lowered for insertion under a load to be lifted and moved.

Infrastructure: The transportation, water, electrical, and telecommunications systems that allow communities and countries to operate.

Initiative: The ability to work without constant supervision and solve problems independently.

Invert: To reverse the order or position of numbers. In fractions, inverting means reversing the positions of the numerator and denominator, so ¾ becomes ⁴⁄₃. When dividing by fractions, one fraction is inverted.

Isosceles triangle: A triangle that has two equal sides and two equal angles.

Italics: Letters and numbers that lean to the right rather than stand straight up.

Jargon: Specialized terms used in a specific industry.

Joint: The intersection of workpiece components where a physical connection is generally made. Joints are often named by the style of connection, such as butt joints and miter joints.

Joists: Lengths of wood or steel that usually support floors, ceilings, or a roof. Roof joists will be at the same angle as the roof itself, while floor and ceiling joists are usually horizontal.

Kerf: The channel created by a saw blade passing through the material, which is equal to the width of the blade teeth.

Lanyard: A cord, rope, or similar material used to limit the travel of two connected objects. Also, a short section of rope or strap, one end of which is attached to a worker's safety harness and the other to a strong anchor point above the work area.

Lath: A thin, flat strip of wood used to form latticework or a foundation for wall plaster.

Leader: In drafting, the line on which an arrowhead is placed and used to identify a component.

Leadership: The ability to set an example for others to follow by exercising authority and responsibility.

Legend: A description of the symbols and abbreviations used in a set of drawings.

Level: Used as an adjective in this context, meaning flat or even with another surface or in relation to the Earth's horizon.

Lifting clamp: A device used to move loads such as steel plates or concrete panels without the use of slings.

Load: The total amount of what is being lifted, including all slings, hitches, and hardware.

Load control: The safe and efficient practice of load manipulation, using proper communication and handling techniques.

Load stress: The strain or tension applied on the rigging by the weight of the suspended load.

Loadbearing: Carrying a significant amount of weight and/or providing necessary structural support. A loadbearing wall is typically carrying some portion of the roof's weight and cannot be removed without risking structural failure or collapse.

Lockout/tagout: A formal procedure for taking equipment out of service and ensuring that it cannot be operated until a qualified person has removed the lock and/or warning tag.

Management system: The organization of a company's management, including reporting procedures, supervisory responsibility, and administration.

Masonry bit: A drill bit with a carbide tip designed to penetrate materials such as stone, brick, or concrete.

Mass: The quantity of matter present.

Master link: The main connection fitting for chain slings.

Material cart: Four-wheeled device used to transport materials around a job site.

Maximum intended load: The total weight of all people, equipment, tools, materials, and loads that a ladder can hold at one time.

Mechanical plans: Engineered drawings that show the mechanical systems, such as motors and piping.

Memo: Informal written correspondence. Another term for memorandum (plural: *memoranda*).

Methamphetamine: A highly addictive crystalline drug, derived from amphetamines, that affects the central nervous system.

Metric scale: A straightedge measuring device divided into centimeters, with each centimeter divided into 10 millimeters. Usually used for architectural drawings and sometimes referred to as a metric architect's scale.

Midrail: Mid-level, horizontal board required on all open sides of scaffolds and platforms that are more than 14 inches (35 cm) from the face of the structure and more than 10 feet (3.05 m) above the ground. It is placed halfway between the toeboard and the top rail.

Mission statement: A statement of how a company does business.

Miter joints: Joints made by fastening parts together with the ends cut at a similar angle, usually 90 degrees.

Mixed number: A combination of a whole number with a fraction or decimal. Examples of mixed numbers are $3^7/_{16}$, 5.75, and $1^1/_4$.

Mushroomed: Refers to the head of a tool that is struck or used for striking when the head becomes wider and distorted, like the head of a mushroom.

Negative numbers: Numbers less than zero. For example, −1, −2, and −3 are negative numbers.

Nonverbal communication: All communication that does not use words. This includes tone of voice, appearance, personal environment, use of time, and body language.

Not to scale (NTS): Describes drawings that show relative positions and sizes only, without scale.

Numerator: The part of a fraction above the dividing line. For example, the 1 in ½ is the numerator. It is equivalent to the dividend in a long division problem.

Obtuse angle: Any angle greater than 90° and less than 180°.

Occupational Safety and Health Administration (OSHA): An agency of the US Department of Labor. Also refers to the Occupational Safety and Health Act of 1970, a law that applies to more than 111 million workers and 7 million job sites in the country. Also, an agency of the US Department of Labor with the mission to establish and enforce safe working conditions for the US workforce.

One-rope lay: The lengthwise distance it takes for one strand of a wire rope to make one complete turn around the core.

Opiates: A narcotic painkiller derived from the opium poppy plant or synthetically manufactured. Heroin is the most commonly used opiate.

Opposite angles: Two angles that are formed by two straight lines crossing. They are always equal.

Overstriking: When a hammer misses the head of a nail, with the head going past it and allowing the hammer handle to impact the nail head instead. Overstriking is a common cause of hammer damage.

Pallet jack: A device used to lift and move heavy or stacked pallets; also known as a pallet truck.

Paraphrase: Express something heard or read using different words.

Peening: The process of bending, shaping, or cutting metal by striking it with a tool.

Perimeter: The distance around the outside of a closed shape, such as a rectangle, circle, square, or any irregular shape.

Permit: A legal document that allows a task to be undertaken.

Permit-required confined space: A confined space that has been evaluated and found to have actual or potential hazards, such as a toxic atmosphere or other serious safety or health hazard. Workers need written authorization to enter a permit-required confined space. Also see *confined space*.

Personal protective equipment (PPE): Equipment or clothing designed to prevent or reduce injuries.

Pi: A mathematical constant value approximately equal to 3.14 (or $^{22}/_7$) used to determine the area and circumference of circles. It is sometimes symbolized by the Greek letter π.

Pipe mule: A two-wheeled device used to transport medium-length pieces of pipe, tubing, or scaffolding; sometimes referred to as a tunnel buggy.

Pipe transport: A wheeled device similar to a pipe mule, but used to move larger pieces of pipe.

Piping and instrumentation drawings (P&IDs): Schematic diagrams of a complete piping system.

Place value: The value that a digit represents in a whole number, determined by its place within the whole number or by its position relative to the decimal point. In the number 124, the number 2 represents 20, since it is in the tens position.

Plane: A surface in which a straight line joining two points lies wholly within that surface.

Plane geometry: The mathematical study of two-dimensional (flat) shapes.

Planked: Having pieces of material 2 inches (5 cm) thick or greater and 6 inches (15 cm) wide or greater used as flooring, decking, or scaffold decks.

Plumb: Perfectly vertical, meaning a surface is at a right angle (perpendicular) to the horizon or other surface used for comparison.

Plumbing isometric drawing: A type of three-dimensional drawing that depicts a plumbing system.

Plumbing plans: Engineered drawings that show the layout for the plumbing system.

Pneumatic: Powered by air pressure, such as a pneumatic tool.

Positive numbers: Numbers greater than zero. For example, 1, 2, and 3 are positive numbers. Except for zero, any number without a negative (–) sign in front of it is treated as a positive number.

Powered wheelbarrow: A vehicle similar to a manual wheelbarrow, but powered by an electric or gas motor; also known as a power buggy.

Product: The result of a multiplication problem. For example, the product of 6×6 is 36.

Professionalism: Integrity and work-appropriate manners.

Project manager: Oversees the planning and delivery of construction projects on time and within budget.

Project owner: The initiator of the project who usually finances the building endeavor.

Proximity work: Work done near a hazard but not actually in contact with it.

Punch list: A written list that identifies deficiencies requiring correction at completion.

Qualified person: A person who, by possession of a recognized degree, certificate, or professional standing, or by extensive knowledge, training, and experience, has demonstrated the ability to solve or prevent problems relating to a certain subject, work, or project. Also, a person who, through the possession of a recognized degree, certificate, or professional standing, or one who has gained extensive knowledge, training, and experience, has successfully demonstrated his or her ability to solve problems relating to the subject matter, the work, or the project.

Quotient: The result of a division problem. For example, when dividing 6 by 2, the quotient is 3.

Radius: The distance from a circle's center point to any point on the curved line, or half the width (diameter) of the circle.

Rafter: One of a series of beams that extend from the perimeter of a building to the high point of a roof, providing structural support for the roof above.

Rated capacity: The maximum load weight a sling or piece of hardware or equipment can hold or lift; also referred to as the working load limit (WLL).

Reciprocating: Moving backward and forward on a straight line.

Rectangle: A four-sided shape with four 90° angles. Opposite sides of a rectangle are always parallel and the same length. Adjacent sides are perpendicular and may or may not be equal in length.

Reference: A person who can confirm to a potential employer that you have the skills, experience, and work habits that are listed in your resume.

Registers: HVAC components that direct air from the heating or cooling system into a room. A damper in the register is used to control the airflow.

Rejection criteria: Standards, rules, or tests on which a decision can be based to remove an object or device from service because it is no longer safe.

Remainder: The amount left over in a division problem. For example, in the problem $34 \div 8$, 8 goes into 34 four times ($8 \times 4 = 32$) with 2 left over, so the remainder is 2.

Respirator: A device that provides clean, filtered air for breathing, no matter what is in the surrounding air.

Revolutions per minute (rpm): The rotational speed of a motor or shaft, based on the number of times it rotates each minute.

Rigging hook: An item of rigging hardware used to attach a sling to a load.

Right angle: An angle that measures 90°. The two lines that form a right angle are perpendicular to each other. This is the angle most used in the crafts.

Right triangle: A triangle that includes one 90° angle.

Roller skids: A device that includes a surface table and two, three, or four roller skids. Materials that are to be moved are placed on the table surface and then pushed on the skids.

Roof plan: A drawing of the view of the roof from above the building.

Rough terrain forklift: Similar to an industrial forklift, but designed to be used on rough surfaces. Rough terrain forklifts are characterized by large pneumatic tires, usually with deep treads that allow the vehicle to grab onto the roughest of roads or ground cover without sliding or slipping.

Safety culture: The culture created when the whole company sees the value of a safe work environment.

Safety data sheet (SDS): A document that must accompany any hazardous substance. The SDS identifies the substance and gives the exposure limits, the physical and chemical characteristics, the kind of hazard it presents, precautions for safe handling and use, and specific control measures.

Scaffold: An elevated platform for workers and materials.

Scale: The ratio between the size of a drawing of an object and the size of the actual object.

Scalene triangle: A triangle with sides of unequal lengths.

Schematic: A one-line drawing showing the flow path for electrical circuitry or the relationship of all parts of a system.

Section drawing: A cross-sectional view of a specific location, showing the inside of an object or building.

Sector: A category of construction distinguished by specific types of construction, materials, equipment, and skills. The four sectors of construction are residential, commercial, industrial, and heavy civil/infrastructure.

Self-feed bit: A bit that uses a lead screw to pull the larger rear portion of the bit into the wood, making it simple to drill larger holes.

Self-presentation: The way a person dresses, speaks, acts, and interacts with others.

Sexual harassment: A type of discrimination that results from unwelcome sexual advances, requests, or other verbal or physical behavior with sexual overtones.

Shackle: Coupling device used in an appropriate lifting apparatus to connect the rope to eye fittings, hooks, or other connectors.

Shank: The smooth part of a drill bit that fits into the chuck.

Sheave: A grooved pulley-wheel for changing the direction of a rope's pull; often found on a crane.

Shielding: A structure used to protect workers in trenches.

Shoring: A support system designed to prevent a trench or excavation cave-in.

Signaler: A person who is responsible for directing a vehicle when the driver's vision is blocked in any way.

Silicosis: A serious lung disease resulting from the inhalation of crystalline silica particles.

Site superintendent: Manages day-to-day activities on a construction site and supervises work performed by the subcontractors.

Six-foot rule: A rule stating that platforms or work surfaces with unprotected sides or edges that are 6 feet (≈2 m) or higher than the ground or level below it require fall protection.

Sling: Wire rope, alloy steel chain, metal mesh fabric, synthetic rope, synthetic webbing, or jacketed synthetic continuous loop fibers made into forms, with or without end fittings, used to handle loads.

Sling legs: The parts of the sling that reach from the attachment device around the object being lifted.

Sling reach: A measure taken from the master link of the sling, where it bears weight, to either the end fitting of the sling or the lowest point on the basket.

Solid geometry: The mathematical study of three-dimensional shapes.

Specifications: Precise written presentation of the details of a plan.

Splice: To join together.

Spoil: Material such as earth removed while digging a trench or excavation.

Spotter: A person who walks in front of another worker who is carrying or transporting a long load to ensure there is a clear, unobstructed path.

Square: (1) A special type of rectangle with four equal sides and four 90° angles. (2) The product of a number multiplied by itself. For example, 25 is the square of 5; 16 is the square of 4. Also, used as an adjective in this context, to have the shape of a square where two components are at precise 90-degree angles (perpendicular) to each other.

Square knot: A knot made of two reverse half-knots and typically used to join the ends of two ropes of similar diameters; also called a reef knot.

Standing end: The end of a rope that is not being knotted.

Standing part: The portion of a rope that is between the standing end and the working end.

Step drill bit: A bit that includes steps that enable the user to cut holes in metal to a desired diameter.

Straight angle: A 180° angle or flat line.

Strand: A group of wires wound, or laid, around a center wire, or core. Strands are laid around a supporting core to form a rope.

Stripping: Damaging or distorting the head or threads of a fastener.

Structural plans: A set of engineered drawings used to support the architectural design.

Stud: A vertical support inside a wall to which the wall finish material attaches. The base of a stud rests on a horizontal baseplate, and a horizontal cap plate rests on top of a series of studs.

Subcontractor/specialty contractor: Typically employs craft professionals who have specialized skills needed to complete a particular part of the construction project (e.g., HVAC, electrical, plumbing).

Sum: The result of an addition problem. For example, in the problem 7 + 8 = 15, 15 is the sum.

Sustainability: The practice of minimizing negative impacts to Earth's climate and natural environment.

Symbols: Drawings that represent a material or component on a plan.

Synthetic drugs: A drug with properties and effects similar to known substances but having a slightly altered chemical structure. Such drugs are often not illegal since they are somewhat different than well-defined restricted or illegal substances. The two typical categories are cannabinoids (lab-produced THC or marijuana substitutes) and cathinones, which are designed to mimic the effects of cocaine or methamphetamines.

Table: A way to present important text and numbers so they can be read and understood at a glance.

Table of contents: A list of book chapters or sections, usually located at the front of the book.

Tactful: Being aware of the effects of your statements and actions on others.

Tag line: Rope that runs from the load to the ground. Riggers hold on to tag lines to keep a load from swinging or spinning during the lift.

Tang: Tapered, pointed end of a file designed for insertion into a file handle.

Tardiness: Habitually showing up late for work.

Tattle-tail: Cord attached to the strands of an endless loop sling. It protrudes from the jacket. A tattle-tail is used to determine if an endless sling has been stretched or overloaded.

Tempered: Treated with heat to enhance or restore steel hardness.

Threaded shank: A connecting end of a fastener, such as a bolt, with a series of spiral grooves cut into it. The grooves are designed to mate with grooves cut into another object in order to join them together.

Title block: A part of a drawing sheet that includes some general information about the project.

Toeboard: A vertical barrier at floor level attached along exposed edges of a platform, runway, or ramp to prevent materials and people from falling.

Top rail: A top-level, horizontal board required on all open sides of scaffolds and platforms that are more than 14 inches (36 cm) from the face of the structure and more than 10 feet (3 m) above the ground.

Torque: The rotating or twisting force applied to an object such as a fastener or the shaft of a motor. Also, the turning force produced by the drill. As the bit's turning speed increases, torque decreases, and vice versa.

Trench: A narrow excavation made below the surface of the ground that is generally deeper than it is wide, with a maximum width of 15 feet (4.6 m). Also see *excavation*.

Triangle: A closed shape that has three sides and three angles.

Trigger lock: A small lever, switch, or part that can be used to activate a locking catch or spring to hold a power tool trigger in the operating mode without finger pressure.

Unit: A standardized quantity used for a particular kind of measurement.

Unstranding: Describes wire rope strands that have become untwisted. This weakens the rope and makes it easier to break.

Vertex: The point at which two or more lines or curves come together.

Virtual reality (VR): Provides a 3D model of a construction site and allows users to directly immerse themselves into the virtual space. VR enables multiple people to envision the final product, feel what it is like to be in the space, and identify any issues before work begins.

Voltage rating: A measurement of the power the tool can deliver.

Volume: The amount of space contained within a three-dimensional space.

Warning yarn: A component of the sling that shows the rigger whether the sling has suffered too much damage to be used.

Welding: The process of heating two or more metal workpieces to their liquid state and fusing them to create a very strong joint.

Welding curtain: A protective screen set up around a welding operation designed to safeguard workers not directly involved in that operation.

Wheelbarrow: A one- or two-wheeled vehicle with handles at the rear that is used to carry small loads.

Whip check: A safety attachment used to prevent whiplashing in hoses when they are inadvertently uncoupled.

Whole numbers: Complete number units without fractions or decimals.

Wind sock: A cloth cone open at both ends mounted in a high place to show which direction the wind is blowing.

Wire rope: A rope made from steel wires that are formed into strands and then laid around a supporting core to form a complete rope; sometimes called cable.

Work ethic: Work habits that are the foundation of a person's ability to do his or her job.

Work zone: The area in which a forklift may come in contact with objects or people, either with the rear end or the front forks.

Working end: The end of a rope that is being used to tie a knot.

Zero tolerance: The policy of applying laws or penalties to even minor infringements of a code in order to reinforce its overall importance, typically related to drug and alcohol abuse when applied to the workplace.

REFERENCES

00101 Basic Safety

Basic Construction Safety and Health. Fred Fanning. 2014. CreateSpace Independent Publishing Platform.

Construction Project Safety. John Schaufelberger, Ken-Yu Lin. 2013. John Wiley & Sons, Inc.

Management of Construction Projects: A Constructor's Perspective. John Schaufelberger, Len Holm. Second Edition. 2017. Routledge.

US Occupational Safety and Health Administration. Numerous safety videos are available online at **www.osha.gov/video**.

00102 Introduction to Construction Math

Mathematics for Carpentry and the Construction Trades. Alfred P. Webster, Kathryn E. Bright. 2012. Upper Saddle River, NJ: Prentice Hall.

Mathematics for the Trades: A Guided Approach. Hal Saunders, Robert A. Carman. 2018. Upper Saddle River, NJ: Pearson Learning.

Using Math in Construction: Math You Will Actually Use. Colin Wilkinson. 2017. Rosen Central.

00103 Introduction to Hand Tools

Hand Tool Basics. Steve Branam. 2018. Des Moines, IA: Popular Woodworking Books.

Essential Guide to the Steel Square: How to Figure Everything Out with One Simple Tool, No Batteries Required. Ken Horner. 2016. Mount Joy, PA: Fox Chapel Publishing.

The Tool Book: A Tool Lover's Guide to More Than 200 Hand Tools. Phil Davy, Luke Edwardes-Evans, Jo Behari, Matthew Jackson. 2018. New York, NY: DK Publishing.

00104 Introduction to Power Tools

29 CFR 1926, OSHA Construction Industry Regulations. Latest Edition. Washington, DC: Occupational Safety and Health Administration, US Department of Labor, US Government Printing Office.

Build Stuff with Wood: Make Awesome Projects with Basic Tools. Asa Christiana. 2017. Newtown, CT: Taunton Press.

Power Tool Institute, Inc. 1300 Sumner Avenue, Cleveland, OH 44115-2851, USA. **www.powertoolinstitute.com**.

Woodworking with Power Tools: Tools, Techniques, and Projects. 2019. Newtown, CT: Taunton Press.

00105 Introduction to Construction Drawings

Autodesk, 1 Market St., Suite 500, San Francisco, CA 94105, USA; 3D design, engineering, and entertainment software and parent company of the AutoCAD software suite. **www.autodesk.com**.

Blueprint Reading for the Construction Trades. Peter A. Mann. 2005. Ontario, Canada: Micro-Press.com.

Datacad, P.O. Box 815, Simsbury, CT 06070, USA. Windows-based CADD solutions. **www.datacad.com**.

Printreading for Residential Construction. Leonard P. Toenjes. 2021. Orland Park, IL: American Technical Publishers.

Reading Architectural Plans for Residential and Commercial Construction. Ernest R. Weidhaas. 2001. Englewood Cliffs, NJ: Prentice Hall Career & Technology.

Reading Architectural Working Drawings. Edward J. Muller, Phillip A. Grau III. 2004. Upper Saddle River, NJ: Prentice Hall.

00106 Introduction to Basic Rigging

Bob's Rigging and Crane Handbook. Bob De Benedictis. Third Edition. 2012. Leawood, KS: Pellow Engineering Services, Inc.

Rigging. James Headley. 2012. Sanford, FL: Crane Institute of America, Inc.

Rigging Handbook. Jerry A. Klinke. 2016. Stevensville, MI: ACRA Enterprises, Inc.

Rigging Math Made Simple. Delbert L. Hall. 2019. Johnson City, TN: Spring Knoll Press.

Rigging Safety in Construction Environments DVD Kit. Port Washington, NY: Global Industrial.

00107 Basic Communication Skills

Effective Communication at Work: Speaking and Writing Well in the Modern Workplace. Vicki McLeod. 2020. Emeryville, CA: Rockbridge Press.

The Elements of Style. William Strunk, Jr. 2015. Grammar, Inc.

The English Grammar Workbook for Adults: A Self-Study Guide to Improve Functional Writing. Michael DiGiacomo. 2020. Emeryville, CA: Rockbridge Press.

How to Win Friends and Influence People. Dale Carnegie. 2013. New York, NY: Simon & Schuster.

Submitting a Winning Bid: Guide to Making Construction Bidding with Examples. Gustavo Cinca. 2020.

Tools for Success: Critical Skills for the Construction Industry. NCCER. 2009. Hoboken, NJ: Pearson Education, Inc.

00108 Basic Employability Skills

Knock 'em Dead Resumes: A Killer Resume Gets More Job Interviews! Martin Yate. 2014. Avon, MA: Adams Media.

Knock 'em Dead: The Ultimate Job Search Guide. Martin Yate. 2014. Avon, MA: Adams Media.

The 7 Habits of Highly Effective People: Powerful Lessons in Personal Change. Stephen R. Covey. 2013. New York, NY: Simon & Schuster.

Starting Your Career as a Contractor: How to Build and Run a Construction Business. Claudiu Fatu. 2015. New York, NY: Skyhorse Publishing – Allworth Press.

00109 Introduction to Material Handling

Heavy Equipment Operations. NCCER. 2020. Hoboken, NJ: Pearson Education, Inc.

Knots: The Complete Visual Guide. Des Pawson. 2012. New York, NY: DK Publishing.

Manufacturing Facilities Design & Material Handling. Matthew P. Stevens, Fred E. Meyers. 2013. West Lafayette, IN: Purdue University Press.

Materials Handling Handbook. The American Society of Mechanical Engineers (ASME) and the International Material Management Society (IMMS), Raymond A. Kulwiec, Editor-in-Chief. 1985. New York, NY: Wiley-Interscience.

Simple Solutions: Ergonomics for Construction Workers. US Centers for Disease Control and Prevention, National Institute for Occupational Safety and Health. Last modified August 2014. **http://www.cdc.gov/niosh/docs/2007-122/**.

INDEX

A

Abbreviations, construction drawings, 270–273

Abrasive, 250

   cutoff saws, 247

Absenteeism, 375–378

Accidents

   at-risk work habits, 40–41

   categories, 36–37

   causes, 38–40

   communication and prevention of, 340

   costs, 37–38

   defined, 35–36

   intentional acts, 42

   unsafe acts, 42

Active listening, 343–345

Acupuncture, 43

Acute angle, 168

Addition, 124–125

   decimals, 142–143

   fractions, 133–135

Adjacent angle, 168

Adjustable wrenches, 194–195

Alcohol and drug abuse, 383–384

   accidents and incidents and, 41

Allen® (hex), 191

Alloy steel chain slings, 314, 318–319

Alternating current (AC), 222

Aluminum ladder, 63, *63*

American Public Works Association (APWA), 78

Amphetamines, 383

Amp hour (Ah), 226

Anchor points, 54–55

Angle grinder, 250–253

Angles, 168

Appendices, 350–351

Apprenticeship/craft training, 12

Arbor, 239

Architects, 270

Architect's scale, 275–278

Architectural plans, 279

Arc welding

   eye protection and, 88–89

   hazards, 105

Area, of shapes, 171–172

Arrowhead, 269

Asbestos, 93–95

Ashley Book of Knots, The (Ashley), 401

Assigned protection factor (APF), breathing filtration, 93

Associated General Contractors of America (AGC), 3

Atmospheric hazards, confined space, 115

At-risk work habits, 40–41

Auger bit, 225

Automation, forklifts, 411

B

Backsaw, 208

Ball peen hammer, 185

Band saws, 246

Banner Bank Building (Boise), 375

Barbiturates, 383

Bar clamp, 211

Barriers and barricades, 52–53

Base (triangle), 170

Basket hitch, 332–333

Batteries, recycling of, 231

Battery packs, cordless drills, 226

Beam anchor, 54–56

Beam grab, 59–61

Beams, 285

Benardos, Nikolay, 18

Bench grinder, 250–253

Benchmark, construction drawings, 279, 281

Benefits, high costs of, 366

Bevel, 187

Big four high hazard areas, 37

Billings, Pat, 7

Biodiesel, 411

Biological hazards, 101

Birdcaging, wire rope slings, 321

Bisect, 170

Bits, power drills, 224–225, 228–229

Block and tackle, 327–328

Bloodborne pathogens, 98

Blue collar jobs, 364

Blueprints, 266, 283

Body language, 343–345

Border (construction drawing), 268

Boredom, listening and, 344

Bowline knot, 397–399

Brazing, 105–106

Break line (construction drawings), 270

Bricklayers, 203

Bridle, 314

Building information modeling (BIM), 8–9, 267

   energy-efficient design and, 279

Bull ring, 330

Bullets, 356

Bullying, 382

Burj Khalifa, 8

C

Cannabinoids, 384

Capsize, 397–398

Carabiner, 57

Carbide, 232

Carbon footprint, 358

Carbon monoxide, 74

Career advancement and mobility, construction industry and, 12–14

Career and technical education (CTE), 26–28

Career categories in construction, 16–25

Career exploration, 27

Carpenter, 16–17

Cat's paw, 187

Caught-in or caught-between accidents, 37, 37

hazards, 74–80

Cave-ins, 75–76

Cell phones, recycling, 346

Celsius temperature scale, 165–166

Chain hoists, 328–329

Change orders, 350

reading and understanding, 350

Charge cycles, 226

Cheater bar, 196

Chemical splashes, 98–99

Chiropractic care, 43

Chisel bar, 187

Chisels, 189–191

Choker hitch, 330–332

Chronic obstructive pulmonary disease (COPD), 92–93

Chuck, 225, 228–229

keyless, 229

Chuck key, 225

Circle, 171

Circles, 168, 171

Circular saws, 239–242

Circumference, 171

Civil plans, 279–280

Clamps, 210–212

Claw hammers, 183

Clean rooms, 93

Clove hitch knot, 397, 400–401

Clutch-drive screwdriver, 191

Coating thickness, measurements of, 144

Codebooks, 349

Code of Federal Regulations (CFR), 428

Cold stress, 103

Combustible materials, 44, 109

Commercial construction, 14

Common Ground Alliance Initiative, 78

Communication

active listening, 343–345

basic skills, 337–348

constructive criticism, 380–381

cultural interpretation, 347

emails, 357–360

failure of, 39–40

Hazard Communication (HAZCOM) Standard, 44

listening and speaking skills, 341–343

nonverbal, 339–341

overview, 338

process, 339–343

reading and writing, 348–360

speaking on the job, 345–347

telephone calls, 346–347

texting, 359–360

Community colleges, construction training at, 29

Competent person

defined, 51, 319

wire rope sling inspection, 319–321

Compressed gas, 106

Compromise, conflict resolution and, 379

Computer-aided design (CAD), 8–9

Computer-aided drafting (CAD), 266

Computers, in construction industry, 6

Concave, 200

Concrete anchor, 54–56

Concrete mule, 407–408

Confidentiality, 377

Confined space

accidents in, 37

safety measures for, 113–116

Conflict resolution, 345, 378–379

with supervisors, 379

Construction

defined, 2–3

estimates, 126

famous examples, 8

history of, 7–9

Construction drawings

abbreviations, symbols, and keynotes, 270–273

care of, 273

civil plans, 279

components, 266–269

dimensions and drawing scale, 274–276

elements, 269–274

elevations, 283

legality, 269

regional and company differences, 269

types of, 278–304

Construction industry

career benefits and opportunities, 10–12

career path in, 15

common myths concerning, 4–7

education requirements, 11–12

employment opportunities in, 364–366

entry-level jobs, 366–369

growth of, 3

mission statement, 366

organization chart, 368

overview of, 1–9

sectors, 14–15

types of careers, 16–25

Construction mathematics

area of shapes, 170–172

conversion of decimals, fractions and percentages, 145–148

decimal addition, subtraction, multiplication and division, 140–145

decimals, 137–140

estimates, 126

fractions, 129–132

fractions, addition and subtraction, 133–135

fractions, multiplication and division, 135–136

geometry, 167–175

improper fractions and mixed numbers, 132–133

length measurement, 152–156, 158–160

metric and inch-pound measurement systems, 157–167

multiplication table, 432

numerators and denominators, 127

overview, 122

place values of whole numbers, 123–124

shapes, 169–171

temperature units, 164–166

volume measurement, 162–164, 172–176

weight measurement, 161–162

whole number addition and subtraction, 124–125

whole number multiplication and division, 125–128

Constructive criticism, 380–381

Container labeling, 99

Contour lines, 279

Contractor, 14

Conversion factors and formulas, 145–148, 160–166, 432–436

Coordinate system, construction drawings, 279, 281

Coping saw, 208–209

Cordless tools and batteries, 225–226

Core, 312

Core failure, wire rope slings, 321

Costs

direct and indirect, 150

incidents and accidents, 37–38

Countersink, 224

Counting systems, 143

Coupling guards, 78–79

Co-workers, conflict resolution with, 379

Craft professionals

benefits of career as, 10

decimal measurements and, 146

defined, 2

demand for, 3–4

salary survey, 11

training and apprenticeships, 11–12, 28–29

Craft/trade, defined, 2

Crane operator, 22–23

hand signals, 332

Cranes, safety procedures, 79–80

Crew leader, 14

Critical thinking

barriers to, 370–371

problem solving and, 371–373

Criticism, giving and receiving, 379–381

Cross-bracing (scaffolds), 71

Crosscut, 210

Cube, 162

Cubic measurements, 162–164

three-dimensional, 173

Culture, communication and, 347

Cut lines (construction drawings), 269

Cutoff saws, 247

Cylinders

carts, 402, 404

measurement of, 174

D

Danger, in construction industry, 6

Decimals, 122, 137–139

addition and subtraction, 140

division, 141–142

dots and commas in, 140

fractions converted to, 146–147

inches converted to, 147–148

percentages conversion to, 145–146

rounding, 138

whole numbers and, 138

Degree (angle), 168

Delays in scheduling, 375

Demolition tools, 183–185

Denominators, 127–132

Destructive criticism, 379

Detail drawings, 284, 285

Detail grinder, 250–253

Diagonal lines, 169, 172

Diameter, 171

Diamond blade guard, 254

Diet, work safety and, 42–43

Difference, 124

Digit, 123

Digital level, 202

Digital thermometers, 166

Dig Safely card, 77–78

Dimensional units, 125

    nominal and true dimensions, 154

Dimension line, 269

Dimensions, construction drawings, 274–276

Direct costs, 150

Direct current (DC), 222

Discrimination, 385

Distractions, listening and, 345

Dividend (division), 126

Division

    decimals, 141–142

    fractions, 135–136

    whole numbers, 125–128

Divisor, 126

Double-locking snap hooks, 58

Dowel, 183

Drawing area, construction drawing, 268

Drawing revision control, 267

Dress, 340

Drills

    bits, 224–225

    cordless, 225–226

    electromagnetic presses, 230–232

    hammer drills/impact drivers, 232–236

    parts, 224–225

    pneumatic, 236–238

    power drills, 222–232

    right-angle, 228

    rotary hammer, 93–94

    safety, 227

D-rings, 54–56, 59

Drones, in construction industry, 6

Drop-testing, safety nets, 62

Drug abuse, 41, 383–384

Drum cart, 403–405

Drum dolly, 403–404

Drywall, 185

Drywall hammer, 185

Drywall saw, 208–209

E

Earmuffs, 91

Earn as You Learn training model, 12

Earplugs, 91

Ego, listening and, 345

Eiffel Tower, 8

Electrical plans, 297–303

Electrical safety

    cord safety, 85

        guidelines, 80–84

        shock accidents, 37

Electrician, 17–18

Electric tools, 222

Electrocution, 81

Electromagnetic drill presses, 230–232

Electronic signatures, 357–360

Elevated work, fall protection, 51–52

Elevation (EL), 284

Elevation drawing, 284

Emails, 357–360

Emergency Stop signal, 332

Emery cloth, 210

Emotion, listening and, 344

Empire State Building, 8

Employability skills, overview, 364–366

Employee rights and responsibilities, 429

End grinder, 250–253

Energy-efficient design, building information modeling (BIM) and, 279

Energy release hazards, 80

Engineers, 270

Engineer's scale, 275, 278

English ruler, 152–153

Environment

    construction industry and, 7, 10

    extreme environments, safety in, 101–104

Equations, 127

Equilateral triangle, 170

Equipment

    material handling, 402–413

    selection and scheduling, 374

Equipment guards, 78–79

Equivalent fractions, 129–130

Essential services, construction as, 10

Evacuation, 101

Excavations, 52

    hazards, 74–78

Exercise, work safety and, 42–43

Extension ladders, 65–68

Eye and face protection, 88–89

Eyebolts, 323–324

F

Face shields, 88–89

Facial expression, communication and, 340

Fahrenheit temperature scale, 165–166

Failure to communicate, 38–39

Falling objects, 72

Falls
  accidents and incidents, 37
  arrest of, 54–62
  types and hazards, 52–54
Fastening systems, power-actuated, 257–258
Fatal four accidents/incidents, 38
Fatalities
  accidents and incidents, 37
  cave-in, 76
  excavations, 75
  struck-by fatalities, 73
Feet, decimal equivalents in, 147–148
Fiberglass ladders, 63
Files, 212–215
Finish carpenter, 16–17
Fire extinguishers, 110–113
Fire hazards and fire fighting, 109–113
Fire protection plan, 304
Fire triangle, 109
Flammable materials, 40
Flaring cup guard, 254
Flash burn, 105
Flash point, 109
Flats (wrenches), 193
Floor plan, 279
Flush, 183
Flying objects, 72–73
Fonts, 353
Foot and leg protection, 90–91
Force, 161
Foreman, 14
Forklifts
  industrial forklift, 409–410
  rough terrain forklift, 411–412
Formula, 171
Forstner bit, 225
Foundation plan, 285, 288–289
Fractions, 122, 129–136
  addition and subtraction, 133–135
  decimal conversion to, 146–147
  equivalent fractions, 129–130
  improper fractions, 132–133
  lowest common denominator, 130–132
  multiplication and division, 135–136
  reduction of, 130
Freight elevator, 407–408
Frostbite, 103
Fuel cell technology, 410

G

Gas cylinders, safety management of, 106–107
Geobond®, 7
Geometry, 167–175
  angles, 168
  shapes, 169–175
  square, 205
Global Harmonization System (GHS), 99–100
Globally Harmonized System of Classification and Labeling of
    Chemicals, 44, 99–100
Glossary, 350–351
Gloves, 89–90
Golden Gate Bridge, 2, 8
GOST standards, 36
Graphs, 350
Great Wall of China, 8
Green-collar jobs, 369
Green construction, 7
  biodiesel, 411
  city governments, 149
  drawings, 274–275
  fuel cell technology, 410
  job opportunities, 369
  materials recycling, 397
Green employers, 387–388
Green technologies, 149
Gridlines, construction drawings, 274
Grilles, 289
Grinders, 250–255
  attachments and accessories, 254–255
Grinding hazards, 104–105
Grit, 250
Grooming, 340
Ground and grounding, electrical safety and, 81–82
Ground fault circuit interrupter (GFCI), 37, 82–83, 227
Ground fault protection, 227
Guarded areas, 52
Guardrails, 61
Guards, tool, machine and equipment, 78–79

H

Haberle, Bruce, 28
Habitat for Humanity, 28
Hacksaw, 208–209
Half hitch knot, 397, 399
Half-measures, accidents and incidents, 38
Hallucinogens, 383
Hammer drills/impact drivers, 232–236
Hammers, 183–185
Hand line, 65
Hand protection, 89–90
Handsaw, 208–210
Hand-screw clamp, 211
Hand signals, material handling, 413
Hand tools
  chisels and punches, 189–191

Hand tools (*Continued*)
 clamps, 210–212
 files and utility knives, 212–215
 hammers, 183–185
 levels and layout tools, 201–207
 measuring tools, 199–201
 nail pullers, 186–187
 overview, 182–183
 pliers and wire cutters, 196–198
 saws, 208–210
 screwdrivers, 191–192
 shovels and picks, 216–217
 sledgehammers, 185–186
 wrenches, 193–196
Hand truck, 402–404
Harassment, 382
Hard hats, 88
Harnesses, 55–58
Hazard Communication (HAZCOM) Standard, 44, 97, 99–100
Hazards
 biological, 101
 caught-in or caught-between hazards, 74–80
 fire hazards and fire fighting, 109–113
 hot work hazards, 104–108
 job-site exposure hazards, 97–101
 radiation, 100
 recognition, 34–36, 44–51
 respiratory hazards, 92–96
 struck-by hazards, 72–74
 vehicle and roadway, 73–74
Hearing protection, 91
Heat cramps, 101–102
Heat exhaustion, 102
Heating, ventilating, and air conditioning (HVAC) plan, 289, 294–295
Heat stress, 101
Heat stroke, 102–103
Heavy civil/infrastructure construction, 14
Heavy equipment
 operator, 20–21
 safety procedures, 79–80
Heavy soil, 76
Height work, 396–397
Hidden line (construction drawings), 270
High-efficiency particulate arrestance (HEPA) filters, 93–94
High-temperature systems, 85
Historic restoration, 3
Hitches, 329–333
 basket, 332–333
 choker, 330–332
 defined, 309

 emergency stop, 333
 vertical, 330
Hoists, 308, 327–329
Hooks (rigging), 326–327
Hoover Dam, 8
Hot work hazards, 104–108
Hot work permits, 107–108, 354–355
Humor, communication and, 341
HVAC technician, 23–24
Hydraulic energy, 45
Hydraulic jacks, 260–261
Hydraulic tools, 222
Hypothermia, 103–104

I

Impact drivers, 232, 235–236
Impact wrenches, 236–238
Improper fractions, 132–133
Inches, decimal equivalents in, 147–148
Inch-pound measurement system, 152–153
 length measurements, 157–159
 volume measurements, 163
 weight measurements, 161
Incidents
 at-risk work habits, 40–41
 categories, 36–37
 causes, 38–40
 costs, 37–38
 defined, 35–36
 intentional acts, 42
 reporting of, 48
 unsafe acts, 42
Indexes, 350–352
Indirect costs, 150
Industrial construction, 14
Industrial forklift, 407, 409–410
Infrastructure
 construction sector, 14
 investment in, 3
Initiative, 377
Injuries, reporting of, 48
Inspection
 alloy steel chain slings, 318–319
 of ladders, 67
 of scaffolds, 70
 safety, 430
 slings, 316–321
 wire rope slings, 319–321
Intentional acts, incidents and accidents and, 42
Internet
 information on, 372
 job applications and, 367

Invert, 135
Ironworker, 23
Isosceles triangle, 170
Italics, 353

J

Jacks, hydraulic, 260–261
Jargon, 339
Jigsaws, 242–244
Job growth estimates
   carpenter, 17
   crane operator, 22
   electrician, 18
   heavy equipment operator, 21
   HVAC technician, 23
   ironworker, 23
   pipefitters, 20
   project manager, 25
   welders, 19
Job safety analysis (JSA), 46–47
Job-site exposure hazards, 97–101
Joint, 185
Joists, measurement for, 154–155
Journalist's questions, 357

K

Kelvin temperature scale, 165–166
Kerf, 209, 241
Keyed chucks, 225, 228–229
Keyhole saw, 208
Keyless chuck, 229
Keynotes, construction drawings, 270–273
Kinking, wire rope slings, 320
Knots, material handling, 397–401

L

Labor, selection and scheduling of, 374–375
Ladders, 62–68
   duty ratings and load capacities, *62*
   extension, 65–68
   inspection, 67
   positioning, 64–65
   safety foot, 64
   securing of, 65
   stepladders, 67
   straight ladders, 63–65
   types of, *63*
Landscape plan, construction drawings, 279, 282
Lanyard, 183
Lanyards, 52, 58–60, 196
   tool lanyards, 74
Laser level, 202–203

Laser measuring tools, 200–201
Lath, 185
Layout tools, 201–207
Lead, 97–98
Leader (construction drawings), 269–270
Leadership in Energy and Environmental Design (LEED), 7
Leadership skills, 386–388
   construction industry, 12–13
Legal issues, construction drawings, 269
Legend, construction drawing, 289
Length measurements, 152–156
   metric units, 159–160
Level, 184, 201–207
Lifelines, 59–61
Lifting and lowering techniques, 393–395
Lifting clamps, 324–325
Lineman's pliers, 196–197
Lines of construction, 269–270
Listening skills, 341–343
   active listening, 343–345
   barriers to, 344–345
Load, 309
Loadbearing walls, measurement of, 154–155
Load control, 329
Load stress, shackles, 322–323
Locking C-clamp pliers, 211
Locking pliers, 196–197
Lockout/tagout (LOTO) devices, 80, 84–85
Long-nose pliers, 196–197
Lowest common denominator, 130–132

M

Machine guards, 78–79
Maintenance, 251–252
   circular saws, 240–242
   construction drawings, 273
   hammer drills, 233–234
   hand tools, 188–189
   hoists, 328–329
   hydraulic jacks, 260–261
   impact drivers, 235–236
   miter and cutoff saws, 247
   oscillating multi-tools (OMT), 255–256
   pneumatic tools, 237–238
   power nailers, 258
   saws, 243–246
Management system, failure, 38, 44
Masonry bit, 225
Mass, 161
Massage therapy, 43
Master link, 314

Material cart, 402–403

Material handling, 391–411

   equipment, 402–413

   hand signals, 413

   height locations, 394–395

   knots, 397–401

   lifting and lowering procedures, 393–395

   pre-task planning, 392

   problem-solving and scheduling for, 373

   stacking and storing, 395–396

Maximum intended load, 62

Maximum noise levels, 91

MDAS (My Dear Aunt Sally) memory aid, 127–128

Measurement

   length, 152–156

   safety and maintenance of tools, 201

   temperature units, 164–167

   tools, 199–207

   volume, 162–164

   weight, 161–162

Measuring tape, 154–156, 200

Mechanical plans, 289–295

Memos, 353

Mercury, 206

Meritens, Auguste de, 18

Metal, drilling techniques, 230

Methamphetamine, 383

Metric scale, construction drawings, 276–278

Metric system

   conversion, 160, 162, 164, 166, 432–436

   as global standard, 162, 164

   measurement with, 157–158

   measuring tape, 154–155

   prefixes, 158

   ruler, 153

   units of length, 159–160

   units of weight, 161

   visualization, 156

   volume measurements, 163–164

Midrail, 61

Minor injuries, accidents and incidents, 36

Mission statement, 366

Miter joints, 208

Miter saws, 246–247

Mixed numbers, 132–133

Mobile devices, construction drawings on, 273

Motorized material handling equipment, 406–413

Multiplication

   decimals, 140–141

   fractions, 135–136

   whole numbers, 125–128

Mushroomed heads, 191

N

Nailers, 256–259

Nail pullers, 186–187

Nails, screws vs., 192

National Association of Women in Construction (NAWIC), 5

National Center for Construction Education and Research (NC-CER)

   salary survey, 11

   training programs, 13–14, 16

Near-miss incidents, 36

   reporting of, 48

Negative numbers, 123

Nominal measurements, 131, 154

Non-adjustable wrenches, 193–194

Non-motorized material handling equipment, 402–406

Non-shock absorbing lanyard, 59

Nontechnical skills training, 27

Nonverbal communication, 339–341

North arrow, construction drawing, 269

Not to scale (NTS), construction drawings, 275

Numeral systems, 125

Numerators, 127–128, 129

O

Obtuse angle, 168

Occupational Safety and Health Administration (OSHA)

   construction safety and, 6

   Hazard Communication (HAZCOM) Standard, 44

   incident and accident investigations, 35–36

   mission and standards, 428

   scaffold standards, 70

   silica standard, 92–93

One-Call Notification System, 78

One-rope lay, wire rope slings, 320

On-the-job learning (OJL), 28–29

Opiates, 383

Opposite angle, 168

Orthographic drawings, 276

Oscillating multi-tools (OMT), 255–256

Overstriking, 183

Oxyfuel cutting, welding and brazing, 105–106

P

Pallet jacks, 403, 406

Paraphrasing, 344

Peening, 185

PEMDAS (Please Excuse My Dear Aunt Sally) memory aid, 127–128

Percentages, decimal conversion to, 145–146

Perimeter, 169

Permanent anchor, 54–56, 55

Permissible exposure limits (PELs), 97

Permits, 350

   confined space requirements, 114–116

   hot work permits, 107–108, 354–355

   reading and understanding, 350

Personal fall arrest system (PFAS), 54–62

Personal habits, 376–377

Personal protective equipment (PPE), 8, 87–91

   accidents and incidents, 42

   material handling, 393

Personal skills, 376–378

Phillips screwdriver, 191

Physical distance, communication and, 341

Picks, 216–217

Pictograms, container labeling, 99–100

Pi measurements, 171, 174

Pins, shackles and, 324–325

Pipe clamp, 211

Pipefitters, 19–20

Pipe mule, 403, 405–406

Pipe transport, 403, 406

Pipe welder, 18–19

Piping and instrumentation drawings (P&IDs), 289, 293

Place values, 123

Plane geometry, 168

Planked scaffolds, 71

Plan locations, construction drawings, 274

Planning problems, 373–375

Plan north, 269

Plane, 312

Pliers, 196–198

Plumb, 199

Plumb bob, 206

Plumbing and piping plans, 296–297

Plumbing isometric drawing, 296–297

Pneumatic drills, 236–238

Pneumatic energy, 45

Pneumatic nailers, 256–259

Pneumatic tools, 222

Portable band saw, 244–246

Positive numbers, 123

Post level, 202

Posture and gestures, communication and, 340

Power-actuated tools, 222

Power drills/drivers, 222–232

   cordless tools, 225–226

   operation, 230

   parts, 224–225

   procedures for, 228–229

   safety, 227

Powered wheelbarrow, 405, 407

Power screwdrivers, 257

Power sources, for power tools, 234

Power tools

   cordless tools, 225–226

   drills/drivers, 222–232

   grinders, 250–255

   hammer drills, 232–234

   hydraulic jacks, 260–261

   impact drivers, 235–236

   impact wrenches, 236–238

   nailers, 256–259

   pneumatic drills, 236–238

   power sources for, 234

   safety guidelines, 223

   saws, 239–249

Precision measuring tools, 201

Pressurized systems, safety guidelines, 85

Prism, triangular, 174–175

Problem solving

   barriers to, 370–371

   critical thinking and, 371–373

   planning and scheduling problems, 373–375

Product (multiplication), 125–128

Professionalism, 377

Project management software, 373–374

Project manager, 14, 24–25

   digging and excavation, 77–78

Project owner, 14

Property damage, accidents and incidents, 36

Property lines, 269

Proximity work, energized electrical equipment, 83–84

Punches, 190–191

Punch lists, 353

Q

Qualified person, 110

   rigging operations, 308

Quotient, 126

R

Radiation hazards, 100

Radius, 171

Rafter, 204–205

Rankine temperature scale, 165–166

Rasps, 212–213

Rated capacity, 309

Reading skills

   codebooks, 349

   communication and, 348–360

   on the job, 349–353

Reality capture, 304

Real-life application, training and education, 26

Rechargeable batteries, recycling of, 231

Reciprocating saws, 242–244

Rectangle, 169, 171

   three-dimensional, 173

Recycling

   cell phones, 346

   materials, 397

   rechargeable batteries, 231

References, 367, 446–447

Registers, 289

Rejection criteria, sling, 316–318

Relationship skills, 376–388

Remainder, 126

Request for information (RFI), 273

Residential construction, 14

Respirators, 40, 95–96

Respiratory hazards, 92–96

Resume, 367

Retractable fall arrest lanyard, 58–60

Revision block, construction drawing, 268

Revolutions per minute (rpm), 230

Rigging

   basic equipment, 308

   hardware, 321–327

   hitches, 329–333

   hoists, 327–329

   hooks, 326–327

   slings, 308–321

Rigging hook, 326

Right angle, 168

Right-to-Know requirement, 44

Right triangle, 169

Risk, rationalization, 43

Risk assessment, 46, 48

Roadway hazards, 73–74

Robertson® (square), 191

Robotic welder, 18

Robots, in construction industry, 6

Rockwell hardness scale, 232

Roller skids, 403, 405

Rolling warehouse ladder, 63

Roman tools, 212

Roof plan, 279, 284

Rope grab, 59–61

Rope stretchers, 169

Rough carpenter, 16–17

Rough terrain forklift, 407, 411–412

Rulers, 152–156

   English ruler, 152–153

   metric ruler, 153

   steel rule, 199

   wooden folding rule, 200

S

Saddles, 58

Safety. *See also* personal protective equipment (PPE)

   caught-in and caught-between hazards, 74–80

   chisels and punches, 190–191

   circular saws, 240–242

   clamps, 212

   Code of Federal Regulations, 428

   compliance, 431

   confined space, 113–116

   in construction industry, 6

   electrical safety guidelines, 80–84

   employee rights and responsibilities, 429

   environmental extremes, 101–104

   fall arrest, 54–62

   fall types and hazards, 52–54

   files and knives, 215

   fire hazards and fire fighting, 109–113

   General Duty Clause, 428–429

   grinders, 251–252

   hammer drills, 233–234

   hand tools, 188–189

   hazard recognition, 34–36, 44–51

   hoists, 328–329

   hot work hazards, 104–108

   hydraulic jacks, 260–261

   impact drivers, 235–236

   incident and accident causes, 38–44

   inspections, 430

   job-site hazards, 97–101

   ladders and stairs, 62–68

   levels, 203

   material handling, 393–397

   miter and cutoff saws, 247

   oscillating multi-tools (OMT), 255–256

   OSHA mission and standards, 428

   personal protective equipment, 86–91

   pliers and wirecutters, 198

   pneumatic tools, 237–238

   poor housekeeping and, 44

   power nailers, 258

   power tools, 222, 227

   respiratory hazards, 92–96

   risk rationalization, 43

   saws, 210, 243–246

   scaffolds, 68–71

   screwdrivers, 192

   shovels and picks, 216–217

struck-by hazards, 72–74

table saw, 248–249

unsafe conditions, 43

violations, 430–431

wrenches, 195–196

Safety culture, 35

Safety Data Sheets (SDSs), 39, 48–51

Safety nets, 61–62

Safety shoes, 90–91

Safety training, 36

Salaries in construction

carpenter, 17

crane operator, 22

electrician, 18

heavy equipment operator, 21

HVAC technician, 23

ironworker, 23

pipefitters, 20

project manager, 25

survey, 11

welders, 19

Saws, 208–210, 239–249

band saw, 246

circular, 239–242

cutoff saws, 247

jigsaws, 242–244

miter saws, 246–247

portable band, 244–246

reciprocating, 243–244

table saws, 248–249

Scaffolds, 52, 68–71

Scale

construction drawings, 274–276

measurement of, 276–278

mechanical plans, 289

Scalene triangle, 170

Scheduling problems, 373–375

Schematic drawings, 276, 289

Scientific notation, 142

Screwdrivers, 191–192

power screwdrivers, 257

Screws, types of, 192

SDS technology, 233

Section cut (construction drawings), 269, 283

Section drawing, 284

Sectors in construction industry, 14–15

Self-care, safety and, 42–43

Self-contained breathing apparatus (SCBA), 40, 95–96

Self-feed bit, 225

Self-presentation, 376

Self-supporting scaffolds, 68–71

Serious/disabling injuries, accidents and incidents, 37

Sexual harassment, 382

Shackles, 320–323

defined, 321

Shank, 228

Shapes, volume, 172–175

Shapes (geometric)

area, 171–172

types, 170–171

Sheaves, 315

Shielding, 76–77

Shock-absorbing lanyard, 58–60

Shock injuries, 84

Shoring, 76

Shovels, 216–217

Shredder rasps, 213

Signaler, 74

Silica standard, 92–93

Silicosis, 92

Simulators, in construction industry, 6

Site superintendent, 14

digging and excavation, 77–78

Six-foot rule, 52

Sketchpad system, 9

Skilled labor, construction as, 4

Skills requirements

carpenter, 17

crane operator, 22

electrician, 18

heavy equipment operator, 21

HVAC technician, 24

ironworker, 23

pipefitters, 20

project manager, 25

safety and, 41–42

welders, 19

Sledgehammers, 185–186

Sleep, work safety and, 42–43

Slings, 308–321

alloy steel chain, 314, 318–319

cut protection, 316–318

inspection, 316–321

jackets, 318

synthetic slings, 310–314

tagging requirements, 309–310

wire rope, 315–316

Sling legs, 310

Sling reach, 310

Slip-joint pliers, 196–197

Slotted screwdriver, 191

Smoking, respiratory hazards and, 93–94

Social issues, in workplace, 381–384

Social skills, 376–378

Socket wrenches, 195

Software, in construction industry, 6

Soil

    conditions, 75–78

    types, 75

Solid geometry, 168

Speaking skills, 341–343

    on the job, 345–346

    telephone calls, 346–347

Specialty contractor, 14

Specifications

    construction drawings, 266

    electrical plans, 297, 299

Specific qualifications

    carpenter, 17

    crane operator, 23

    electrician, 18

    heavy equipment operator, 21

    HVAC technician, 24

    ironworker, 23

    pipefitters, 20

    project manager, 25

    welders, 19

Spell-checking, 353, 356

Spirit levels, 201–202

Splice, 318

Spoil piles, 77

Spotter, 403

Spring clamp, 211

Spud wrenches, 194

Square, 169, 171, 205

Square knot, 397–398

Squares, 204–205

Stacking and storing materials, 395–396

Stainless steel welder, 18

Stairways, 67–68

Standard measure, 156

Standing end knot, 397–398

Standing part knot, 397–398

Statista website, 4

Steel rule, 199

Step drill bit, 225

Stonemasons, 189

Straight angle, 168

Straight ladder, 63–65

Strands, 312

Stress effects, 41

Stripping, 191

Struck-by accidents, 37

    hazards, 72–74

Structural plans, 284, 287

Stud spacing, measurement for, 154–155

Subcontractor, 14

Subtraction, 124–125

    decimals, 142–143

    fractions, 133–135

Sullivan, Louis Henry, 29

Sum, 124

Suspension scaffolds, 68–71

Suspension trauma strap, 57

Sustainability, construction industry and, 7

Sutherland, Ivan, 9

Symbols, construction drawings, 270–273

Synthetic drugs, 383–384

T

Table of contents, 350

    reading and understanding, 350

Tables and graphs, 350–351

Table saw, 248–249

Tactful, 378

Tags and signs

    safety tags, 39–40, 85

    scaffolds, 70

Tag line, 330

Tang, 213

Tape measure, 154–156, 200

Tardiness, 375, 377–378

Task safety analysis (TSA), 46

Tattle-tail, 313

Teamwork, 384–388

Techcrunch.com website, 5–6

Technology

    in construction industry, 5–6

    job applications and, 369

Teeth per inch (TPI), saws, 245–246

Telephone calls, 346–347

Temperature

    extremes, 104

    measurement units, 164–167

Tempered, 190

Tempered steel, 190

Terrorism, incidents and accidents and, 42

Texting, 359–360

Text presentation, 351–352

Thomas, Holley, 5

Threaded shank, 323–324

Three-dimensional shapes, volume measurement, 172–175

3-4-5 rule, 175

Tie-back lanyard, 54–56

Time, communication and, 340

Title block, 267–268

Toeboards, 54

Tongue-and-groove pliers, 196–197

Tool belts, 57–58

Tool blades, 44

Toolbox Talks, 39

Tool guards, 78–79

Tool lanyards, 74, 196

Tools, selection and scheduling, 374

Top rails (scaffolds), 71

Torque, 191

impact drivers, 235–236

wrenches, 195

Torx®, 191

Trade, defined, 2

Transportation Security Administration, 368

Transportation Worker Identification Credential, 368

Trenches

hazards, 74–78

sloped and bent trenches, 76

Trench sites, accidents in, 37

Triangle, 170–171

Triangular prism, 174–175

Trigger lock, 223

True measurements, 154

Tugs, 408

Tuition debt, construction industry and, 11–12

TWIC® NexGen, 368

U

Underwater welder, 18

Units

in length measurement, 158–161

in weight measurement, 161

Universities, construction training at, 29

Unsafe acts, 42

Unsafe conditions, 43

Unstranding, wire rope slings, 321

US Green Building Council (USGBC), 7

Utility knives, 213–215

V

Vehicle hazards, 73–74

Vertex, 168

Vertical hitch, 330

Villages Charter School, The (TVCS), 28

Virtual reality (VR), 9

Voltage rating, 226

Volume measurements, 162–164

inch-pound system, 163

shapes, 172–175

W

Wages, in construction industry, 4–5

Walt Disney World, 7

Warning yarns, 311

Water, gram weight, 162

Web (car, band) strap, 211

Weight measurements

conversion, 162, 432–436

inch-pound system, 161

metric units, 161

Welder, 18–19

Welding, 185

Welding curtains, 105

Weld inspector, 18

Wheelbarrow, 403, 405, 407

Whip check, 237

White collar jobs, 364

Whole numbers, 122–128

addition and subtraction with, 124–125

multiplication and division, 125–128

order of operations, 127–128

place values, 123

Window schedule and detail, 283, 286

Wind sock, 101

Wire cutters, 196–198

Wire rope, 309

slings, 315–316, 319–321

Women, in construction industry, 5

Wooden folding rule, 200

Work ethic, 376–377

Working end knot, 397–398

Workplace, social issues in, 381–384

Work zone, 409

Worm-drive saw, 242

Worsham, Boyd, 13

Wrecking bar, 187

Wrenches, 193–196

impact wrenches, 236–238

Writing skills

communication and, 348, 353–360

emails, 356–350

on the job, 353, 356

resumes, 367–368

texting, 359–360

text presentation, 351–352

Y

Y-configure shock absorbing lanyard, *59*

Z

Zero tolerance policy, 383